Proofs in Competition Math: Volume 2

Dennis Chen, Freya Edholm, Alex Toller

Version 1.1: April 2020

Contents

IV Applied Discrete Math   1

10 Beginning Discrete Math   3
10.1 Introduction to Set Theory 3
10.2 Introduction to Statistics 19
10.3 Introduction to Probability 32
10.4 Combinatorics and Probability: Part 1 44

Chapter 10 Review Exercises   58

11 Intermediate Discrete Math   60
11.1 Introduction to Game Theory and Rational Decision-Making 60
11.2 Combinatorics and Probability: Part 2 75
11.3 Graph Theory: The Massive Web of Mathematics 84

Chapter 11 Review Exercises   92

12 Advanced Discrete Math   94
12.1 Combinatorics and Probability: Part 3 94
12.2 Graph Theory: Part 2 104
12.3 Overview of Operations Research 112

Chapter 12 Review Exercises   117

V Number Theory   119

13 Beginning Number Theory   121
13.1 Factors and Divisibility 121
13.2 Bases Other than Base 10 130
13.3 Modular Arithmetic: Part 1 137
13.4 Studying Prime Numbers 144
13.5 Introduction to Diophantine Equations 151
13.6 Special Functions in Number Theory 156

Chapter 13 Review Exercises   163

14 Intermediate Number Theory   165
14.1 Modular Arithmetic: Part 2 165
14.2 Primality Theorems 174
14.3 Symmetry and Quadratic Residues/Reciprocity 183

Chapter 14 Review Exercises   193

15 Advanced Number Theory — 194
- 15.1 Lifting the Exponent (LTE) 194
- 15.2 Continued Fraction Representations and the Pell Equation 200
- 15.3 Order and Cyclotomic Polynomials 211

Chapter 15 Review Exercises — 223

VI Open Problems in Mathematics and Fun Proofs! — 224

16 Through the Looking Glass into Some Open Problems — 225
- 16.1 A Brief Introduction to the Millennium Prize Problems 225
- 16.2 Hilbert's 23 Problems 252
- 16.3 The Twin Prime Conjecture 261
- 16.4 The Collatz Conjecture 263
- 16.5 Fermat's Last Theorem: Not an Open Problem (Anymore), But... 265

17 Just for Fun... — 267
- 17.1 The Sea of Pythagorean Theorem Proofs 267
- 17.2 Paradoxes and Bogus "Proofs": The Art of Mathematical Deception 276

VII End Matter — 285

Closing Remarks — 286

Closing Exam — 287

Solutions to Exams — 291

Appendix A: Counting by Infinities — 313

Appendix B: Studying the Roots of nth-Degree Polynomials for $n \geq 3$ — 314

Appendix C: Proving the Fundamental Theorem of Algebra — 315

Appendix D: The Many, Many Centers of a Triangle — 316

Appendix E: Exploring Projective Geometry — 317

VIII Solutions to Exercises — 323

Part 4: Applied Discrete Math — 324

Chapter 10: Beginning Discrete Math — 324
- Section 10.1: Introduction to Set Theory 324
- Section 10.2: Introduction to Statistics 330
- Section 10.3: Introduction to Probability 333
- Section 10.4: Combinatorics and Probability: Part 1 337
- Chapter 10 Review Exercises 340

Chapter 11: Intermediate Discrete Math — 344
    Section 11.1: Introduction to Game Theory and Rational Decision-Making 344
    Section 11.2: Combinatorics and Probability: Part 2 349
    Section 11.3: Graph Theory: The Massive Web of Mathematics 353
    Chapter 11 Review Exercises 356

Chapter 12: Advanced Discrete Math — 361
    Section 12.1: Combinatorics and Probability: Part 3 361
    Section 12.2: Graph Theory: Part 2 365
    Chapter 12 Review Exercises 368

Part 5: Number Theory — 371

Chapter 13: Beginning Number Theory — 371
    Section 13.1: Factors and Divisibility 371
    Section 13.2: Bases Other than Base 10 374
    Section 13.3: Modular Arithmetic: Part 1 377
    Section 13.4: Studying Prime Numbers 381
    Section 13.5: Introduction to Diophantine Equations 384
    Section 13.6: Special Functions in Number Theory 387
    Chapter 13 Review Exercises 390

Chapter 14: Intermediate Number Theory — 393
    Section 14.1: Modular Arithmetic: Part 2 393
    Section 14.2: Primality Theorems 397
    Chapter 14 Review Exercises 400

Chapter 15: Advanced Number Theory — 404
    Section 15.1: Lifting the Exponent 404
    Section 15.2: Continued Fraction Representations and the Pell Equation 407
    Section 15.3: Order and Cyclotomic Polynomials 409
    Chapter 15 Review Exercises 411

Bibliography — 414

Glossary — 416

Index — 432

About the Authors — 444

Part IV

Applied Discrete Math

Why Study Discrete Math?

In our daily lives, we mostly think and operate in terms of *continuous mathematics:* mathematics that works on a spectrum of infinite length and consisting of an infinite number of infinitely small intervals. Time and space are continuous - passing us by with each passing second and inch. They do not have a smallest unit of measure, nor a largest; they are representations of a quantity that is "smooth" in that it is never broken or separated.

Yet, it is equally important to be able to think in terms of *discrete mathematics:* the branch of mathematics that deals with finite quantities that are defined in terms of isolation from each other. For instance, whereas a continuous definition of time would operate under the notion that time is an ever-flowing fabric, a discrete definition would split time into days, hours, minutes, and seconds, and treat seconds as the "smallest" interval, thereby separating each moment from the others. Similarly, if space is thought of as a continuum, there is no such thing as the "smallest distance,"[1] but from a discrete point of view, distance can be thought of in terms of kilometers, meters, centimeters, millimeters, and increasingly small units of measure, but would always have separation between any two objects.

More formally, a *discrete* mathematical object is defined in terms of *countable sets* - which we will delve into throughout the chapter. As opposed to continuous sets, it is important to study countable sets in particular. Their versatility in fields such as computer science and algorithms, cryptography, and software development and engineering, is one that simply does not carry over to continuous mathematical objects. Discrete math also has several connections to logic and rationality, which are invaluable tools in conquering the challenges of everyday society. As it turns out, the answer to the oft-asked student question, "When will we ever use this in the real world?" is a rather surprising, "All the time." Especially with discrete math!

---

[1] However, the *Planck length* is the distance at which the laws of quantum physics begin to take effect over the laws of classical physics, and as such, it is often misconstrued as a hypothetical "smallest length."

Chapter 10

Beginning Discrete Math

10.1 Introduction to Set Theory

As the study of countable, isolated sets, discrete math naturally invites itself, first and foremost, to the study of *set theory:* the analysis of collections of objects that have several intriguing, unique, and elegant properties.

At its core, set theory revolves around just one mathematical object: the *set,* which is loosely defined as any collection of objects.

> **Definition 10.1: Definition of a Set**
>
> A set is a well-defined collection of distinct objects, and is itself considered a mathematical object.

For example, $\{1, 3, 5\}$ is considered a set, but not $\{1, 3, 3\}$ (although both are examples of objects in and of themselves). This definition, however, could use some additional rigor:

> **Definition 10.2: A More Rigorous Definition of a Set**
>
> Any set must abide by the axioms laid out in *axiomatic set theory,* which posit the following:
>
> - A set consists of either one or more *elements,* which are objects within the set, or zero elements (in which case it is known as the *empty set,* denoted by $\emptyset$).
> - As aforementioned, all sets must not have duplicate elements.
> - Two sets are *equal* if and only if each element of one set is also an element of the other.
> - If a collection of objects should contain two or more copies of the same element, it is equal to the set containing only one copy of that element. This is known as the *extensionality property of sets.*

10.1.1 Defining and Working with Sets and Set Notation

Sets are most commonly denoted by capital letters; if there are multiple sets in a sequence of sets, the sets will almost always be named in alphabetical order. Otherwise, S is a very common name for a singular set (alongside A).

To define a set and its elements, we use notation such as the following:

$$A = \{1, 2, 3, 4\} \tag{10.1}$$

$$B = \text{the set of all rational numbers with a denominator less than 10} \tag{10.2}$$

$$C = \{1, 2, 3, 4, \ldots, 2019\} = \text{the set of all positive integers up to 2019} \tag{10.3}$$

As in equation (10.3), if we want to define a set with many elements that obey a clear pattern, we use ... (ellipses) to indicate that the elements continue to obey that pattern. Often, however, ellipses do not suffice to indicate a pattern, especially an unclear/ambiguous one, one that ceases after a certain point, or one that switches to another pattern midway through. In these cases, mathematicians use *set-builder notation* to characterize a set, such as in the following:

$$D = \{x^2 + x + 1 \mid x \in \mathbb{N}\} \tag{10.4}$$

$$E = \left\{ \frac{x+1}{x^2+1} \mid x \in \mathbb{R} \right\} \tag{10.5}$$

In set-builder notation, the range of the set is indicated by the expression to the left of the vertical line, and the domain of the set is indicated by the expression to the right of the vertical line.

When working with multiple sets, we must define additional operations:

Definition 10.3: Operations on Multiple Sets

Let A and B be two sets.

- The *union* of A and B, denoted $A \cup B$, is the set consisting of all elements that are either in A only, in B only, or in both A and B.

- The *intersection* of A and B, denoted $A \cap B$, is the set consisting of all elements that are in both A and B.

- The *sum* of A and B (also known as the *Minkowski sum*, denoted $A + B$, is the set $C = \{a + b \mid a \in A, b \in B\}$; i.e. the set consisting of all elements that are the sum of one element from A and an element from B.

The *difference* of A and B, denoted $A - B$ or $A \backslash B$, is unlike the Minkowski sum, however, in that it denotes the set A with any elements that are also in B removed; $A - B = \{x \mid x \in A, x \notin B\}$.

For three or more sets, the union of those sets is the set consisting of elements that are in *any* of the sets, and the intersection of those sets is the set consisting of elements that are in *all* of the sets.

Shorthand notations for the union/intersection of sets $S_1, S_2, S_3, \ldots, S_n$ are $\cup_{k=1}^n S_k$ and $\cap_{k=1}^n S_k$ respectively (similar to the sum/product shorthand notations, $\sum$ and $\prod$).

Furthermore, we say that A and B are *disjoint* if $A \cap B = \emptyset$ (i.e. $A \cup B = C$ where C is a set consisting of all of the elments in A and in B).

In addition, each set has a size, or *cardinality*:

Definition 10.4: Cardinality of a Set

The cardinality of a finite set S, denoted $|S|$, is the number of elements it contains. For infinite sets, we can define the cardinality in terms of injectivity and bijectivity:

1. Of two sets A and B, we say that $|A| = |B|$ if there exists a bijection from A to B. In this case, A and B are *equipotent* or *equinumerous*.

2. We say that $|A| \leq |B|$ if there is an injective function from A to B. The inequality is strict if there is an injective function, but no bijective function, from A to B.

Furthermore, we use the following standard notations to denote special, commonly used sets: $\mathbb{N}$ for the natural numbers (positive integers; i.e. the set $\{1, 2, 3, 4, \ldots, \}$), $\mathbb{Z}$ for the integers, $\mathbb{Q}$ for the rationals (and, by extension, $\mathbb{R}\backslash\mathbb{Q}$ for the irrationals), $\mathbb{R}$ for the real numbers, and $\mathbb{C}$ for the complex numbers.

10.1.2 Subsets and Supersets

Sets can be defined not only in relation to English vocabulary, counting numbers or functions with a domain and range, but also in relation to other sets! Sets can be *subsets* or *supersets* of other sets:

> **Definition 10.5: Subsets and Supersets**
>
> Let A and B be two sets.
>
> - A is a *subset* of B (denoted $A \subseteq B$) if all of the elements in A are also contained in B; that is, $\forall x \in A$, $x \in B$. We write $A \subset B$ (without the line at the bottom) if $\exists b \in B$ such that $b \notin A$.
>
> - Similarly, A is a *superset* of B (contains B, denoted $A \supseteq B$) if all of the elements in B are also contained in A. If $\exists a \in A$ such that $a \notin B$, we write $A \supset B$ strictly.
>
> A specific type of subset is the *complement*, which is defined for a set A in terms of a superset B of A. The complement A^c of A is given by $A^c = \{x \mid x \in B, x \notin A\}$. (Here, B is commonly known as a *universal set*. Another common notation for this is U.)

Sets can even be members of other sets! For instance, consider $A = \{1\}$ and $B = \{1, \{1\}\}$. Even though the two elements in set B look similar, they are completely different, and by extension, set B is completely different from set A.

10.1.3 Power Sets

Another important object associated with a set is its *power set*:

> **Definition 10.6: Power Set**
>
> The power set $\mathcal{P}(A)$ (sometimes 2^A) of a set A is the set consisting of all subsets of A.

> **Corollary 10.0: Cardinality of the Power Set**
>
> If $|A| = n$, then $|\mathcal{P}(A)| = 2^n$.

Proof. For each element of A, we can choose to either include it or not include it. There are n elements in A, and 2 choices for each element. Each set of choices corresponds to a unique element in $\mathcal{P}(A)$, so $|\mathcal{P}(A)| = 2^n$. $\square$

Example 10.1.3.1. *What is the power set of $A = \{\emptyset, 1, \{2\}\}$?*

Solution 10.1.3.1. *The power set is the set of subsets of A, or*

$$\mathcal{P}(A) = \{\emptyset, \{\emptyset\}, \{1\}, \{\{2\}\}, \{\emptyset, 1\}, \{\emptyset, \{2\}\}, \{1, \{2\}\}, \{\emptyset, 1, \{2\}\}\}$$

Be very careful here; the set $\{\emptyset\}$ is not empty! It contains one element: $\emptyset$. As a sanity check, we note that A has three elements ($\emptyset$, 1, and $\{2\}$ (not 2)), and so $\mathcal{P}(A)$ should have $2^3 = 8$ elements, which is indeed how many it has.

Example 10.1.3.2. *How many subsets does the set $\{-4, -3, -2, -1, 0, 1, 2, 3, 4\}$ have?*

Solution 10.1.3.2. *A set with n elements has 2^n subsets, since we can pick whether or not any given element is included in the subset. For $n = 9$, the number of subsets is $\boxed{512}$.*

Exercise 1. Let $S_1 = \{1\}$, $S_2 = \{1, \mathcal{P}(S_1)\}$, $S_3 = \{1, \mathcal{P}(S_2)\}$, and so forth. What is S_n in terms of n?

10.1.4 Open and Closed Sets

The idea of *open* and *closed sets* is a generalization of open and closed intervals on the real line. Defining open sets allows us in turn to define continuity, as well as the concepts of *connectedness* and *compactness* in a space.[2] To be precise, we define an *open set* as follows:

> **Definition 10.7: Definition of an Open Set in $\mathbb{R}^n$**
>
> Where a *set* is defined as a subset S of $\mathbb{R}^n$, S is open if, for any point $x \in S$, there exists a *ball* $B_\epsilon(y)$ of radius ϵ around x that is contained entirely within S; in other words, there exists an $\epsilon > 0$ such that, for any point $y \in \mathbb{R}^n$ with $|x - y| < \epsilon$, $y \in S$. Otherwise, S is closed iff $\mathbb{R}^n \backslash S$ is open, and neither open nor closed otherwise. (*Note that every set does not have to be open or closed! In fact, most sets fall into neither category.*)

We can define a ball in $\mathbb{R}^n$ more intuitively as follows. Consider the circular disk $\mathcal{C}$ in $\mathbb{R}^2$ with general equation $x^2 + y^2 \leq r^2$ (a circle with center $(0,0)$ and radius r, and its interior filled in):

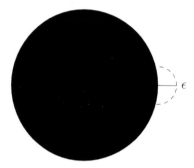

Figure 10.1: The circular disk with equation $x^2 + y^2 \leq r^2$ with closed boundary is an example of a closed set in $\mathbb{R}^2$.

In order for $\mathcal{C}$ to be open, for any point $p = (x, y) \in \mathcal{C}$, any ball $B_\epsilon(p)$ with radius $r > 0$ around p should be contained entirely within $\mathcal{C}$. However, if we take $p = (r, 0)$ (or any other point $p = (x, y)$ such that $x^2 + y^2 = r^2$), then $B_\epsilon(p)$ would extend to outside the circle, which contradicts the definition of an open set. Hence $\mathcal{C}$ is closed. However, the slightly different set $\mathcal{C}_r := \{(x, y) \in \mathbb{R}^2 \mid x^2 + y^2 < r^2\}$ would be open.

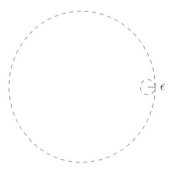

Figure 10.2: The circular disk with equation $x^2 + y^2 < r^2$ with open boundary is an example of an open set in $\mathbb{R}^2$. (Note the dashed lines on the boundary; the interior of the circle has not been colored, to aid with visibility.)

Proof. To prove this, we take an arbitrary point $p = (x_0, y_0) \in \mathbb{R}^2$ such that $x_0^2 + y_0^2 < r^2$. Let $r_1 > 0$, and let $x \in B_{r_1}(p)$, so that $|x - p| < r$ (where $|x - p|$ denotes the distance between x and p in $\mathbb{R}^2$). Take

---

[2] We study these concepts further in §16, "Poincaré Conjecture."

$r_2 = \frac{r_1 - |x-p|}{2}$ (in fact, this works for any constant $c > 1$ in the denominator), and let $y \in B_{r_2}(x)$. Then

$$|y - p| \leq |y - x| + |x - p| \leq \frac{r}{2} + \frac{|x - p|}{2} < r$$

by the Triangle Inequality, so $B_{r_2}(x) \subseteq B_{r_1}(p)$. Therefore, $B_{r_1}(p)$ is open. $\square$

10.1.5 de Morgan's Law

Logic and set theory are perhaps two of the most closely intertwined areas of mathematical study. Indeed, set theory is essentially a manifestation of all logical thought in mathematics using specialized notation. In this section, we will study one of the most prominent connections between the two disciplines: de Morgan's Law, which governs a crucial logical relationship between sets.

Theorem 10.1: de Morgan's Law

For any two sets A and B, we have the following:

(i) $(A \cup B)^c = A^c \cap B^c$.

(ii) $(A \cap B)^c = A^c \cup B^c$.

Here is a not-so-rigorous proof that nevertheless concisely demonstrates de Morgan's Law for Boolean variables A and B:

Proof.

(i) $(A \cup B)^c = A^c \cap B^c$.

A	B	$(A \cup B)^c$	$A^c \cap B^c$
T	T	F	F
T	F	F	F
F	T	F	F
F	F	T	T

(ii) $(A \cap B)^c = A^c \cup B^c$.

A	B	$(A \cap B)^c$	$A^c \cup B^c$
T	T	F	F
T	F	T	T
F	T	T	T
F	F	T	T

This shows that the laws are true when A and B are either true or false. $\square$

This proof, unfortunately, is considerably more time-consuming for sets of more than one element. Here is the standard, more rigorous proof that applies to all sets A, B:

Proof.

(i) Let $P = (A \cup B)^c$ and $Q = A^c \cap B^c$. Let $x \in P$ be arbitrary; then $x \in P$ implies $x \in (A \cup B)^c$, or $x \notin (A \cup B)$. Then $x \notin A$ and $x \notin B$, and so $x \in A^c$ and $x \in B^c$. This is synonymous with saying that $x \in (A^c \cap B^c)$, or $x \in Q$. Hence, $P \subseteq Q$.

Henceforth, let $y \in Q$ with y arbitrary. Then $y \in Q \implies y \in (A^c \cap B^c)$, or $y \in A^c$ and $y \in B^c$. Thus, $y \notin A$ and $y \notin B$, or $y \notin (A \cup B)$. It follows that $y \in (A \cup B)^c$, or $y \in P$. Therefore, $Q \subseteq P$. Combining this with $P \subseteq Q$ yields $P = Q$, or $(A \cup B)^c = A^c \cap B^c$, as desired.

(ii) Similarly, let $R = (A \cap B)^c$ and $S = A^c \cup B^c$. Let $x \in R$; $x \in R \implies x \in (A \cap B)^c$, or $x \notin (A \cap B)$. Hence, $x \notin A$ or $x \notin B$, which means that $x \in A^c$ or $x \in B^c$. Therefore, $x \notin A$ or $x \notin B$, from which it follows that $x \in A^c$ or $x \in B^c$, and so $x \in (A^c \cup B^c) \implies x \in S$. We then have $R \subseteq S$ as a direct result.

As before, let $y \in S$; then $y \in S \implies y \in (A^c \cup B^c)$, which implies $y \in A^c$ or $y \in B^c \implies y \notin A$ or $y \notin B$. Then $y \notin (A \cap B)$, or $y \in (A \cap B)^c = R$. Since $S \subseteq R$ as a result of this, $R = S$, or $(A \cap B)^c = A^c \cup B^c$ as desired.

This completes the proof of both parts of the theorem. $\square$

10.1.6 The Venn Diagram: A Visual Representation of Logical Set Theory

There are countless ways to represent set theory beyond the abstract notation, but by far the most familiar is the *Venn diagram*, which depicts the relationships between the cardinalities of specific subsets of a universal set. Venn diagrams consist of two or three circles that overlap with each other; these circles are usually labeled with the letters of the English alphabet (or whatever the sets happen to be named). (For $n \geq 4$, we cannot draw an n-set Venn diagram with circles. The proof of this is in §11.3, "Graph Theory: The Massive Web of Mathematics.") The intersections of the circles (the areas of overlap) represent the intersections of two sets, and the central intersection area of all circles is the intersection of all sets in the universal set.

In the standard diagram below, there are three subsets of a universal set U: A, B, and C:

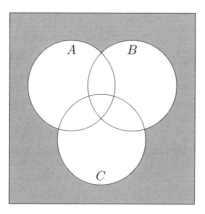

Figure 10.3: A standard Venn diagram, with $(A \cup B \cup C)^c$ shaded.

We can also depict the union and intersection of only specific sets by shading those parts of the Venn Diagram. For this part, we will limit our discussion to two-set diagrams (larger-set diagrams are similar).

The union of sets A and B in some universal set U can be depicted by shading in the circles corresponding to sets A and B:

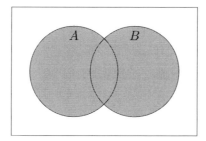

Figure 10.4: A two-set Venn diagram with $A \cup B$ shaded.

Similarly, the intersection of sets A and B can be depicted by shading in the area of overlap between circles A and B:

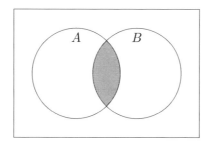

Figure 10.5: A two-set Venn diagram with $A \cap B$ shaded.

Exercise 2. *How many distinguishable Venn diagrams (in the sense that one can tell them apart visually based on which parts of the diagrams are shaded) are there that consist of 2 sets? What about 3 sets?*

10.1.7 The Relationship Between Set Theory and Cartesian Geometry

Set theory has numerous deeply-rooted connections with not only logic, but also other branches of math altogether. Namely, geometry links to set theory in intriguing and unconventional ways, one of which is the notion of the *Cartesian product*.

The Cartesian Product

To best understand the Cartesian product, first recall that $\mathbb{R}^2$, or the familiar two-dimensional Cartesian coordinate plane, is the set of all points with two real coordinates, namely $\mathbb{R}^2 = \{(x, y) \mid x, y \in \mathbb{R}\}$. (Similarly, $\mathbb{R}^n = \{(x_1, x_2, x_3, \ldots, x_n) \mid x_i \in \mathbb{R}, \forall i \in [1, n]\}$.) We can say that $\mathbb{R}^2$ is the "product" of the x- and y-axes, according to the general definition:

Definition 10.8: Cartesian Product

The *Cartesian product* of two sets A and B is given by $A \times B = \{(a, b) \mid a \in A, b \in B\}$.

Remark. Note that the Cartesian product is *not commutative!* In general, $A \times B \neq B \times A$. This is because $(a, b) \neq (b, a)$; that is, n-tuples are *ordered*, unlike sets, which are *unordered*.

Example 10.1.7.1. *If $A = \{1, 2, 3\}$ and $B = \{4, 5\}$, then*

$$A \times B = \{(1, 4), (1, 5), (2, 4), (2, 5), (3, 4), (3, 5)\}$$

We can write $\mathbb{R}^2 = \{(x, y) \mid x \in \mathbb{R}, y \in \mathbb{R}\}$, which shows that $\mathbb{R}^2 = \mathbb{R} \times \mathbb{R}$.

> **Theorem 10.2: Cardinality of the Cartesian Product**
>
> For any two sets A and B, $|A| \cdot |B| = |A \times B|$.

Here are some examples that may aid with the intuition:

Example 10.1.7.2. *If we roll two fair six-sided dice simultaneously, there are 6^2 possible outcomes in the Cartesian product, and there are 6 possible outcomes for each individual die.*

Example 10.1.7.3. *Suppose that we flip three fair coins in succession. There are 2 possible outcomes for each of the coins C_1, C_2, and C_3, namely H (heads) and T (tails). Thus, we may write $C_i = \{H, T\}$ for $i = 1, 2, 3$. Then $C_1 \times C_2 \times C_3 = \{(H,H,H), (H,H,T), (H,T,H), (H,T,T), (T,H,H), (T,H,T), (T,T,H), (T,T,T)\}$, which has $8 = 2^3$ elements.*

Example 10.1.7.4. *For each of the following, state with justification whether the statement is true or false.*

(a) $\left(0, \dfrac{1}{2}\right) \in \mathbb{Z} \times \mathbb{Q}$.

(b) $\left(2, \dfrac{2}{3}\right) \in \mathbb{Q} \times \mathbb{Z}$.

(c) $(\sqrt{2}, 3) \in (\mathbb{R}\backslash\mathbb{Q} \times \mathbb{R}\backslash\mathbb{Q})^c$.

(d) $(\sqrt{-2}, i^4) \notin (\mathbb{C} \times \mathbb{C})^c$.

Solution 10.1.7.1.

(a) True. $0 \in \mathbb{Z}$, and $\dfrac{1}{2} \in \mathbb{Q}$.

(b) False. $2 \in \mathbb{Q}$, but $\dfrac{2}{3} \notin \mathbb{Z}$ (it is not an integer).

(c) True. $(\mathbb{R}\backslash\mathbb{Q} \times \mathbb{R}\backslash Q)^c$ is the set $\mathbb{C}^2 - \{(x,y) \mid x, y \in \mathbb{R}\backslash\mathbb{Q}\}$, i.e. the set of ordered pairs whose coordinates are not both irrational. Since $3 \in \mathbb{Q}$, the statement is true (even though $\sqrt{2} \in \mathbb{R}\backslash\mathbb{Q}$).

(d) True. Not only are both $\sqrt{-2}$ and i^4 in $\mathbb{C}$ (even though $i^4 = 1 \in \mathbb{R}$ as well), but even if they weren't, the set $(\mathbb{C} \times \mathbb{C})^c$ is empty (remember that $\mathbb{C}^2$ is the universal set).

Balls in Generalized Metric Spaces

In this section, we'll briefly revisit the idea of a *ball*, and attempt to formalize it at least a little bit more. First we need to define the concept of a *metric space*:

> **Definition 10.9: Metric Spaces**
>
> A *metric space* is an ordered pair (M, d) where M is a set and d is a *metric*, often called a *distance function*, on M. d is defined to be a function from $M \times M$ to $\mathbb{R}$ such that, for any $x, y, z \in M$, we have
>
> (a) $d(x, y) \geq 0$, and $d(x, y) = 0$ iff $x = y$.
>
> (b) $d(x, y) = d(y, x)$.
>
> (c) $d(x, y) + d(y, z) \geq d(x, z)$ (Triangle Inequality).

Exercise 3. *Show that the first part of condition (a) follows from the second part of condition (a), and conditions (b) and (c).*

Example 10.1.7.5. *The following are examples of metric spaces:*

1. *The real line;* $d(x,y) = |x-y|$.

2. $\mathbb{R}^2$ *equipped with its usual "distance function":* $d((x_1,y_1),(x_2,y_2)) = \sqrt{(x_2-x_1)^2 + (y_2-y_1)^2}$, *often called the 2-metric* d_2.

3. *The taxicab metric* $d((x_1,y_1),(x_2,y_2)) = |x_2-x_1| + |y_2-y_1|$, *often called the 1-metric* d_1.

4. *The p-metric* d_p *for any* $p \in \mathbb{N}$ *(essentially the Distance Formula in p dimensions).*

5. *The supremum metric or maximum metric* d_∞, *given by* $d((x_1,y_1),(x_2,y_2)) = \max(|x_2-x_1|,|y_2-y_1|)$.

6. *The discrete metric:* $d(x,y) = 0$ *if* $x=y$, *and* $d(x,y) = 1$ *otherwise.*

Exercise 4. *Verify that all of the above are examples of metric spaces.*

Once we have defined the notion of a metric space,[3] we can proceed to contextualize our current understanding of balls within it.

Earlier, we studied the nature of open and closed sets, or, more generally, balls, in $\mathbb{R}^2$, all the while assuming the standard distance formula, the distance function (or metric) $d_2 := d((x_1,y_1),(x_2,y_2)) = \sqrt{(x_2-x_1)^2 + (y_2-y_1)^2}$. However, in general, the definition of a ball can be extended to any distance metric as follows:

> **Definition 10.10: Definition of Open and Closed Balls in General Metric Spaces**
>
> If (M,d) is a metric space, $a \in M$, and $r > 0$, then the open ball $B_r(a)$ with radius r, centered at a, is the set $\{x \in M \mid d(x,a) < r\}$.
>
> Likewise, the closed ball $\overline{B}_r(a)$ is defined by the set $\{x \in M \mid d(x,a) \leq r\}$ (the difference being non-strict equality rather than strict equality).

---

[3] We traditionally also define *complete* metric spaces, otherwise known as *Banach spaces*, if they are also *normed vector spaces*(of which metric spaces are a superset), but this is a topic usually covered in a university-level functional analysis course, and so will not be touched upon here. Defining a Banach space first requires that we define Cauchy sequences, which are rooted in firm understanding of calculus topics.

10.1.8 Relations Within a Set

We can also define *binary relations* over a set: the *reflexive, symmetric,* and *transitive* relations, as well as the *equivalence relation*. First we must define a binary relation:

> **Definition 10.11: Binary Relations Over a Set**
>
> A binary relation R on a set S is a set of ordered pairs of elements of S; it is a subset of S^2. (Similarly, a binary relation on $S \times T$, where S and T are sets, is a subset of $S \times T$. Here $\times$ denotes the Cartesian product.) It is also called a *correspondence*.
>
> If s and t are elements of S and T respectively, then we write sRt if s is related to t in the manner described by the binary relation R. For example, if R is defined by $\geq$ (greater than or equal to), $s = 6$, and $t = 5$, then sRt, but if $s = 5$ and $t = 6$, then sRt does not hold.

> **Theorem 10.3: Types of Binary Relations Over a Set**
>
> Let R be a binary relation on a set X. Then R may have the following properties:
>
> (i) Reflexivity: For all $x \in X$, xRx.
>
> (ii) Symmetry: For all $x, y \in X$, xRy iff yRx.
>
> (iii) Transitivity: For all $x, y, z \in X$, if xRy and yRz, then xRz.
>
> Finally, R is an *equivalence relation* if it is a reflexive, symmetric, and transitive relation.

Example 10.1.8.1. *Is the binary relation "divides" over $X = \{1, 2, 4, 8, \ldots, 512\}$ reflexive, symmetric, transitive, an equivalence relation, or none of those?*

Solution 10.1.8.1. *The "divides" relation is reflexive, since any $x \in X$ divides itself (this is true of any real number). However, it is not symmetric, since for any $x, y \in X$, $x > y$, x does not divide y (even if y divides x). In addition, the relation is transitive; note that all of the elements in X are powers of 2. If $\exists x, y \in X$ such that $x \mid y$, then we necessarily have $x \leq y$, so the $x > y$ issue is moot. Since $x, y \in \{2^k \mid k \in \mathbb{N} \cup \{0\}\}$, $x \mid y$ whenever $x \leq y$, and so for any $x, y, z \in X$ (not necessarily distinct), $(xRy) \wedge (yRz) \implies xRz$. Finally, since R is not symmetric on X, it is not an equivalence relation.*

Remark. We can also represent binary (as well as ternary, quaternary, quinary, and n-ary) relations discretely as either graphs or matrices.

Equivalence relations also give rise to *equivalence classes;* disjoint partitions $[x]$ of a set X such that, for all $x \in X$, $[x] = \{y \in X \mid x \sim y\}$, which is the set of elements $y \in X$ equivalent to x. (The notation $x \sim y$ denotes that x is equivalent to y.) In equivalence classes, all of the following hold:

> **Theorem 10.4: Equalities of Equivalence Classes**
>
> (a) For all $x \in X$, $x \in [x]$.
>
> (b) $X = \bigcup_{x \in X} [x]$.
>
> (c) $x \sim y \iff [x] = [y]$.
>
> (d) $x \nsim y \iff [x] \cap [y] = \emptyset$.

Proof.

(a) This follows immediately from the fact that equivalence is reflexive.

(b) This, too, follows directly from $x \in [x]$ for all $x \in X$.

(c) For any $z \in [x]$, $x \sim z$. Since $x \sim y$, $y \sim x$ by the symmetric property of an equivalence relation, and so $y \sim z$ by the transitive property. Thus, $z \in [y]$, which in combination with $z \in [x]$, yields $[x] \subset [y]$. Since $\sim$ is symmetric, $[y] \subset [x]$, which implies $[x] = [y]$.

(d) Assume the contrary; namely, that $z \in [x] \cap [y]$. Then $x \sim z$ and $y \sim z \implies z \sim y$, which in turn implies $x \sim y$ by the transitive property, but this is a contradiction.

This proves the properties of all equivalence classes. $\square$

Example 10.1.8.2. *An example of an equivalence class over $\mathbb{Z}$ is $[x] = \{y \in \mathbb{Z} \mid x \equiv y \bmod 3\}$. $[x]$ can be partitioned into three disjoint subsets: namely, $[0] = [3k]$, $[1] = [3k+1]$, $[2] = [3k+2]$ for $k \in \mathbb{Z}$.*

10.1.9 Examples

Example 10.1.9.1. *Let $A = \{1, 2, 3, 4, 5\}$, $B = \{3, 4, 5, 6, 7\}$. Compute $A + B$, $A - B$, $A \times B$, $|A \times B|$, $|\mathcal{P}(A \cap B)|$, and $|\mathcal{P}(A \cup B)|$.*

Solution 10.1.9.1. *$A + B = \{4, 5, 6, \ldots, 12\}$, since every positive integer from 4 to 12, inclusive, can be obtained by summing an element from A and an element from B. $A - B = A \backslash B = \{1, 2\}$ (the set with elements in A that are not also in B). $A \times B = \{(1,3), (1,4), (1,5), (1,6), (1,7), (2,3), \ldots, (5,7)\}$, so $|A \times B| = 25 = 5^2$. $\mathcal{P}(A \cap B)$ is the power set of a set with three elements, so its cardinality is $2^3 = 8$. Similarly, $A \cup B$ has 7 elements, so $|\mathcal{P}(A \cup B)| = 2^7 = 128$.*

Example 10.1.9.2. *(2017 AIME II) Find the number of subsets of $\{1,2,3,4,5,6,7,8\}$ that are subsets of neither $\{1,2,3,4,5\}$ nor $\{4,5,6,7,8\}$.*

Solution 10.1.9.2. *Let $U = \{1,2,3,4,5,6,7,8\}$, $A = \{1,2,3,4,5\}$ and $B = \{4,5,6,7,8\}$. Since $A \cup B = U$, any subset S of U will have at least one element that is also in A, B, or both. Hence, we must consider those subsets S with elements in both $A - B = \{1,2,3\}$ and $B - A = \{6,7,8\}$.*

Optionally, we may chose to include a subset of $C = \{4,5\}$ as well, since $C \subset A$ and $C \subset B$; including 4 or 5 will not change the fact that a subset $S \subset U$ is not also a subset of A or B. There are 8 subsets of A, and 8 subsets of B; however, 1 of each of those is the empty set, so we have $7^2 = 49$ choices of non-empty subsets of A and B. We then multiply by $2^2 = 4$ to account for the choices from $C = \{4,5\}$ to obtain 196 valid subsets of U that are subsets of neither A nor B.

Example 10.1.9.3. *Show that the set $S = \bigcup_{n \in \mathbb{N}} n^{-1}$ is neither open nor closed.*

Solution 10.1.9.3. *Note that $0 = \lim_{n \to \infty} n^{-1}$ is a limit point of the set, but $0 \notin S$, so S cannot be closed. (That S is closed iff it contains all of its limit points is a corollary of the definition of an open set.) Furthermore, S is not open; if we take the ball $B_r(a)$ for any $a \in S$, with $r > 0$, the ball will necessarily contain points $p \notin S$, so S is not open. (We can also use a similar argument with $B_r(1)$ necessarily containing points $p > 1$ to show that S is not open.) Therefore, S is neither open nor closed.*

Example 10.1.9.4. *Draw the Venn diagram representing the following scenario:*

"At a summer cook-off, there are 150 people, and 100 people who can make at least one of hamburgers, ribs, or potato salad. There are 55 people can cook hamburgers, 38 people can cook ribs, 49 people can make potato salad, 22 people who can make both hamburgers and ribs, 12 people who can make both ribs and potato salad, 15 people who can make both hamburgers and potato salad, and 9 people who can make all three dishes."

Solution 10.1.9.4. *Here is a standard diagram, with $A =$ "hamburgers," $B =$ "ribs," and $C =$ "potato salad":*

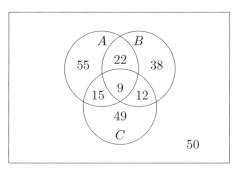

Example 10.1.9.5. *Give an example of a set X, and a binary relation R on X, satisfying each of the following, or show that no such ordered pair (X, R) exists:*

(a) R is an equivalence relation.

(b) R is reflexive, but not symmetric or transitive.

(c) R is symmetric, but not reflexive or transitive.

(d) R is transitive, but not reflexive or symmetric.

(e) R has exactly two of the properties.

(f) R has none of the three properties.

Solution 10.1.9.5.

(a) On the set $X = \mathbb{R}$, $\leq$ (as well as $\geq$) is an equivalence relation. It is reflexive, since $x \leq x$ for all $x \in \mathbb{R}$; it is symmetric, as $x \leq y \iff y \leq x$, and it is transitive, since if $x \leq y$ and $y \leq z$, then $x \leq z$ as well.[4]

(b) On the set $X = \mathbb{R}$, for $x, y \in X$, xRy iff $x < y + 1$ is reflexive (since $x < x + 1$ for all $x \in X$), but not symmetric (as it does not directly follow from $x < y + 1$ that $y < x + 1$), and not transitive for $x, y, z \in X$ (take $x = 1, y = 0.1, z = -0.8$).

(c) On the set of a group of people $X = \{Amy, Brody, Cathy\}$, $R =$ "is a mutual friend of" is symmetric, but not reflexive, since a person cannot be his/her own friend, and not transitive, since it may be the case that Amy and Brody are friends, and Brody and Cathy are friends, but not Amy and Cathy. (Another R that would work is $R =$ "are of different genders.")

(d) On the set $X = \mathbb{R}$, for $x, y \in X$, xRy iff $x < y$ is neither reflexive nor symmetric ($x \not< x$ for all $x \in X$, and $x < y$ does not imply $y < x$ (in fact, this is always false)). However, R is transitive, as for all $x, y, z \in X$, if $x < y$ and $y < z$, then necessarily, $x < z$.

(e) On the set of three siblings $X = \{Dan, Evelyn, Frank\}$, $R =$ "is a sibling of" is symmetric and transitive, but not reflexive (a person cannot be his or her own sibling).

(f) On the set $X = \{Rock, Paper, Scissors\}$, the binary relation $R =$ "beats" given by xRy if $(x, y) = (Rock, Scissors), (Paper, Rock), (Scissors, Paper)$ is non-reflexive, as no choice beats itself, non-symmetric, as, for instance, Rock beats Scissors but Scissors loses to Rock, and non-transitive; Rock beats Scissors, and Scissors beats Paper, yet Rock loses to paper. (Another R that would work would be a binary relation for the variant "Rock Paper Scissors Lizard Spock,"[5] or any similar variant with any number of choices.)

Another such relation is $R =$ "is a parent of." It is not reflexive (it is, in fact, irreflexive; i.e. $\forall x \in X$, $x \not\sim x$), since one cannot be one's own parent, not symmetric (as if x is the parent of y, then y is certainly not the parent of x), and not transitive (if x is the parent of y, and y is the parent of z, then x is not the parent of z, but rather the grandparent).

---

[4]This is just one example of many, and to be honest, not the most exciting one, but it is simple and familiar. We leave it to you to try to find as many examples as you can!

[5]Invented by Sam Kass and Karen Bryla, and popularized by the television show *The Big Bang Theory*.

10.1.10 Exercises

Exercise 5. *Each of the following sets is either described in English, or has its (first few) elements listed; write each in set-builder form.*

(a) $A = \{$the set of letters in the word "mathematics"$\}$

(b) $B = \{1, 1, 2, 4, 7, 13, 24, \ldots\}$

(c) $C = \{1, 2, 4, 8, 16, 32, 64, \ldots\}$

(d) $D = \{$the set of all perfect squares that have exactly 3 positive integral divisors$\}$

(e) $E = \{$the set of positive integers that can be written as $1 + 2 + 3 + \ldots + k$ for some positive integer $k\}$

(f) $F = \{2, 3, 5, 7, 11, 13, 17\}$

Exercise 6. *Using a Venn Diagram, demonstrate both of De Morgan's Laws.*

Exercise 7. *Show that $S = T$ iff $S \subseteq T$ and $S \supseteq T$.*

Exercise 8. *Prove that if $S \cap T = S$, then $S \subseteq T$.*

Exercise 9. *Prove that $(A^c)^c = A$ for all sets A in some universal set U.*

Exercise 10.

(a) *Prove that the interval $(1, 2)$ is open.*

(b) *Prove that the interval $[0, 1]$ is closed.*

(c) *Prove that the interval $(0, 2]$ is neither open nor closed.*

Exercise 11.

(a) *Prove that the union of closed sets is also closed.*

(b) *Prove that the intersection of open sets is also open.*

Exercise 12. *For any two sets A and B, show that $\mathcal{P}(A \cap B) = \mathcal{P}(A) \cap \mathcal{P}(B)$. Is it necessarily true that $\mathcal{P}(A \cup B) = \mathcal{P}(A) \cup \mathcal{P}(B)$?*

Exercise 13. *Is it possible for a set to be both open and closed? If not, provide a disproof. If so, provide an example of such a set.*

Exercise 14.

(a) *Prove or disprove the following: $A \times (B \cup C) = (A \times B) \cup (A \times C)$ for all sets A, B, C.*

(b) *Prove or disprove the following: $A \times (B \cap C) = (A \times B) \cap (A \times C)$ for all sets A, B, C.*

Exercise 15. *For each of the sets X and binary relations R on X, list whether or not R is a reflexive, symmetric, transitive, or equivalence relation on X, and justify your answers.*

(a) (i) $R = $ "is a spouse of," $X = \{$Michael, Steven, Nadia, Francis, Julia$\}$. Nadia is married to Francis, and Julia is married to Michael.

(ii) $R = $ "is a husband of," $X = \{$Michael, Steven, Nadia, Francis, Julia$\}$. Nadia is married to Francis, and Julia is married to Michael.

(b) $R = $ "is relatively prime with," $X = \mathbb{N}$

(c) (i) $R = $ "is at least 1 (Cartesian coordinate) unit away from," $X = \mathbb{N}^2$

(ii) $R = $ "is at most 1 (Cartesian coordinate) unit away from," $X = \mathbb{R}^2$

Extension Exercise 1. *Prove the cardinality relationships in Definition 10.4.*

Extension Exercise 2. *(2015 Berkeley Math Tournament) Find two disjoint sets N_1 and N_2 with $N_1 \cup N_2 = \mathbb{N}$, so that neither set contains an infinite arithmetic progression.*

Extension Exercise 3. *Prove that the intersection of two equivalence relations on a nonempty set X is also an equivalence relation.*

Extension Exercise 4. *Let $f : \mathbb{R} \mapsto \mathbb{R}$. Define a binary relation R on f by xRy iff $x = y \iff f(x) = f(y)$. Show that R is an equivalence relation. What are the equivalence classes for $f(x) = x$, $f(x) = x^2$, $f(x) = x^2 + 20x + 18$, and $f(x) = \sin x$? (Here, x is in radians, not degrees.)*

Extension Exercise 5. *Let $n \in \mathbb{N}$, and let $x, y \in \mathbb{Z}$. The equivalence relation $x \sim y$ of modulo n has an associated quotient set, $\mathbb{Z}_n = \mathbb{Z}/\sim = \{[0], [1], [2], \ldots, [n-1]\}$. For an integer $a \in \mathbb{Z}$, define $f_a : \mathbb{Z}_n \mapsto \mathbb{Z}_n$, $f_a([x]) = [ax]$. Prove that f_a is a bijection iff a and n are relatively prime.*

10.2 Introduction to Statistics

The everyday world is flooded with all sorts of data - from governmental census and tax data, to race, gender, and socioeconomic data collected in cities, states, and nations, to financial data in the stock market and in our own bank accounts, to the frightening contents of our performance evaluations. At times, we are so busy dealing with the stress of life that we take for granted that everything can work so seamlessly due to the power of *statistics*.

Statistics is the glue that binds masses of data into something cohesive and meaningful - it is the synthesis of data via data collection, organization, interpretation, and presentation. Statistical analysis is much like the brain's methods of processing stimuli. In order to process, and make sense of, the waves of stimuli that they are constantly confronted with, our brains must first receive the information from our sensory organs, which transform physical stimuli into electro-chemical signals (the *collection* of data). These signals must then travel through an *attention filter*, which decides which signals are most important and deserving of the brain's immediate attention; the most important signals head directly to the brain for organization.

From there, the brain encodes these signals into either sensory, working, or long-term memory (the *organization* of data). Once the information is encoded, it must then be maintained and analyzed. At this stage, the brain decides how it will handle the influx of information (the *analysis* of data). Once it assimilates the data, the brain decides how to react to the stimulus (the *presentation* of data). For example, if one touches a hot stove, the transmission of heat signals will lead the brain to emit pain signals that urge the immediate removal of the stimulus - taking one's hand off the stove.

If any one of these steps is skipped over, the brain cannot adequately process information, leading to potentially disastrous scenarios. Similarly, in statistics, if we fail to collect the data, we will obviously have nothing to work with; if we fail to organize the data, analysis will be vastly complicated; if we fail to analyze the data properly, the end result will likely be meaningless; and if we fail to present the data, we cannot use it to effect any change (much like keeping one's hand on the stove - just like nerve cells will begin to die, a real-world problem that may easily be resolved with statistics may end up being exacerbated beyond the point of no return). Statistics is an extremely important field of study, just like our brain is, by far, the most important part of our body. (Is statistics the most important field of study in *all* of mathematics? Debatable, but most versatile in its real-world applications? Statistics, no contest!) For now, we will be venturing into the world of *descriptive statistics*, the subfield of statistics that deals primarily with summative descriptions of a data set.

10.2.1 Statistical Terminology

With all of this in mind, how do we use sets to categorize data and interpret the world around us? The answer lies in the application of statistics to set theory. First, we relate a given set to an array of statistical terms:

> **Definition 10.12: Statistical Terminology**
>
> Consider a set $S = \{x_1, x_2, x_3, \ldots, x_n\}$ of data points.
>
> - The *mean* of the elements of S is equal to $\dfrac{x_1 + x_2 + x_3 + \ldots + x_n}{n}$. (The mean is also referred to as the *arithmetic mean*, as opposed to the *geometric mean*, $\sqrt[n]{x_1 x_2 x_3 \cdots x_n}$.)
>
> - The *median* of the elements of S is their "middle" element, when they are arranged from smallest to largest. If the x_i are already in that order, the median depends on the parity of n. If n is odd, the median is $x_{\left(\frac{n+1}{2}\right)}$; if n is even, the median is $\dfrac{x_{\left(\frac{n}{2}\right)} + x_{\left(\frac{n}{2}+1\right)}}{2}$ (i.e. the mean of the middlemost two terms).
>
> - The *mode* of the elements of S is the most frequent element. If S has no most frequent element (i.e. if it has no duplicate elements), then S does not have a mode. If there are n elements with the same number of instances in S, then all of them are modes of S. We say S is *bi-modal*, *tri-modal*, or more generally, *n-modal*.
>
> - The pth *percentile* of S ($0 \leq p < 100$) is the (smallest) value[a] which exactly $p\%$ of the values in S fall under.[b]
>
> - The *quartiles* of S are the *first*, *second* (which is also the median), and *third*, which are the elements at the 25th, 50th, and 75th percentiles of S, respectively. (If there is no element exactly at those percentiles, the corresponding quartile is defined to be the mean of the two elements immediately above and below the percentile in question.)[c]
>
> - The *interquartile range* (IQR) of S is the positive difference between the first and third quartiles of S.
>
> - The *five-number summary* of S consists of the minimum, first quartile, median, third quartile, and maximum.

- The *standard deviation* of S has two variants: the *sample standard deviation* and the *population standard deviation*. The former is calculated by summing the squares of the differences of each element from the mean, dividing by $n-1$, and then taking the square root of the entire fraction; the latter is calculated similarly except we divide by n rather than $n-1$. The formula for the SSD is $\sigma = \sqrt{\dfrac{\sum_{i=1}^{n}(x_i - \bar{x})^2}{n-1}}$ and the formula for the PSD is identical except for the denominator, where σ denotes the SSD/PSD and $\bar{x}$ denotes the mean of S.

---

[a] We must make a distinction here between discrete and continuous sets; for discrete sets, we assign to the pth percentile the smallest value above $p\%$ of the data points, but for continuous sets, the pth percentile is one specific value.

[b] It is also of note that some texts require that $p\%$ of the values are less than *or equal to* the pth percentile, so that $0 < p \leq 100$ (the "optimistic" definition; ours is the "pessimistic" definition). Others make a compromise and state that the percentile associated to a data point is the mean of the percentage of values strictly less than the data point, and the percentage of values less than or equal to the data point (so that $0 < p < 100$, with strict inequality on both sides). For instance, the element 2 in the set $S = \{1, 2, 3, 4, 5\}$ would be at the 20th percentile using the pessimistic definition in this book, the 40th percentile using the optimistic definition, and the 30th percentile using the compromise definition.

[c] Using the compromise definition here, as opposed to the pessimistic or optimistic definitions, allows us to conveniently define the median as the 50th percentile when there are an odd number of elements, say $2k+1$, in a set. Under the pessimistic definition, the percentile of the median is $\left(100 \cdot \dfrac{k}{2k+1}\right)\%$, and under the optimistic definition, the percentile of the median is $\left(100 \cdot \dfrac{k+1}{2k+1}\right)\%$. Indeed, the mean of these is $\left(100 \cdot \dfrac{k+0.5}{2k+1}\right)\% = 50\%$.

Let's solidify these concepts with a computational example:

Example 10.2.1.1. *Let $S = \{n^2 - 2n \mid n \in [1, 10], n \in \mathbb{N}\}$. Compute the mean, median, mode, SSD, PSD, and five number summary of S.*

Solution 10.2.1.1. *The mean of the elements in S is equal to* $\dfrac{1}{10} \cdot \left(\sum_{k=1}^{10} k^2 - 2k\right) = \dfrac{1}{10} \cdot \left(\sum_{k=1}^{10} k^2 - \sum_{k=1}^{10} 2k\right) = \dfrac{1}{10} \cdot \left(\dfrac{10 \cdot 11 \cdot 21}{6} - 2 \cdot \dfrac{10 \cdot 11}{2}\right) = \dfrac{1}{10} \cdot (385 - 110) = \dfrac{1}{10} \cdot 275 = 27.5$. *Since the number of elements of S is even, the median of S is the mean of the fifth and sixth elements, or* $\dfrac{(5^2 - 2 \cdot 5) + (6^2 - 2 \cdot 6)}{2} = \dfrac{15 + 24}{2} = 19.5$. *Furthermore, S has no mode, since all of its elements are distinct. The SSD of S is* $\sigma = \sqrt{\dfrac{\sum_{i=1}^{10}(x_i - 27.5)^2}{9}} = \sqrt{\dfrac{(-1 - 27.5)^2 + (0 - 27.5)^2 + \ldots + (80 - 27.5)^2}{9}} = \sqrt{\dfrac{7210.5}{9}} = \sqrt{\dfrac{4807}{6}} \approx 28.305$, *and the PSD is* $\sqrt{\dfrac{7210.5}{10}} = \sqrt{721.05} \approx 26.852$. *The five-number summary of S is* $\{-1, 3, 19.5, 41.5, 80\}$.

10.2.2 Visualizing Statistics

Statistics is a very visual field of study; we can interpret and represent real-world quantities and relationships through a vast array of lenses. We will begin our exploration of the various methods of depicting data with introducing the types of graphs customarily used to illustrate data.

Types of Graphs

A *line graph*, or *frequency polygon*, represents data points pictorially via plotting them as points in the xy-plane. We can then draw a line through each pair of consecutive data points to obtain a line graph. This line may pass through all of the data points in one clean stroke, or it may be jagged and mountainous, requiring several line segments of small length to accurately depict the trends in the data. With the exception of graphs of a specialized nature, most line graphs deal only with non-negative x- and y-values, so the origin and the first quadrant are part of a standard line graph. The variable in the data represented by the x-axis is the *independent variable*, since it is independent of any prior input; the variable represented by the *y*-axis is the *dependent variable,* as its value is dependent on that of the independent variable. As a direct consequence, the curve plotted in a line graph must always be a function (since no two *y*-values can be mapped to a single *x*-value, the curve must pass the Vertical Line Test). We can also plot multiple lines in the same graph; in fact, this is most often done to illustrate the relationships between two or more sets of data in the same context.

Line graphs have their most versatile real-world applications in determining trends over time; the *x*-axis in this case most often is broken up into months or years. The *y*-axis, in turn, is usually labeled with some relevant real-world quantity, such as revenue in (thousands or millions of) dollars, temperature in degrees Fahrenheit or Celsius, the percentage probability of an event occurring, or time in minutes or hours.

An example of a line graph is as follows, using the following three data sets:

$$A = \{(2,3), (5,6), (15,7), (17,8)\}$$

$$B = \{(2.5, 1.5), (3, 2.5), (6, 5)\}$$

$$C = \{(1,6), (1.5, 7), (4.5, 8.5), (7, 9), (12, 10)\}$$

where the *x*-axis represents the time in hours spent studying for a test for students A, B, and C, and the *y*-axis represents their grades out of 10:

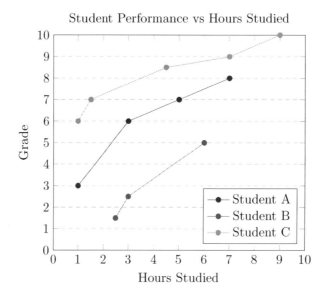

Figure 10.6: The line graph corresponding to the three data sets above. Note that the title in any graph usually follows the format "*y* vs *x*," where *x* and *y* are the independent and dependent variables, respectively.

Whereas a line graph depicts data by connecting them via a line in the positive *xy*-plane, a *bar graph*, or *histogram*, illustrates the quantities of data according to specific intervals that they fall under. The *x*-axis is separated into several brackets that represent ranges for the representation of data, and the *y*-axis shows the quantity of data within each bracket. The more of the data is present in a bracket, the taller its bar will be. Similarly to a line graph, the axes of a bar graph usually begin at zero and extend into positive territory. (However, there are exceptions in which the starting point can be negative.) On the next page is a bar graph that represents the studying time/grade data for student A.

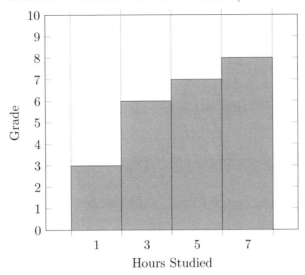

Figure 10.7: A bar graph representing Student A's studying time/grade data.

A *pie chart,* much like a line or a bar graph, is a two-dimensional representation of data, but is even more visual in that it depicts larger quantities of data as occupying larger portions of a "pie," or a 360° circular arc. The value of each data point in a set is depicted as proportional to the total value.

As an example, consider the set $S = \{\text{red}, \text{blue}, \text{blue}, \text{green}, \text{green}, \text{green}\}$ of marble colors in a bag. The sum of the elements in S is 6, which we set equal to the full 360° circular arc. Let x, y, and z be the portions, in degrees, associated to red, blue, and green respectively. Then $x = \dfrac{y}{2} = \dfrac{z}{3}$ and $x + y + z = 360$, so $x = 60$, $y = 120$, and $z = 180$.

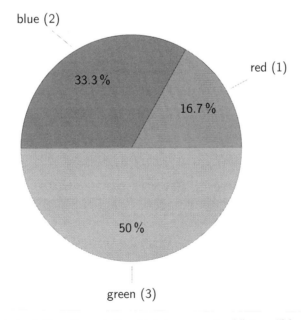

Figure 10.8: A pie chart depicting the set S, which has one "red," two "blues," and three "greens."

Scatter plots are perhaps the "most discrete" of the discrete visualizations: they imagine each data point as a separate entity in the positive xy-plane. They are used to identify the relationships, if there are any, between the independent and dependent variables in the data. As opposed to a bar graph, line graph, or pie chart, a scatter plot has a *line of best fit*. Informally speaking, this line is "closest," on average, to all of the data points. Formally speaking, the sum of the squares of the smallest distances from each point to the line is minimized (via *linear regression;* this produces the *least squares regression line (LSRL)*.[6]

Scatter plots indicate the *correlation* between the data variables, as precisely measured by the *correlation coefficient* r, which is a real number between -1 and 1, inclusive. A coefficient of 0 indicates no correlation, a positive coefficient indicates that the independent and dependent variable increase or decrease in tandem, and a negative coefficient indicates that the independent and dependent variables are in disagreement; i.e. one increases while the other decreases. The higher the absolute value of the coefficient, the stronger the agreement or disagreement. For instance, a correlation of -0.90 is stronger of a negative correlation than a correlation of $+0.40$ is a positive correlation. In this sense, the correlation coefficient can roughly be thought of as the "slope" of the data, or perhaps slightly more formally, as an approximation of the slope of the LSRL.

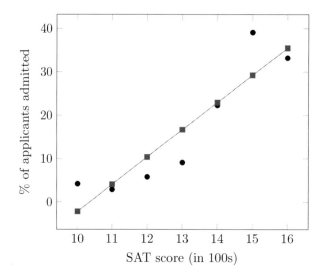

Figure 10.9: A scatter plot of college admission rate (in %) vs. SAT scores (in 100s) with correlation coefficient $r = 0.912$. (Disclaimer: this is purely fictitious data used solely for the purpose of illustrating a mathematical concept. Any resemblance to actual college admissions data is purely coincidental.)

In the scatter plot above, a noteworthy point is that the (imaginary) students who scored in the 1000s on the SAT actually had a higher acceptance rate to this (equally imaginary) college than did students who scored in the 1100s! This may be explained by the former group of students working extra hard in their classes and attaining higher GPAs to compensate for their lower test scores, or perhaps by athletic scholarships that the students scoring in the 1100s lacked.

Similarly, the students with perfect 1600s were actually admitted in lesser numbers than the students with 1500s. This may be accounted for by complacency on the part of the perfect scorers in keeping high GPAs, since they may (mistakenly) feel that their perfect test score will impress admissions officers and somehow automatically grant them admission despite potentially mediocre extracurriculars and essays. Another factor may be the determination of the not-quite-perfect SAT scorers to "beat out" their 1600-scoring competitors with stellar GPAs, awards and honors, and extracurricular activities, glowing letters of recommendation, and extraordinary personal statements.

This shows that, although the data undeniably has a very strong positive correlation (meaning that, the higher a student's SAT score, the higher his or her chances of admission *in general*), there are exceptions. Thus, the correlation coefficient is less than 1 (albeit only slightly in this example).

---

[6]The precise algorithm for computing the LSRL relies on methods traditionally taught in a post-secondary linear algebra course, which we have covered in §6.8.

To compute the correlation coefficient, we use the following formula:

$$r = \frac{\sum_{i=1}^{n}(x_i - \bar{x})(y_i - \bar{y})}{\sqrt{\sum_{i=1}^{n}((x_i - \bar{x})(y_i - \bar{y}))^2}}$$

where n is the number of data points. An alternate formula is

$$r = b_1 \cdot \sqrt{\sum_{i=1}^{n}\left(\frac{x_i - \bar{x}}{y_i - \bar{y}}\right)^2}$$

where b_1 is the estimated slope of the LSRL.

Proof. Let X and Y be the independent and dependent variable, respectively. The formula is an expanded form of the more concise formula

$$r = \frac{\text{cov}(X, Y)}{\sigma_X \cdot \sigma_Y}$$

where $\text{cov}(X, Y)$ denotes the *covariance* of X and Y, and σ_X and σ_Y denote the (population) standard deviations of X and Y, respectively. The covariance, $\text{cov}(X, Y)$, is defined by $\mathbb{E}[(X - \mathbb{E}[X])(Y - \mathbb{E}[Y])]$, where $\mathbb{E}[X]$ denotes the expected value of the variable X. (By the linearity property of expectation, this can be simplified to $\mathbb{E}[XY] - \mathbb{E}[X]\mathbb{E}[Y]$.)

We show that $|r| \leq 1$. We have

$$|\text{cov}(X, Y)| = |\mathbb{E}[(X - \mathbb{E}[X])(Y - \mathbb{E}[Y])]|$$

by definition. By the Cauchy-Schwarz Inequality,

$$|\mathbb{E}[(X - \mathbb{E}[X])(Y - \mathbb{E}[Y])]| \leq \sqrt{\mathbb{E}[(X - \mathbb{E}[X])^2]}\sqrt{\mathbb{E}[(Y - \mathbb{E}[Y])^2]} = \sqrt{\text{Var}(X)}\sqrt{\text{Var}(Y)}$$

It follows that

$$\left|\frac{\text{cov}(X, Y)}{\text{Var}(X)\text{Var}(Y)}\right| \leq 1$$

as desired. $\square$

We can also define the *coefficient of determination* r^2, which is the proportion of variance in the dependent variable that can be predicted from the flux of the independent variable. More so than the correlation coefficient r, it is primarily used in statistical models to test hypotheses and predict future outcomes. As $r \in [-1, 1]$, $r^2 \in [0, 1]$.

The Box-and-Whisker and Stem-and-Leaf Interpretations of Mean, Median, and Mode

Using descriptive statistics, we can relate the measurements of mean and median to another type of data visualization: the *box-and-whisker plot*. In addition, we can use the *stem-and-leaf plot* to concisely illustrate the general distribution of data points in given brackets.

A *box-and-whisker plot,* or simply a *box plot,* is a pictorial representation, either horizontal or vertical, of the five-number summary of a data set, either discrete or continuous (or a slightly modified variant thereof that distinguishes outliers from the rest of the data). It marks the first, second, and third, quartiles, and the minimum and maximum, as straight lines, has lines extending from the first quartile to the minimum and from the third quartile to the maximum, and has a box indicating the IQR, as in the following example:

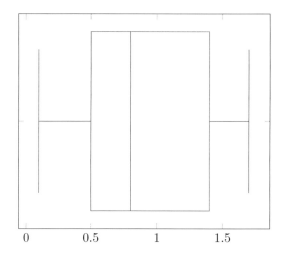

Figure 10.10: A typical box plot (horizontal).

Here, the minimum is at $x = 0.1$, the first quartile is at $x = 0.5$, the median is at $x = 0.8$, the third quartile is at $x = 1.4$, and the maximum is at $x = 1.7$ (with the box having length $1.4 - 0.5 = 0.9$). This is the most common type of box plot; some variants choose to omit the minimum and maximum in favor of outliers (marked as small dots) below the 9th or 2nd percentiles, and above the 91st or 98th percentiles. Other variants choose to mark any data point that is not within 1.5 IQR of the respective quartile as an outlier.

There is also the *stem-and-leaf* plot, which is considerably less visual but no less emblematic of the intriguing properties of data. A stem-and-leaf plot displays data according to the bracket that they fall in, arranged from smallest to largest within those brackets. In this sense, it can be thought of as a histogram laid on its side, except with the bars replaced by the data points.

As a concrete example, consider the set $S = \{1, 4, 9, 10, 15, 21, 24, 25, 27, 28, 29, 31,$
$32, 33, 36, 38, 51, 56, 60, 69, 72, 78\}$ representing the ages, in years, of people at a (very small) wedding. The following would be the corresponding stem-and-leaf plot for S:

0	1 4 9
1	0 5
2	1 4 5 7 8 9
3	1 2 3 6 8
4	
5	1 6
6	0 9
7	2 8

Table 10.1: Stem plot of ages at wedding, Key: $1|2 = 12$ years

If you were to tilt your head 90° clockwise, the stem-and-leaf plot would take the shape of a histogram, with each successive data point in a bracket contributing to the height of the bar in that bracket! In these plots, the mode(s) can be easily identified, and the median is not much harder to scan for, given the separation between elements in different brackets. To see the difference, try hunting for the median of a long set (like S in this example, which has 22 elements) in list form - it would take significantly longer![7]

10.2.3 Introduction to Inferential Statistics

As a natural extension of descriptive statistics, *inferential statistics* interprets data that is part of a larger *sample* or *population*, as opposed to just one data set. Inferential statistical methods make predictions and inferences about the larger sample from the smaller data set, rather than just confining its observations to the smaller set. They are often used when it is not feasible to use statistical methods on all of a population; for instance, when trying to draw conclusions about the population of the United States. Instead, we must simply ensure that our sample is a *representative sample*; i.e. has the same general characteristics as the whole population.

Sampling Strategies

The process of ensuring that the sample is representative is known as *sampling*, and takes place via a set of *sampling strategies*:

Definition 10.13: Some Sampling Strategies

- Probability sampling
  - Simple random sampling
  - Systematic random sampling
  - Stratified random sampling
- Non-probability sampling
  - Convenience sampling
  - Self-selection sampling
  - Quota sampling

Probability sampling refers to the class of sampling methods that select a sample from a population at random *using probabilistic methods*, while *non-probability sampling* refers to the class of sampling methods that use *non-probabilistic methods*; i.e. methods that are based on the subjective judgment of the researcher or experimenter. Namely, simple random sampling, the selection of a sample under the assumption of an equal chance of each member of the population being selected, is a probability sampling method, as are systematic random sampling, which depends on the probabilistic method used to select the first member to select the

---
[7] It's 30, by the way. See how stem-and-leaf plots help with the intuition?

rest of the sample in the same way, and stratified random sampling, which splits the population into several *strata* (groups, such as male/female and college major).

On the other hand, non-probability sampling methods, such as convenience sampling, self-selection sampling, and quota sampling, rely on the researcher's deliberate choices. In convenience sampling, the researcher picks the members of the population that are the easiest to access throughout the experiment. In self-selection sampling, the members of the population choose to participate in the experiment on their own accord. (This is often done via a questionnaire or advertising vehicle; the candidates are then filtered through the requirements of the study.) In quota sampling, the researcher picks out enough members of the population from each stratum so as to satisfy some set of predetermined conditions.

Hypothesis Testing

In any meaningful scientific study, before any sampling or experimentation can occur, there must be a well-formed *hypothesis* that conforms to all tenets of the scientific method.

> **Definition 10.14: The Hypothesis**
>
> Rigorously speaking, a *hypothesis* is an *a priori* supposition that is either verified or disproven through further analysis and investigation.

The hypothesis is traditionally not based on any previous findings within the study, save for the cursory overview of the requisite background. Scientists and mathematicians alike take special care to distinguish a *hypothesis* from a *theory*. Although these words may be interchangeable in everyday vernacular,[8] they are certainly not synonymous in the scientific sphere! Whereas a hypothesis is but an untested conjecture that is not supported by empirical data and may well turn out to be completely false, a theory is a scientifically accepted and well-substantiated explanation of a natural phenomenon, based on a body of facts that has been confirmed through repeated observation and experimentation. That is, a theory is not merely a conjecture; it does not arise out of thin air.

This is why *hypothesis testing* is such an important part of the research process! In order to form a generally-accepted and plausible scientific theory, we must run the hypothesis through a gamut of checks. First, we formulate the *null hypothesis* H_0, which posits that the observations of the experiment are the result of pure chance, and the *alternative hypothesis* H_a, which posits that the observations can be traced back to a non-random cause with reasonable certainty.[9] Second, we identify a *test statistic*, which is the main variable used to determine whether the null hypothesis H_0 is true. Third, we compute the *p*-value, which is a measure of the likelihood that a test statistic at least as significant as was observed would be obtained given the null hypothesis. Finally, we compare this *p*-value to a threshold to determine whether or not the observed effect is *statistically significant;* that is, most likely a result of a real effect and not random chance. The most common threshold (as mentioned into footnote 53) is $p < 0.05$ (which corresponds to a 5% probability that the result is *not* due to a real effect). If $p < 0.05$, or whatever threshold α we are using, then we reject the null hypothesis and conclude that the observed effect is statistically significant. If $p \geq \alpha$, however, then we accept the null hypothesis and cannot conclude that the observed effect is due to anything more than pure chance.[10]

---

[8]Even Google's dictionary feature lists "theory" as a synonym for "hypothesis." For shame!

[9]"Reasonable certainty" has been conventionally established as greater than 95% certainty; i.e. in the scientific literature, a *p*-value of less than $\alpha = 0.05$.

[10]One mnemonic for this is to think about a school exam. Let's say your teacher is particularly strict, and only gives A's to students who miss fewer than 5% of the questions. If your *p*-value (for *p*lease review) is less than 0.05, you got an A, for H_a, the *a*lternative hypothesis. Otherwise, you got a B or worse, and must accept H_0, for "*O*h no, I'm in trouble now!"

Errors in Hypothesis Testing

Nothing is impervious to human error - not even hypothesis testing. Sometimes, we mistakenly reject the null hypothesis when it is true (when the *p*-value is higher than the set threshold), or mistakenly accept the null hypothesis when, in fact, the alternative hypothesis is true (when the *p*-value is below the set threshold). In the first case, we have made a *type I error*, or a *false positive*; in the second, we have made a *type II error*, or a *false negative*. This may seem counter-intuitive at first glance, but remember that the null hypothesis is itself the rejection of a real effect (as it chalks experimental observation up to chance). Thus, if we make a type I error, we are saying that an effect is taking place when it isn't, and if we make a type II error, we are saying that observations are the result of random chance, when in fact there is an effect driving them.

H_0	True	False
Accept	correct	Type I
Reject	Type II	correct

Table 10.2: The two types of hypothesis testing errors. The top row represents reality, and the leftmost column represents the results of the research.

This table represents a concrete example: whether or not a certain drug is effective on an experimental group.

H_0: drug isn't effective	Non-effective	Effective
Accept: placebo effect	correct	Type I
Reject: drug is effective	Type II	correct

Table 10.3: A concrete application of Table 9.2 above.

10.2.4 Examples

Example 10.2.4.1. *The median of the set $\{2, 5, x, 15, 18\}$ is x, where x is a positive integer. What is the sum of the possible values for x?*

Solution 10.2.4.1. *In order for x to be the median of the set, $5 \leq x \leq 15$. The sum of the possible values of x is $5 + 6 + \cdots + 15 = \dfrac{15 \cdot 16}{2} - \dfrac{4 \cdot 5}{2} = \boxed{110}$.*

Example 10.2.4.2. *Let $S = \{1, x, x^2\}$. If the mean of S is M and the median of S is m, and $3M + m = 100$, compute the sum of the elements of S.*

Solution 10.2.4.2. *Let S be the sum of the elements of S; notice that $S = 3M$, so $S + m = x^2 + x + 1 + m = 100$. Then m must be a number that makes $x^2 + x + 1 + m$ a perfect square; in fact, if $m = x$, then $x^2 + x + 1 + m = x^2 + 2x + 1 = (x+1)^2$. Then $(x+1)^2 = 100$ and $x + 1 = 10$, so $x = 9$ (note that we must pick the positive value of x; otherwise, the median would be 1 instead of x). It follows that $S = 100 - 9 = \boxed{91}$.*

Example 10.2.4.3. *What is the IQR of the set $\{1, 2, 4, 8, 16, \ldots, 2048\}$?*

Solution 10.2.4.3. *The set has 12 elements (all the powers of 2 from $2^0 = 1$ to $2^{11} = 2048$), so the first quartile is the mean of 4 and 8, or 6, and the third quartile is the mean of 2^8 and 2^9, or 384. Hence, the IQR is $384 - 6 = \boxed{378}$.*

Example 10.2.4.4. *Suppose that a mammogram reports that a woman has breast cancer (which occurs with frequency 1 in 10 women) when she doesn't 25% of the time, and reports that a woman doesn't have breast cancer when she does 5% of the time. Otherwise, it will correctly detect breast cancer (or its absence). Let E_0 be the expected number of women in a sample of 1,000 women that obtain the correct result, E_1 the expected number of women subject to a Type I error by the mammogram machine, and E_2 the expected number of women subject to a Type II error by the mammogram machine. What are E_0, E_1, and E_2?*

Solution 10.2.4.4. *In the sample, 100 women have breast cancer, and 900 do not. Of the 100 with breast cancer, 5 of the women will receive false negatives (Type II error), and of the 900 without breast cancer, 225 of them will receive false positives (Type I error). Thus, $E_0 = 770$ will receive a correct result, with $E_1 = 225$ and $E_2 = 5$.*

10.2.5 Exercises

Exercise 16. *A researcher selects 50 freshmen from a pool of 500 high school students, who attend the same high school, in order to evaluate their athletic capabilities. What types of sampling is this an example of?*

Exercise 17. *For each of the following situations, determine, with explanation, whether there has been a Type I error or a Type II error.*

(a) *In a village during the dark of night, a boy cries out, "Wolf, wolf!" but amidst all the ensuing panic and commotion, there is no wolf to be found (and the people lose trust in the boy as a result).*

(b) *A few minutes later, a wolf barges through the gates out of nowhere and begins viciously attacking the villagers! However, this time, the boy has gone home. He is unaware of the wolf's presence, and so does not report anything.*

Exercise 18.

(a) *What is the five-number summary of $S = \{2, 2.5, 3.5, 6, 8, 10, 12\}$?*

(b) *Determine any outliers (elements below the 9th and above the 91st percentiles, using the "compromise" definition). Are these any different from outliers obtained from the > 1.5 IQR definition?*

Exercise 19. *A standardized test is graded with a real number ranging from 0 to 6. The mean score is 4.8, and the standard deviation is 0.9. To the nearest whole number, in what percentile does a score of 5.4 fall?*

Exercise 20. *In the fictitious country of Mathlandia, there are two candidates for President: Sophie Noether and Emmy Germain. 51% of the voting population in one of Mathlandia's provinces, Pascalia, have a Noetherian party affiliation, while 49% of the Pascalian voting population have a Germainian party affiliation. In another Mathlandian province, Pythagorea, 52% of the voting population side with Noether, while the other 48% of the population side with Germain. A survey is conducted on a simple random sample of 100 people from each of Pascalia and Pythagorea. What is the probability that the survey will indicate that Pascalia has a smaller percentage of Noetherian voters than does Pythagorea? (Hint: Use the formula $\sigma_d = \sqrt{\dfrac{P_1(1-P_1)}{n_1} + \dfrac{P_2(1-P_2)}{n_2}}$ for the standard deviation of the difference (where P_1 and P_2 denote the proportions of Noetherian voters in Pascalia and Pythagorea respectively, and n_1 and n_2 denote the sample populations of Pascalia and Pythagorea respectively))*

Exercise 21. *Mike is a participant in a study (of thousands of participants) where the researchers test his reaction time to a stimulus via a computer program. He is given 5 different stimuli to react to (and does not know when the stimulus will be triggered). His reported result will be the mean of his 5 reaction times.*

(a) *Given that the mean reaction time is 215 ms, the standard deviation of the sample population is 50 ms, and Mike's reported mean reaction time is 205 ms, approximately in what percentile is Mike's result? (Lower reaction times have lower percentiles; lower percentiles are considered "better" results.)*

(b) *Suppose that the researchers' hypothesis is that Mike's reaction time is the result of chance (and that he is actually perfectly average). What is the p-value associated with the hypothesis? Interpret this in terms of statistical significance.*

(c) *What is the interval of reported mean reaction times that would be considered statistically insignificant?*

(d) *What problems, if any, does this hypothetical study have that may affect the validity of the results? What parts of the study are properly conducted?*

Extension Exercise 6.

(a) *Prove that, if X and Y are variables in a linear relationship, then the correlation coefficient of X and Y is ± 1.*

(b) *Prove the converse of the statement above.*

Extension Exercise 7. *Let X and Y be independent random variables, and let a and b be constants. Compute $Var(aX + bY)$.*

10.3 Introduction to Probability

10.3.1 Defining Probability

Mathematicians define *probability* in terms of *probability experiments:* procedures that can be infinitely repeated and have associated to them a set of potential outcomes, known as a *sample space*. Within this sample space, there are several elements that represent *events*, or combinations of outcomes.

Let S be a sample space, and let $E \subseteq S$ be an event in S. The probability that the event E occurs is given by $P(E) = \dfrac{n(E)}{n(S)}$, where $n(E)$ is the number of possibilities with E occuring, and $n(S)$ is the total number of possibilities in the sample space S. This formula depends on the number of elements of the sample space and event space being countable; i.e. them being discrete metrics. (If they are continuous metrics, then we can still compute probability geometrically.)

Multiple events $E_1, E_2, E_3, \ldots, E_n$ in the sample space can be either *independent* or *dependent*. (For now, let's work with only two elements, E_1 and E_2.) If the likelihood of E_2 occurring does not depend on the outcome of E_1, then the events are said to be independent. Otherwise, they are dependent. Intuitively, if an object in a probability experiment cannot "remember" its previous outcomes, such as a die or coin, its associated events are independent.

We now define two crucial probability laws:

> **Theorem 10.5: Additive and Multiplicative Laws of Probability**
>
> - Additive law: For two independent events E_1 and E_2, $P(E_1 \cup E_2) + P(E_1 \cap E_2) = P(E_1) + P(E_2)$.
> - Multiplicative law: For two independent events E_1 and E_2, $P(E_1 \cap E_2) = P(E_1) \cdot P(E_2)$. We can henceforth combine these into one compact law: If E_1 and E_2 are independent events, then $P(E_1 \cup E_2) = P(E_1) + P(E_2) - P(E_1) \cdot P(E_2)$.

Example 10.3.1.1. *If the probability of E_1 occurring is 0.7 and the probability of E_2 occurring is 0.4, then the probability that either occurs is $0.7 + 0.4 - 0.7 \cdot 0.4 = 0.82$. The probability that both occur is $0.7 \cdot 0.4 = 0.28$, and the probability that neither occurs is $(1 - 0.3) \cdot (1 - 0.4) = 0.18$. (Thus, the probability that exactly one of the two events occurs is $0.82 - 0.28 = 0.54$. We can verify this by computing $0.7 \cdot (1 - 0.4) + 0.4 \cdot (1 - 0.7) = 0.7 \cdot 0.6 + 0.4 \cdot 0.3 = 0.54$.)*

The proofs of these two laws will follow after we introduce the idea of conditional probability.

Example 10.3.1.2. *Three fair coins are flipped, one after another.*

(a) What is the sample space S?

(b) Find the probability that not all of the coins come up the same.

Solution 10.3.1.1.

(a) The sample space S consists of all possible outcomes of flipping a fair coin three times in a row, namely
$S = \{HHH, HHT, HTH, HTT, THH, THT, TTH, TTT\}$.

(b) The probability of HHH occurring is $\left(\dfrac{1}{2}\right)^3 = \dfrac{1}{8}$, by the multiplicative law. The probability of TTT occurring is also $\dfrac{1}{8}$ (since T and H have the same probability of occurring; the coin is fair). Hence, the probability that neither outcome occurs is $1 - \dfrac{1}{8} - \dfrac{1}{8} = \dfrac{3}{4}$.

Example 10.3.1.3. *Two fair six-sided dice are rolled simultaneously.*

(a) Describe the sample space S of the sums of the dice rolls.

(b) Compute the probability of rolling a sum of exactly 2, 11, or 12.

(c) Compute the probability of rolling a sum of 7 or higher.

Solution 10.3.1.2.

(a) *The sample space S is the set of positive integers between 2 and 12, inclusive. The smallest possible sum of the dice rolls is $1 + 1 = 2$, and the largest possible sum is $6 + 6 = 12$. Every positive integer in between is also attainable.*

(b) *The probability of rolling a sum of 2 is $\frac{1}{6} \cdot \frac{1}{6} = \frac{1}{36}$; similarly, the probabilities of rolling sums of 3, 11, and 12 are $\frac{2}{36}, \frac{2}{36},$ and $\frac{1}{36}$ respectively by the multiplicative law of probability.*

(c) *The probability of rolling a 7 is $\frac{6}{36}$, and the probability of rolling a sum of n for $8 \leq n \leq 12$ is $\frac{13-n}{36}$ (by symmetry; the probability of rolling a sum of n is the same as that of rolling a sum of $13 - n$). The sum of the probabilities is $\frac{21}{36} = \frac{7}{12}$.*

10.3.2 Expected Value

Intuitively, we can define the *expected value* of a variable as its average value; i.e. the weighted mean of its outcomes according to their frequencies. Formally, we define expected value as follows:

Definition 10.15: Expected Value

Let X be a discrete random variable with possible values $\{x_1, x_2, x_3, \ldots, x_n\}$, and a *frequency* f_i associated to each distinct value x_i. Then the expected value of X is equal to

$$\mathbb{E}(X) = \frac{x_1 f_1 + x_2 f_2 + x_3 f_3 + \ldots + x_n f_n}{f_1 + f_2 + f_3 + \ldots + f_n}$$

Note that *expected value* should not be taken literally to mean "the value that is expected." There are plenty of examples in which the expected value is not a value that is "expected" at all in the common sense, due to its low probability of actually happening: consider a rigged 20-die with 18 faces having 1 printed on them, 1 face having 2 printed on it, and the last face having 20 printed on it. Though the expected value is indeed equal to $\frac{18 \cdot 1 + 1 \cdot 2 + 1 \cdot 20}{20} = 2$, 2 is certainly not the value one would reasonably expect from this die (as it only has a 5% chance of coming up). Instead, it would be 1 (which has a 90% chance of coming up). Yet, 2 is the mathematically-defined expected value, as it is the limit of the mean value of the dice rolls as the number of rolls tends to infinity.

In fact, much of the time, the expected value is not actually attainable at all! Consider the same die from before, except with 19 faces having 1 printed on them and the last face having 20 printed on it. This time, the expected value is $\frac{1 \cdot 19 + 20 \cdot 1}{20} = 1.95$, but this is not even a possible outcome, let alone the "expected" one! This is a prime example of why we must be careful with what words we use to describe certain mathematical objects.

Example 10.3.2.1. *A tetrahedral die is rigged to come up with n ($1 \leq n \leq 4$) with a probability in direct proportion with the value of n. Compute the expected value of a dice roll using this rigged die.*

Solution 10.3.2.1. *Since $1 + 2 + 3 + 4 = 10$, the value of n will come up $\frac{n}{10}$ of the time. If X is the value of the roll, then $\mathbb{E}(X) = \frac{1^2 + 2^2 + 3^2 + 4^2}{10} = 3$.*

This is probably the closest to an "expected" value (in the informal sense) that an actual mathematical expected value will get (other than, of course, the trivial case of having a fixed "random" variable). But sometimes expected values can come seemingly out of the blue, as in the d20 example above.

Worse yet, they may not even be defined at all. Consider the following famous example of a non-existent expected value:

Example 10.3.2.2. *One day, a stranger on the street comes up to you with a fair coin in his hand. He offers to give you 2^n dollars to flip n heads until the first tail appears. So if you flip T, you earn \$1; if you flip HT, you earn \$2, if you flip HHT, you earn \$4, and so forth. How many dollars should you pay so that the game is fair for both parties?*

Solution 10.3.2.2. *If X is the random variable representing the number of dollars earned, then $\mathbb{E}(X) = 1 \cdot \frac{1}{2} + 2 \cdot \frac{1}{4} + 4 \cdot \frac{1}{8} + \ldots = \frac{1}{2} + \frac{1}{2} + \frac{1}{2} + \ldots \to \infty.$*

But most people would not be willing to pay an infinite amount of money to play a game that has a 50% chance of netting them nothing more than a fast food value meal and an abysmal chance of netting them anything over \$10; in fact, hardly anyone would be willing to fork over more than \$5 (and rightfully so). Hence the paradox!

We can skirt around this paradox by introducing the concept of *utility*, which is also a cornerstone idea in economics:

> **Definition 10.16: Utility**
>
> A *utility* function $U : \mathbb{R}^+ \mapsto \mathbb{R}$ gives a *utility level*, or perceived benefit to the consumer/recipient, mapped to each unit of a consumed good and measured in *utils*.[a]
>
> ---
> [a]Note that we have set the domain as $\mathbb{R}^+$ rather than $\mathbb{R}$, since consumed good quantities must be non-negative; in addition, the domain is not $\mathbb{N}$, since the utility curve is continuous and not discrete. Consumed good quantities can be non-integers, such as 0.5 liters of soda or 1.25 hours spent waiting for a flight, but they cannot be negative. Furthermore, the utility level can be negative, if the consumed good is detrimental to the consumer. The 0.5 liters of soda may yield, say, +2 utils, but the time wasted waiting for the flight may come at −5 utils.

As such, we can define a gross income of \$n to have a utility level of $f(n)$ utils as opposed to 2^n utils, where $f(n)$ is any function from $\mathbb{N}$ to $\mathbb{R}^+$ such that f is non-decreasing over $\mathbb{N}$ and $\sum_{k=1}^{\infty} \frac{f(k)}{2^k} < \infty$. (Here, the codomain is $\mathbb{R}^+$ rather than $\mathbb{R}$; after all, who *wouldn't* want to gain more money?) Specifically, we can set $f(n) = \sqrt{n}$, yielding the finite sum

$$\sum_{k=1}^{\infty} \frac{\sqrt{k}}{2^k} \approx 1.347$$

or even $f(n) = n$, which yields

$$\sum_{k=1}^{\infty} \frac{k}{2^k} = 2.$$

Both of these values are quite reasonable, and make intuitive sense, since the payment scheme is naturally skewed towards the bottom end of possible outcomes. There is a 50% chance of earning only \$1, and a 25% chance of earning just \$2, but for even \$16 or better, the likelihood is a lowly 6.25%.

Essentially, we are simply re-defining the benefit of playing this game - the consumer's raw income in dollars - by converting it into an equivalent quantity of *utils* based on the consumer's value system. Indeed, the model of diminishing returns is a more realistic real-life model, as consumers often get bored of a good if they have a great surplus of it. In fact, with certain goods (medicine, for instance), an excess amount is actively harmful, thus incurring *negative* utils. So yes, it is possible to have too much of a good thing!

10.3.3 Conditional Probability

Conditional probability is a measure of the probability of an event in a sample space given that some other event has already occurred. If we want to compute the probability that event E_1 occurs given that event E_2 has already occurred, we denote this probability $P(E_1 \mid E_2)$, $P(E_1/E_2)$, or $P_{E_2}(E_1)$.

> **Definition 10.17: Conditional Probability Formula**
>
> For events E_1, E_2, $P(E_1 \mid E_2) = \dfrac{P(E_1 \cap E_2)}{P(E_2)}$.

Proof. Let a random experiment be repeated n times, and let E_1 and E_2 be events in the sample space. Let E_1 occur n_1 times, E_2 occur n_2 times, and $E_1 \cap E_2$ occur n_3 times. Then the relative frequency of E_1 occurring given E_2 is $\dfrac{n_3}{n_2}$. We can write this as $\dfrac{\left(\frac{n_3}{n}\right)}{\left(\frac{n_2}{n}\right)} = \dfrac{P(E_1 \cap E_2)}{P(E_2)}$. □

Example 10.3.3.1. *The probability that a student passes a test is 30%, but if she studies beforehand (with probability 50%), the probability is 80%. 80% is the probability that the student passes the test given that she studies.*

> **Theorem 10.6: Independent Events in Conditional Probability**
>
> If E_1 and E_2 are events in a sample space S, $P(E_1) = P(E_1 \mid E_2)$, then E_1 and E_2 are independent.

Example 10.3.3.2. *In the example above, if E_1 is the event of the student passing the test, and E_2 is the event of the student studying, then $P(E_1 \mid E_2) = 0.8$, where $P(E_2) = 0.5$. $P(E_1 \mid E_2) = 0.8 \neq 0.3 = P(E_1)$; therefore, E_1 and E_2 are not independent. That is, studying affects the probability that the student passes her test.*

Proof. If $P(E_1) = P(E_1 \mid E_2)$, then $P(E_1) = \dfrac{P(E_1 \cap E_2)}{P(E_2)} \implies P(E_1) \cdot P(E_2) = P(E_1 \cap E_2)$, which is the definition of the multiplicative law *for independent events only*. Hence the statement is proved. □

In general, $P(E_1 \mid E_2) \neq P(E_2 \mid E_1)$!

Example 10.3.3.3. *Suppose we are conducting a probability experiment to determine the accuracy rate of a machine that diagnoses terminal cancer. Let E_1 represent the event that a person has cancer, and let E_2 represent the event that the machine diagnoses cancer. We might have $P(E_1 \mid E_2) = 0.05$, but $P(E_2 \mid E_1) = 0.9$. (A person only has a 5% probability of actually having cancer if diagnosed with cancer by the machine, but the machine will correctly detect cancer in someone who has it 90% of the time.)*

A common fallacy, known as the *base rate fallacy*, lies in falsely stating that the two probabilities are equal. We can use *Bayes' Theorem*, however, to correctly determine the relationship between $P(E_1 \mid E_2)$ and $P(E_2 \mid E_1)$.

10.3.4 Bayes' Theorem

> **Theorem 10.7: Bayes' Theorem**
>
> For any two events E_1 and E_2, $P(E_1 \mid E_2) = \dfrac{P(E_2 \mid E_1) \cdot P(E_1)}{P(E_2)}$ where $P(E_2) \neq 0$.

Proof. We have $P(E_1 \cap E_2) = P(E_1) \cdot P(E_2 \mid E_1)$, and also $P(E_1 \cap E_2) = P(E_2) \cdot P(E_1 \mid E_2)$. Thus, $P(E_1) \cdot P(E_2 \mid E_1) = P(E_2) \cdot P(E_1 \mid E_2)$, which simplifies to $P(E_1 \mid E_2) = \dfrac{P(E_1) \cdot P(E_2 \mid E_1)}{P(E_2)}$. □

Example 10.3.4.1. *Consider the cancer example above. Given that 1% of the people in a sample population have cancer, and 18% of all results given by the machine are positive, what is the probability that a randomly selected individual with a negative result actually has cancer, and was mis-diagnosed?*

Solution 10.3.4.1. *One way to solve the problem is to use a direct application of Bayes' Theorem, where $P(E_1) = 0.01$, $P(E_2) = 0.18$, and $P(E_2 \mid E_1) = 0.9$. We immediately obtain $P(E_1 \mid E_2^c) = \dfrac{(1-0.9) \cdot 0.01}{0.82} = \dfrac{1}{820}$.*

Alternatively, we can construct a table of members of the sample population in four categories: cancer, diagnosed; cancer, not diagnosed (false negative, Type II error); no cancer, diagnosed (false positive, Type I error); and no cancer, not diagnosed. (Here we assume that the sample has 1000 people, 10 of which have cancer. + denotes a diagnosis of "cancer," and − denotes a diagnosis of "no cancer.")

	Cancer	No cancer
+	9	171
−	1	819

Table 10.4: Table of cancer diagnoses in a sample of 1,000 people, 10 of which have cancer.

Thus, the probability that an individual not diagnosed with cancer actually has it is a mere $\dfrac{1}{820}$, which is good news for the medical industry! (Or maybe not. But it could always be worse...)

10.3.5 Proving the Additive and Multiplicative Rules of Probabilities

Proving the Additive Rule. The sets $E_1, E_2 \subseteq S$ of event spaces in the sample space S are the unions of the two disjoint sets $E_1 \backslash E_2, E_1 \cap E_2$ and $E_2 \backslash E_1, E_2 \cap E_1$ respectively. Thus $P(E_1) = P(E_1 \backslash E_2) + P(E_1 \cap E_2)$ and $P(E_2) = P(E_2 \backslash E_1) + P(E_2 \cap E_1)$. Next, observe that $(E_1 \backslash E_2) \cap (E_2 \backslash E_1) = \emptyset$, and so

$$P(E_1) + P(E_2) = P(E_1 \backslash E_2) + P(E_2 \backslash E_1) + 2P(E_1 \cap E_2)$$
$$= P((E_1 \backslash E_2) \cup (E_1 \cap E_2) \cup (E_2 \backslash E_1)) + P(E_1 \cap E_2)$$
$$= P(E_1 \cup E_2) + P(E_1 \cap E_2)$$

This completes the proof. □

Proving the Multiplicative Rule for independent events. Let there be n independent events $E_1, E_2, E_3, \ldots, E_n$. Starting with the result $P\left(\bigcap_{i=1}^{n} E_i\right) = P(E_1) \cdot P(E_2 \mid E_1) \cdot P(E_3 \mid E_1 \cap E_2) \cdots P\left(E_n \mid \bigcap_{i=1}^{n-1} E_i\right)$, we repeatedly apply the definition of conditional probability:

$$P(E_2 \mid E_1) = \frac{P(E_2 \cap E_1)}{P(E_1)}$$

$$P(E_3 \mid E_1 \cap E_2) = \frac{P(E_3 \cap E_1 \cap E_2)}{P(E_1 \cap E_2)}$$

$$\vdots$$

$$P\left(E_n \mid \bigcap_{i=1}^{n-1} E_i\right) = \frac{P\left(E_n \cap \left(\bigcap_{i=1}^{n-1} E_i\right)\right)}{P\left(\bigcap_{i=1}^{n-1} E_i\right)}$$

This yields

$$P(E_1) \cdot P(E_2 \mid E_1) \cdot P(E_3 \mid E_1 \cap E_2) \cdots P\left(E_n \mid \bigcap_{i=1}^{n-1} E_i\right) = P\left(E_n \cap \left(\bigcap_{i=1}^{n-1} E_i\right)\right) = P\left(\bigcap_{i=1}^{n} E_i\right)$$

Therefore, we have

$$P\left(\bigcap_{i=1}^{n} E_i\right) = P(E_1) \cdot P(E_2 \mid E_1) \cdot P(E_3 \mid E_1 \cap E_2) \cdots P\left(E_n \mid \bigcap_{i=1}^{n-1} E_i\right)$$

as desired. □

10.3.6 Probability Distribution Functions

In an experiment dealing with continuous values, the random variable has a *probability density function* in the sample space, which is a function that maps each outcome in the sample space to the relative probability that it occurs. Here, the already-important distinction between the discrete and continuous definitions of probability pops out. Under the discrete definition, the probability that any given unit of a sample space would be chosen is zero, but the continuous definition guarantees that the sum of all probabilities under the PDF is 1.[11]

Definition 10.18: Types of Distributions

- We have the aforementioned *probability density/distribution function* (PDF) of a *continuous* random variable X, which is a function of an outcome in the sample space S that outputs the probability of that outcome occurring.

- We also have the *cumulative distribution function* (CDF) of a *continuous* random variable X, which plots the probabilities for each $x \in S$ (where S is the sample space) that X will assume a value less than or equal to x. Thus, at the upper bound of S, the CDF will always be 1. (Here, "maximum," "limit supremum," and "upper bound" are synonymous; the CDF plots probabilities for $X \leq x$, not $X < x$.)

- Finally, we have the *probability mass function* (PMF) of a *discrete* random variable X, which gives the probability that X is exactly equal to some $x \in S$ (where S is the sample space). The outputs of the PMF must always sum to 1.

We will begin visualizing these distributions after we introduce a few important statistical concepts.

The z-Score

In a PDF of a continuous random variable, the plot has an x-axis that is split by *z-scores*, or distance from the mean in terms of the standard deviation σ.

Definition 10.19: z-Score

Let S be a sample space, let $\bar{x}$ be the mean of the elements in S, and let σ be the sample standard deviation of the elements in S. Then for all $x \in S$, the z-score of x is given by $z_x = \dfrac{x - \bar{x}}{\sigma}$.

In addition, we can show that the z-scores across the entire standard normal distribution have a mean of 0, and a variance of 1.

---

[11] This relates to the ideas of the limit and Riemann integration in calculus; while the value of a function at a specific point may be equal to zero, the area under the curve can still be non-zero. An example is $f(x) = \dfrac{1}{x^2}$, whose limit as x tends to infinity is zero, but whose area under the curve from $x = 1$ to $x = \infty$ is 1.

Proof. To show that the mean of the z-scores is 0, it suffices to show that their sum is 0. We have

$$\sum z_x = \sum \frac{x - \bar{x}}{\sigma}$$

$$= \frac{1}{\sigma} \cdot \sum (x - \bar{x})$$

$$= \frac{1}{\sigma} \cdot (x - \sum \bar{x})$$

$$= \frac{1}{\sigma} \cdot \sum \bar{x} - \sum \bar{x} = 0$$

where $x = \sum \bar{x}$ by the definition of the mean.

To show that the variance (i.e. σ^2) of the z-scores is 1, it suffices to show that $\sum z_x^2 = n - 1$ (where n is the number of elements in the sample space S). We have

$$\sum z_x^2 = \sum \left(\frac{x - \bar{x}}{\sigma}\right)^2$$

$$= \frac{1}{\sigma^2} \cdot \sum (x - \bar{x})^2$$

$$= \frac{1}{\sigma}^2 \cdot (\sigma^2 \cdot (n - 1)) = n - 1$$

where the second-to-last step follows from the fact that $\sum (x - \bar{x})^2 = (n - 1) \cdot \sigma^2$ (which itself follows from the definition of the standard deviation σ). $\square$

The Normal Distribution

By far the most ubiquitous probability distribution function is the *Gaussian distribution*, more commonly referred to as the *normal distribution* and informally known as the *bell curve* (especially in educational or psychological contexts).[12] The normal distribution is a perfectly symmetrical distribution that is vital in fields such as the natural and social sciences, since distributions of random variables are often unknown in such contexts.

There are actually multiple types of bell-shaped distributions that are called *normal*; infinitely many, in fact. The specific shape that we usually refer to as a normal distribution is the *standard normal distribution*, which has standard deviation and variance both equal to 1.

The PDF of the normal distribution with μ (mean) and σ (standard deviation) arbitrary is given by

$$f(x \mid \mu, \sigma^2) = \frac{1}{\sqrt{2\pi\sigma^2}} e^{-\frac{(x-\mu)^2}{2\sigma^2}}$$

The PDF of the familiar *standard* normal distribution, with $\mu = 0$ and $\sigma = 1$, is given by

$$\phi(x) = \frac{1}{\sqrt{2\pi}} e^{-\frac{1}{2}x^2}$$

Of note is that

$$\int_{-\infty}^{\infty} \phi(x) \, dx = 1$$

(the area under the standard normal distribution curve over $\mathbb{R}$ is 1, indicating that the sum of all possible probabilities of the function is 1).

---
[12] There are actually several other bell-shaped curves, including the *Student's t*, *Cauchy*, and *logistic* distributions, but this one is the most commonly-used and well-known.

The following are the PDF and CDF of the *standard* normal distribution:

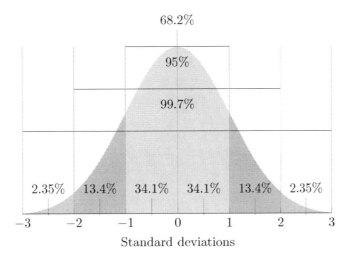

Figure 10.11: The probability density function of the standard normal distribution.

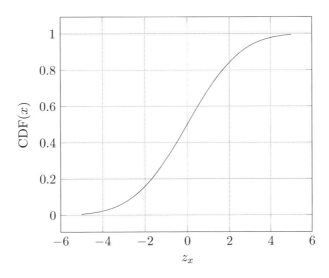

Figure 10.12: The cumulative distribution function of the standard normal distribution.

If we want to obtain the PDF and CDF of *any* normal distribution, we can simply apply a scale factor of σ (dilation) and a translation factor of μ (where the distribution is centered; where it has its peak, and where the rate of change of the CDF is maximized).

Computing an Exact p-Value

In order to determine whether a data observation is statistically significant, we need to compute an exact p-value. (Note that this procedure is only valid under a set of precise conditions: namely, that a sample is randomly selected, that the sample is sufficiently smaller than the population it was drawn from, and that the variable in the experiment has a perfectly normal distribution.) In turn, we must know the actual population mean μ, the mean according to the null hypothesis μ_0, the population standard deviation σ, the sample mean $\bar{x}$, and the sample size n.

Furthermore, we must determine what kind of p-value we are computing. Recall that, by definition, the p-value is a measure of the probability, under the null hypothesis, that a result would be observed as or more extreme than what was actually observed. But are we viewing the event corresponding to this as a left-tailed, right-tailed, or two-tailed event? If the event is single-tailed (either left-tailed or right-tailed), then the p-value corresponding to that event is given by $P(X \leq x)$ or $P(X \geq x)$ for a left-tailed and right-tailed event, respectively. If the event is double-tailed, then the p-value is given by $2\min(P(X \leq x), P(X \geq x))$ (as we want the probability on both sides that an observation under H_0 is *at least* as extreme as the observation). If we know all of these values, along with what kind of p-value we seek, we can compute the z-statistic exactly:

$$z = \frac{\sqrt{n} \cdot (\bar{x} - \mu_0)}{\sigma} \qquad (10.6)$$

This value is the number of standard deviations a sample is away from the population mean. Then we can use a z-score table (for either one or two tails as appropriate) to determine the p-value - the likelihood that a sample is more extreme than what was observed. As an example, consider the following scenario:

Example 10.3.6.1. *Suppose that we want to test whether a certain drug that treats high cholesterol is effective, with a significance threshold of $\alpha = 0.05$. We have a sample of 9 people with high cholesterol, a population mean of 200 mg/dL, which is also the mean by the null hypothesis, an SD of $\frac{360}{7} \approx 51.43$ mg/dL, and a sample mean of 230 mg/dL. Then our p-value is $\dfrac{3 \cdot (230 - 200)}{\left(\dfrac{360}{7}\right)} = 1.75$.*

Here, H_0 is the hypothesis that the drug is ineffective, and that the elevated cholesterol levels of the sample population are due to pure chance, whereas H_a is the alternative hypothesis that the drug is effective, and that there is a driving effect behind the higher sample mean of 230 mg/dL. To determine whether H_0 or H_a should be accepted, we use a z-score table to convert a z-score of 1.75 to a p-value. If we use a one-tailed table, then we obtain a p-value of 0.96, which indicates that $1 - 0.96 = 0.04$ is the proportion of the standard normal distribution that lies above the observed data. Since $0.04 < 0.05$, the data is statistically significant, and so we reject H_0 in favor of H_a. However, if we are using a two-tailed table, we instead obtain a p-value of 0.08, and so we accept the null hypothesis (effectively, chalk the observed data up to chance).

The Central Limit Theorem

The *Central Limit Theorem* is what makes the normal distribution work the way it does: it ensures that, as more and more independent random variables are added to an experimental setting, their normalized sum tends towards a normal distribution, even if the variables themselves were not normally distributed. Its name is actually dual in meaning: it is not only "central" in that it centralizes distribution to the normal distribution, but also "central" to the field of statistics as a whole, in that it allows probabilistic methods that work for normal distributions to also work for other types of distributions.

> **Theorem 10.8: Central Limit Theorem (CLT)**
>
> Let n be the number of independent random variables introduced into an experiment, and let $f_X(n)$ be the probability density function of the normalized sum of the independent random variables. Then
>
> $$\lim_{n \to \infty} f_X(n) = \phi(n) = \frac{1}{\sqrt{2\pi\sigma^2}} e^{-\frac{(n-\mu)^2}{2\sigma^2}}$$
>
> for μ, σ arbitrary.

The field of statistics is replete with other concepts that have vast real-world applications, but these are all of the basic statistical concepts that most significantly aid our intuitive understanding of probability and combinatorics. In the following sections, we will shift our focus from rigorous statistics to developing a computational foundation of combinatorics.

10.3.7 Examples

Example 10.3.7.1. *(2018 Stanford Math Tournament) A number is formed using the digits $\{2, 0, 1, 8\}$, using all 4 digits exactly once. Note that $0218 = 218$ is a valid number that can be formed. What is the probability that the resulting number is strictly greater than 2018?*

Solution 10.3.7.1. The number can either begin with a 2 or an 8; and if it begins with 2, then all arrangements of 018 other than 018 itself are valid (giving $3! - 1 = 5$ possibilities). With 8, there are $3! = 6$ possibilities, so out of all $4! = 24$ permutations of the digits in 2018, the desired probability is $\boxed{\dfrac{11}{24}}$.

Example 10.3.7.2. *(1985 AJHSME) Assume every 7-digit whole number is a possible telephone number except those that begin with 0 or 1. What fraction of telephone numbers begin with 9 and end with 0?*

Solution 10.3.7.2. The fraction that begin with 9 is $\dfrac{1}{8}$ the total, and the fraction that end with 0 is $\dfrac{1}{10}$ the total; multiplying the two yields $\boxed{\dfrac{1}{80}}$.

Example 10.3.7.3. *Two integers a, b are randomly selected from the interval $[1, 100]$. What is the probability that, when a and b are concatenated (strung together) with a followed by b, they form an integer less than or equal to 2019?*

Solution 10.3.7.3. Clearly, a and b must be be two digits (or 100). If $a \in [1, 9]$, this is guaranteed for all b. If $a \in [10, 19]$, $b \in [1, 99]$. If $a = 20$, $b \in [1, 19]$. If $a \in [21, 100]$, $b \in [1, 9]$. There are $9 \cdot 100 + 10 \cdot 99 + 19 + 80 \cdot 9 = 2629$ possibilities out of $100^2 = 10{,}000$ total, so the desired probability is $\boxed{\dfrac{2629}{10{,}000}}$.

10.3.8 Exercises

Note: Since probability, in particular, is such an important and heavily-used topic in competition math, we have included a greater amount of exercises accordingly.

Exercise 22. What is the probability that, if you select two elements with replacement from the set $\{1, 2, 3, 4, 5\}$, at least one will be prime?

Exercise 23. Richard and Harvey both pick two random numbers between 18 and 20, inclusive. What is the probability that Richard's number is at least twice Harvey's number, minus 21?

Exercise 24. Three unfair coins, A, B, and C, are flipped in that order. Coin A lands on heads $\frac{2}{3}$ of the time, coin B lands on tails $\frac{3}{4}$ of the time, and coin C lands on heads $\frac{9}{10}$ of the time if coins A and B both landed on heads, and $\frac{3}{10}$ of the time otherwise. What is the probability that at least two coins land on tails?

Exercise 25. (2017 AMC 12B) Real numbers x and y are chosen independently and uniformly at random from the interval $(0, 1)$. What is the probability that $\lfloor \log_2 x \rfloor = \lfloor \log_2 y \rfloor$, where $\lfloor r \rfloor$ denotes the greatest integer less than or equal to the real number r ?

Exercise 26. (2013 Stanford Math Tournament) Nick has a terrible sleep schedule. He randomly picks a time between 4 AM and 6 AM to fall asleep, and wakes up at a random time between 11 AM and 1 PM of the same day. What is the probability that Nick gets between 6 and 7 hours of sleep?

Exercise 27. (2006 AMC 10B) Bob and Alice each have a bag that contains one ball of each of the colors blue, green, orange, red, and violet. Alice randomly selects one ball from her bag and puts it into Bob's bag. Bob then randomly selects one ball from his bag and puts it into Alice's bag. What is the probability that after this process the contents of the two bags are the same?

Exercise 28. Compute the probability that a randomly chosen positive integral factor of 10^{2019} is not also a factor of 10^{2017}.

Exercise 29. (2016 AIME II) There is a 40% chance of rain on Saturday and a 30% chance of rain on Sunday. However, it is twice as likely to rain on Sunday if it rains on Saturday than if it does not rain on Saturday. The probability that it rains at least one day this weekend is $\frac{a}{b}$, where a and b are relatively prime positive integers. Find $a + b$.

Exercise 30. An unfair coin has probability p of landing heads and $1 - p$ of landing tails. Given that the probability of it landing on heads twice and tails three times out of five flips is equal to p^2, what is the value of p?

Exercise 31. (2006 AIME II) When rolling a certain unfair six-sided die with faces numbered 1, 2, 3, 4, 5, and 6, the probability of obtaining face F is greater than $\frac{1}{6}$, the probability of obtaining the face opposite is less than $\frac{1}{6}$, the probability of obtaining any one of the other four faces is $\frac{1}{6}$, and the sum of the numbers on opposite faces is 7. When two such dice are rolled, the probability of obtaining a sum of 7 is $\frac{47}{288}$. Given that the probability of obtaining face F is $\frac{m}{n}$, where m and n are relatively prime positive integers, find $m + n$.

Extension Exercise 8. A giant parking lot has 2018 consecutive parking spaces, each with the same width. 2000 cars park in a random space. Then, a 2-space-wide RV rolls into the parking lot. What is the probability that it can park?

Extension Exercise 9. (2016 ARML Individual Round #9) An n-sided die has the integers between 1 and n (inclusive) on its faces. All values on the faces of the die are equally likely to be rolled. An 8-sided die, a 12-sided die, and a 20-sided die are rolled. Compute the probability that one of the values rolled is equal to the sum of the other two values rolled.

Extension Exercise 10. *Andrew and Beth both flip a certain amount of fair coins. Andrew flips n coins in succession, while Beth flips $n+1$ coins in succession. Show that Andrew always has exactly a $\frac{1}{2}$ chance to flip at least as many heads as Beth, independently of the value of n.*

Extension Exercise 11. *(2016 ARML Team Round #7) Real numbers x, y, and z are chosen at random from the unit interval $[0,1]$. Compute the probability that $\max(x,y,z) - \min(x,y,z) \leq \frac{2}{3}$.*

Extension Exercise 12. *(1992 AHSME) An "unfair" coin has a $\frac{2}{3}$ probability of turning up heads. If this coin is tossed 50 times, what is the probability that the total number of heads is even?*

Extension Exercise 13.

(a) *Compute the probability that a random 10-digit string of 0's and 1's will not contain more than 2 consecutive 0's or 1's.*

(b) *In general, compute the probability that a random d-digit string of 0's and 1's will not contain more than 2 consecutive 0's or 1's.*

Extension Exercise 14. *Compute the probability that a solution $(x,y,z) \in \mathbb{R}^3$, $0 < x, y \leq 3$ to the equation $\log_4 x + \log_4 y = \log_{16} z$ is also a solution to the equation $x^2 + y^2 + z \leq 3$.*

10.4 Combinatorics and Probability: Part 1

In the study of *combinatorics*, we focus on the art of *counting*. At first, this seems trivial; even ludicrous. Surely counting isn't a real area of study - we've all learned how to do it by first grade! Indeed we have, but what combinatorics actually studies is the shorthand of counting set cardinalities, as in the sets of potential outcomes of one or multiple events. We must also clarify what we mean by "outcome" - are we counting outcomes that are distinguishable to the naked eye, the total of all physical outcomes, or the number of outcomes that have some other property (such as a certain sum of dice rolls)?

The laws governing probability computations also apply here. For instance, we can use the additive law to solve problems where we are concerned with the number of ways to select elements from a set $A \cup B$, where A and B are disjoint sets:

Example 10.4.0.1. *If an ice cream shop offers 20 flavors, then adds 11 more for the summer season, then it will have 31 flavors, since all of the flavors are in the same set (the set of flavors offered by the shop).*

Furthermore, we can apply the multiplicative law to determine the number of ways to select elements from the set $A \times B$, where A and B are sets and $\times$ denotes the Cartesian product:

Example 10.4.0.2. *How many ways are there to pick out a pair of pants, a shirt, a pair of socks, a pair of shoes, a jacket, and a tie from a wardrobe with 6 pairs of pants, 3 shirts, 4 differently-colored pairs of socks, 2 pairs of shoes, 3 jackets, and 2 ties?*

Solution 10.4.0.1. *By the multiplicative law, there are $6 \cdot 3 \cdot 4 \cdot 2 \cdot 3 \cdot 2 = 864$ choices.*

10.4.1 Introduction to Casework

Casework is an oft-used technique in combinatorics to efficiently sort through cases that present different restrictions to the problem at hand. It is a type of *proof by exhaustion*, in which we go through all possible cases to ensure that we have fully solved the problem. When using casework, we must add the results of the different cases, since they are disjoint from each other.

Example 10.4.1.1. *(2005 AMC 10A) How many three-digit numbers satisfy the property that the middle digit is the average of the first and the last digits?*

Solution 10.4.1.1. *This solution will be an exhaustive 9-case-long, case-by-case check (affectionately a casework bash).*

First note that the middle digit (henceforth M for brevity; we will similarly denote the first and last digits by F and L respectively) cannot be 0, since this would force the sum of the other digits to be 0 as well (and a positive integer cannot begin with a 0). So there are obviously 0 cases here.

If $M = 1$, then $F + L = 2$; this forces either $(F, L) = (1, 1)$ or $(2, 0)$, so 2 cases. If $M = 2$, then $1 \leq F \leq 4$; if $M = 3$, then $1 \leq F \leq 6$, and if $M = 4$, then $1 \leq F \leq 8$. This yields 20 cases so far.

For $M \geq 5$, F can still be no higher than 9, so $L = 0$ is no longer a valid possibility. Hence, $1 \leq F \leq 9$ gives 9 cases. For $M = 6$, $F + L = 12$, so $3 \leq F \leq 9 \implies$ 7 cases. This continues until $M = 9$, which yields just 1 case: $(F, L) = (9, 9)$. Our total in the collection of cases where $M \geq 5$ is $9 + 7 + 5 + 3 + 1 = 25$ cases.

Thus, the final count is $20 + 25 = 45$ cases.

The previous problem is a prime example of how lengthy a casework bash can get; the official solution provided by the MAA for this problem uses the fact that twice the middle digit should be the sum of the first and last digits (and so that sum must be even).

Here is an example of a casework problem with just two cases to check:

Example 10.4.1.2. *(2014 AIME I) An urn contains 4 green balls and 6 blue balls. A second urn contains 16 green balls and N blue balls. A single ball is drawn at random from each urn. The probability that both balls are of the same color is 0.58. Find N.*

Solution 10.4.1.2. *This problem is much less casework-heavy, but still a good demonstration of the general concept and an opportunity to combine previous concepts with casework.*

If both balls are green, then the probability of drawing a green marble from the first urn is $\frac{2}{5}$, and the probability of drawing a green marble from the second urn is $\frac{16}{16+N}$. Similarly, if both balls are blue, the probability of drawing a blue marble from the first urn is $\frac{3}{5}$, and the probability of drawing a blue marble from the second urn is $\frac{N}{16+N}$. Multiplying the probabilities together for each case yields $\frac{32}{80+5N}$ and $\frac{3N}{80+5N}$, respectively. Adding these together yields $\frac{32+3N}{80+5N}$, and setting this equal to $\frac{29}{50}$ yields $50(32+3N) = 29(80+5N)$, or $N = 144$.

10.4.2 The Principle of Inclusion-Exclusion (PIE)

When we were discussing Venn diagrams and de Morgan's Law, we were actually taking advantage of the *Principle of Inclusion-Exclusion,* (or Inclusion-Exclusion Principle, PIE for short[13]) which gives a formula for the cardinality of a union of multiple sets in terms of the cardinalities of each set, as well as those of intersections of n-tuples of sets. For instance, in a 2-circle Venn diagram, the area of overlap represents the intersection of the two sets (Figure 9.5), and the area in one of the circles represents the union of the two sets (Figure 9.4). The area inside exactly one of the circles is either $A\setminus B$ or $B\setminus A$, and the area outside the circles but inside the universal set is $(A \cup B)^c = A^c \cap B^c$.

Theorem 10.9: Inclusion-Exclusion Principle

For two sets A and B, $|A \cup B| = |A| + |B| - |A \cap B|$.

For three sets A, B, and C, $|A \cup B \cup C| = |A| + |B| + |C| - |A \cap B| - |A \cap C| - |B \cap C| + |A \cap B \cap C|$.

For n sets $S_1, S_2, S_3, \ldots, S_n$,

$$\left|\bigcup_{i=1}^{n} S_i\right| = \sum_{i=1}^{n} |S_i| - \sum_{j>i} |S_i \cap S_j| + \sum_{k>j>i} |S_i \cap S_j \cap S_k| + \ldots + (-1)^n \cdot \left|\bigcap_{i=1}^{n} S_i\right|$$

Proof. There are two routes we can take here. We will first prove the formula by induction.

Our base case is $n = 2$: $|A \cup B| = |A| + |B| - |A \cap B|$. (For $n = 1$, the formula is trivial: $|A| = |A|$.) This is true by the same reasoning as used to prove additive rule of probabilities (see §9.3.4).

Next, we formulate our induction step; namely, we show that

$$\left|\bigcap_{i=1}^{n+1} S_i\right| = \sum_{i=1}^{n+1} |S_i| - \sum_{j>i} |S_i \cap S_j| + \sum_{k>j>i} |S_i \cap S_j \cap S_k| + \ldots + (-1)^{n+1} \cdot \left|\cap_{i=1}^{n+1} S_i\right|$$

using the n-set generalization of PIE. This can be accomplished by observing that

$$\left|\bigcup_{i=1}^{n+1} S_i\right| = \left|\bigcup_{i=1}^{n} S_i \cup S_{n+1}\right|$$

---

[13] Not related to π, the world's (arguably) most famous mathematical constant.

$$= \left| \bigcup_{i=1}^{n} S_i \right| + |S_{n+1}| - \left| \bigcup_{i=1}^{n} S_i \cap S_{n+1} \right|$$

Henceforth, consider

$$\left| \bigcup_{i=1}^{n} S_i \cap S_{n+1} \right|$$

This can be written as

$$\left| \bigcup_{i=1}^{n} (S_i \cap S_{n+1}) \right|$$

(where $S_i \cap S_{n+1}$ is considered as one entity, instead of as the union of the S_i intersecting with S_{n+1}) by the following lemma:

Lemma 10.9: Intersection Distributes Over Union

For any three sets A, B, and C, $A \cap (B \cup C) = (A \cap B) \cup (A \cap C)$.

Proof. Let $x \in A \cap (B \cup C)$. Then $x \in A \wedge (x \in B \vee x \in C)$, which implies that either $(x \in A) \wedge (x \in B)$, $(x \in A) \wedge (x \in C)$, or both. In turn, $x \in ((A \cap B) \cup (A \cap C))$, so $A \cap (B \cup C) = (A \cap B) \cup (A \cap C)$, as desired. $\square$

(It is also true that the union distributes over the intersection; proving this is left as an exercise to the reader.)

From Lemma 9.9 above and the induction hypothesis, it follows that

$$\left| \bigcup_{i=1}^{n} (S_i \cap S_{n+1}) \right| = \sum_{i=1}^{n} |S_i \cup S_{n+1}| - \sum_{j>i} |S_i \cap S_j \cap S_{n+1}|$$

$$+ \sum_{k>j>i} |S_i \cap S_j \cap S_k \cap S_{n+1}| + \ldots + (-1)^{n+1} \cdot \left| \bigcup_{i=1}^{n} S_i \cap S_{n+1} \right|$$

In general, consider the term in the formula for i intersections:

$$(-1)^{i+1} \cdot \sum_{|P|=i} \left| \bigcap_{i \in P} S_i \right| - (-1)^i \cdot \sum_{|Q|=i-1} \left| \bigcap_{i \in Q} S_i \cap S_{n+1} \right|$$

where $1 \leq i \leq n$. This is equal to

$$(-1)^{i+1} \cdot \sum_{|R|=i} \left| \bigcap_{i \in R} S_i \right|$$

where R is a subset with i elements from the set of $n+1$ sets $S_1, S_2, S_3, \ldots, S_{n+1}$. Indeed, this is what we desire in the expansion of $|\bigcap_{i=1}^{n} S_i|$.

To finish off the proof, we can verify that the first term, $\sum_{i=1}^{n} |S_i| + |S_{n+1}|$, is equal to $\sum_{i=1}^{n+1} |S_i|$. (Since $|\bigcup_{i=1}^{n} S_i \cap S_{n+1}|$ is the intersection of multiple sets, it does not contribute to the first term.) In addition, the last term can be verified to be equal to $(-1)^n \cdot |\bigcap_{i=1}^{n} S_i|$. We have verified the formula for $n+1$, completing the proof. $\square$

Alternatively, we can prove the Inclusion-Exclusion Principle using the Binomial Theorem:

Alternate proof. Let $x \in \bigcup_{i=1}^{n} S_i$ belong to exactly s of the sets S_i. Then all intersections of more than s sets are empty, as are any intersections including sets that do not contain x. The number of 2-set intersections including x is $\binom{s}{2}$, the number of 3-set intersections including x is $\binom{s}{3}$, and so forth; the number of s-set intersections including x is $\binom{s}{s} = 1$. Then we have

$$\binom{s}{1} - \binom{s}{2} + \binom{s}{3} - \binom{s}{4} + \ldots + (-1)^{s+1} \cdot \binom{s}{s} = 1$$

since each intersection is counted only once. This simplifies to

$$1 - \binom{s}{1} + \binom{s}{2} - \binom{s}{3} + \ldots + (-1)^s \cdot \binom{s}{s} = 0 \implies (1-1)^s = 0$$

by the Binomial Theorem. Hence the statement is proven. □

10.4.3 The Pigeonhole Principle/Dirichlet Box Principle

The *Pigeonhole Principle*, sometimes called the *Dirichlet Box Principle*,[14] is yet another deceptively simple theorem that seems at first to be an extension of common sense:

> **Theorem 10.10: Pigeonhole Principle (Simplest Case)**
>
> If there are $n+1$ objects (pigeons) and n boxes (holes), at least one of the boxes will necessarily contain two or more of the objects.

Seems simple enough to visualize in practice. And indeed, the proof is a fairly straightforward contradiction argument:

Proof. Assume the contrary - that each box contained only zero or one objects. If each box contained exactly one object, there would be a total of n objects in the boxes, not $n+1$; contradiction. Therefore one box must contain two or more objects. □

We can also generalize the Pigeonhole Principle to a more rigorous form:

> **Theorem 10.11: Generalized Pigeonhole Principle**
>
> If there are n pigeons and k holes, then at least one of the holes will necessarily contain at least $\lceil \frac{n}{k} \rceil$ pigeons (where $\lceil x \rceil$ denotes the *ceiling* of x; i.e. the smallest integer greater than or equal to x).

Proof. Again, we can use a contradiction argument. Suppose that each hole only contains up to $\lceil \frac{n}{k} \rceil - 1$ pigeons. Then the holes will, in total, contain $k\left(\lceil \frac{n}{k} \rceil - 1\right)$ pigeons; $k\left(\lceil \frac{n}{k} \rceil - 1\right) < k\left(\frac{n}{k}\right) = n$, which gives the desired contradiction. □

Here is a set of somewhat standard examples:

Example 10.4.3.1.

(a) In Armand's drawer, there are 5 pairs of blue socks, 6 pairs of red socks, 7 pairs of black socks, and 3 pairs of white socks. How many pairs of socks must Armand draw to guarantee that he draws two or more pairs of different colors?

(b) What if he wanted to draw three or more pairs of different colors?

(c) How many pairs of socks must Armand draw to guarantee that he draws two or more pairs of the same color?

---
[14] Mostly by those who like to sound fancy; almost everyone I know calls it Pigeonhole.

(d) *What if he wanted to draw five or more pairs of the same color?*

Solution 10.4.3.1.

(a) *Armand could draw 7 pairs of black socks, but once he draws an eighth pair, he will be forced to pick out a different color, so the answer is 8 by the simplest case of Pigeonhole. (Here, the drawing of a pair of socks represents a pigeon and the number of same-colored socks represents a hole.)*

(b) *Armand could draw 7 black pairs and 6 red pairs, but his fourteenth pair must be a different color (by generalized Pigeonhole).*

(c) *There are 4 different colors of socks Armand can draw, but his fifth pair is guaranteed to be a different color (with 5 pigeons and 4 holes).*

(d) *Armand could draw 4 pairs of each color, except he only has 3 pairs of white socks! Uh-oh. What do we do here? Simple. Armand can still draw 4 pairs of the other-colored socks, but he is limited to drawing 3 pairs of white socks - that's 15 pairs. His sixteenth pair will necessarily be a fifth duplicate of the same color.*

Example 10.4.3.2. *Prove that if eleven or more distinct numbers are selected from the set $\{1, 2, 3, \ldots, 19\}$, then at least two of them will sum to 20.*

Solution 10.4.3.2. *(In this example, the pigeons are the selected numbers, and the holes are the sets of pairs of elements of the set summing to 20.) There are ten "holes": $\{1, 19\}, \{2, 18\}, \{3, 17\}, \ldots, \{9, 11\}, \{10\}$. We can select one element from each of these holes without any trouble. If we try to pick an eleventh element, however, we are forced to pick two elements from at least one of the holes, and this forces a sum of 20 in that hole. (This, in itself, invokes the simplest case of the Pigeonhole Principle.)*

You may have already been instinctively aware of this theorem at a sub-conscious level. Perhaps when you were a child, you tried, in vain, to fit a square peg into a round hole at the playground for hours on end, before you found yourself being whisked away to go home for dinner. No matter how much you knew, deep down, that it was impossible, you tried anyway - because you wanted to see it happen, regardless of what the laws of mathematics or physics *said* was possible. You wanted to defy the laws of nature - you wanted to *win*.

As it turns out, this defiant spirit is also what drives many professional mathematicians! Finding counterexamples to some conjectures can be a truly daunting, even impossible, task (and it often is impossible). But mathematicians continue to take a stab at it anyway; they truly live by the motto "never say never." Regardless of how many failed counterexamples they've already been through, mathematicians keep searching for the successful one - the holy grail - that may or may not exist, much like you kept trying to achieve that which may or may not even be possible, regardless of how many hours you spent trying to fit that square peg in that round hole.

10.4.4 Combinations and Permutations

Imagine that you are a middle school student who has forgotten the combination to your lock that you haven't used since your elementary school days. You desperately want to use this cool lock - emblazoned with math-themed decorum and puns - on your new locker, but you can't without trying all the possibilities to see which one works. You're sitting in your room, when your parents step in and see you still at 0400 by your bedtime. Your mother says, "Hey sweetie, maybe it's not worth it. We'll just buy you a new lock tomorrow, okay?" You reluctantly agree, but you refuse to accept defeat, so you keep trying.

Then, out of the blue, around 0200, a thought hits you: "How many of these do I have to try, anyway?" You see that the lock has 4 digits, so you run the math and figure out that you still have another 205 combinations to go! Unfazed as you are, though, you resolve to stay up all night if you have to, since you don't have anything super important to do at school tomorrow anyway. You keep at it for a little while longer, until - voilà - miraculously, you discover the code: 0405!

Exhausted but overjoyed, you roll over onto your back and find yourself in the deep bliss of sleep, instantly relieved from finger-hurting dial turning and the stress of not knowing when you'd finally unveil the elusive combination. The next morning, you excitedly run over to your mom's bedroom to tell her the good news. It's a good thing, too - had the combination been much higher of a number, and had you not run the math beforehand, you would've been stuck there all night!

This little story is perhaps the first real, applicable example of combinatorics for many a reader. As much as the question of real-life relevance is brought up for other areas of math, the non-probability aspects of combinatorics, in particular, strike some disillusionment amongst those students who can't wait to barge out of the classroom, almost reflexively, at the sound of the bell. Though probability's real-world use is somewhat clearer, the versatility of combinations and permutations is no less prevalent.

> **Definition 10.20: Combinations and Permutations**
>
> Consider a set $S = \{x_1, x_2, x_3, \ldots, x_n\}$. Let T be a subset of S.
>
> A *permutation* π of S is a rearrangement of the elements of S with respect to order. (For instance, let $T = \{x_1, x_2, x_3\}$; then $\pi_1(T) = T$ is a different permutation from $\pi_2(T) = \{x_1, x_3, x_2\}$.
>
> A *combination* of S is a selection of a subset T from S without regard for order. (In the example above, $\pi_1(T)$ and $\pi_2(T)$ are both the same combination of S.)

> **Theorem 10.12: Combination/Permutation Formulas**
>
> Let $S = \{x_1, x_2, x_3, \ldots, x_n\}$ as before. Then the number of permutations of S is equal to $n!$, the number of permutations of S with cardinality $k \leq n$ (denoted ^nP_k) is $\frac{n!}{k!}$, and the number of combinations of S with cardinality $k \leq n$, denoted $\binom{n}{k}$, is equal to $\frac{n!}{k!(n-k)!}$. This is frequently referred to as "n choose k," or the *binomial coefficient*. (We define $0! = 1$ to allow these formulas to work.)

> **Corollary 10.12: Rearrangements of a String**
>
> Let W be a *word*; i.e. a string consisting of the letters $L_1, L_2, L_3, \ldots, L_n$. Let $f(L_i)$ denote the *frequency* of the letter L_i; that is, how many times it appears in W. Then the number of distinguishable rearrangements (permutations) of W is equal to
>
> $$\frac{\left(\sum_{i=1}^{n} f(L_i)\right)!}{\prod_{i=1}^{n} f(L_i)!}.$$

Proof.

(a) Permutation formula: Suppose we are choosing k objects, without replacement, from a set of n objects. We have n ways to pick the first object, $n-1$ ways to pick the second object, $n-2$ ways to pick the third object, and in general, $n - k + 1$ ways to pick the kth object. We multiply these together to obtain $^nP_k = (n)(n-1)(n-2) \cdots (n-k+1)$; multiplying by $\frac{(n-k)!}{(n-k)!} = 1$ yields
$$^nP_k = \frac{(n)(n-1)(n-2) \cdots (n-k+1) \cdot (n-k)!}{(n-k)!} = \frac{n!}{(n-k)!} = \frac{n!}{k!} \text{ (since } k! = (n-k)!).$$

(b) Combination/choose formula: From the formula for ^nP_k, it suffices to divide by $k!$, since there are $k \cdot (k-1) \cdot (k-2) \cdots 2 \cdot 1 = k!$ different ways to order a set of k elements. (We have k positions for the first element, $k-1$ positions for the second element, $k-2$ for the third, and so forth.) The formula is thus $\binom{n}{k} = \frac{n!}{k!(n-k)!}$.

This completes the proofs of the permutation and choose formulas for all $n, k \in \mathbb{N}$, $k \leq n$. □

Example 10.4.4.1. *Going back to the combination lock story, we could say that the "combination lock" is actually a "permutation lock," since it only accepts the digits 0, 4, 0, and 5 in the specific order (permutation) 0405. A true combination lock would accept any of the $\frac{4!}{2!} = 12$ 4-digit numbers (allowing for leading zeros) containing the digits 0, 4, 0, and 5. (This would've saved you 360 combinations - or about 15 minutes of stress and hard work!)*

Example 10.4.4.2.

(a) *The number of ways to rearrange the letters in a word with no duplicate letters (such as MATH) is just the factorial of the number of letters, in this case $4! = 24$.*

(b) *The number of permutations of the word PROOF is equal to $\frac{5!}{2!} = 60$. (Note that there are also 3 instances of $1!$ in the denominator for P, R, and F, but these can safely be ignored.)*

(c) *The number of ways to arrange the letters in ABBCCCDDDD is given by $\frac{10!}{4! \cdot 3! \cdot 2! \cdot 1!} = 12,600$.*

Proof. Consider each group of distinct letters in the l-letter word W as a block. Then call the blocks $B_1, B_2, B_3, \ldots, B_n$. In each block B_i, if we instead assume that all of the k letters in B_i were distinct, then it would have $k!$ arrangements. All of the letters are identical, so there is only 1 arrangement for each block. Thus, we must divide $n!$ by $\sum_{i=1}^{n} L(B_i)! = \sum_{i=1}^{n} f(L_i)$ (where $L(B_i)$ denotes the number of letters in the block B_i and $f(L_i)$ denotes the frequency of the letter L_i; hence, $L(B_i) = f(L_i)$). Since $l = \sum_{i=1}^{n} f(L_i)$, the result immediately follows. □

Let's cement our understanding of this fundamentally important idea (used often in contests) with a couple of examples from the most recent AMC 8 contest:[15]

Example 10.4.4.3. *(2018 AMC 8) Professor Chang has nine different language books lined up on a bookshelf: two Arabic, three German, and four Spanish. How many ways are there to arrange the nine books on the shelf keeping the Arabic books together and keeping the Spanish books together?*

Solution 10.4.4.1. There are $2! = 2$ ways to arrange the Arabic books, $3! = 6$ ways to arrange the German books, and $4! = 24$ ways to arrange the Spanish books. Now consider the Arabic books all as one unit A, and the Spanish books as one unit S. Then we want the number of ways to arrange the string $AGGGS$, which is equal to $\frac{5!}{3!} = 20$. The total number of arrangements is therefore $2 \cdot 6 \cdot 24 \cdot 20 = 5,760$.

Example 10.4.4.4. *(2018 AMC 8) Abby, Bridget, and four of their classmates will be seated in two rows of three for a group picture, as shown.*

$$X \quad X \quad X$$
$$X \quad X \quad X$$

If the seating positions are assigned randomly, what is the probability that Abby and Bridget are adjacent to each other in the same row or the same column?

---
[15] As of the publication of the first edition of this book.

Solution 10.4.4.2. *There are $\binom{6}{2} = 15$ ways for Abby and Bridget to position themselves in the seats. We can do some minor casework from here, with Abby in the first row. If she is in the leftmost seat, Bridget can be in two seats: the ones right and down from Abby. The same goes for when Abby is in the middle-top seat. If she is in the upper-right seat, however, there is only one seat for Bridget: the lower-right seat. When Abby is in the second row, Bridget must be to the right of Abby (if she is above her, that is a case that has already been counted), so there are 2 possibilities. The total is 7 possibilities, so the desired probability is $\frac{7}{15}$. (We did not need to consider the positions of the other students, since everything would cancel out in the fraction anyway.)*

10.4.5 Pascal's Triangle and its Combinatorial Connections

Introduction to Pascal's Triangle

Arguably the most famous (or at least readily recognizable) of all objects in combinatorics, *Pascal's Triangle* is a compact visual depiction of the binomial coefficients, constructed in a nice, clean, simple manner.

> **Definition 10.21: Pascal's Triangle**
>
> Pascal's Triangle consists of an infinite number of *rows*, beginning with row 0 (at the very center of the triangle), which consists of a single 1. Row 1, which is immediately below Row 0, consists of two 1's, and from then on, all elements at the edges of the triangle (the left- and right-most points of each row) are 1's, and all other elements (the elements in the triangle's interior) are equal to the sum of the two elements directly above them.

Here are the first six rows of Pascal's Triangle:

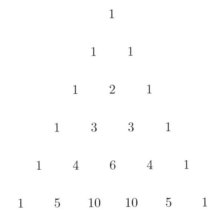

Figure 10.13: The first few rows of Pascal's Triangle.

And here are the first six rows of the triangle with the elements re-written as binomial coefficients:

$$\binom{0}{0}$$
$$\binom{1}{0} \quad \binom{1}{1}$$
$$\binom{2}{0} \quad \binom{2}{1} \quad \binom{2}{2}$$
$$\binom{3}{0} \quad \binom{3}{1} \quad \binom{3}{2} \quad \binom{3}{3}$$
$$\binom{4}{0} \quad \binom{4}{1} \quad \binom{4}{2} \quad \binom{4}{3} \quad \binom{4}{4}$$
$$\binom{5}{0} \quad \binom{5}{1} \quad \binom{5}{2} \quad \binom{5}{3} \quad \binom{5}{4} \quad \binom{5}{5}$$

Figure 10.14: Figure 9.13 with the elements written as binomial coefficients, to illustrate the relationships that will be proven in the following sections.

To fully understand and appreciate why the elements of Pascal's Triangle can all be written as binomial coefficients in the form $\binom{n}{k}$ where n is the row number and k is the position of the element in that row (where 0 is the far left and n is the far right), we must first introduce a crucial combinatorial identity.

Pascal's Identity

Pascal's Identity, named in honor of Pascal himself, is what allows us to show that Pascal's Triangle consists entirely of binomial coefficients. Pascal's Identity is particularly useful in simplifying expressions with several complicated binomial coefficients:

> **Theorem 10.13: Pascal's Identity**
>
> For all $n, k \in \mathbb{N}$,
> $$\binom{n-1}{k-1} + \binom{n-1}{k} = \binom{n}{k}.$$

Proof. Expanding the LHS yields

$$\binom{n-1}{k-1} + \binom{n-1}{k} \tag{10.7}$$

$$= \frac{(n-1)!}{(k-1)! \cdot (n-k)!} + \frac{(n-1)!}{k! \cdot (n-k-1)!} \tag{10.8}$$

$$= (n-1)! \cdot \left(\frac{k}{k! \cdot (n-k)!} + \frac{n-k}{k! \cdot (n-k)!} \right) \tag{10.9}$$

$$= (n-1)! \cdot \left(\frac{n}{k! \cdot (n-k)!} \right) \tag{10.10}$$

$$= \frac{n!}{k! \cdot (n-k)!} = \binom{n}{k} \tag{10.11}$$

as desired. □

Rationale for the Binomial Coefficients

Consider some non-edge element $N = \binom{n}{k}$. (All edge elements are 1, and also equal to $\binom{n}{0}$ for some $n \geq 0$.) Observe that N can be written as $\binom{n-1}{k-1} + \binom{n-1}{k}$, since the two elements above it are both in row $n-1$. By Pascal's Identity, we immediately have $N = \binom{n}{k}$ as desired.

Hockey Stick Identity

The *Hockey Stick Identity*, so named for the appearance of a hockey stick emerging when the relevant terms of Pascal's Triangle are highlighted, is another strikingly elegant combinatorial result owing to the great Pascal:

> **Theorem 10.14: Hockey Stick Identity**
>
> For all $n, r \in \mathbb{N}$, $n > r$,
> $$\sum_{i=r}^{n} \binom{i}{r} = \binom{n+1}{r+1}$$

Proof. Exercise. □

Informally speaking, we start at some element of the triangle and proceed diagonally downwards and to the left, until we reach a point we are happy with. Then we go one down and to the *right*; the sum of the elements on the down-left diagonal is equal to the remaining element. (Two Pascal's Triangles are printed in the pages above for your highlighting convenience.)

10.4.6 Complementary Counting

The art of combinatorics often lies not in counting the cases that satisfy some restriction, but rather the cases that *don't* satisfy it. In situations where the satisfactory cases are too cumbersome to count, then we can use *complementary counting*, which subtracts the unsatisfactory cases from the total. Consider the following basic example:

Example 10.4.6.1. *(2016 AMC 8) An ATM password at Fred's Bank is composed of four digits from 0 to 9, with repeated digits allowable. If no password may begin with the sequence 9, 1, 1, then how many passwords are possible?*

Solution 10.4.6.1. *With no restrictions, there would be $10^4 = 10,000$ possible passwords, since repeated digits are allowed. However, we must subtract off the "unsatisfactory" cases; i.e. those passwords beginning with 911. There are 10 of these, since it does not matter what the last digit is; thus, our total is $10,000 - 10 = 9,990$ acceptable passwords.*

More rigorously, consider some universal set U. If we want to count a subset $S \subseteq U$, then complementary counting entails computing $|S^c|$, rather than $|S|$.

Here is an example of a problem that combines complementary counting with the Inclusion-Exclusion Principle:

Example 10.4.6.2. *(2002 AIME I) Many states use a sequence of three letters followed by a sequence of three digits as their standard license-plate pattern. Given that each three-letter three-digit arrangement is equally likely, the probability that such a license plate will contain at least one palindrome (a three-letter arrangement or a three-digit arrangement that reads the same left-to-right as it does right-to-left) is $\frac{m}{n}$, where m and n are relatively prime positive integers. Find $m + n$.*

Solution 10.4.6.2. *We want to subtract from 1 the probability that a given license plate will not contain a palindrome in either its letters or its numbers. For the letters, we have 26 choices for the letters in the end positions (they must be the same), and 26 choices for the middle letter, so this gives $26^2 = 676$ choices (out of 26^3 total choices, for a probability of $\frac{1}{26}$). For the numbers, there are similarly 10 choices for the end digits and 10 choices for the middle digit, as well as $10^3 = 1000$ choices in total, so the probability that a 3-digit string of numbers is a palindrome is $\frac{1}{10}$.*

From here, we can take one of two solution paths:

(a) *Use complementary counting. The probability that a given three-letter string is not a palindrome is $\frac{25}{26}$, and the probability that a given three-digit string is not a palindrome is $\frac{9}{10}$. Multiplying these together yields $\frac{45}{52}$, and the complement of this is our answer, $\frac{7}{52}$.*

(b) *Use PIE. The probability of at least one of the strings being a palindrome is the sum of the individual probabilities that each is a palindrome, minus the probability that both are palindromes. This is equal to $\frac{1}{26} + \frac{1}{10} - \frac{1}{260} = \frac{7}{52}$, confirming our complementary counting prowess.*

10.4.7 Examples

Example 10.4.7.1. *A fair coin is flipped 9 times, landing on heads every time. What is the probability that the tenth flip will also be heads?*

Solution 10.4.7.1. *Since the coin does not have a "memory" (i.e. all the flips are independent events), the answer is $\frac{1}{2}$. (We can also use conditional probability to show that the probability is $\frac{1}{2}$. Given an event with probability $\left(\frac{1}{2}\right)^9 = \frac{1}{512}$, the probability that an event with probability $\left(\frac{1}{2}\right)^{10}$ occurs is $\frac{1}{2}$.)*

Example 10.4.7.2.

(a) *Your body made a protein with the following amino acid chain: Met-Leu-Pro-Ile-Phe-STOP. If there is 1 way to code up Met, 6 for Leu, 4 for Pro, 3 for Ile, 2 for Phe, and 3 for STOP, in how many ways could the protein chain have been coded up?*

(b) *How many times more ways could the above protein have been coded up if the only order restriction were that Met comes first and STOP comes last?*

Solution 10.4.7.2.

(a) *There are $1 \cdot 6 \cdot 4 \cdot 3 \cdot 2 \cdot 3 = \boxed{432}$ ways to code up the protein by the multiplicative principle.*

(b) *There are 4 proteins other than Met and STOP that can be re-arranged in any order, so the answer is $4! = \boxed{24}$.*

Example 10.4.7.3. *(2017 Berkeley Mini Math Tournament) There are 2000 math students and 4000 CS students at Berkeley. If 5580 students are either math students or CS students, then how many of them are studying both math and CS?*

Solution 10.4.7.3. *Let M be the set of math students and C be the set of CS students. Then this is a straightforward application of PIE, where $|S| = 2000$ and $|C| = 4000$; given that $|C \cup S| = 5580$, $|C \cap S| = |C| + |S| - |C \cup S| = 420$. No rearrangement required!*

Example 10.4.7.4. *(2017 AMC 8) Mrs. Sanders has three grandchildren, who call her regularly. One calls her every three days, one calls her every four days, and one calls her every five days. All three called her on December 31, 2016. On how many days during the next year did she not receive a phone call from any of her grandchildren?*

Solution 10.4.7.4. *Call the grandchildren who call in every 3, 4, and 5 days, A, B, and C respectively. 2017 is not a leap year, so it will have 365 days. During 2017, A will call Mrs. Sanders $\lfloor \frac{365}{3} \rfloor = 121$ days, B will call her $\lfloor \frac{365}{4} \rfloor = 91$ days, and C will call her $\lfloor \frac{365}{5} \rfloor = 73$ days. Furthermore, both A and B call every 12 days, or 30 days out of the year; A and C call every 15 days, or 24 days in the year; B and C call every 20 days, or 18 days in 2017; and finally, all three children call every 60 days, or 6 days in 2017. Using PIE, we get that $|(A \cup B \cup C)^c| = 365 - |A \cup B \cup C| = 365 - (121 + 91 + 73 - (30 + 24 + 18) + 6) = 365 - 219 = 146$.*

(Alternate solution: Since lcm(3, 4, 5) = 60, and 360 is a nice, clean multiple of 60, consider only the first 360 days. A will not call on $\frac{2}{3}$ of those days, B will not call on $\frac{3}{4}$ of those days, and C will not call on $\frac{4}{5}$ of those days. Multiplying these together yields $360 \cdot \frac{2}{3} \cdot \frac{3}{4} \cdot \frac{4}{5} = 144$. Finally, accounting for the last 5 days, none of the children call in on days 361 or 362 (A does call on day 363, B calls on day 364, and C calls on day 365). Thus, our answer is $144 + 2 = 146$.)

Example 10.4.7.5. *Solve the equation $\binom{n+1}{3} - \binom{n}{2} = 7n$ for $n \in \mathbb{N}$.*

Solution 10.4.7.5. *We have $\binom{n+1}{3} = \frac{(n+1)(n)(n-1)}{6}$ and $\binom{n}{2} = \frac{n(n-1)}{2}$, so their difference is equal to $\frac{n^3 - n}{6} - \frac{3n^2 - 3n}{6} = \frac{n^3 - 3n^2 + 2n}{6} = \frac{n(n-1)(n-2)}{6}$. Setting this equal to $7n$ yields $(n-1)(n-2) = 42$, which has solutions $n = 8, -5$. We must have $n > 0$, so $n = 8$ is our only solution.*[16]

Example 10.4.7.6. *(2005 AMC 12B) An envelope contains eight bills: 2 ones, 2 fives, 2 tens, and 2 twenties. Two bills are drawn at random without replacement. What is the probability that their sum is $20 or more?*

Solution 10.4.7.6. *In total, there are $\binom{8}{2} = 28$ possible choices of two distinct bills. Of these, if one of them is a twenty, then the sum of the two bills will automatically be more than $20; this occurs with probability $\frac{\binom{8}{2} - \binom{6}{2}}{\binom{8}{2}} = \frac{13}{28}$. On the other hand, if neither bill is a twenty, the only way for them to sum to at least $20 is for them to sum to exactly $20: two tens, which occurs with probability $\frac{1}{28}$. Hence, the total probability that the sum is at least $20 is $\frac{14}{28} = \frac{1}{2}$.*

Example 10.4.7.7. *Briefly explain why the sum of the elements in any given row of Pascal's Triangle is equal to 2 raised to the power of the row number.*

---

[16] To verify that no other positive integers n work, note that $f(n) = \binom{n+1}{3} - \binom{n}{2}$ grows faster than does $7n$ for all $n > 8$; more rigorously, $f'(n) = \frac{3n^2 - 6n + 2}{6} > 7 = (7n)'$. To show $\frac{3n^2 - 6n + 2}{6} > 7$ for all $n \geq 8$, we simplify to $3n^2 - 6n - 40 > 0$. Indeed, the quadratic equation $3n^2 - 6n - 40 = 0$ has solutions $n = \frac{3 \pm \sqrt{129}}{3}$, so $3n^2 - 6n - 40 > 0$ whenever $n > \frac{3 + \sqrt{129}}{3} < 1 + \frac{12}{3} = 5$. Here, $f'(n)$ denotes the derivative of f.

Solution 10.4.7.7. *Let the row number be n, with elements $\binom{n}{0}, \binom{n}{1}, \binom{n}{2}, \ldots, \binom{n}{n}$. Effectively, we must show that*
$$\sum_{k=0}^{n} \binom{n}{k} = 2^n.$$
This is a consequence of the Binomial Theorem (proven as an exercise).

10.4.8 Exercises

Exercise 32. *Roll two fair six-sided dice. What is the probability that you roll at least one prime number?*

Exercise 33. *At a dance party, there are 7 boys and 5 girls. Each boy is paired with one or zero girls. How many different arrangements are possible?*

Exercise 34. *A modified, single-person variation of the card game Blackjack has the dealer deal to the bettor two cards that are either the Ace, or 2-10 (not including Jacks, Queens, or Kings), and where Aces are always worth 1. A blackjack in this version is a hand with two cards summing to 18. If the sum is less than 18, the bettor wins a dollar amount equal to his sum. If the sum is exactly 18, the bettor wins $100. Otherwise, if the sum is over 18, the bettor busts and wins nothing. Suppose that the dealer is generous and wants to make the game truly fair (and not slightly advantageous to the casino, like in real life). To the nearest dollar, how much should he charge to play this version of Blackjack?*

Exercise 35. *How many ways are there to arrange the letters in the word TARGET in a circle such that the T's are not adjacent?*

Exercise 36.

(a) *Show that, if there are at least 367 people in a room, at least two of them are guaranteed to have the same birthday (assuming they were all born in a leap year).*

(b) *Show that, if there are at least $366n + 1$ people in a room, at least $n+1$ of them are guaranteed to have the same birthday.*

(c) *Let there be p people in a room, $1 \leq p \leq 366$. Find, with proof, the smallest value of p such that the probability of two or more people sharing the same birthday is at least $\frac{9}{10}$.*

(d) *Let there be p people in a room, $1 \leq p \leq 366$. Find, with proof, the smallest value of p such that the probability of two or more people sharing the same birthday is at least $\frac{k-1}{k}$, in terms of k.*

Exercise 37. *(2016 AMC 12A) Each of the 100 students in a certain summer camp can either sing, dance, or act. Some students have more than one talent, but no student has all three talents. There are 42 students who cannot sing, 65 students who cannot dance, and 29 students who cannot act. How many students have two of these talents?*

Exercise 38. *Find the value of*
$$\binom{3}{0} + \binom{3}{1} + \binom{4}{2} + \binom{5}{3} + \ldots + \binom{98}{96} + \binom{99}{97}.$$

Exercise 39. *Prove the Hockey Stick Identity.*

Extension Exercise 15. *We can also prove Pascal's Identity purely combinatorially. How so? (Hint: consider an arbitrary element x in a set with n elements.)*

Extension Exercise 16.

(a) *Let $\{a_i\}_{i=1}^n$ be an integer sequence. Show that there exists an ordered pair of integers (k, m) with $k < m$ and $k + m \leq n$ such that $a_k + a_{k+1} + a_{k+2} + \ldots + a_{k+m}$ is divisible by n.*

(b) *Using the Generalized Pigeonhole Principle, prove the Chinese Remainder Theorem.[17]*

Extension Exercise 17. *(2014 Caltech-Harvey Mudd Math Competition) There is a long-standing conjecture that there is no number with $2n + 1$ instances in Pascal's triangle for $n \geq 2$. Assuming this is true, for how many $n \leq 100,000$ are there exactly 3 instances of n in Pascal's triangle?*

Extension Exercise 18. *Generalize Pascal's Identity to the p-variable case, then prove that generalized form of the identity.*

---

[17] For the statement of the CRT, see §14.

Chapter 10 Review Exercises

Exercise 40. *Roll three fair six-sided dice. What is the probability of getting a triple-digit product?*

Exercise 41. *(1970 AHSME) If the number is selected at random from the set of all five-digit numbers in which the sum of the digits is equal to 43, what is the probability that the number will be divisible by 11?*

Exercise 42. *In a length-k string of characters, define the polar of position p to be position $k - p + 1$ (from left to right). Call a string of A's and B's yin-yang if there is an A in the nth spot whenever the letter in the polar of the nth spot is a B, and vice versa. The probability that a randomly-chosen 10-letter string with 5 A's and 5 B's is yin-yang can be written in the form $\frac{m}{n}$, where m and n are relatively prime positive integers. Find $m + n$.*

Exercise 43. *Let $S = \{-3, -3, -3, -1, 1, 1, 9, 16, 26, 41\}$.*

(a) *What are the mean, median, mode(s), and sample/population standard deviations of S?*

(b) *What is the five-number summary of S?*

(c) *Using each of the following definitions of an outlier, compute the outlier(s) of S.*

   (i) *A data point that lies more than 1.5 IQR away from the mean*

   (ii) *A data point that lies below the 9th percentile or above the 91st percentile (using the "compromise" definition)*

(d) *Suppose that the elements of S are each written down on a slip of paper, and the slips are shuffled and put into a hat. Josie picks three of the slips of paper (with replacement) uniformly at random. She knows that there are 10 slips of paper in the hat, but does not know what the numbers are beforehand. In her three picks, she comes up with -3 every time. Using a two-tailed test, she then concludes that S must contain an unusually large proportion of -3's. Is she correct in her conclusion, or is she making a statistical error? If so, what kind of error is she making, and why? Would her conclusion change if she instead used a single-tailed test? (Use the standard threshold for statistical significance, $\alpha = 0.05$.)*

Exercise 44. *(2017 MathCounts Chapter Sprint) Yu has 12 coins, consisting of 5 pennies, 4 nickels and 3 dimes. He tosses them all in the air. What is the probability that the total value of the coins that land heads-up is exactly 30 cents? Express your answer as a common fraction.*

Exercise 45. *(2003 AMC 12B) Let S be the set of permutations of the sequence 1, 2, 3, 4, 5 for which the first term is not 1. A permutation is chosen randomly from S. The probability that the second term is 2, in lowest terms, is $\frac{a}{b}$. What is $a + b$?*

Exercise 46. *(2015 AIME II) Two unit squares are selected at random without replacement from an $n \times n$ grid of unit squares. Find the least positive integer n such that the probability that the two selected unit squares are horizontally or vertically adjacent is less than $\frac{1}{2015}$.*

Exercise 47. *How many ways can the digits of 112233 be arranged such that no two identical digits are adjacent?*

Exercise 48. *(2015 AIME I) The nine delegates to the Economic Cooperation Conference include 2 officials from Mexico, 3 officials from Canada, and 4 officials from the United States. During the opening session, three of the delegates fall asleep. Assuming that the three sleepers were determined randomly, the probability that exactly two of the sleepers are from the same country is $\frac{m}{n}$, where m and n are relatively prime positive integers. Find $m + n$.*

Exercise 49. *Jackson is taking a Probability class which has three equally weighted tests - two midterms and a final. In this class, an A equates to any percentage score 90 or above, a B to a score at least 80 but less than 90, and so forth, until an F which equates to any score under 60. On the first test, he scores a B, and on the second test, he scores an A. Jackson forgot his exact percentage scores on the first two tests, but wants to secure an A in the class, so he studies long and hard for the final. What is the probability that he will be able to do so, given that he scores an A on the final? (The test scores need not be integers.)*

Exercise 50. A 5-question exam awards 5 points for a correct answer, 1 point for a blank, and 0 points for an incorrect answer. For each attainable score s, let $N(s)$ be the number of distinct answer patterns corresponding to that score. For example, $N(6) = 20$, since the only way to achieve a score of 6 is to answer 1 correctly, leave 1 blank, and answer 3 incorrectly; this can be done in $\frac{5!}{(3! \cdot 1! \cdot 1!)} = 20$ ways. Over all attainable scores s, the average value of $N(s)$ can be written in the form $\frac{p}{q}$, where p and q are relatively prime positive integers. Find $p + q$.

Exercise 51. If we pick a positive integer between 1 and 1000, inclusive, the probability that it is expressible as $x \lfloor x \lfloor x \rfloor \rfloor$ (where $x > 0$ is a real number and $\lfloor x \rfloor$ represents the floor function/greatest integer function) can be written as $\frac{m}{n}$, where m and n are relatively prime positive integers. Find $m + n$.

Exercise 52. *(2018 AIME I)* Kathy has 5 red cards and 5 green cards. She shuffles the 10 cards and lays out 5 of the cards in a row in a random order. She will be happy if and only if all the red cards laid out are adjacent and all the green cards laid out are adjacent. For example, card orders RRGGG, GGGGR, or RRRRR will make Kathy happy, but RRRGR will not. The probability that Kathy will be happy is $\frac{m}{n}$, where m and n are relatively prime positive integers. Find $m + n$.

Exercise 53. A prototype of an email spam filter is $p_1\%$ effective at filtering spam: that is, it will correctly identify a spam email $p_1\%$ of the time. However, it will also incorrectly identify legitimate emails as spam $p_2\%$ of the time. In terms of p_1 and p_2, what is the probability that an email identified as spam by the filter actually is spam?

Extension Exercise 19. If a, b, c are real numbers between 0 and 10 inclusive, compute the probability that $a + 2b + 3c \leq 15$.

Extension Exercise 20. *(2012 HMMT February Combinatorics Round)* A frog is at the point $(0, 0)$. Every second, he can jump one unit either up or right. He can only move to points (x, y) where x and y are not both odd. How many ways can he get to the point $(8, 14)$?

Extension Exercise 21. *(2018 AIME I)* For every subset T of $U = \{1, 2, 3, \ldots, 18\}$, let $s(T)$ be the sum of the elements of T, with $s(\emptyset)$ defined to be 0. If T is chosen at random among all subsets of U, the probability that $s(T)$ is divisible by 3 is $\frac{m}{n}$, where m and n are relatively prime positive integers. Find m.

Extension Exercise 22. *(1972 USAMO P3)* A random number selector can only select one of the nine integers $1, 2, \ldots, 9$, and it makes these selections with equal probability. Determine the probability that after n selections ($n > 1$), the product of the n numbers selected will be divisible by 10.

Extension Exercise 23. *(2018 USAJMO P1)* For each positive integer n, find the number of n-digit positive integers that satisfy both of the following conditions:

- no two consecutive digits are equal, and
- the last digit is a prime.

Chapter 11

Intermediate Discrete Math

11.1 Introduction to Game Theory and Rational Decision-Making

We begin the next part of our foray into the limitless expanse that is discrete math with a cursory overview of *game theory*. Alas, this is not the study of video games,[18] but it is something fairly close! By definition, a *game* is any exchange of strategic moves played amongst two or more *participants*, much like a single-player campaign mode usually entails a real-life human being playing against one or more computer-controlled opponents, and a multi-player mode entails playing a match against (or on a team with) several other players.

Each of these participants makes rationally-based *decisions*, just like in a video game, where a player chooses to fight a brutal monster boss by choosing his or her attack, defense, and healing moves, or chooses how to collaborate with his or her teammates, who do the same to achieve a common goal. Indeed, this is the driving motive of game theory: to maximize the benefit to all players involved in the game.

> **Definition 11.1: Game**
>
> A *game* is any strategic interaction between two or more rational decision-makers.

We say a decision is *rational* if it maximizes the benefit to the decision maker.

11.1.1 Types of Strategies

How do we define a *strategy*? The answer depends on what probabilities are assigned to each strategy being chosen from a *strategy set* (i.e. the set of all possible decisions)[19] available to the player.

> **Definition 11.2: Strategy Types**
>
> A *pure* strategy is a strategy that determines the player's move for any and all situations he or she should face. A *mixed* strategy assigns a probability to each pure strategy in the player's strategy set. A *totally mixed* strategy is a mixed strategy in which all probabilities are non-zero. (A pure strategy can be thought of as a mixed strategy with probability 1 for that strategy, and probability 0 for all other strategies.)

---

[18]That would be *ludology*. Or, more precisely, a specialized subfield of it - ludology as a whole is the psychologically- and culturally-oriented analysis of game design, player interactions, the role that the game fulfills in its niche, and the influence of games on human behavior. A "game" may be just about anything - a board game, card game, dice game, tabletop game, pencil-and-paper game, or even a conversation game (à la *20 Questions*, *Never Have I Ever...*, or *Truth or Dare*) - not necessarily a video game, even though it may seem that way to us 2019 folk!

[19]We say a strategy set is *finite* if the player has a finite number of *discrete* strategies available to them. For instance, the set {rock, paper, scissors} is a finite strategy set, but {the set of times to leave for work so as to minimize commute time from traffic} or {the set of portions of a cake to cut in the cake cutting game} are not finite.

For example, if the player always picks strategy A in a two-decision game, he or she is using a pure strategy. If s/he flips a coin to determine which decision s/he makes, this is an example of a mixed strategy (the coin need not be fair; the probability can be utterly lopsided, as long as the probabilities of the coin coming up heads or tails are both nonzero). If the coin has either heads or tails on both sides, then this is an example of a mixed strategy that is *not* also totally mixed.

11.1.2 Types of Games

Games can be played in a vast multitude of ways, and with no limit on the number of players: they can range from simple two-player games with one decision made by each player, or extremely complex multiplayer games with constant decision-making and decisions that are dependent on others. The most common descriptions, or *forms*, of a game are the *normal form* and the *extensive form*:

> **Definition 11.3: Normal and Extensive Form Games**
>
> A *normal form game* is a representation of a game through a square matrix, with *payoffs*, or benefits (measured in utils, or whatever currency is at stake) as elements in the cells of the matrix. An *extensive form game* is a more graphical representation of a game as a branching tree, with each decision being a branch and the result of each decision being a node.

Games are also classified depending on the nature of their payoffs for each player involved:

> **Definition 11.4: Zero-Sum Game**
>
> A *zero-sum* game is a game in which one group of participants gains that which is collectively lost by the other participants; i.e. the sum of each participant's net gain/loss is zero.

We can also classify games under the general umbrellas of *cooperative* or *non-cooperative*:

> **Definition 11.5: Cooperative and Non-Cooperative Games**
>
> A game is *cooperative* if two or more players form a *coalition* that competes with other coalitions, and the players in each coalition share the payoff. These coalitions can be enforced through such means as contracts set up at the beginning of the game, and can make binding agreements and redistribute the payoff as they see fit. Otherwise, a game is *non-cooperative*; i.e. when there exists no possibly of forging alliances.

> **Definition 11.6: Special Coalitions**
>
> Let $\mathcal{P}$ be the set of players; then $\mathcal{P}$ itself is the *grand coalition*, and $\emptyset$ is the *empty coalition*.

> **Theorem 11.1: Coalitions in a Cooperative Game**
>
> Every cooperative game with N players has 2^N coalitions, including the grand and empty coalitions.

Proof. This amounts to proving that $|\mathcal{P}(S)| = 2^N$ when $|S| = N$, where $\mathcal{P}(S)$ denotes the power set of S. This was proven in §10.1.3. $\square$

11.1.3 I Know You Know I Know...

The strategic nature of a game also depends on what information, if any, its players have access to. Does everyone know everything about everyone else? Is some information classified, or is all of it? Is there any means of espionage; that is, can boundaries imposed on the players be circumvented? The answers of all of these questions determine another classification for a game:

> **Definition 11.7: Perfect Information**
>
> A game has *perfect information* if all players have access to all previous events in the game, including the starting conditions, as well as all players' potential payoffs and the payoff matrix. Otherwise, a game has *imperfect information*.

Example 11.1.3.1. *Examples of games with perfect information are chess, checkers, Go, and tic-tac-toe, since all players can see the entire board at any time, and, in the case of chess and checkers, are aware of which pieces have already been captured. Examples of games with imperfect information are bridge, any variation of poker, and even purchasing a car from a car dealership,[20] since you are unaware of the intrinsic properties and qualities of the car, whereas the seller knows (nearly) everything about it, and how it will benefit you, the consumer.*

Furthermore, games can be *simultaneous* or *sequential*, tying into the concept of perfect and imperfect information:

> **Definition 11.8: Simultaneous and Sequential Games**
>
> A game is *simultaneous* if all players make decisions at the same time. If they do stagger their decisions, then the game cannot have perfect information (i.e. players making decisions later cannot know what decisions earlier players have made). However, not having perfect information is not a sufficient condition for a game of this type to be simultaneous. We can say that such a game requires "perfect *non*-information."
>
> On the contrary, a game is *sequential* if later players have at least some knowledge about the actions of earlier players. Even something as little as knowing that an earlier player did *not* perform one action, out of several, is enough to ensure that a game is sequential and not simultaneous.

11.1.4 Extensive Form Games: Time-Sequenced Decision Trees

In an *extensive form game*, players' decisions/strategies are depicted through a branching tree, with each node representing an opportunity for a player to make a decision, and with each branch representing the "path" to the outcome of that decision.

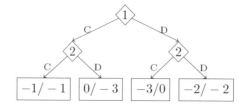

Figure 11.1: An extensive form version of the Prisoner's Dilemma (see §11.1.5 below).

In the game above, players 1 and 2 make their decisions simultaneously, since neither player's move depends on the other's. If player 1 chooses to *cooperate* (C), then player 2 is left with the same decision as player 1.

---
[20] Quite possibly the least fun "game" in the world, if you ask me. But then again, I'm not a car salesman.

If player 2 chooses to *defect* (D), then player 2 will still have the same strategy set at his or her disposal, only the potential consequences will differ.

In this case, both players have perfect information at their disposal; however, this is not always the case. When a game has imperfect information, the extensive form usually represents the information that *is* known by both/all players by connecting it with a dotted/dashed line. For instance, in Figure 11.1, if player 2 lacks information as to player 1's decision (as is the case in the real prisoner's dilemma), then we would draw a dashed line between the "2" nodes.

If, instead, we construct a game in terms of an infinite continuum of decisions (i.e. non-discrete strategy sets), then the extensive form adopts a slightly different notation. Rather than straight lines connecting nodes, we use a circular arc, reaching out to the second node, that adjoins two line segments protruding from the first node. (We write the upper and lower bounds above and below the circular arc, accompanied by a variable used to express the payoffs.)

Backward Induction

In *backward induction,* we essentially deconstruct an extensive form game by looking at which decision is rational for each player at each point in the game, starting from the end of the game. This will determine the sequence of optimal strategies for the players of the game. Backward induction has been around for as long as game theory itself - it was proposed in 1944 in the very book that introduced game theory into the contemporary focus of mathematical study, *Theory of Games and Economic Behavior,* by John von Neumann and Oskar Morgenstern (at least in the case of two-player games).

An introductory example of backward induction lies in the form of the *ultimatum game:* a game in which there is a proposer, who proposes a distribution of a stack of ten $1 bills, and a responder, who either accepts or rejects the distribution proposed by the first player. The proposer chooses to split the money into either a *fair* distribution of $5/$5, or an *unfair* distribution of $8/$2 (keeping the $8 for himself/herself). If the responder accepts the proposer's offer, the money is split between the two players as per the proposal; if the responder rejects, both players receive nothing (the money is effectively discarded).

The following is the extensive form diagram for this game:

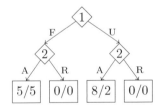

Figure 11.2: The ultimatum game in extensive form.

Starting from the nodes marked "2" (for player 2's decision), which strategy should s/he go with? If s/he rejects the proposal, regardless of whether player 1 (the proposer) has gone with "F" (for *fair*) or "U" (for *unfair*), both players receive a payoff of zero. However, if s/he accepts, then both payoffs will be non-zero (in particular, his or her own payoff will be at least $2). And what about player 1's decision? If player 2 will accept no matter what, then the only rational thing to do is to propose an unfair decision, since doing so will net him/her $8, as opposed to $5. Thus, the ideal strategies for player 1 and player 2 are U (unfair) and A (accept), respectively.

A slightly more complicated example, which uses mixed strategies, invokes the idea of expected value in backwards decision-making. Suppose a homeowner bought a house, knowing that he would keep the house for exactly 10 years. Each year, he has the choice to install either a 1-car or a 2-car garage, or to leave his car parked in the street. Once he installs the garage, he will not be able to change it to something else for the remainder of his home ownership. The utility to the homeowner of having a 1-car garage is +60 utils, but the utility to him of having a 2-car garage is +100 utils, while the utility of having no garage at all is 0 utils. Should the homeowner ever install a 1-car garage, or is it best to install a 2-car garage from the beginning?

In performing backwards induction, we assume that the homeowner has a 50% chance of choosing either strategy when he does install a garage. Just before year 10 of his home ownership (at time $t = 9$), he will gain 100 utils from installing the 2-car garage, but only 60 utils from installing the 1-car garage. Both of these are clearly better than installing no garage, so at $t = 9$, the homeowner should install some kind of garage if he hasn't already. At $t = 8$, the expected total value of waiting is zero at the time, but at $t = 9$, it will become $\frac{100 + 60}{2} = 80$ utils, as opposed to $60 \cdot 2 = 120$ utils from building the 1-car garage and $100 \cdot 2 = 200$ from building the 2-car garage. So he should still install a garage, whether it be 1-car or 2-car.

As for $t = 7$, the expected value of waiting totals to $\frac{200 + 120}{2} = 160$ utils, and building a 1-car garage immediately yields 180 utils (building a 2-car garage yields 300 utils). Still a disadvantageous strategy, but for $t = 6$ and earlier, an interesting phenomenon takes hold. At $t = 6$, the expected value of waiting is then 240 utils, which is exactly equal to the value of building a 1-car garage! The 2-car garage option is still superior, of course, but at this point, it matters not whether the homeowner builds a 1-car garage or chooses not to build.

For $t = 5$ and earlier, waiting it out is actually the optimal course of play: the expected value of waiting at $t = 5$ is $\frac{240 + 400}{2} = 320 > 60 \cdot 5 = 300$. This trend is exacerbated for $t < 5$ as well. So in summary, counter-intuitive as it seems, our homeowner would be better off waiting for an opportunity to install the more expensive 2-car garage, as opposed to accepting the 1-car garage offer! But only if it's not too late. Once he's past the halfway point of his residence, he might as well throw in the towel and just install the cheaper, 1-car garage. That is to say, it is more important to deeply consider a better option if one plans to stick with it for a longer period of time. Think of this like deciding between two job offers: would you want to work a job that you absolutely loathed, with very miserly salary and benefits, for years and years, maybe even the rest of your career? Probably not - if you were to establish a long-term career, you would almost certainly go for the job that would pay more and would be more beneficial and/or convenient to you.

11.1.5 Normal Form Games: Payoff Matrices

In a two-player game, a *payoff matrix* is a table representing the benefit or cost to each player of each potential set of simultaneous decisions. The benefits to the players may be in comma-separated ordered pairs (as in this book), or they may be separated with a slash or some other visual mark.

Example 11.1.5.1. *An example of a payoff matrix for a game with players A and B is as follows:*

A, B	choice 1	choice 2
choice 1	$(1, 2)$	$(3, -3)$
choice 2	$(-2, 4)$	$(-1, 3)$

Table 11.1: A standard two-decision payoff matrix for a two-player game. A's decisions are traditionally in the left column, and B's decisions are traditionally in the top row.

This matrix represents the situation where if A picks choice 1 and B picks choice 1, then A receives a payoff of 1 util and B receives a payoff of 2 utils, and similarly for the other three cells.

For games with three or more players, the payoff matrix becomes vastly more complicated (hence why we usually stick to representing two-player games in this manner). If we wanted to depict the benefit and cost to each player in a three-player game, we could either isolate one of the players and create separate tables for each case, or we could turn the payoff matrix into a payoff "cube," so to speak:

	(1,1,1)	(1,1,2)
(2,1,1)	(2,1,2)	(,2,2)
(2,2,1)	(2,2,2)	

Figure 11.3: A two-decision payoff "cube" for a three-player game. Player 1's decision dictates the z-position of the desired cell, player 2's decision corresponds to the x-axis, and player 3's decision corresponds to the y-axis.

In general, we can represent the payoffs for an n-player game via an n-dimensional hypersphere. As the number of decisions granted to each other grows, so does the "size," or radius, of the hypersphere.

At this point, we can introduce yet another two game types:

> **Definition 11.9: Symmetric and Asymmetric Games**
>
> A game is *symmetric* iff changing which players make the (same) decisions does not affect the payoff matrix. More rigorously stated:
>
> Let $G : S \mapsto P(S)$ (where $S = \{p_1, p_2, p_3, \ldots, p_n\}$ is a set of n players $P_1, P_2, P_3, \ldots, P_n$ and $P(S)$ is the n-variable payoff function associated with the set S of players) be a function representing a game. Then G is *symmetric* if, for any permutation π of S (i.e. a bijective function $\pi : S \mapsto S$), $G(S) = G(\pi(S))$. Else it is *asymmetric*.

The Prisoner's Dilemma

The *prisoner's dilemma* is one of the most classic and widely-known game-theoretic scenarios, in which two convicted and incarcerated prisoners are offered a choice between either remaining silent, or testifying against the other prisoner. They cannot communicate with each other (as they are in solitary confinement), and their sentences will be decided solely by their decisions.

The following is the standard payoff matrix for the prisoner's dilemma:

	cooperate	defect
cooperate	$(-1, -1)$	$(-3, 0)$
defect	$(0, -3)$	$(-2, -2)$

Table 11.2: The payoff matrix for the prisoner's dilemma. Here, to "cooperate" means to remain silent.

The prisoners are perfectly rational, but interestingly, they do not cooperate, even when it seems that it would be to their common benefit for them to do so. Since the prisoners know the payoff matrix, but the game is simultaneous, it only makes logical sense for each prisoner to defect rather than cooperate (since there is the promise of a payoff of 0 utils instead of -1 util). However, if both prisoners defect, then they will both be imprisoned for 2 years (receive a payoff of -2 utils), rather than just 1 year (a payoff of -1 util). If they were to cooperate with each other, on the other hand, their sentences would be half as strict. This is where the dilemma lies - if the prisoners know what the payoffs are, then why would they not both just cooperate and receive the lighter sentence?

The short answer: because humans are greedy by nature. The long answer: Greed and selfishness is the very definition of rational behavior. A rational decision-maker is purely self-interested, so he or she does not consider the payoff for his or her opponent, or his or her opponent's strategies to maximize payoff. Instead, he or she will merely attempt to pursue the maximum payoff, which is a complete pardon. In addition, even if the opponent defects, the worst possible consequence is only 2 years in prison, not 3. Regardless of what the opponent does, defection is the superior strategy, since cooperation yields 0 utils if the opponent cooperates, and -2 utils if the opponent also defects, whereas cooperating yields -1 util if the opponent does the same, but a hefty -3 utils if the opponent defects. As such, defection is a *dominant strategy*: that is, a strategy which beats another strategy for all strategies that the opponent can play. In fact, both prisoners defecting is the *Nash equilibrium* for the prisoner's dilemma (we will cover this in depth in the following section).

We can also generalize the prisoner's dilemma to an infinite number of payoff matrices of a certain form; the payoffs need not be specifically as in Table 10.2. Any payoff matrix of the form in the table below preserves the dominant strategy and Nash equilibrium of the game:

	cooperate	defect
cooperate	(a, a)	(b, c)
defect	(c, b)	(d, d)

Table 11.3: The payoff matrix for the generalized prisoner's dilemma, where $c > a > d > b$ and $2a > b + c$.

In this generalized form, we require $c > a > d > b$,[21] since the dilemma lies in having the Nash equilibrium (both prisoners defecting) provide a worse punishment to both prisoners than if they both cooperated; hence, $a > d$ (i.e. $d < a$, since the values are non-positive). In addition, if one prisoner cooperates and the other defects, the defector will get off scot-free (or at least receive a much more lenient sentence than the cooperator), so $b << c$ and $a < c$. Furthermore, as the common punishment for both prisoners defecting should be less than that of cooperating while the other defects, $d > b$, so the only logical order is $c > a > d > b$. (If we consider the variables as measures of prison time rather than utils, then the order is reversed, so that $c < a < d < b$.) Finally, we must have $2a > b + c$, since the collective payoff for cooperate/cooperate should be better (less harsh) than the payoff for cooperate/defect.

Battle of the Sexes

The *battle of the sexes* is a two-player *coordination game*[22] in a similar vein as the prisoner's dilemma. A married couple has agreed to meet somewhere (either a football game or an opera) one evening, but they have both forgotten where they would go (and both parties are aware of each other's forgetfulness). The husband wishes to attend the football game, while the wife wishes to attend the opera. Both would prefer going to the same place over ending up at different places (even if the husband is at the football game and the wife is at the opera). In addition, they (inexplicably) cannot communicate.[23] Where should the couple go to maximize each spouse's satisfaction level?

A traditional payoff matrix for the battle of the sexes game looks something like this:

	football	opera
football	$(3, 2)$	$(1, 1)$
opera	$(0, 0)$	$(2, 3)$

Table 11.4: A payoff matrix for the battle of the sexes.

---

[21] Some authors use r, s, t, p in place of a, b, c, and d respectively, for "reward," "sucker," "temptation," and "punishment" payoffs respectively.

[22] We'll define this more rigorously in the next section, after we develop the idea of the Nash equilibrium. For now, we can loosely define a coordination game as a game in which players choose the same, or corresponding, strategies.

[23] Though we can handwave this somewhat by saying that the football and opera are somewhat close to each other, and that both the husband and wife are already driving to the general area, but still. Haven't they heard of hands-free communication? ...Oh wait, silly me, this game was invented in 1944! Even then, you'd think they would've arranged all this beforehand...

Note that, though the payoff is 2 utils, as opposed to 3, for the spouse who ends up at the wrong event but with his or her significant other, it is still higher than the payoff of 1 util for the spouse who ends up at the correct event, but alone. This indicates that the couple's togetherness is more important than the events that they attend. However, the choices of events *do* matter: the payoffs for the husband attending the football game and the wife attending the opera are both 1 util, but the payoffs for both attending their undesired events are both zero.

Like the prisoner's dilemma, the battle of the sexes presents an apparent contradiction in rationality, except in a slightly different way. If the husband goes to the football game, then it would be in the wife's best interest to go to the football game, even though she would normally prefer the opera. Likewise, if the wife goes to the opera, then the husband would get a higher payoff from going to the opera as well.

This is an instance in which there are two combinations of pure strategies that provide the greatest payoffs to both spouses; we will study the details in the following section.

11.1.6 Analyzing Nash Equilibria

The inarguable cornerstone of game theory, or at least of the study of normal form games, lies in the analysis of *Nash equilibria* (NEs), named for the late John Forbes Nash, who made crucial contributions to game theory upon which our current understanding of the field and focus on future projects rests. A Nash equilibrium of a non-cooperative game in normal form, simply put, is a solution in which neither player gains anything by unilaterally changing his or her own strategy. Each player knows the equilibrium strategies of all of the other player(s).

> **Definition 11.10: Formal Definition of a Nash Equilibrium**
>
> Let G be an n-player game, which we will alternately denote (S, f) (where $S = S_1 \times S_2 \times S_3 \times \cdots \times S_n$ is the Cartesian product of the strategy sets S_i for $i \in [1, n]$ and $f = f_1 \times f_2 \times f_3 \times \cdots \times f_n$ is the Cartesian product of the payoff functions f_i ($i \in [1, n]$). For each strategy $x \in S$, let x_i be the strategy profile (or strategy combination) of player i and let x_{i^C} be the strategy profile of all other players than player i. For each $i \in [1, n]$, when player i chooses strategy $x_i \in S_i$, then the payoff will be $f_i(x)$ for that player. Let $x = (x_1, x_2, x_3, \ldots, x_n) = x^* \in S$; then x^* is an NE if and only if
>
> $$\forall i, x_i \in S_i : f_i(x_i^*, x_{i^C}^*) \geq f_i(x_i, x_{i^C}^*)$$
>
> In other words, a unilateral change in strategy from the NE by one player (player i) will always affect the profitability of the strategy negatively.

(Note: If the equality in the definition above is strict, then the NE is known as a *strict NE*. If there is equality for x_i^* and some other strategy in S, then the NE is a *weak NE*.)

A game may have either a pure strategy NE or mixed strategy NE.

> **Theorem 11.2: Nash's Existence Theorem**
>
> If mixed strategies are allowed, then every game with a finite number of players, and a finite number of *pure* strategies in the strategy set of each player, has at least one Nash equilibrium.[a]
>
> [a]Recall that mixed strategies are divisions of the probability of occurrence amongst multiple pure strategies.

Proof. A comprehensive proof with definitions can be found in the paper "A Tutorial on the Proof of the Existence of Nash Equilibria."[24] □

---

[24]https://www.cs.ubc.ca/ jiang/papers/NashReport.pdf.

> **Definition 11.11: Coordination Game**
>
> If a game has multiple pure strategy NEs that involve all players picking the same (or corresponding) strategies, it is a *coordination game*.

Games with an infinite continuum of strategies for some coalition of players need not have any Nash equilibria. For instance, if we were to set up a game in which two players pick a random positive integer, and the player who picks the larger number wins, then this game would not have an NE, since both players would have everything to gain from simply making their numbers larger (provided that the other player keeps his or her number constant). However, if the strategy set is compact (i.e. both closed[25] and bounded), then there necessarily exists an NE for the game.

Computing pure strategy NEs relies on determining the dominant strategies for each player in a two-player game; computing mixed strategy NEs, however, requires that we determine *p-mix* and *q-mix* options.

Consider the general 2-player payoff matrix below (with Rose and Colin as the row and column players, respectively):

Rose/Colin	A	B
A	(a,b)	(c,d)
B	(e,f)	(g,h)

Let Rose play strategy A with probability p; then she plays strategy B with probability $1-p$. Similarly, let Colin play strategy A with probability q, so that he plays strategy B with probability $1-q$. We then compute their respective p-mix and q-mix values by multiplying their payoffs by p and $1-p$ for Rose and q and $1-q$ for Colin, then summing the two for each player:

Rose/Colin	A	B	q-mix
A	(a,b)	(c,d)	$aq + c(1-q)$
B	(e,f)	(g,h)	$eq + g(1-q)$
p-mix	$bp + f(1-p)$	$dp + h(1-p)$	

The p-mix values for each of Colin's decisions represent the expected value of his mixed strategy, and the q-mix values for each of Rose's decisions represent the expected value of her mixed strategy. In order to equalize the payoff that the opponent will receive from playing either strategy, we set the p-values equal to each other, and the q-values equal to each other. Doing so yields $bp + f - fp = dp + h - hp$, or $p = \dfrac{h-f}{b-d+h-f}$. Similarly, we obtain $q = \dfrac{g-c}{a-e+g-c}$.

Example 11.1.6.1. *Find all Nash equilibria (in pure and mixed strategies) of the game whose normal form is given by the payoff matrix below.*

Rose/Colin	A	B
A	$(2,2)$	$(1,2)$
B	$(0,1)$	$(4,-1)$

Solution 11.1.6.1. *A pure strategy NE (in fact, the only one) is $(2,2)$ (also denoted (A,A) or (up, left)), since neither Rose nor Colin would unilaterally switch his or her strategy to gain an advantage. $(B,B) = (4,-1)$ is not a pure strategy NE, since Colin has great incentive to switch strategies (in fact, -1 is not only Colin's lowest payoff, but the lowest payoff in the entire game!) A mixed strategy NE, on the other hand, can be computed by assigning to Rose and Colin q-mix and p-mix values, respectively. The q-mix value is*

---

[25] See §10.1.4, "Open and Closed Sets."

$2q + 1(1-q) = q + 1$ for strategy A and $4 - 4q$ for strategy B, yielding $q + 1 = 4 - 4q \implies q = 0.6$. Thus, Colin should play strategy A $\frac{3}{5}$ of the time and strategy B $\frac{2}{5}$ of the time. As for Rose, her expected payoff is $\frac{(2+0) \cdot 0.6 + (1+4) \cdot 0.4}{2} = 1.6$.

11.1.7 March to 30 Septendecillion: Strategic Modeling in *Universal Paperclips*

In the 2017 web browser game *Universal Paperclips*,[26] the player assumes control of ClipMaker, an artificial intelligence machine "protagonist" (you'll see why that's in quotes shortly) programmed with only one task: make paperclips. The game harkens back to the *paperclip maximizer* thought experiment (originally proposed in 2003 by Swedish philosopher Nick Bostrom): if an AI is programmed with a single task, but otherwise left to its own devices, will it know when to stop, or will it just keep going on forever, consuming all of Earth's (and, eventually, the universe's) resources and establishing utter tyranny over the human race[27] in the process?

In the game, the player at one point unlocks *strategic modeling* capabilities: the ability to run "tournaments" consisting of one or multiple rounds. In each round of a tournament, the player can select a strategy that will determine which decisions the AI goes with. There are eight different strategies offered to the player (once he/she has unlocked them all over the course of the game): RANDOM, A100, B100, GREEDY, GENEROUS, MINIMAX, TIT FOR TAT, and BEAT LAST.[28]

Here (assuming that the player has unlocked all eight strategies), the opponent runs through each of its eight strategies, pairing each against all of the protagonist AI's strategies, thus making each "full" tournament $8^2 = 64$ rounds long (or if the player has unlocked n of the strategies, n^2 rounds long).

Which strategies are the best (in that they provide the most utils to the AI) in the long run? To answer this, we must first define the util in context. The game uses, alongside exorbitant quantities of the humble paperclip, a currency called *Yomi*,[29] which correlates positively with the payoffs from the rounds in a strategic modeling tournament. (The payoff matrix differs from tournament to tournament, but the amount of Yomi earned from a certain sum of payoffs from game to game is adjusted accordingly.) Here, our definition of a util is simple enough: one util is equal to one Yomi, and so "util" and "Yomi" are synonyms for the same general idea: a measure of benefit from the payoff matrix.[30]

As it turns out, given that the AI has access to all eight strategies, the hierarchy of strategies, from best to worst, is BEAT LAST, GREEDY, GENEROUS, A100/B100 (tie), MINIMAX, TIT FOR TAT, RANDOM.[31] Let's take a (quick) look at why this should be the case.

We'll begin with the worst strategy of all: RANDOM (which is, appropriately, the first strategy that is available to the player before upgrades allow him or her to unlock the other strategies). Despite its seemingly promising appearance and tendency to outperform the A100 and B100 strategies in the early game, once all eight strategies are unlocked, its shortcomings are quickly exposed. The RANDOM strategy, despite its simplicity, completely fails to account for the opponent's move, or to adjust its choice based on the ac-

---

[26] Designed by Frank Lantz of New York University, and programmed by Lantz and Bennett Foddy (of *QWOP* fame).

[27] And any extraterrestrial life, too.

[28] The RANDOM strategy selects each decision with a probability of exactly 50%; the A100 and B100 strategies, respectively, select the "A" and "B" choices exclusively; the GREEDY strategy selects the strategy with the highest *potential* payoff; the GENEROUS strategy selects the strategy with the highest potential payoff for the AI's opponent; the MINIMAX strategy selects the strategy which does not contain the maximum payoff for the opponent (thereby always choosing the opposite strategy as GENEROUS); the TIT FOR TAT strategy selects the strategy that the opponent chose in the previous round; and the BEAT LAST strategy selects the strategy that would beat the opponent's strategy in the previous round.

[29] A Japanese term which roughly translates to "reading the mind of [one's] opponent and planning ahead accordingly."

[30] In-game, aside from strategic modeling, Yomi has diminishing returns, but we will ignore this for the purposes of our strategic modeling analysis.

[31] The approximate average amounts of Yomi earned under each strategy were obtained from https://universalpaperclips.gamepedia.com/Strategic_Modeling. In addition, this assumes the player has *not* unlocked upgrades that increase Yomi even further.

tual payoffs themselves, unlike the BEAT LAST, GREEDY, GENEROUS, MINIMAX, and TIT FOR TAT strategies. Thus, it has a much higher chance of being bested by the opponent (who will run through all eight strategies; the BEAT LAST and GREEDY strategies, in particular, stand out as advantageous against the RANDOM strategy, as BEAT LAST is liable to make a decision that beats a decision made by pure chance, and GREEDY explicitly targets the row that has the highest value in the payoff matrix. Hence, RANDOM is not part of an optimal play strategy!

Next on our list is TIT FOR TAT; this strategy plays the same move that the opponent played in the previous round. While this strategy beats RANDOM considerably, it is still far from ideal. The opponent tends to go for a payoff that maximizes its gains while minimizing your AI's, so playing that same decision will likely lead to a situation where the opponent counters with that decision as well, causing it to benefit majorly and your AI to lose that round. This is not to say that this strategy is without merit, however; though it is lacking compared to the best strategies, it is still possible to win a round if the opponent has used the GREEDY or GENEROUS strategies, since this would ensure that it has picked a column with a cell that has the highest payoff for it. It is somewhat likely that this cell will also contain the highest payoff for the protagonist AI, especially in payoff matrices where the payoffs in the diagonal cells are high and the payoffs in the off-diagonal cells are low. However, usually this is not a risk that is worth taking (most matrices tend to be fairly balanced across their cells around 4-7 Yomi for both players; TIT FOR TAT is optimized when there are payoffs of 1-2 and 9-10).

Similar in overall benefit to the AI is the MINIMAX strategy, which is effectively the opposite of the GENEROUS strategy; it denies the opponent the highest-scoring cells in the payoff matrix at all costs, even if it means consigning the AI to a lower payoff. Evidently this strategy is not optimal either, and GENEROUS just barely beats it out, as expected (since it does allow for higher payoffs for the AI, but more often than not, the cells containing high payoffs for the opponent will also contain low payoffs for the AI, and vice versa). That is, MINIMAX and GENEROUS strategies can be highly effective in rarer cases where the payoff matrix is greatly unbalanced, but otherwise, they are not the best strategies overall.

Moving forward, we have the A100 and B100 strategies, which are tied (or at least have any differences amounting to nothing more than natural variation in the balance of the payoff matrices between A and B decisions; the Law of Large Numbers says A100 and B100 should ultimately be equally profitable). The opponent's winning strategy here depends on the balance of the payoff matrix, and most often, the matrix is balanced; other times, it will be skewed in favor of the AI, or it will be equally skewed in favor of the opponent. These are significantly better odds than other strategies such as RANDOM or TIT FOR TAT, but A100 and B100 are fixed strategies that cannot be tailored to the game, and cannot adapt to changing circumstances, unlike BEAT LAST; as such, they lie squarely in the middle of the pool of strategies in terms of profitability. GREEDY is the second-best strategy (and, in fact, the best strategy after the late-game "Strategic Attachment" upgrade, which grants the player bonus Yomi if s/he picks a strategy in the top three) due to its tendency to pursue the highest possible payoffs at all costs. While the opponent may counter with a MINIMAX or even a GREEDY strategy, the negative effects will be mostly mitigated, since the minimum of the two potential payoffs is guaranteed.

Finally, BEAT LAST is the best strategy of the eight, and usually what constitutes "optimal play" in terms of maximizing utility/Yomi. If we have a decently balanced payoff matrix, then the opponent's move should maximize its payoff while minimizing the AI's; "beating" this move would require that the AI's payoff is higher than the opponent's from that move.

The design of the tournament component of *Universal Paperclips* exemplifies both the breadth and real-world versatility of applied game theory; namely, how the simplicity of the rational decision-making premise can be extended to even the most ludicrous scenario: an AI turning everything in the universe - even itself - into paperclips!

11.1.8 Example

Example 11.1.8.1. *(2013 AMC 12B) Barbara and Jenna play the following game, in which they take turns. A number of coins lie on a table. When it is Barbaras turn, she must remove 2 or 4 coins, unless only one coin remains, in which case she loses her turn. What it is Jennas turn, she must remove 1 or 3 coins. A coin flip determines who goes first. Whoever removes the last coin wins the game. Assume both players use their best strategy. Who will win when the game starts with 2013 coins and when the game starts with 2014 coins?*

Solution 11.1.8.1. *Consider the 2 cases with 2013 and 2014 coins. In the case of 2013 coins, note that, with 5 coins left, whoever goes first will automatically lose (they must leave an amount that the other person can just take). If Jenna goes first, she can take 3 coins; then whenever Barbara takes coins, Jenna can just take an amount that makes the number of coins a multiple of 5. Then it will be Barbaras move at 5 coins, so she will lose. But if Barbara goes first, Jenna can take an amount that makes the total coins taken 5 during that round. Then Barbara will be left with 3 coins, so she takes 2 and Jenna takes the remaining coin and win. Thus, Jenna has a winning strategy for 2013 coins.*

In the case of 2014 coins, if Jenna goes first, she can remove 1 coin and win that way. If Barbara goes first, she can take 4 coins, which leaves 2010. But then Barbara has a winning strategy, so whoever goes first will win with 2014 coins.

11.1.9 Exercises

Exercise 54. *The essay portion of a standardized test is graded as follows. Two graders grade the essay simultaneously, and cannot collude with each other at any point during the grading process. Each grader assigns the essay an integer score from 1 to 6, inclusive. Then those scores are added together to obtain the student's final score.*

(a) Represent this scenario as a normal form game.

(b) Determine all Nash equilibria for this game.

The payoff function for a sum score of n ($2 \leq n \leq 12$, $n \in \mathbb{Z}$) is given by $P(n) = |(n-7)^2 - 49|$, so that the payoff is maximized for a sum score of 7 (at 49 utils, or \$49 in the graders' pockets) and minimized for a sum score of 2 or 12 (at a measly 24 utils, or \$24).[32]

Exercise 55. *The city of Forbes has an infamously dangerous intersection, so the city planners are considering adding some traffic lights to the intersection. Each year during their 10-year-long tenure, they may elect to have zero, one, or two traffic lights installed. Once a traffic light has been installed, it cannot be removed, and the intersection can only have up to two lights (any more would be prohibitively expensive!) The payoff of having no lights is 0 utils, the payoff of having one light is +50 utils, and the payoff of having two lights is +80 utils (where 1 util represents 10 annual accidents prevented). When should the Forbes city planners add the first and second traffic lights? Assume the city planners have a 50% chance of choosing to add one or two lights each year.*

Exercise 56. *On a math test with 30 multiple-choice questions (each with 5 answer choices), participants earn 4 points for each correct answer, 0 points for blanks, and lose 1 point for each wrong answer. Suppose that the passing score is 60 points out of a possible 120. What is the optimal strategy to pass this test (assuming each answer choice is equally likely to be correct)? Describe the best course of action for < 15, 15, 16, 17, and ≥ 18 confident answers.*

Exercise 57. *Describe an example of a symmetric two-player, two-decision game that is also a zero sum game and a generalized prisoner's dilemma, or show that no such game exists.*

Exercise 58. *(2016 Caltech-Harvey Mudd Math Competition) A gambler offers you a \$2 ticket to play the following game. First, you pick a real number $0 \leq p \leq 1$. Then, you are given a weighted coin with probability p of coming up heads and probability $1p$ of coming up tails, and flip this coin twice. The first time the coin comes up heads, you receive \$1, and the first time it comes up tails, you receive \$2. Given an optimal choice of p, what is your expected net winning?*

Exercise 59. *Show that the mixed-strategy equilibrium for Rock Paper Scissors against an unknown, unpredictable adversary is $\left(\frac{1}{3}, \frac{1}{3}, \frac{1}{3}\right)$, where (x, y, z) denotes the proportions x, y, z played of Rock, Paper, Scissors respectively.*

The "Bernardo and Silvia" series of problems from the AMC competitions:

Exercise 60. *(2010 AMC 10A) Bernardo randomly picks 3 distinct numbers from the set $\{1,2,3,4,5,6,7,8,9\}$ and arranges them in descending order to form a 3-digit number. Silvia randomly picks 3 distinct numbers from the set $\{1,2,3,4,5,6,7,8\}$ and also arranges them in descending order to form a 3-digit number. What is the probability that Bernardo's number is larger than Silvia's number?*

Exercise 61. *(2012 AMC 12B) Bernardo and Silvia play the following game. An integer between 0 and 999 inclusive is selected and given to Bernardo. Whenever Bernardo receives a number, he doubles it and passes the result to Silvia. Whenever Silvia receives a number, she adds 50 to it and passes the result to Bernardo. The winner is the last person who produces a number less than 1000. Let N be the smallest initial number that results in a win for Bernardo. What is the sum of the digits of N?*

---

[32] Yes, in this imaginary universe, the graders receive bonuses according to the quality of the essays that they have scored! Seems preposterous, but fear not - each essay is also scanned by the higher-ups, and if they disagree excessively with the graders, the graders forfeit their bonuses and risk termination (so the graders can't just hand out 6's (or 1's) like candy). This (quite Orwellian, and thankfully hypothetical) testing bureaucracy wants most students to have average scores, even if their essays are all top-notch. Why? Because otherwise, they would be accused of making the test too easy or too hard (or just plain unfair)!

Exercise 62. *(2016 AMC 12A)* Let k be a positive integer. Bernardo and Silvia take turns writing and erasing numbers on a blackboard as follows: Bernardo starts by writing the smallest perfect square with $k+1$ digits. Every time Bernardo writes a number, Silvia erases the last k digits of it. Bernardo then writes the next perfect square, Silvia erases the last k digits of it, and this process continues until the last two numbers that remain on the board differ by at least 2. Let $f(k)$ be the smallest positive integer not written on the board. For example, if $k = 1$, then the numbers that Bernardo writes are $16, 25, 36, 49, 64$, and the numbers showing on the board after Silvia erases are $1, 2, 3, 4,$ and 6, and thus $f(1) = 5$. What is the sum of the digits of $f(2) + f(4) + f(6) + \cdots + f(2016)$?

Extension Exercise 24. *(2014 Stanford Math Tournament)* A one-player card game is played by placing 13 cards (Ace through King in that order) in a circle. Initially all the cards are face-up and the objective of the game is to flip them face-down. However, a card can only be flipped face-down if another card that is '3 cards away' is face-up. For example, one can only flip the Queen face-down if either the 9 or the 2 (or both) are face-up. A player wins the game if they can flip all but one of the cards face-down. Given that the cards are distinguishable, compute the number of ways it is possible to win the game.

Extension Exercise 25. *(2013 Stanford Math Tournament)* Robin is playing notes on an 88-key piano. He starts by playing middle C, which is actually the 40th lowest note on the piano (i.e. there are 39 notes lower than middle C). After playing a note, Robin plays with probability $\frac{1}{2}$ the lowest note that is higher than the note he just played, and with probability $\frac{1}{2}$ the highest note that is lower than the note he just played. What is the probability that he plays the highest note on the piano before playing the lowest note?

Extension Exercise 26. *(2004 USAMO P4)* Alice and Bob play a game on a 6 by 6 grid. On his or her turn, a player chooses a rational number not yet appearing on the grid and writes it in an empty square of the grid. Alice goes first and then the players alternate. When all squares have numbers written in them, in each row, the square with the greatest number is colored black. Alice wins if she can then draw a line from the top of the grid to the bottom of the grid that stays in black squares, and Bob wins if she can't. (If two squares share a vertex, Alice can draw a line from on to the other that stays in those two squares. Find, with proof, a winning strategy for one of the players.

Extension Exercise 27. *(1999 USAMO P5)* The Y2K Game is played on a 1×2000 grid as follows. Two players in turn write either an S or an O in an empty square. The first player who produces three consecutive boxes that spell SOS wins. If all boxes are filled without producing SOS then the game is a draw. Prove that the second player has a winning strategy.

Extension Exercise 28. *(2014 USAMO P4)* Let k be a positive integer. Two players A and B play a game on an infinite grid of regular hexagons. Initially all the grid cells are empty. Then the players alternately take turns with A moving first. In his move, A may choose two adjacent hexagons in the grid which are empty and place a counter in both of them. In his move, B may choose any counter on the board and remove it. If at any time there are k consecutive grid cells in a line all of which contain a counter, A wins. Find the minimum value of k for which A cannot win in a finite number of moves, or prove that no such minimum value exists.

Extension Exercise 29. *(2016 USAMO P6)* Integers n and k are given, with $n \geq k \geq 2$. You play the following game against an evil wizard.

The wizard has $2n$ cards; for each $i = 1, ..., n$, there are two cards labeled i. Initially, the wizard places all cards face down in a row, in unknown order.

You may repeatedly make moves of the following form: you point to any k of the cards. The wizard then turns those cards face up. If any two of the cards match, the game is over and you win. Otherwise, you must look away, while the wizard arbitrarily permutes the k chosen cards and turns them back face-down. Then, it is your turn again.

We say this game is winnable if there exist some positive integer m and some strategy that is guaranteed to *win in at most m moves, no matter how the wizard responds.*

For which values of n and k is the game winnable?

11.2 Combinatorics and Probability: Part 2

Combinatorics and probability can be taken to an entirely new level by considering how we can make the ideas motivating them much more streamlined, and how we can make them resemble more of a big-picture perspective. Rather than myopically taking a look at detailed cases, we can instead turn to methods that make use of recursive methods, identities involving binomial coefficients, and Diophantine equations. These sorts of approaches to combinatorics also make use of important techniques and constructs in analysis, such as bijections and limits, to make our existing foundation of applied discrete math that much stronger.

11.2.1 Stars-and-Bars/Balls-and-Urns

In solving particular types of Diophantine (integer) equations, we may not be able to solve the equations efficiently for all variables, or perhaps at all. Done by brute force, it may take minutes, or even hours, to determine all solutions and ensure that none are missed. Clearly, this is not an optimal strategy! Say, for instance, that we wanted to find all solutions in positive integers to the equation $a+b+c+d = 19$. We could perform casework, starting from $a = 1$, $b = 1$, $c = 1$, and $d = 16$, decreasing the value of d and increasing c until we have c and d as close as possible, then repeating the process for a, b. However, this is horribly inefficient. We require another algorithm for determining the number of solutions, which will also reveal the nature of the solutions themselves.

As it turns out, this particular equation has an inseparable combinatorial link! Due to how we can represent a Diophantine equation of this form as a partition of a set of n indistinguishable objects, or *stars*, into $k+1$ bins with k *bars*, hence the name "stars and bars." (Alternately, we can think of this general problem as that of placing the objects (balls) into labeled urns, hence the alternate moniker "balls-and-urns.")

> **Theorem 11.3: Stars-and-Bars**
>
> For any positive integers n and k, the number of k-tuples of positive integers summing to exactly n is equal to $\binom{n-1}{k-1}$.
>
> If our positive integers are $a_1, a_2, a_3, \ldots, a_k$, and we additionally require that $a_i \leq a_j$ for all $a < j$, then the number of k-tuples of positive integers with $\sum_i a_i = n$ is given by $\dfrac{n+k-1}{k-1}$.

> **Corollary 11.3: Using the Hockey-Stick Identity**
>
> We can sum over all positive integers n to obtain a variant of the Hockey-Stick Identity.

Proof. If we have n "stars," then we can place a partition, or "bar," between the stars in $n-1$ positions (since there are $n-1$ gaps between n stars). Then, if we wish to place $k-1$ bars in the $n-1$ positions, this can be done in $\binom{n-1}{k-1} = \binom{n+k-1}{k-1}$ ways.

To prove the form of the theorem with non-decreasing positive integers, consider the general form with n indistinguishable stars and k bars. Consider inserting $k-1$ bars strictly between the stars, so as to divide the stars into k groups. (Note that, in such a grouping, some bars may be adjacent.) There are a total of $n+k-1$ positions, since n are stars and $k-1$ are bars. Thus, we can select n positions among $n+k-1$ spaces, which can be done in $\binom{n+k-1}{n} = \binom{n+k-1}{k-1}$ ways. $\square$

Note that, using this method, we are essentially proving the existence of a bijection between two sets - the first set being the set of positive integer k-tuples that satisfy a Diophantine equation, and the second set being the set of positions that the bars can assume between the n stars.

11.2.2 Introduction to Recursive Probability

In some probability problems, we cannot work directly with probability at face value, but instead must work with a recurrence relation. An example of this comes in the form of a problem where outcomes are *dependent*: i.e. contingent on the outcome of a previous event(s). If we do not know how one variable will behave, how could we possibly know how others that depend on it will behave?[33]

Our answer? *Recursion.* In most cases, we can understand dependent events in non-probabilistic terms; that is, we can relate them to regular recurrence relations. Consider the following illustrative example:

Example 11.2.2.1. *In terms of n, how many sequences of n fair coin flips do not have any two consecutive flips of heads?*

Solution 11.2.2.1. *Note that whether two consecutive flips are heads depends on whether the previous flip was heads. So we can't just use the probability of flipping two H's $\left(\frac{1}{4}\right)$ to solve the problem.*

Instead, we can use induction on n. Let a_n be this number of sequences for any positive integer n. Then $a_1 = 2$ and $a_2 = 3$ by inspection. But what about much larger values of n? Clearly, it will not do to test every one of the possible 2^n sequences! But what we can do is to show inductively that $a_n = a_{n-1} + a_{n-2}$ (which is very Fibonacci-like, we might add, except with a different starting point).

Consider a sequence with no HH of length n. The sequence starts with either a T or an H. If it starts with T, the rest of the sequence is a similar sequence with length $n - 1$. If it starts with H, then its second flip must be T, which yields an HH-free sequence of length $n - 2$. Therefore, $a_n = a_{n-1} + a_{n-2}$, as desired.

In this particular case, note that the number of HH-free sequences is, in fact, given by the Fibonacci sequence! In the Advanced Number Theory sections, we learn how to obtain a closed form for the Fibonacci series, among other such recursive series - namely $\left(\frac{1+\sqrt{5}}{2}\right)^n + \left(\frac{1-\sqrt{5}}{2}\right)^n$ for $F_n = F_{n-1} + F_{n-2}, F_1 = F_2 = 1$ here.

Studying the Random Walk/Gambler's Ruin

If we recall our (brief) venture into game-theoretic territory earlier this chapter, we may remember that we used decision trees to navigate our way through a problem, or "game," in order to make the best decisions and receive the largest payoff (profit). Though we can make decision trees for a wide variety of decision problems, there are just as many (if not more) cases in which this approach hits a brick wall. Such problems are often in the form of *random walks,* which are exactly as their name implies: random, in that the parameter in question may fluctuate between two extremes and tend towards the middle, yet, paradoxically, we wish to determine when the parameter is expected to reach an extreme (with it becomes less likely as time goes on).

> **Definition 11.12: Random Walk**
>
> In a probabilistic context, a *random walk*, also known as a *stochastic process* or a *random process*, is any series of random events with the outcomes of previous events dictating the probabilities that future events occur.

Most commonly, random walks are special types of *Markov chains* or processes (which we will study in the following chapter). A simplified example of a random walk is as follows:

Example 11.2.2.2. *An ant begins at $(0,0)$ in the Cartesian plane. Each minute, it moves one unit up, right, down, or left, each direction with equal probability. Compute the probability that, after four minutes, the ant ends back up at $(0,0)$.*

---

[33]This question is actually the basis for *chaos theory*, the most well-known example of which is the butterfly effect - the study of how very slight changes in initial conditions can lead to major, perhaps devastating, effects in the system.

This is a (rare) example of a random walk problem that can be solved entirely computationally (using casework). Most of these types of problems entirely reinforce the point of a random walk problem in that *we cannot (realistically) weasel our way out of them with casework!* So what exactly should our approach be? Let's take a look at a famous random walk problem to gain a better perspective.

In a famous problem known as *gambler's ruin*, there are two gamblers (whom we will call Alpha and Beta) who play a gambling game repeatedly. In each round of the game, Alpha earns \$1 with probability p, and loses \$1 with probability $1-p$. Beta earns \$1 with probability $1-p$ and loses \$1 with probability p. The game ends when either Alpha or Beta runs out of money (and then the other gambler wins by default). Assuming each round is independent of the previous rounds, and if we are given the initial dollar amounts that Alpha and Beta have (namely \$$d$ for Alpha and \$$N-d$ for Beta, where N is some positive integer), what is the probability of each gambler winning?

Problems like these, contrary to what first appearances might suggest, can be solved quite effectively using recursive methods. Note that, after the first round, Alpha can end up with either $d+1$ or $d-1$ dollars (and, similarly, Beta can end up with either $N-d+1$ or $N-d-1$ dollars). We consider, then, the probability p_i that Alpha wins with a starting balance of i dollars:

$$p_i = P(A \mid A_1)P(A_1) + P(A \mid A_1^C)P(A_1^C) = p_{i+1}p + (1-p)p_{i-1}$$

where $P(A)$ denotes the probability that Alpha wins the entire game and $P(A_r)$ denotes the probability that Alpha wins round r.

Hence, it follows that $p_{i+1}p = p_i + (p-1)p_{i-1}$. To solve this recursive equation, it suffices to find the values of p_i for two different i. Since $p_0 = 0$, we really only need one more value of i. Clearly, $p_N = 1$ (since Beta will then have nothing to his name, and will automatically lose).

As it turns out (see the following chapter for further details on obtaining the closed form), the characteristic polynomial of this recurrence is $px^2 - x + (1-p) = 0$, which has roots $r_1 = 1$ and $r_2 = \dfrac{1-p}{p}$ (if $p \neq 1-p \implies p \neq \dfrac{1}{2}$; otherwise, $p_i = \alpha + \beta i = \dfrac{1}{N}$). These roots yield the closed form $p_i = \dfrac{1 - \left(\dfrac{1-p}{p}\right)^i}{1 - \left(\dfrac{1-p}{p}\right)^N}$.

To summarize, the solution to this problem demonstrates that gambling is, in fact, a losing proposition in the long run! The house, mathematically, is rigged to give itself a slight advantage, which is why it's best to quit while you're ahead. If we were to suppose that $p = \dfrac{1}{2}$ (i.e. the game is completely and utterly fair), then Alpha's (and Beta's) chance of winning would be $\dfrac{i}{N}$, but $N > i$, so eventually, raising this to a high power will yield a limit of zero (i.e. both Alpha and Beta lose all their money to the house over a long enough period of time).

These types of conundrums are prime examples of *random walks:* probability problems that cannot be solved by the traditional methods we have learned thus far, as outcomes are dependent on previous ones. (In particular, the gambler's ruin problem is a prime example of a *one-dimensional* random walk, in that there is only one "dimension" in which the parameter can travel: along the number line as a representation of currency remaining.) However, we can apply recursion to determine the closed form of the desired probability.

In particular, we can consider an offshoot of gambler's ruin that has a slightly less convoluted final outcome. In fact, we can show that the final outcome is just as intuitive as it would seem in these cases!

> **Theorem 11.4: Probability of Exhaustion**
>
> Let Alpha and Beta have d and $N - d$ dollars, respectively. Assuming both players have a $\frac{1}{2}$ chance of having to surrender one dollar to the other player per turn, the probability that Alpha will go bankrupt before Beta is $\frac{N-d}{N}$, and the probability that Beta will go bankrupt before Alpha is $\frac{d}{N}$.

Proof. (Sketch) This is a cumulative result from the work of Christiaan Huygens, who used the recursive relation

$$p_i = \begin{cases} 0 & i = 0 \\ 1 & i = N \\ p \cdot p_{i+1} + (1-p)p_{i-1} & 1 \leq i \leq N-1 \end{cases}$$

In particular, one of the gamblers (WLOG Alpha) has already lost for $i = 0$, and has already won for $i = N$ (as aforementioned). To elaborate on the somewhat cursory form of the result we listed in introducing the problem, let D_i denote the number of dollars Alpha has at time $t = i$, let A represent the event of Alpha winning the game, and let W represent the event of Alpha winning round 1. Between $i = 0$ and $i = N$, we have

$$p_i = P(A \mid D_0 = N)$$
$$= P(A \wedge W \mid D_0 = N) + P(A \wedge W^C \mid D_0 = N)$$
$$= P(W \mid D_0 = n)P(A \mid W^C \wedge D_0 = n) + P(W^C \mid D_0 = n)P(A \mid W^C \mid D_0 = n)$$
$$= p \cdot P(A \mid D_1 = n+1) + (1-p) \cdot P(A \mid D_1 = n-1)$$
$$= p \cdot P(A \mid D_0 = n+1) + (1-p) \cdot P(A \mid D_0 = n-1)$$
$$p \cdot p_{i+1} + (1-p) \cdot p_{i-1}$$

The limits of the probabilities of Alpha and Beta bankrupting correspond to the roots of the characteristic polynomial, namely $\frac{1-p}{p}$ and 1 (taking the limit as $N \to \infty$ might make this more clear). $\square$

11.2.3 Vandermonde's Identity

Vandermonde's Identity provides an efficient and neatly simplified way of condensing the product of several binomial coefficients into one tidy expression. In combo problems, its main use is similar to that of the Hockey-Stick Identity for addition:

> **Theorem 11.5: Vandermonde's Identity**
>
> For all $r, m, n \geq 0$,
> $$\binom{m+n}{r} = \sum_{k=0}^{r} \binom{m}{k}\binom{n}{r-k}.$$

We can prove this identity algebraically or combinatorially:

Algebraic proof. Consider the binomial expansion of $(1+x)^{m+n}$:

$$(1+x)^{m+n} = \sum_{k=0}^{m+n} \binom{m+n}{k} x^k$$

Also observe that $(1+x)^{m+n} = (1+x)^m \cdot (1+x)^n$, and the Binomial Theorem on both expressions separately yields

$$\left(\sum_{i=0}^{m} \binom{m}{i} x^i\right)\left(\sum_{j=0}^{n} \binom{n}{j} x^j\right)$$

$$= \sum_{r=0}^{m+n} \left(\sum_{k=0}^{r} \binom{m}{k}\binom{n}{r-k} \right) x^r$$

By comparing coefficients of the x^k terms, we can see that

$$\sum_{k=0}^{r} \binom{m}{k}\binom{n}{r-k} = \binom{m+n}{r}$$

as desired. $\square$

Combinatorial proof. Say we have m men and n women in a group, and we must pick exactly r of these people for a committee. This can clearly be done in $\binom{m+n}{r}$ ways. Now observe that this is equivalent to choosing 0 men and r women - which can be done in $\binom{m}{0}\binom{n}{r}$ ways - 1 man and $r-1$ women - which can be done in $\binom{m}{1}\binom{n}{r-1}$ ways - and so forth, until r men and no women - which can be done in $\binom{m}{r}\binom{n}{0}$ ways. Thus, we have

$$\sum_{k=0}^{r} \binom{m}{k}\binom{n}{r-k} = \binom{m+n}{r}$$

which is Vandermonde's Identity. $\square$

Indeed, in the same vein as the combinatorial proof above, we can also obtain a geometric proof (which is really combo in disguise):

"Geometric" proof. Consider a grid of $(m+n-r) \times r$ squares. There are $\binom{m+n}{r}$ ways to traverse the grid in $m+n$ movements up or to the right. If we let the bottom-left vertex (the starting point) be at $(0,0)$, and the top-right vertex (the ending point) be at $(m+n-r, r)$, then there will be $\binom{m}{k}$ paths beginning at $(0,0)$ and ending at $(m-k, k)$, and then $\binom{n}{r-k}$ paths beginning at $(m-k, k)$ and ending at $(m+n-r, r)$. Hence, there are $\binom{m}{k}\binom{n}{r-k}$ paths in total, which is equal to $\binom{m+n}{r}$, as claimed. $\square$

Generalizing Vandermonde's Identity

We can generalize Vandermonde's Identity to work for all p-tuples of integers summing to some positive integer n, as opposed to just singular sums. In this way, we are multiplying out p polynomials, rather than just two.

> **Theorem 11.6: Generalized Vandermonde's Identity**
>
> $$\sum_{\sum_{i=1}^{p} k_i = m}^{m} \prod_{j=1}^{p} \binom{n}{k_j} = \binom{pn}{m}$$
>
> for all $m, n, p \geq 0$ and k_i ($1 \leq i \leq p$) non-negative integers.

Proof. We can consider p bags, with each bag having n balls (total of pn balls). We are selecting m of these balls in total, which clearly yields $\binom{pn}{m}$ choices. Similarly as in the proof of the two-polynomial case, we can select k_i balls from bag i, with $\sum k_i = m$. Hence, we have

$$\prod_{j=1}^{p} \binom{n}{k_j} = \binom{pn}{m}$$

which is what we claimed. □

Despite initial impressions of the identity, which may lead one to believe that it only works for integer arguments, there actually exists a full generalization to non-integers, called the *Chu-Vandermonde identity*. Unlike Vandermonde's Identity, which dates back to the early 1300s in its earliest form, the Chu-Vandermode form is much more recent, and is traditionally proven using probablistic methods (such as the hypergeometric distribution) and/or more abstract constructs such as the gamma function.

> **Theorem 11.7: The Chu-Vandermonde Identity**
>
> For all $s, t \in \mathbb{C}$, and $n \geq 0$,
> $$\binom{s+t}{n} = \sum_{k=0}^{n} \binom{s}{k}\binom{t}{n-k}$$

In fact, we can take this a step further to the statement of the *Rothe-Hagen identity*, which extends Vandermonde's Identity to *all* triples of complex numbers (a, b, c), with $n, k \geq 0$:

> **Theorem 11.8: Rothe-Hagen Identity**
>
> For all $a, b, c \in \mathbb{C}$,
> $$\sum_{k=0}^{n} \frac{a}{a+bk}\binom{a+bk}{k}\binom{c-bk}{n-k} = \binom{a+c}{n}$$

11.2.4 The Hypergeometric Probability Distribution

Despite being more of a continuous probability-based construct than a discrete one, this particular type of probability distribution - the *hypergeometric distribution* - warrants special mention here, as it can be supplemented considerably by Vandermonde's Identity.

In the statement of Vandermonde's Identity, we can perform a surprisingly simple manipulation: simply divide both sides of the equation by $\binom{m+n}{r}$, so that the LHS becomes 1; then we have license to interpret the RHS as a sum of probabilities. Namely,

$$1 = \sum_{k=0}^{r} \frac{\binom{m}{k}\binom{n}{r-k}}{\binom{m+n}{r}}$$

$$= \sum_{k=0}^{r} \frac{\binom{m}{k}\binom{n}{r-k}}{\binom{m+n}{r}}$$

$$= \frac{\binom{m}{0}\binom{n}{k}}{\binom{m+n}{k}} + \frac{\binom{m}{1}\binom{n}{k-1}}{\binom{m+n}{k}} + \frac{\binom{m}{2}\binom{n}{k-2}}{\binom{m+n}{k}} + \ldots$$

Each term in this sum can be interpreted as the probability of, say, a certain number of red balls being drawn - without replacement and in r draws - from a bag with n red and m blue balls.

11.2.5 Examples

Example 11.2.5.1. *How many ways are there to select four letters from the 26 letters of the English alphabet, with each successive letter further in the alphabet than the previous?*

Solution 11.2.5.1. *The key insight here is that this is really just equivalent to choosing any four distinct letters period, since there is only one possible order for the chosen letters to go in. Our answer, then, is* $\boxed{\binom{26}{4}}$.

Example 11.2.5.2. *(2014 MathCounts State Sprint) Seven different prizes are to be distributed among three contest winners such that each winner receives at least one prize and each of the prizes goes to one of the three winners. In how many different ways can the prizes be distributed among the three winners?*

Solution 11.2.5.2. *Prizes can be distributed in the combinations 1, 1, 5; 1, 2, 4; 1, 3, 3; 2, 2, 3. If the combination is 1, 1, 5 we can pick the lucky person in 3 ways, then the 5 presents in $\binom{7}{5} = 21$ ways, then the 2 other prizes in 2 ways $\implies$ 126 ways. The same logic applies for the other combos. Stars-and-bars may also be used to expedite the casework process. At any rate, we should arrive at an answer of* $\boxed{1806}$.

Example 11.2.5.3. *(Drunkard's Walk) A drunk man finds himself wandering near the edge of a cliff. With one more step forward, he will fall off the cliff. At any given step, the probability that he steps forward is $\frac{1}{3}$, and the probability that he steps backward is $\frac{2}{3}$. What is the probability that he survives?*

Solution 11.2.5.3. *From reading the section, we might be able to immediately deduce that the answer should be $\frac{1-\frac{2}{3}}{\frac{2}{3}} = \boxed{\frac{1}{2}}$. This is quite a surprising result, as even though the man is twice as likely to step away from the cliff as he is to step towards it, he is still just as likely to ultimately fall off as he is to avoid falling off. By this formula, we can further deduce that, for any $p \leq \frac{1}{2}$ (where p is the probability of the man stepping away from the cliff), the man is guaranteed to fall off the cliff! "Don't drink and derive," indeed! (Or, perhaps, "Don't drink and stand near the edges of cliffs.")*

Example 11.2.5.4. *(2016 AMC 12B) Tom, Dick, and Harry are playing a game. Starting at the same time, each of them flips a fair coin repeatedly until he gets his first head, at which point he stops. What is the probability that all three flip their coins the same number of times?*

Solution 11.2.5.4. *Flipping the coins once occurs with probability $\left(\frac{1}{2}\right)^3 = \frac{1}{8}$. Flipping twice requires that none of the men flip heads (i.e. they all flip tails) on their first turns, i.e. with probability $\frac{1}{64}$. We can continue indefinitely in this fashion, finding that the probability is given by the infinite geometric series $\frac{1}{8} + \frac{1}{64} + \frac{1}{512} = \boxed{\frac{1}{7}}$.*

11.2.6 Exercises

Exercise 63. *(1997 AHSME)* Two six-sided dice are fair in the sense that each face is equally likely to turn up. However, one of the dice has the 4 replaced by 3 and the other die has the 3 replaced by 4. When these dice are rolled, what is the probability that the sum is an odd number?

Exercise 64. *(2015 Caltech-Harvey Mudd Math Competition)* There are twelve indistinguishable blackboards that are distributed to eight different schools. There must be at least one board for each school. How many ways are there of distributing the boards?

Exercise 65. *(2015 AMC 12B)* An unfair coin lands on heads with a probability of $\frac{1}{4}$. When tossed n times, the probability of exactly two heads is the same as the probability of exactly three heads. What is the value of n ?

Exercise 66. Compute the number of non-decreasing sequences of positive integers with length 7, whose last term does not exceed 9.

Exercise 67. *(2018 ARML Team Round #4)* A fair 12-sided die has faces numbered 1 through 12. The die is rolled twice, and the results of the two rolls are x and y, respectively. Given that $\tan(2\theta) = \dfrac{x}{y}$ for some θ with $0 < \theta < \dfrac{\pi}{2}$, compute the probability that $\tan \theta$ is rational.

Exercise 68. *(1983 AHSME)* Three balls marked 1, 2, and 3, are placed in an urn. One ball is drawn, its number is recorded, then the ball is returned to the urn. This process is repeated and then repeated once more, and each ball is equally likely to be drawn on each occasion. If the sum of the numbers recorded is 6, what is the probability that the ball numbered 2 was drawn all three times?

Exercise 69. *(2016 AMC 12A)* Jerry starts at 0 on the real number line. He tosses a fair coin 8 times. When he gets heads, he moves 1 unit in the positive direction; when he gets tails, he moves 1 unit in the negative direction. The probability that he reaches 4 at some time during this process is $\frac{a}{b}$, where a and b are relatively prime positive integers. What is $a + b$? (For example, he succeeds if his sequence of tosses is $HTHHHHHH$.)

Exercise 70. *(1999 AHSME)* A tetrahedron with four equilateral triangular faces has a sphere inscribed within it and a sphere circumscribed about it. For each of the four faces, there is a sphere tangent externally to the face at its center and to the circumscribed sphere. A point P is selected at random inside the circumscribed sphere. What is the probability that P lies inside one of the five small spheres?

Exercise 71. How many ordered size-18 sets with elements from $\{1, 2, 3\}$ have elements summing to 50 or greater?

Exercise 72. *(2017 HMMT February Combinatorics Round)* How many ways are there to insert $+$s between the digits of 111111111111111 (fifteen 1s) so that the result will be a multiple of 30?

Exercise 73. *(2013 AIME I)* Melinda has three empty boxes and 12 textbooks, three of which are mathematics textbooks. One box will hold any three of her textbooks, one will hold any four of her textbooks, and one will hold any five of her textbooks. If Melinda packs her textbooks into these boxes in random order, the probability that all three mathematics textbooks end up in the same box can be written as $\dfrac{m}{n}$, where m and n are relatively prime positive integers. Find $m + n$.

Exercise 74. *(1998 AIME)* Let n be the number of ordered quadruples (x_1, x_2, x_3, x_4) of positive odd integers that satisfy $\sum_{i=1}^{4} x_i = 98$. Find n.

Exercise 75. Let $a, b, c \in [-k, k] = \mathcal{I}$, $a, b, c, k \in \mathbb{R}$. In terms of k, determine the probability that the quadratic equation $ax^2 + bx + c = 0$ has a real solution, assuming a, b, c are selected at random from the interval $\mathcal{I}$.

Exercise 76. *(2018 AIME II) A real number a is chosen randomly and uniformly from the interval $[-20, 18]$. The probability that the roots of the polynomial*

$$x^4 + 2ax^3 + (2a-2)x^2 + (-4a+3)x - 2$$

are all real can be written in the form $\frac{m}{n}$, where m and n are relatively prime positive integers. Find $m+n$.

Extension Exercise 30. *(2015 AIME I) Consider all 1000-element subsets of the set 1, 2, 3, ... , 2015. From each such subset choose the least element. The arithmetic mean of all of these least elements is $\frac{p}{q}$, where p and q are relatively prime positive integers. Find $p+q$.*

Extension Exercise 31. *(2015 Berkeley Math Tournament) Evaluate*

$$\sum_{k=0}^{37} (-1)^k \cdot \binom{75}{2k}.$$

Extension Exercise 32. *(2014 Stanford Math Tournament) A composition of a natural number n is a way of writing it as a sum of natural numbers, such as $3 = 1+2$. Let $P(n)$ denote the sum over all compositions of n of the number of terms in the composition. For example, the compositions of 3 are 3, $1+2$, $2+1$, and $1+1+1$; the first has one term, the second and third have two each, and the last has 3 terms, so $P(3) = 1+2+2+3 = 8$. Compute $P(9)$.*

Extension Exercise 33. *(2018 HMMT February Combinatorics Test) How many ways are there for Nick to travel from $(0,0)$ to $(16,16)$ in the coordinate plane by moving one unit in the positive x or y direction at a time, such that Nick changes direction an odd number of times?*

Extension Exercise 34. *(2010 USAMO P2) There are n students standing in a circle, one behind the other. The students have heights $h_1 < h_2 < \ldots < h_n$. If a student with height h_k is standing directly behind a student with height h_{k-2} or less, the two students are permitted to switch places. Prove that it is not possible to make more than $\binom{n}{3}$ such switches before reaching a position in which no further switches are possible.*

Extension Exercise 35. *(2018 Berkeley Math Tournament) Compute*

$$\sum_{i=0}^{\infty} \sum_{j=0}^{\infty} \binom{i+j}{i} \cdot 3^{-(i+j)}.$$

Extension Exercise 36. *The odds of San Holo successfully navigating his way through an asteroid field are $3720 : 1$. As $n \to \infty$, compute the limit of the probability that San navigates his way through at least n out of $3721n$ asteroid fields.*

11.3 Graph Theory: The Massive Web of Mathematics

On the extremely popular social media website Facebook, each registered user has a set of *friends*, who have special privileges such as viewing each others' news feeds and posting updates exclusive to one's specific crowd of friends. One person's friend might be friends with another person, who turns out to be friends with the first person, but another friend of the first person may not be friends with the third person (or, for that matter, anyone at all other than the first person). In addition, friendship on Facebook is mutual; that is, if Alice is friends with Bob, then Bob is necessarily friends with Alice as well.

This is a prime example of *graph theory* in action. In graph theory, we study a single general mathematical object - the *graph* - in an enormous level of detail, analyzing all of its properties, its sub-types, its upshots and offshoots, and its applications throughout several fields, such as computer science, AI and linguistics, and sociology (perhaps the most pertinent and topical example lies in the analysis of famous actors' relationships with each other, in the form of the *(Kevin) Bacon number*, as well as the related *Erdős number* and *Erdős-Bacon number*[34]), to name just a few.

11.3.1 Fundamental Definitions and Properties of Graphs

As our very first foray into the subject, how do we define a graph? In the graph-theoretic context, we do not refer to many of the traditional objects as graphs, as the term has a very specialized meaning. We do not refer to the representation of a function on the coordinate plane, nor do we refer to a visual depiction of statistical data. Rather, a *graph* is a set of vertices connected to each other in some fashion, as in the Facebook example that opened the section:

> **Definition 11.13: Graph**
>
> A *graph* G is an ordered pair (V, E) that consists of a set of vertices V (depicted as nodes) and a set E of edges (depicted as line segments connecting two vertices).

Essentially, graphs are representations of pairwise relationships (the presence or absence of edges) between objects (the vertices). Graphs may be either *directed*, meaning that each edge passes strictly from one vertex to another, or *undirected*, meaning that each edge connects the vertices in both directions. Selective permeability of a cell membrane is a perfect example of this concept; the membrane allows certain substances to pass through it, but not others (and determines which substances may pass through in both directions as well). The edge represents whether or not a substance may pass through the membrane (in either direction; going from outside the membrane to in, or from inside the membrane to out).

> **Definition 11.14: Order**
>
> The number of vertices in a graph is called the *order* of the graph.

> **Definition 11.15: Degree**
>
> The *degree* of a given vertex $V \in G$, denoted $\deg(V)$, is the number of edges connected to V.

One may notice that the term *degree* is also used to describe the highest exponent in a polynomial. This is no accident - graphs and polynomials are very much interrelated! (We'll take a look at this in depth later.)

---

[34] The Bacon number of an actor represents the number of degrees of separation of that actor from Kevin Bacon; i.e. the number of movies "linking" that actor and Kevin Bacon. For instance, Tom Cruise has a Bacon number of 1, since he starred in *A Few Good Men* with Bacon. It is hypothesized that every movie actor has a Bacon number less than or equal to 6 - an idea known as the *six degrees of separation*. Similarly, one's Erdős number represents the number of academic papers that separate one from Paul Erdős (in that co-authoring a paper with someone else adds a degree of separation), and one's Erdős-Bacon number is the sum of one's Bacon and Erdős numbers.

Putting all of this together, here is an example of a graph annotated with an explanation of each of its components:

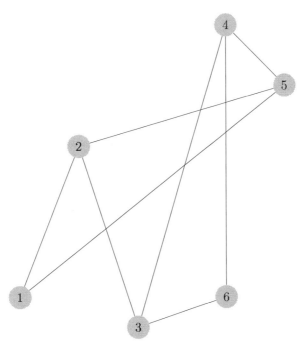

Figure 11.4: An example of an undirected graph with 6 vertices and 7 edges. Vertices 2, 4, and 5 have degree 3, vertices 1 and 3 have degree 2, and vertex 6 has degree 1.

Going back to the Facebook example from the beginning of the section, we could draw the conclusion that Facebook user 1 has two friends (namely, users 2 and 5), user 2 has friends 1, 3, and 5, and so forth, with user 6 having users 3 and 4 as friends. (Note that, if friendship were not mutual, but instead uni-directional, we would draw a directed graph, and the edges would have arrows pointing in the direction of the connection/friendship.)

We can use this example to introduce another crucial definition:

Definition 11.16: Path

A *path* in graph G is a route from one vertex of G to another that travels only along the edges in G. The *length* of the path P is the number of vertices in G that P passes through.

In fact, we can reduce this definition to being a subgraph of G (see §11.3.3). We usually write the path P traveling through vertices $v_0, v_1, v_2, \ldots, v_n$ as $P = v_0 v_1 v_2 \cdots v_n$.

Definition 11.17: Cycle

A *cycle* of G is a path P in G such that $P = v_0 v_1 v_2 \cdots v_n v_0$ for some n.

Similarly, the length of a cycle $P \in G$ is the number of vertices/edges in P, namely $n + 1$.

Finally, we define a *matching* of a graph G:

> **Definition 11.18: Matching**
>
> A *matching* of a graph G is a subgraph of G (see §11.3.3) is a set of edges with no common vertices.

11.3.2 Classifying Graphs

Different classifications of graphs have remarkably unique properties; we devote a large amount of focus and analysis to each separately for good reason! These types of graphs are not mutually exclusive, nor are they common to all graphs, which makes them special and worthy of investigation. These classifications serve as the basis for much of the contemporary interest in graph-theoretic research, including in solving NP-hard problems such as the traveling salesman problem. Here are some of the fundamental definitions in classifying some of the countless varieties of graphs.

Firstly, we say a graph is *regular* if each vertex has the same degree. That is, each vertex has the same number of neighbors. In the case where $\deg v = d$ for all $v \in V$, $G = (V, E)$ is *d-regular*.

We say a graph is *connected* if there exists a path from every vertex to every other vertex:

> **Definition 11.19: Connected Graph**
>
> A graph G is connected if, for every pair of vertices $(v_i, v_j) \in G$, $i \neq j$, there exists a path P connecting v_i and v_j.

In loose terms, G is connected in the intuitive sense of the word: that is, every pair of distinct vertices is strung together by an edge. There is no vertex that is completely isolated from the rest; no vertex has degree zero. In addition, graphs that have all vertices with degree 1 or higher are necessarily connected. (Note that a graph consisting of a single vertex - and no edges - is connected by definition; an graph with multiple vertices and no edges is necessarily disconnected.)

The *complete graph* is one of the single most iconic symbols of graph theory, along with one of the centerpieces of research interest in the field. As it lends itself extraordinarily well to many of the other intrinsically interesting properties of graphs in general, it proves an excellent object of study:

> **Definition 11.20: Complete Graph**
>
> A complete graph is an undirected graph in which every pair of distinct vertices is connected by a unique edge. A *complete digraph* is a directed graph in which every pair of distinct vertices is connected by an edge in each direction.

Remark. The complete graph on n vertices is customarily denoted by K_n.[35]

By drawing the first few K_n, we observe a pattern that becomes abundantly clear once we list out some of its more concrete properties. We notice that K_n has $\frac{n(n-1)}{2} = 1+2+3+\ldots+(n-1)$ edges, and is furthermore a regular graph of degree $n-1$. Additionally, K_n represents the edges of the $(n-1)$-simplex; that is, K_3 forms the edges of a triangle, K_4 forms the edges of a tetrahedron, and so forth. (Note that, as the key takeaway for this section, and as food for thought, observe that as $n \to \infty$, the overall appearance of K_n tends toward that of the circle. This heavily suggests that K_n is a manifestation of the regular n-gon. Can you prove this?)

---

[35] It is unknown what the letter "K" represents here; some say that it stands for the German *komplett*, while others contend that it is honorary of the Polish mathematician Kazimierz Kuratowski.

Next, we proceed to cover the *bipartite graph,* a variety of graph that involves a splitting, or partition, of its vertices in a manner that is likely reminiscent of combinatorics:

> **Definition 11.21: Bipartite Graph**
>
> A *bipartite graph* G (or *bigraph* for short) is a graph that can be decomposed into two disjoint sets of vertices such that no two vertices in the same set are adjacent in G.

The bipartite graph is in fact a special case of the more general k-partite graph G, which is similar in that it can be decomposed into k disjoint sets of vertices with no two vertices in the same subset adjacent in G. It is worthy of note that a graph is bipartite iff it is 2-colorable (we rigorously prove this in §11.3.4).

It is also noteworthy that the broad classification of a bipartite graph opens up entirely new branches of classification into at least tens of other graph types: for instance, acyclic graphs (graphs with no cycles, connected graphs of which are also trees - see below), book graphs, cycle graphs with an even order (i.e. graphs that consist of a single cycle through all their vertices), and a plethora of other graph sub-types!

For now, however, we will narrow our focus to discuss an important category of graphs: the *tree*.

> **Definition 11.22: Trees**
>
> A *tree* is an undirected graph in which any pair of vertices is connected by exactly one path. More rigorously, a tree is a graph G satisfying any of the following conditions (all of which are equivalent):
>
> - G is both connected and acyclic (containing no cycles).
> - G is acyclic, but a simple cycle[a] is formed by adding an edge to G.
> - Any two vertices in G can be connected by a simple path (a path with no repeated vertices).
> - G is connected and would become disconnected by removing a single vertex.
>
> Assuming the order of G is finite, the following statements are also equivalent:
>
> - G has $n - 1$ edges, and is connected.
> - G is connected, and every subgraph $H \in G$ has at least one vertex with degree 0 or 1.
> - G has no simple cycles.
>
> ---
> [a]A cycle with no repeated vertices.

Using this "tree" analogy, we can also define a *forest* and *leaf:*

> **Definition 11.23: Forest**
>
> A *forest* is an undirected graph such that all of its connected components[a] are trees. Id est, a forest is a disjoint union of trees, and also an undirected cyclic graph.
>
> ---
> [a]A *connected component* of an undirected graph G is a subgraph of G such that any two vertices are connected by a path, and which is disconnected from all other vertices in G.

> **Definition 11.24: Leaf**
>
> A *leaf* of a tree is a vertex of degree 1.

11.3.3 Subgraphs

From any graph, we can construct its *subgraphs*, just as from any set, we can construct its subsets. Similarly to a subset of a set of elements, a subgraph consists of the elements in its parent graph:

> **Definition 11.25: Subgraph**
>
> A *subgraph* H of a graph $G = (V, E)$ is a graph formed from the subsets of the vertices and edges of G.
>
> Furthermore, we say $H \in G$ is a *vertex-induced*, or just an *induced*, subgraph, if it consists of both a subset of the vertices of G and a set of edges connecting any pair of vertices in the subset. (That is, each edge must have both its endpoints in the subset.) We say that H is an *edge-induced* subgraph of G if it is a subset of the set of edges of G, along with any vertices that are the endpoints of those edges.

Remark. Much like every set is a subset of itself, so every graph is a subgraph of itself.

Analogously, we may define *induced paths, induced cycles,* and *induced matchings*. These are all induced subgraphs of G that also happen to be paths, cycles, and matchings of G, respectively. The definition of an induced path is then conducive to setting up another famous problem in game theory, with a combinatorial link in higher dimensions: the *snake-in-a-box problem*. In this problem, we seek the longest path between two distinct vertices in a hypercube. Beginning at one corner of the hypercube, we desire to travel along as many edges as possible, and reach as many corners as possible. After we reach a new corner, it, together with all of its immediate neighbors, can no longer be revisited. Rigorously speaking:

> **Definition 11.26: The Snake-in-a-Box Problem**
>
> In the snake-and-a-box problem, we define a *snake* to be a connected path in the hypercube with each node in the snake has exactly two neighbors also in the snake (with the exception of the head (starting vertex) and tail (ending vertex)).

It is particularly helpful in this problem to understand a hypercube not strictly geometrically, but combinatorially as well. If we take the hypercube in n dimensions to be the collection of 2^n vertices such that each vertex is associated with a binary string of length n, and the strings of neighboring vertices always differ in exactly one place, then it becomes significantly easier to make the logical leap from making random - but guessably correct - connections to determining an optimal solution (and, therefore, an induced path of the hypercube).

11.3.4 Coloring Arguments and the Chromatic Polynomial

With graphs, especially more complicated ones, labeling in some fashion is of the utmost importance. Traditionally, on a non-rigorous level, we tend to label graph vertices with numbers or letters or some other concrete indicator, but much more often, mathematicians turn to colors to label graphs. This is especially manifest in some of the most intriguing open problems related to graph theory, namely the *four-color problem* (which was proven in 1976, thereby becoming the four-color *theorem*). Such construction as known as *graph colorings*, and are the key elements in coloring arguments that show that a graph adheres to certain key properties that allow us to work with it in a set of ways that would otherwise be inaccessible. In this regard, a key piece of terminology is the *chromatic number* of a graph:

> **Definition 11.27: Chromatic Number**
>
> The *chromatic number* $\chi(G)$ of a graph G is the smallest k such that G is k-colorable (i.e. can be colored with k vertices such that no two adjacent vertices are the same color).

Upon toying around a bit with some types of graphs using this newfound tool, we may come to an especially keen (and significant) observation regarding bipartite graphs in particular:

> **Theorem 11.9: Bipartite Graphs are 2-Colorable**
>
> A graph G is bipartite iff $\chi(G) = 2$.

Proof.

($\implies$) This direction is rather trivial: it suffices to take a construction with all the vertices in one of the disjoint subsets one color, and all the vertices in the other subset the other color (since the vertices in the subsets are non-adjacent).

($\impliedby$) Let G be 2-colorable, and let f be a color mapping onto G consisting of color classes[36] C_1, C_2. G has no edges in C_1 or C_2, and so it must be bipartite with partition into C_1, C_2. $\square$

> **Corollary 11.9: k-partite Graphs are k-Colorable**
>
> The graph G is k-partite iff $\chi(G) = k$.

In a similar vein, and as an extension of this concept, we can also define the *chromatic polynomial* of a graph. Initially devised by George David Birkhoff in an attempt to solve the four color problem, the chromatic polynomial counts the number of colorings of a graph with a given number of vertices:

> **Definition 11.28: Chromatic Polynomial**
>
> For a graph G with order k the chromatic polynomial of G, denoted $P(G, k)$ or $\chi_G(k)$, is the number of k-vertex colorings of G.

Example 11.3.4.1. *As an example, consider the linear graph G on 3 vertices with k colors. We have k choices for the color of the first vertex, $k-1$ choices for the color of the second vertex, and $k-1$ choices for the color of the third vertex (since it is disconnected from the first vertex). Thus, $\chi_G(k) = k(k-1)^2$.*

11.3.5 Proving that Circular Venn Diagrams are Non-Constructible for $n \geq 4$

In this section, we prove a claim that we made in §10.1.6:

> **Theorem 11.10: Venn Diagram Non-Constructibility for $n \geq 4$**
>
> For $n \geq 4$, it is impossible to draw an n-set Venn diagram with only circles.

Proof. In a Venn diagram D_n consisting of n circles, consider the intersection points of the circles. Via this approach, we claim that, for $n \geq 4$, the graph of D_n is not planar.

Each circle in D_n overlaps another, creating two intersections with each overlap. It follows that, for each pair of circles, there are at most two vertices, and so the total number of vertices is bounded above by $2\binom{n}{2}$. As vertices are at the intersections of two or more distinct circles in D_n, each vertex v has an even degree d_v of at least 4. By the formula $2e = \sum_{v \in V} d_v$,[37] we have $e \geq 2v = 4\binom{n}{2}$ edges, where e is the number of edges in the 3-dimensional representation of the graph. As D_n represents all possible intersections of n

---
[36] A *color class* of G is a set of edges of G which all have the same color.
[37] This follows from the fact that each edge connects two vertices, and counting all the edges double-counts the vertices of the graph. There is a *much* more complicated proof of this fact, however, which we will not go over here.

sets, we have 2^n faces in the 3-dimensional representation. Substituting these results into Euler's formula $F + E - V = 2$ yields $2^n - 2\binom{n}{2} \leq 2$, a contradiction for all $n \geq 4$. Hence a Venn diagram D_n with n circles can only be drawn in $\mathbb{R}^2$ for $n \leq 3$. $\square$

11.3.6 Examples

Example 11.3.6.1. *(2019 AMC 12B) On a 2×3 rectangular grid of unit squares, showing 12 cities and 17 roads connecting certain pairs of cities, Paula wishes to travel along exactly 13 of those roads, starting at the upper-left corner and ending at the bottom-right corner, and without traveling along any portion of a road more than once. (Paula is allowed to visit a city more than once.) In how many ways can she do this?*

Solution 11.3.6.1. We have $\boxed{4}$ paths; try making a path go through two vertices either once or twice (forming a shape that looks like a rigid version of this: δ, like a 6 in reverse, or a 9 flipped upside-down). Then on the right side of the path, have a 9 in reverse or 6 upside-down).

Example 11.3.6.2. *Let K_{2018} be the complete graph on 2018 vertices. Compute the probability that a random partition of K_{2018} into two disjoint sets of 1009 vertices is bipartite.*

Solution 11.3.6.2. This is $\dfrac{\binom{2018}{1009}}{2^{2018}}$, as we can form 2^{2018} partitions by choosing in which group each of the vertices will be a member of, but only $\binom{2018}{1009}$ of these are actually bipartite (choosing any 1009 vertices to be a member of one group, which forces all other vertices to be a member of the other group).

Example 11.3.6.3. *Prove that for a graph G with order n, the chromatic polynomial is monic with degree n.*

Solution 11.3.6.3. We proceed by induction on the number of edges. Base case: assume that G has order n and no edges. Then each of the n vertices can be colored in k ways, so the chromatic polynomial is k^n; hence, proved.

For the inductive step, we assume that the chromatic polynomial has degree n and is monic for all G with $< m$ edges; then consider G which has m edges. Let $e \in E$; then we can write $\chi_G(k) = \chi_{G-e}(k) - \chi_{G/e}(k)$. As $G-e$ is a subgraph of G with n vertices and $m-1$ edges, we have $\chi_{G-e}(k) = k^n + P(k)$, where $\deg P(k) \leq n-1$.

Also note that G/e is a subgraph with $n-1$ vertices, $m-1$ edges, so $\chi_{G/e}(k)$ has degree $n-1$ and is monic. Hence $\chi_G(k) = k^n + P(k) - \chi_{G/e}(k)$. As k^n has the largest degree, we conclude that $\deg \chi_G(k) = n$, and moreover, that $\chi_G(k)$ is monic.

11.3.7 Exercises

Exercise 77. What is the chromatic number of the complete graph K_n on n vertices in terms of n? What about the chromatic polynomial of K_n?

Exercise 78. Determine the chromatic polynomial for a cycle of order 4.

Exercise 79. How many cycles does K_4 have?

Exercise 80. (2012 AMC 10A/12A) Adam, Benin, Chiang, Deshawn, Esther, and Fiona have internet accounts. Some, but not all, of them are internet friends with each other, and none of them has an internet friend outside this group. Each of them has the same number of internet friends. In how many different ways can this happen?

Exercise 81. (2016 Berkeley Math Tournament) How many graphs are there with 6 vertices that have degrees 1, 1, 2, 3, 4, and 5?

Exercise 82. Show that every tree has at least two leaves.

Exercise 83. Show that a graph G is a tree if and only if G has no cycles but the addition of any edge (between two existing vertices) would create a cycle.

Exercise 84. There are n people at a meeting. Among any group of 4 participants, there is one who knows the other three members of the group. Prove that there is a person who knows everyone else.

Exercise 85. Prove that in a group of 18 people, there is either a group of 4 mutual friends or a group of 4 mutual strangers.

Exercise 86. There are n people at a meeting, some of whom are mutual friends. Demonstrate how we can split them into two groups such that each person has at least half of his/her friends in a different group.

Exercise 87. Show that, if $v \in G$ has $\deg(v_1)$ odd, then there exists a path $P = v_1 v_2$ such that $\deg(v_2)$ is also odd.

Exercise 88. At a chess tournament, there are 18 competitors. In each round, the contests are paired off, and play against each other once. Show that, after 8 rounds, there are three contestants, no pair of which has played against each other.

Extension Exercise 37. Let there be n points in the plane such that the distance between every pair of points is at least 1. Prove that there are at most $3n$ pairs of points with a distance of exactly 1.

Extension Exercise 38. Find the smallest positive integer n such that, in any set of n irrational numbers, there exist three numbers with all pairwise sums of those numbers irrational.

Extension Exercise 39. In Podunk School District, there are 3 schools, each of which has n students. Every student knows exactly $n+1$ students from the other two schools. Prove that we can find one student from each of the schools such that the three students all know each other (pairwise).

Extension Exercise 40. Prove that any coloring of the edges of K_6 using only the colors red and blue will necessarily result in there being two monochromatic triangles (triangles with all three edges the same color).

Chapter 11 Review Exercises

Exercise 89. *(2018 HMMT November Guts Round)* Consider an unusual biased coin, with probability p of landing heads, probability $q \leq p$ of landing tails, and probability $\frac{1}{6}$ of landing on its side (i.e. on neither face). It is known that if this coin is flipped twice, the likelihood that both flips will have the same result is $\frac{1}{2}$. Find p.

Exercise 90.

(a) *(2014 HMMT November Guts Round)* If you flip a fair coin 1000 times, what is the expected value of the product of the number of heads and the number of tails?

(b) If you flip a fair coin k times, where $k \in \mathbb{N}$, what is the expected value of the product of the number of heads and the number of tails?

(c) If you choose a positive integer n uniformly at random from $[1, 2019]$, and then flip a fair coin n times, what is the expected value of the expected value of the product of the number of heads and the number of tails?

Exercise 91. Call an ordered triple of positive integers (a, b, c) *quadratically satisfactory* if the quadratic equation $ax^2 + bx + c = 0$ has two real solutions. Let $a, b, c \in \{1, 2, 3, 4, 5\}$. Compute the probability that (a, b, c) is quadratically satisfactory.

Exercise 92. *(1990 USAMO P1)* A certain state issues license plates consisting of six digits (from 0 through 9). The state requires that any two plates differ in at least two places. (Thus the plates $\boxed{027592}$ and $\boxed{020592}$ cannot both be used.) Determine, with proof, the maximum number of distinct license plates that the state can use.

Exercise 93.

(a) How many cycles does the complete graph K_6 have?

(b) In terms of n, compute the number of cycles in the complete graph K_n.

Exercise 94. *(2012 HMMT November Guts Round)* Alice generates an infinite decimal by rolling a fair 6-sided die with the numbers 1, 2, 3, 5, 7, and 9 infinitely many times, and appending the resulting numbers after a decimal point. What is the expected value of her number?

Exercise 95. A restaurant has a 4.2 star average rating, based on 15 reviews, on Yowl, the hot new online review service. Ratings are integers from 1 to 5, inclusive. If it received no 1- or 2-star ratings, how many distributions of star ratings are possible? (The order in which the reviews were left is irrelevant.)

Exercise 96. *(2018 Fall OMO)* Patchouli is taking an exam with $k > 1$ parts, numbered Part $1, 2, \ldots, k$. It is known that for $i = 1, 2, \ldots, k$, Part i contains i multiple choice questions, each of which has $(i + 1)$ answer choices. It is known that if she guesses randomly on every single question, the probability that she gets exactly one question correct is equal to 2018 times the probability that she gets no questions correct. Compute the number of questions that are on the exam.

Exercise 97. A string of l letters has exactly 360 distinguishable permutations. Compute all possible values of l, and prove that no others exist.

Exercise 98. *(2018 Berkeley Math Tournament)* Alice and Bob play a game where they start from a complete graph with n vertices and take turns removing a single edge from the graph, with Alice taking the first turn. The first player to disconnect the graph loses. Compute the sum of all n between 2 and 100 inclusive such that Alice has a winning strategy. (A complete graph is one where there is an edge between every pair of vertices.)

Exercise 99. *(1990 AHSME)* A subset of the integers $1, 2, \cdots, 100$ has the property that none of its members is 3 times another. What is the largest number of members such a subset can have?

Exercise 100. *(2017 ARML Local Individual Round #5)* $ABCDEFG$ is a pyramid with a hexagonal base. Compute the number of distinct ways all seven vertices of $ABCDEFG$ can be colored one of either red, blue, or green such that no vertices that share an edge are identically colored.

Exercise 101. *(2014 AMC 12B)* The number 2017 is prime. Let $S = \sum_{k=0}^{62} \binom{2014}{k}$. What is the remainder when S is divided by 2017?

Exercise 102. *(2017-2018 USAMTS Round 1 P2)* After each Goober ride, the driver rates the passenger as 1, 2, 3, 4, or 5 stars. The passenger's overall rating is determined as the average of all of the ratings given to him or her by drivers so far. Noah had been on several rides, and his rating was neither 1 nor 5. Then he got a 1 star on a ride because he barfed on the driver. Show that the number of 5 stars that Noah needs in order to climb back to at least his overall rating before barfing is independent of the number of rides that he had taken.

Extension Exercise 41. *(1964 IMO P4)* Seventeen people correspond by mail with one another - each one with all the rest. In their letters only three different topics are discussed. Each pair of correspondents deals with only one of these topics. Prove that there are at least three people who write to each other about the same topic.

Extension Exercise 42. *(1967 IMO P6)* In a sports contest, there were m medals awarded on n successive days ($n > 1$). On the first day, one medal and $\frac{1}{7}$ of the remaining $m - 1$ medals were awarded. On the second day, two medals and $\frac{1}{7}$ of the now remaining medals were awarded; and so on. On the nth and last day, the remaining n medals were awarded. How many days did the contest last, and how many medals were awarded altogether?

Extension Exercise 43. *(2018 Stanford Math Tournament)* Consider a game played on the integers in the closed interval $[1, n]$. The game begins with some tokens placed in $[1, n]$. At each turn, tokens are added or removed from $[1, n]$ using the following rule: For each integer $k \in [1, n]$, if exactly one of $k - 1$ and $k + 1$ has a token, place a token at k for the next turn, otherwise leave k blank for the next turn. We call a position static if no changes to the interval occur after one turn. For instance, the trivial position with no tokens is static because no tokens are added or removed after a turn (because there are no tokens). Find all non-trivial static positions.

Extension Exercise 44. *(2017 ARML Local Individual Round #10)* Let S be a set of 100 points inside a square of side length 1. An ordered pair of not necessarily distinct points (P, Q) is bad if $P \in S$, $Q \in S$ and $|PQ| < \frac{\sqrt{3}}{2}$. Compute the minimum possible number of bad ordered pairs in S.

Extension Exercise 45. *(2018 HMMT February Guts Round)* Michael picks a random subset of the complex numbers $\{1, \omega, \omega^2, \ldots, \omega^{2017}\}$ where ω is a primitive 2018th root of unity and all subsets are equally likely to be chosen. If the sum of the elements in his subset is S, what is the expected value of $|S|^2$? (The sum of the elements of the empty set is 0.)

Extension Exercise 46. *(1975 USAMO P5)* A deck of n playing cards, which contains three aces, is shuffled at random (it is assumed that all possible card distributions are equally likely). The cards are then turned up one by one from the top until the second ace appears. Prove that the expected (average) number of cards to be turned up is $\frac{n+1}{2}$.

Extension Exercise 47. *(2014 IMO P2)* Let $n \geq 2$ be an integer. Consider an $n \times n$ chessboard consisting of n^2 unit squares. A configuration of n rooks on this board is peaceful if every row and every column contains exactly one rook. Find the greatest positive integer k such that, for each peaceful configuration of n rooks, there is a $k \times k$ square which does not contain a rook on any of its k^2 squares.

Chapter 12

Advanced Discrete Math

12.1 Combinatorics and Probability: Part 3

Combinatorics and probability are perhaps the disciplines in discrete math with the *second*-most diverse set of applications (statistics being the crowning champion, of course). Not only can we apply it immediately to real-world situations, such as those in which we assess reward and risk (as in the gambler's ruin and drunken walk examples from Combo/Probability Part 2, §11.2), but we can also apply it (less) immediately to more pure, abstract purposes, such as those studied in group theory and linear algebra. In this concluding chapter of combinatorics, we will explore several fascinating topics that admittedly may not seem instantly useful in many contexts, but provide even more motivation for the operations that we may, on occasion, take for granted.

12.1.1 Advanced Recursion: States/Markov Chains

Returning to our study of the applications of recursion to probability once more leads us to the study of *states* problems, so named as they represent the likelihood of a random variable assuming a certain set of values that depends on its previous states.

The general idea motivating the recursive aspect is the notion that the expected value of a dependent event should be equal to the sum of the products of the probabilities of the previous outcomes and their expected values; roughly speaking, it should consist of prior outcomes that are "weighted" according to their likelihoods of occurring.

In states problems, we usually consider some expected value function $E(i)$ (or, more traditionally, E_i) and define this function recursively such that $E_i = \alpha E_{i-1} + \beta E_{i-2} + \ldots$ for constants $\alpha, \beta, \ldots$ Then we can set up a system of equations by substituting each value of E_i into each successive equation, until we reach the final value, and then solve the system as normal to obtain all expected values at each parameter i (whether it be time, number of moves, et cetera).

Let's try an extended example (wording modified from original) to illustrate the concept, courtesy of an old HMMT contest.

Example 12.1.1.1. *(2009 HMMT November Team Round) Mario is saving Princess Peach from the clutches of the evil Bowser. When he enters Peach's castle, he finds himself in a room with 4 doors. This room is the first of 2 (indistinguishable) rooms. In each room, one door leads to the next room (or, for the second room, into Bowser's lair), while the other 3 doors lead to the first room.*[38]

    (a) Suppose that, in each room, Mario randomly picks a door to walk through. What is the expected number of doors (not including Marios initial entrance to the first room) through which Mario will pass before he reaches Bowsers lair?

---

[38]This isn't quite how it works in the real *Mario* games, but close enough, I guess...

(b) *In general, if there are d doors in every room (with 1 door in each room that leads to the next room) and r rooms, the last of which leads into Bowser's lair, determine the expected number of doors through which Mario will pass before he reaches Bowser's lair.*

Solution 12.1.1.1.

(a) Denote by E_i the expected number of doors through which Mario must still pass before reaching Bowser's lair if he is currently in room i $(1 \leq i \leq 3)$. Mario must pass through at least 1 door to leave any room, he has probability $\frac{3}{4}$ of returning to room 1 where he started, and probability $\frac{1}{4}$ of advancing to room $i+1$, where we invoke the expected value E_{i+1}. Hence, $E_i = 1 + \frac{3}{4}E_1 + \frac{1}{4}E_{i+1}$, which yields $E_1 = 1 + \frac{3}{4}E_1 + \frac{1}{4}E_2 \implies \frac{1}{4}E_1 = 1 + \frac{1}{4}E_2 \implies E_2 = E_1 - 4$, along with $E_2 = 1 + \frac{3}{4}E_1 + \frac{1}{4}E_3 \implies E_1 - 4 = 1 + \frac{3}{4}E_1 + \frac{1}{4}E_3 \implies \frac{1}{4}E_1 = 5 + \frac{1}{4}E_3$. But since $E_3 = 0$ (as the third room is Bowser's lair), we have $E_1 = \boxed{20}$.

(b) More generally, let $E_1, E_2, E_3, \ldots, E_{r+1}$ be defined as above (with $E_{r+1} = 0$). As Mario has probability $\frac{d-1}{d}$ of returning to room 1, and probability $\frac{1}{d}$ of advancing to the next room, we have $E_i = 1 + \frac{d-1}{d}E_1 + \frac{1}{d}E_{i+1}$. Then we can list out the equations in the system and solve (for large i, it will help considerably to use computing software).

In general, the recursive element of combinatorics (leading to characteristic equations that themselves lead to closed forms for probabilities, as in the gambler's ruin scenario) relates to the idea of a *Markov chain*, which is essentially a stochastic process: a memoryless process that involves dependent events.

> **Definition 12.1: Markov Chain**
>
> A *Markov chain*, or *stochastic process,* is a model that describes a sequence of events in which the probability of each event is solely determined by the outcome/state of the previous event (hence the moniker *states*). In other words, the associated Markov process is memoryless; i.e. it has the *Markov property*.

(A simplified definition of memorylessness is that, given the present, the future does not depend at all on the past.)

To expand on the concept of memorylessness, say that we are trying to guess a forgotten phone password. Assuming there are 10,000 possibilities, if we guess a password, then we have eliminated one possibility, with 9,999 left to go. After 10,000 attempts, one is guaranteed to have inputted the correct password, so this is *not* a memoryless dynamic. On the other hand, if we have 10,000 different phones,[39] then inputting an incorrect password would not alter one's chances of guessing the correct password, so this would be a memoryless phenomenon.

More precisely, we can phrase the definition of memorylessness of a random variable in terms of its probability distribution:

> **Definition 12.2: Markov Property**
>
> A *discrete* random variable X with values in the non-negative integers has the Markov property (i.e. memorylessness) iff, for any $x, y \in \{0, 1, 2, \ldots\}$, we have $P(X > x + y \mid X > x) = P(X > y)$. (If X is

---

[39] But who would even have *ten* phones, let alone ten *thousand?* Either we're the richest person in the world or we've somehow gotten our hands on 3D printing technology very early!

> *a continuous* random variable, then x and y need not be non-negative integers, just non-negative real numbers.)

Remark. A common misunderstanding of the Markov property is that $P(X > x + y \mid X > x) = P(X > x + y)$, which is true only if the events are independent (as they often are not)!

It can actually be shown that the only probability distributions that are truly memoryless are the geometric distributions (either of the probability distributions representing the number of Bernoulli trials necessary to obtain a single success, on either the set $\mathbb{N}$ or the set $\mathbb{N} \cup \{0\}$). All other probability distributions have a "memory" to them; i.e. the outcomes of future events do depend, to some extent, on the past. We will leave this to the reader for further exploration.

12.1.2 The Monte Carlo Simulation

The phrase "Monte Carlo" may ring a bell for those who are familiar with the idea of uncertainty in random sampling. The ideas behind Monte Carlo uncertainty manifest in problems like the famous *Monty Hall problem*,[40] which we can actually verify via a *Monte Carlo simulation* (albeit non-rigorously).

The Monte Carlo simulation essentially uses randomness to crack ordinarily deterministic problems so as to determine optimization scenarios, determine draws from a probability distribution, and carry out numerical integration where precise integration through standard means is extremely difficult,[41] for instance. In using a Monte Carlo model, we seek to integrate the Markov chain into a probability distribution that is configured such that the limit (given by the Law of Large Numbers) of the expected value of a random variable is given by the empirical mean of random samples of the variable.

The primary, overarching method by which a Monte Carlo model is usually developed involves four key steps:

> **Definition 12.3: Monte Carlo Simulation**
>
> First, we define a domain of inputs. Second, we randomly select input values from the domain given a certain probability distribution. Third, we evaluate the ratio of "success" deterministically, and finally, we aggregate the results together into one coherent picture. (The Law of Large Numbers ensures that the values selected over time will converge to the expected value according to the probability distribution).

Note that the Monte Carlo approximation falters *if the points in the domain are not uniformly distributed, or if the number of sample points is too small*. This also reflects a similar tendency in statistically-driven studies to be unreliable if the sample size is insufficient, or if the pool is large enough, but inadequately spread out over the domain of conditions that are being researched.

We can use this general method to approximate the values of several mathematical constants, such as π (by way of inscribing a quarter-circle within a circle, then scattering random points throughout the square and determining the ratio of the number of points inside the quarter-circle to the number of points inside the square) and e (using many of the properties in which e is involved, such as in the derangement formula,

---

[40] For those unfamiliar, here's a quick summary. Imagine that you're a contestant on a game show, and the host gives you a choice of one of three doors, two of which have goats behind them, and one of which has a brand-new car behind it. The host has you pick a door, and then reveals that one of the doors that you haven't picked has a goat behind it. He then asks you whether or not you would like to switch doors. As it turns out, the probability that you win by switching is $\frac{2}{3}$, whereas the probability of winning by sticking with your original choice is just $\frac{1}{3}$, contrary to what initial impressions would suggest (namely, that both probabilities are $\frac{1}{2}$, since there are only two potential winning doors after one of the goat doors is revealed)!

[41] Or even impossible, as it is with almost all functions - indeed, any function that is not the composition of elementary functions.

and its expression as the infinite sum of the reciprocals of factorials).

But this is actually just a basic, cursory overview of the Monte Carlo method as a whole; in fact, there is little consensus as to what constitutes a Monte Carlo simulation proper. Many authors make the distinction between a Monte Carlo *method* and a Monte Carlo *simulation,* citing the fictitious, "sandbox"-like nature of a *simulation* (as opposed to the empirical, experimental reality of a *method*). Others proclaim that the *Monte Carlo* part of the nomenclature is what makes all the difference: for a *simulation* - sans the *Monte Carlo* - consists of a single choice of (pseudo)-random variable from the unit interval, whereas the *Monte Carlo* simulation specifically entails drawing a sufficiently large number of random variables from the unit interval (either simultaneously or at a large number of different, staggered times).

Regardless of such distinctions, however, the statistical occurrences and checks that comprise the Monte Carlo process prove to be absolutely invaluable in multiple disciplines, both pure and applied. For instance, in computational physics, the Monte Carlo process can be used to carry out critical calculations in quantum chromodynamics, along with molecular dynamics and particle physics (in the form of designing particle detectors). In engineering and environmental science, Monte Carlo methods can be used in the analysis of both analog and digital circuits, the (predicted) energy output of a windmill during its life of service, and the predicted impacts of pollution from fossil fuel combustion. They are also largely instrumental in computer graphic design (via the method of path tracing), as well as in applied statistics, the development of AI in modern game theory, and in finance (in the way of predicting stock trades, interest rate flux, and supply and demand, for example) and law. And perhaps more immediately and relatably to the common consumer, Monte Carlo methods can forecast climate change, through the probability density function of *radiative forcing;* i.e. the difference between sunlight absorbed by the Earth's atmosphere and the energy radiated back into space.

Going back to the Monty Hall example, applying the Monte Carlo process will demonstrate (at least on an intuitive level) that the probabilities of winning from switching/staying tend towards $\frac{2}{3}$ and $\frac{1}{3}$, respectively. This can be done using pseudocode, computer simulation, or even direct, simple experimentation via hands-on methods (but this may take hours - even days - if one were to carry it out in real life, so we don't recommend this!)

12.1.3 Symmetry, Orbit, and Burnside's Lemma

This section will engage with some concepts from group theory: namely, the inherent symmetry of a group.

As a refresher going forward, the following provides the rigorous criteria for a set of elements, together with some binary operation, to be considered a *group:*

> **Definition 12.4: Definition of a Group**
>
> A group G is a set of elements equipped with a binary operation $\cdot : G \times G \mapsto G$ such that G is associative (i.e. $\forall a, b, c \in G$, $(ab)c = a(bc)$), G has an identity element I with respect to $\cdot$, and for any $g \in G$, $\exists g^{-1} \in G$ such that $gg^{-1} = g^{-1}g = I$.

Remark. Groups often arise from considering the permutations or symmetries of objects. In fact, the underlying concept symmetry is inextricably tied with groups as a whole, as we shall soon see.

We begin by examining a property of a group known as its *orbit*. Loosely speaking, the orbit of a group is part of a set upon which the group acts. More specifically:

> **Definition 12.5: Orbit**
>
> Let G be a group, and let S be a G-set. The *orbit* of $x \in S$ is the set Gx, which is the set of $y \in S$

such that there exists $a \in G$ with $ax = y$.

In addition, we must take care to define what we mean by "acts." If a group G *acts* upon a set, it invokes a group action on the set; that is, a group action of G on S is an assignment of a bijection $S \mapsto S$ to each $g \in G$. (Then G acts on S.)

From here, we can define the *stabilizer* of a set S of a group G:

> **Definition 12.6: Stabilizer**
>
> Let S be a set in group G, and let $T \subseteq S$. The *stabilizer* of T, denoted $\mathrm{stab}(T)$, is the set of $g \in G$ for which $g(T) \subseteq T$. The *strict stabilizer* of T is the set of $g \in G$ for which $g(T) = T$ exactly.

The stabilizer is not *quite* associative, but it does have a property that closely resembles associativity:

> **Theorem 12.1: Multiplicativity of the Stabilizer**
>
> Let a group G act on a set S. Then for all $s \in S$, $g \in G$, $\mathrm{stab}(gs) = g\mathrm{stab}(s)g^{-1}$.

Proof. For any $a \in \mathrm{stab}(s)$, we have $(gag^{-1})gs = gas = gs$. We then have $g\mathrm{stab}(s)g^{-1} \subseteq \mathrm{stab}(gs)$. Substituting gs in place of s, and g with g^{-1}, yields $\mathrm{stab}(gs) \subseteq g\mathrm{stab}(s)g^{-1}$, as desired. $\square$

These definitions culminate in the *orbit-stabilizer theorem*:

> **Theorem 12.2: Orbit-Stabilizer Theorem**
>
> Let G be a group acting upon a finite set S. Let $s \in S$, and denote the orbit and stabilizer of s by $\mathrm{orb}(s)$ and $\mathrm{stab}(s)$ respectively. Then $|\mathrm{orb}(s)| = \dfrac{|G|}{|\mathrm{stab}(x)|}$ where $|\mathcal{S}|$ denotes the cardinality of the set $\mathcal{S}$.

Proof. Consider the mapping $f : G \mapsto \mathrm{orb}(s)$, $f(g) = g \cdot s = gs$. As G acts upon all $s \in S$, f is surjective *a priori*. Furthermore, for all $g, h \in G$, we have $gs = hs \implies g^{-1}h \in \mathrm{stab}(s)$:

Proof. From $hs = gs$, we have
$$g^{-1}(hs) = g^{-1}(gs)$$
$$\implies (g^{-1}h)s = (g^{-1}g)s$$
$$\implies (g^{-1}h)s = I \cdot s = s$$
$$\implies g^{-1}h \in \mathrm{stab}(s)$$

as desired. $\square$

Thus, $g \equiv h \pmod{\mathrm{stab}(s)}$. It follows that we have a bijection from $G/\mathrm{stab}(s)$ to $\mathrm{orb}(s)$, namely $g\mathrm{stab}(s) \mapsto gs$. Then $|g\mathrm{stab}(s)| = |\mathrm{orb}(s)|$ and the result follows. $\square$

Having equipped ourselves with these tools as our base, we can now introduce and discuss at length *Burnside's Lemma*, known humorously as the *lemma that is not Burnside's*.[42] This lemma, in particular, has strong applications in combinatorics (and, by extension, probability), as its greatest uses lie in counting arrangements and permutations of a group with regard to symmetry of an object:

---

[42]Though this lemma is named for William Burnside, other contributors to the foundational ideas of the theorem include George Pólya, Cauchy - who originally stated the theorem in 1845 - and Frobenius. The theorem was mis-attributed to Burnside due to his proof of it in 1897.

Theorem 12.3: Burnside's Lemma

Let G be a group acting on a set S. For $g \in G$, denote by fix(g) the set of fixed points of g (i.e. fix$(g) = \{s \in S \mid gs = s\}$). Then we have

$$|G| \cdot |S/G| = \sum_{g \in G} |\text{fix}(g)|.$$

where $|\cdot|$ denotes the number of orbits of $\cdot$ and S/G denotes the quotient of S and G.

Remark. In the case where G is infinite, we must write this form instead of the alternate form of the theorem, $|S/G| = \frac{1}{|G|} \sum_{g \in G} |\text{fix}(g)|$ (which is equally valid when G is finite).

As a point of clarification before we proceed with the proof: $|\text{fix}(g)|$ denotes the number of ordered pairs $(g, s) \in G \times S$ such that $g(s) = s$.

Proof. We have

$$\sum_{g \in G} |\text{fix}(g)|$$

$$= |\{(g, s) \in G \times S \mid g(s) = s\}|$$

$$= \sum_{s \in S} |\text{stab}(s)|$$

$$= \sum_{H \in S/G} \sum_{h \in H} |\text{stab}(h)|$$

Henceforth, assume WLOG that G operates on S from the left. Let $g \in G$ such that $g(x) = y$, where x and y are elements of the same orbit. Then $a \mapsto gag^{-1}$ is a bijection from stab(x) to stab(y), and thus, from the orbit-stabilizer theorem, for any $h \in H$,

$$\sum_{h \in H} |\text{stab}(h)| = |G|$$

and, therefore,

$$\sum_{g \in G} |\text{fix}(g)| = \sum_{H \in S/G} \sum_{h \in H} |\text{stab}(h)|$$

$$= \sum_{H \in S/G} |G| = |G| \cdot |S/G|$$

This completes the proof. □

We now turn to a more concrete instance in which Burnside's Lemma proves to be extremely useful:

Example 12.1.3.1. *In how many ways can we color the four edges of a square with either black or white if rotations are considered indistinguishable?*

Solution 12.1.3.1. *Note that, if rotations were considered distinguishable from each other, then this problem would be trivial; the answer would simply be $2^4 = 16$. However, this is not the case, as black, white, black, white (clockwise from the topmost edge) is considered identical to white, black, white, black (differing by 90° of rotation in either direction). To solve this deceptively difficult variant, we can use Burnside's Lemma.*

Consider the group G, which consists of four elements: rotation by 0°, 90°, 180°, or 270° (clockwise). The set S consists of $2^4 = 16$ colorings, but we must henceforth consider its fixed points. Namely, we have the following:

- *In the first case - rotation by 0° - there are clearly $2^4 = 16$ fixed points.*

- In the case where we rotate by 90°, the top side will become the right side (and rotation is analogous for the other three sides). There are 2 fixed points, since we require that, post-rotation, each side maintain its color. This is synonymous with saying that all sides must be the same color, which can be done in two ways.

- For the 180° case, notice that the top/bottom and left/right pairs of sides must be of the same color before and after the rotation, so there are $2^2 = 4$ fixed points.

- The final case - the 270° case - is identical to the 90° case (just counter-clockwise rather than clockwise), hence there are 2 fixed points.

We therefore have 24 fixed points over all cases. By Burnside's Lemma, there are $\frac{24}{4} = \boxed{6}$ distinct configurations.

Essentially, Burnside's does boil down to casework in principle, and for simpler variants of the general group symmetry problem such as the example above, it is indeed entirely possible to use pure casework. However, in other cases - which are more common - the casework technique quickly falters, and Burnside's is left as the much more viable option.

12.1.4 Examples

Example 12.1.4.1. In a tic-tac-toe board, how many final stalemate configurations (with player 1 going first by marking an X) are there if rotations are considered identical?

Solution 12.1.4.1. We require 5 X's and 4 O's for a filled board, which can be done in $\binom{9}{5} = 126$ ways. But a stalemate cannot have any row or column occupied with all X's or O's; WLOG consider just the X's. Each row, column, and diagonal must contain either 1 or 2 X's. There are $\boxed{3}$ rotationally distinct configurations: XXO/OOX/XOX, OXO/XXO/XOX, OXO/XOX/XOX.

Example 12.1.4.2. *(2019 HMMT February Combinatorics Test)* Contessa is taking a random lattice walk in the plane, starting at $(1,1)$. (In a random lattice walk, one moves up, down, left, or right 1 unit with equal probability at each step.) Points of the form $(6m, 6n)$ for $m, n \in \mathbb{Z}$ are blue, but points of the form $(6m+3, 6n+3)$ for $m, n \in \mathbb{Z}$ are red. What is the probability that she lands on her first blue point before her first red point?

Solution 12.1.4.2. Denote by $P(m,n)$ the probability that Contessa reaches a blue point before a red point, starting from (m,n). Then, clearly, $P(6m, 6n) = 1$ and $P(6m+3, 6n+3) = 0$. At other points, we have $4P(m,n) = P(m-1,n) + P(m+1,n) + P(m,n-1) + P(m,n+1)$. By symmetry, we also have $P(m,n) = P(m+6a, n+6b)$ and $P(m,n) = P(n,m)$, as well as $P(m,n) = P(-m,n)$, with $P(m,n) = 1 - P(m+3, n+3)$. As such, $4P(1,1) = 2P(0,1)+1$, $4P(0,1) = P(0,2)+2P(1,1)+1$, and $4P(0,2) = P(0,1) + \frac{3}{2}$. Solving gives $P(1,1) = \boxed{\frac{13}{22}}$.

Example 12.1.4.3. *(2018 PUMaC Division A Combinatorics Test)* How many ways are there to color the 8 regions of a three-set Venn Diagram with 3 colors such that each color is used at least once? Two colorings are considered the same if one can be reached from the other by rotation and reflection.

Solution 12.1.4.3. For three sets, there are six possible symmetries, namely two in the class

$$(A, B, C)(AB, BC, CA)(ABC)(\emptyset)$$

three in the class

$$(A, B)(C)(AB)(AC, BC)(ABC)(\emptyset)$$

and one in the class

$$(A)(B)(C)(AB)(AC)(BC)(ABC)(\emptyset)$$

Throughout each of these classes, Burnside's says that the total number of arrangements is $\dfrac{3^8 + 3^6 + 2 \cdot 3^4}{6} = 1485$ (since there are 3 group classes, and so we divide by $3! = 6$). To account for the rotation/reflection constraints, we subtract all ways that involve coloring with 1 or 2 colors. Again applying Burnside's, we have $\dfrac{2^8 + 3 \cdot 2^6 + 2 \cdot 2^4}{6} = 80$ in the 2-color case, multiplied by $\binom{3}{2} = 3 \implies 240$. But we must also add back the 3 colorings with a single color, giving $1485 - 240 + 3 = \boxed{1248}$ distinct colorings.

12.1.5 Exercises

Exercise 103. *A cube is painted on all six faces, with the color of each face chosen from six coats of paint: red, blue, yellow, orange, teal, and violet. Two color arrangements are considered to be identical if the cube can be rotated or reflected to obtain the other. Find the number of distinct color arrangements.*

Exercise 104. *(2016 HMMT February Combinatorics Test) Let $V = \{1, \ldots, 8\}$. How many permutations $\sigma : V \mapsto V$ are automorphisms of some tree? (Authors note: An automorphism of a tree T is a permutation $\sigma : V \mapsto V$ such that vertices $i, j \in V$ are edge-connected iff $\sigma(i)$ and $\sigma(j)$ are.)*

Exercise 105. *(2019 AIME I) A moving particle starts at the point $(4,4)$ and moves until it hits one of the coordinate axes for the first time. When the particle is at the point (a,b), it moves at random to one of the points $(a-1,b)$, $(a,b-1)$, or $(a-1,b-1)$, each with probability $\frac{1}{3}$, independently of its previous moves. Compute the probability that it will hit the coordinate axes at $(0,0)$ is $\frac{m}{3^n}$.*

Exercise 106. *Show that for any sequence of integers $a_1, a_2, \ldots, a_n$,*
$$\prod_{1 \leq i < j \leq n} \frac{a_i - a_j}{i - j} \in \mathbb{Z}$$

Exercise 107. *Suppose we wish to color the edges of a tetrahedron. There are n colors available, and each color may be used any number of times. Up to rotation of the tetrahedron, how many different colorings are possible?*

Exercise 108. *(2018 Berkeley Math Tournament) Consider a $2 \times n$ grid where each cell is either black or white, which we attempt to tile with 2×1 black or white tiles such that tiles have to match the colors of the cells they cover. We first randomly select a random positive integer N where N takes the value n with probability 2^{-n}. We then take a $2 \times N$ grid and randomly color each cell black or white independently with equal probability. Compute the probability the resulting grid has a valid tiling.*

Extension Exercise 48. *(2000 USAMO P4) Find the smallest positive integer k such that, if k squares of an $m \times n$ chessboard are painted black, then there will be three black squares whose centers form a right triangle with sides parallel to the edges of the board.*

Extension Exercise 49. *(2018 Spring OMO) Let m and n be positive integers. Fuming Zeng gives James a rectangle, such that $m-1$ lines are drawn parallel to one pair of sides and $n-1$ lines are drawn parallel to the other pair of sides (with each line distinct and intersecting the interior of the rectangle), thus dividing the rectangle into an $m \times n$ grid of smaller rectangles. Fuming Zeng chooses $m+n-1$ of the mn smaller rectangles and then tells James the area of each of the smaller rectangles. Of the $\binom{mn}{m+n-1}$ possible combinations of rectangles and their areas Fuming Zeng could have given, let $C_{m,n}$ be the number of combinations which would allow James to determine the area of the whole rectangle. Given that*
$$A = \sum_{m=1}^{\infty} \sum_{n=1}^{\infty} \frac{C_{m,n} \binom{m+n}{m}}{(m+n)^{m+n}}$$
then find the greatest integer less than $1000A$.

Extension Exercise 50. *(2014 IMO Shortlist C5) Consider $n \geq 3$ lines in the plane such that no two lines are parallel and no three have a common point. These lines divide the plane into polygonal regions; let S be the set of regions having finite area. Prove that it is possible to colour $\lceil \sqrt{\frac{n}{2}} \rceil$ of the lines blue in such a way that no region in S has a completely blue boundary.*

Extension Exercise 51. *(2016 IMO Shortlist C3) Let n be a positive integer relatively prime to 6. We paint the vertices of a regular n-gon with three colours so that there is an odd number of vertices of each colour. Show that there exists an isosceles triangle whose three vertices are of different colours.*

Extension Exercise 52. *(2016 IMO Shortlist C1) The leader of an IMO team chooses positive integers n and k $(n > k)$, and announces them to the deputy leader and a contestant. The leader then secretly tells the deputy leader an n-digit binary string, and the deputy leader writes down all n-digit binary strings which differ from the leaders in exactly k positions. (For example, if $n = 3$ and $k = 1$, and if the leader chooses 101, the deputy leader would write down 001, 111 and 100.) The contestant is allowed to look at the strings written by the deputy leader and guess the leaders string. In terms of n and k, what is the minimum number of guesses needed to guarantee the correct answer?*

12.2 Graph Theory: Part 2

In this section we tie up some loose ends in our discussion of graph theory. As competitions primarily use the basics in graph theory that we have covered thus far, this (briefer) section serves to extend our ideas from the first Graph Theory section, as well as the rest of the chapter, to provide some food for thought to the reader and hopefully make the connections (pun intended) to other disciplines.

12.2.1 Isomorphisms and Homeomorphisms of Graphs

Recall that two objects are *isomorphic* if they have an invertible mapping between them. We can similarly extend this concept to graph theory. Graphs $G_1 = (V_1, E_1)$ and $G_2 = (V_2, E_2)$ are *isomorphic* if they have the same number of vertices that are connected in the same way. More formally:

> **Definition 12.7: Graph Isomorphism**
>
> We say two graphs $G_1 = (V_1, E_1)$ and $G_2 = (V_2, E_2)$ are *isomorphic* if $|V_1| = |V_2|$, and furthermore, there exists a permutation π of $V = V_1 = V_2$ such that $\{u, v\} \in E_1 \iff \{\pi(u), \pi(v)\} \in E_2$.

That is, an isomorphism between graphs - as with any other objects - is a bijection, this variant commonly called an *edge-preserving bijection* (in that an isomorphism in general preserves structure). Studying isomorphisms between graphs allows us to better understand the underlying "structure" of graphs, as we can establish fundamental truths regarding the isomorphisms that we know to exist between certain classes of graphs, that we can then apply to graphs in general.

Similarly, two objects are *homeomorphic* to each other if they have a continuous function as a mapping between them that also has a continuous inverse. Essentially, homeomorphism is a stronger form of an isomorphism. In similar fashion, we have the following condition for the homeomorphism of graphs:

> **Definition 12.8: Graph Homeomorphism**
>
> The graphs $G_1 = (V_1, E_1)$, $G_2 = (V_2, E_2)$ are *homeomorphic* iff there is an isomorphism from some subdivision of G_1 to a subdivision of G_2.

> **Definition 12.9: Subdivision of a Graph**
>
> A *subdivision* of a graph $G = (V, E)$ is a graph that results from subdividing at least one of its edges. The subdivision of an edge $e \in E$ with endpoints v_i and v_j yields two new edges, $e_1 = \{v_i, v_k\}$ and $e_2 = \{v_k, v_j\}$, where v_k is a node in between v_i and v_j.

We can actually use the homeomorphism of graphs to determine which finite graphs are planar:

> **Theorem 12.4: Kuratowski's Theorem**
>
> A finite graph G is planar iff it does *not* contain a subgraph homeomorphic to the complete graph K_5 on five vertices or $K_{3,3}$ (the complete bipartite graph on six vertices).

> **Corollary 12.4: Kuratowski Subgraphs**
>
> A graph that is homeomorphic to K_5 or $K_{3,3}$ is known as a *Kuratowski subgraph*.

Proof. We first show that K_5 and $K_{3,3}$ themselves are not planar. First we show that K_5 is non-planar. This is trivial, however, using the fact that $e \leq 3v - 6$ (proving this is left as a section exercise), where e and v are the edge and order, respectively. Here we have $e = 10$ and $v = 5$, but $10 \not\leq 9$, so we have arrived at the desired contradiction. Next we show that $K_{3,3}$ is also non-planar. This arises from the fact that the upper bounds on e and f also contradict Euler's formula $f + v - e = 2$. Assume for contradiction that $K_{3,3}$ is planar, so that $v = 6$, $e = 9$, and we subsequently require $v = 5$. However, $K_{3,3}$ is bipartite, so it contains no 3-cycles (indeed, no odd cycles whatsoever), and thus at least 4 edges must bound each face of the embedding of $K_{3,3}$. Finally, each edge is counted twice in the embedding, and so $f \leq \dfrac{e}{2} = \dfrac{9}{2} < 5$. Hence Euler's formula cannot hold, and $K_{3,3}$ cannot be planar.

To prove the remainder of the theorem, note that, if G is planar, every subgraph of G is also planar, and furthermore, G is planar iff every subdivision of G is also planar (both left as end-of-section exercises). This covers the reverse direction; as for the forward direction, we want to show that any non-planar graph contains a subdivision of either K_5 or $K_{3,3}$. For contradiction, assume the existence of a minimal counterexample G, which would have to be 2-connected and not have any vertices with degree 2. (To prove that it must be 2-connected, observe that G must contain a non-planar block, but if it were a proper subgraph, then it would be a non-planar graph not containing a subdivision of K_5 or $K_{3,3}$, hence contradicting the minimality of G. To prove that it cannot have any vertex of degree 2, assume the existence of such a vertex v, whose neighbors are u and w. If u is adjacent to w, then we can remove v from G to obtain a new graph H, which is then planar. In an embedding of H, place the path $P = uvw$ next to the edge $\{u,w\}$. This forms a plane embedding of G, which is a contradiction. Otherwise, should u and w be non-adjacent, remove v from G, and then replace it with $\{u,w\}$ to form H as before. H must be planar, so we can subdivide $\{u,w\}$ to form G. Then G is planar, which is a contradiction.)

Henceforth, let H be the subgraph of G obtained from removing $\{u,v\}$ (the edge connecting u and v, but not the vertices themselves). Since G is minimal, H is planar, as well as 2-connected, so H has at least one cycle including both u and v. Then create a plane embedding of H with a cycle C such that C contains u and v, namely $C = v_0 v_1 v_2 \cdots v_{k-1} v_k \cdots v_l v_0$. $k \geq 2$ (as u and v are not adjacent in H), so there is no path between any two vertices v_i ($0 \leq i \leq k$) outside C. (To prove this, note that, assuming the contrary, we would be able to form a cycle $C' = v_0 v_1 v_2 \cdots v_j v_{j+1} \cdots v_0$, but this would contradict the maximality of C.)

We then must have a path P outside C connecting a vertex v_i ($1 \leq i \leq k-1$) to a vertex v_j ($k \leq j \leq l$). No vertex in P is adjacent to a vertex in C, and we cannot place P inside C, as otherwise, we could include the edge $\{u,v\}$ outside C. Thus, there must exist some "chord" inside C through which P cannot pass. No matter how this chord is configured, if we add back the edge $\{u,v\}$, we will end up with a subdivision of K_5 or $K_{3,3}$, hence reaching the desired contradiction. □

12.2.2 Route Problems

7 Bridges of Königsberg

In 1736, Euler's negative resolution of the problem of the *seven bridges of Königsberg* was effectively the first major landmark in the development of graph theory, as well as a milestone in the progression of ideas regarding topology, since this problem showed that objects need not be thought of as rigid shapes (and indeed, it is most often better *not* to think of them that way). In the formerly-Prussian city of Königsberg (now Kalinigrad in Russia), there were seven bridges between the two islands of Kneiphof and Lomse, arranged as follows:

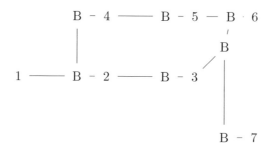

Figure 12.1: A (rough) sketch of the 7 Bridges of Königsberg (where B represents a bridge). Euler showed that it was impossible to traverse all seven bridges without either straying off the path or traversing at least one bridge twice. Each town is the area boxed in by the bridges (and to the very right), and each number is an edge.

The intended solution required a crossing of all seven bridges that neither strayed off the edges, or visited a numbered vertex twice or more. As Euler proved, this is impossible due to the nature of the graph to which the problem simplifies. Specifically, we can reduce the seven bridges and towns to the figure on the following page (Figure 12.2).

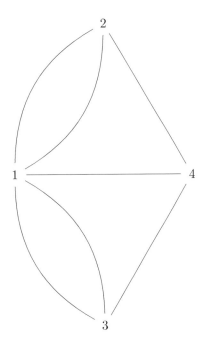

Figure 12.2: The reduced/simplified form of the 7 bridges problem.

As only the sequence of bridges crossed is pertinent, the shape of the graph may be distorted without changing the fundamental properties of the graph itself. As such, Euler could demonstrate that only the existence of an edge between nodes is relevant.

To accomplish this, he in turn observed that, when we enter a vertex via a bridge, we also leave the vertex through a bridge. Thus, we enter a vertex the same number of times as we depart it. If we cross each bridge exactly once, it follows that the number of bridges touching each town must be even. However, both of the towns have an odd number of bridges, so we have a contradiction.

Euler's revelation that the number of bridges - and their endpoints - uniquely determined the non-existence of the solution - was arguably the single breakthrough that ultimately led to the development of modern topology. In this way, this is perhaps one of the most landmark problems - and solutions - in all of mathematics!

The Chinese Postman Problem/Shortest Path Problem

In the *Chinese postman problem,* also known as the *route inspection problem,* we do just that - inspect the set of possible routes to determine the shortest one.[43]

The *shortest path problem,* unlike the route inspection problem, actually assumes a less literal approach. Rather than find the shortest path in terms of actual distance traveled (measured in terms of number of edges), we want to find the path that has the smallest sum of numbers assigned to the edges traversed by the path. To better understand why this distinction is important, consider the graph (on the following page) with numbers assigned to each edge, along with weights specified in a table:

---

[43] Note that, as we will see in the following section, it is important to distinguish between the different contexts of the word "shortest." Here, we mean "traversing through the fewest number of edges," whereas in the problem literally known as the *shortest path problem,* we aim to minimize the sum of assigned weights of each edge that we traverse.

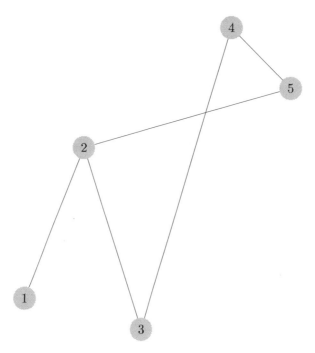

Figure 12.3: An example of a graph where the "shortest" path from vertex 1 to 5 (fewest edges, $1 \to 2 \to 5$) is *not* actually the shortest (in terms of the sum of weights, $1 \to 2 \to 3 \to 4 \to 5$)!

between vertices:	weight
1, 2	5
2, 3	2
2, 5	13
3, 4	3
4, 5	6

Remark. Though the figure above is an undirected graph, we can solve the problem for directed and mixed (some edges directed, others undirected) graphs as well.

We henceforth define the shortest path problem in rigorous terms (for the undirected case). Recall that a *path* between two vertices in a graph G is a sequence of vertices $P = v_1 v_2 v_3 \cdots v_n \in V^n$ with v_i adjacent to v_{i+1} for all $1 \leq i \leq n-1$. Then let $e_{i,j}$ be the edge connecting vertices v_i, v_j. If we define a real-valued *weight function* $f : E \mapsto \mathbb{R}$ (where E is the set of edges in G), the shortest path from v to v' in G is the path $P = v_1 v_2 v_3 \cdots v_n \in G$ such that, over all n, the sum

$$\sum_{i=1}^{n-1} f(e_{i,i+1})$$

is minimal. (Note that this does indeed become equivalent to the route inspection problem iff all edges have equal weight.)

To solve the shortest path problem, we may apply a wide variety of algorithmic solution techniques, such as Dijkstra's algorithm, as well as the considerably lesser-known tools such as the Bellman-Ford algorithm (which generalizes solutions to cases where edge weights are negative) and the A^* heuristic search algorithm. In particular, we will focus on Dijkstra's Algorithm, which begins at the initial node, and defines the *distance* of a node to be the distance (sum of weights) from that node to the initial node.

In Dijkstra's algorithm, we begin by setting all nodes as "unvisited," and by subsequently adding them to the set of *unvisited nodes*. We then consider each node's *tentative distance* from the initial node as being 0 if it is the initial node, and ∞ otherwise. Then, for the current node, we consider all of its neighbors in the unvisited set. We then calculate the distance of each from the current node; the smallest of these becomes the new tentative distance of the node. (For larger values, keep the current, smaller value.) When we have considered all neighbors of the current node of consideration in the unvisited set, we mark the current node as visited (thereby removing it from the unvisited set, never to be looked at again). If the final node (the *destination* node) is visited, or if the smallest tentative distance in the unvisited set is ∞, then stop. Otherwise, the node in the unvisited set with the smallest tentative distance becomes the current node, and the algorithm repeats from considering the unvisited neighbors of the new current node.

The Traveling Salesman Problem/Hamiltonian Path Problem

Imagine that you are a cross-country salesman tasked with delivering goods to all 50 states. This is quite the tall order, and you're not sure that you'll be able to meet your tight deadline. Fortunately, you've studied a little bit of graph theory, and remember the *traveling salesman problem:* the optimization problem of passing through multiple destinations (or nodes in a graph) in the shortest distance possible (including the return trip to the point of origin at the end).

This is one of the more famous examples of an *NP-hard* problem: in simplest terms, a problem that is easy to verify computationally, but difficult to solve in polynomial time. (We discuss this in greater detail and rigor in §16.) It is the epitome of an optimization problem - the class of problems that forms the basis for operations research and queue theory - and operates in a similar vein as the Chinese postman and shortest path problems. In the traveling salesman problem, we seek to find a Hamiltonian cycle with the least weight.

Often, however, we will not be concerned with optimization with regard to a graph, but rather whether or not a path that loops back around to the initial node (a *Hamiltonian path*) *exists at all*. This also extends to the problem of proving the existence of (along with actually finding) a *Hamiltonian cycle* (whether the graph be directed or undirected). Both of these similar problems are NP-complete, or belonging to both the NP and NP-hard classes.

The reasoning behind the difficulty of solving this problem *forwards* (and not merely verifying it) lies in the slowness of the brute-force approach. In a graph G with n vertices, there are $n!$ possibilities to check, and what with the almost exponential-like growth of $n!$ for large n, this process will naturally slow to a crawl. Instead, we must apply other enumerative algorithms to narrow the number of possibilities down to a more reasonable value. One such algorithmic method is *Martello's algorithm,* as described in his paper "An Enumerative Algorithm for Finding Hamiltonian Circuits in a Directed Graph."[44]

To briefly summarize this algorithm, consider a digraph $G = (V, E)$, and a positive integer $h \geq 1$. The most common cases are $h = 1$ (one Hamiltonian path/circuit) and $h = \infty$ (with ∞ being used as a stand-in for the supremum of h, namely 2^n, where $|V| = n$ - perhaps an abuse of notation). Assume WLOG that $(v_i, v_i) \notin G$ for any $1 \leq i \leq n$. We employ a branching strategy in which we extend a path $P = v_{s_1} v_{s_2} v_{s_3} \cdots v_{s_{k-1}} v_{s_k}$. In each iteration of this branching, we backtrack by removing the edge $v_{s_{k-1}} v_{s_k}$, add to P a new edge from $v_{s_{k-1}}$, and prune all edges that will never be part of the path, as well as add to a set I of *implied edges* those edges that will, at some point, be included as part of the path P.

---

[44]Martello 1983, "An Enumerative Algorithm for Finding Hamiltonian Circuits in a Directed Graph," pp. 131-138.

12.2.3 Examples

Example 12.2.3.1. *Define the distance $d(v_1, v_2)$ between two vertices v_1, v_2 to be the length of the shortest path between them. Demonstrate that the distance function satisfies the triangle inequality ($d(u,v) + d(v,w) \geq d(u,w)$).*

Solution 12.2.3.1. *Connect the path from u to v to the path connecting v to w; then we have constructed a path with length $d(u,v) + d(v,w)$. Since we are seeking the path of minimal length, the triangle inequality will be satisfied.*

Example 12.2.3.2. *(2017 MOSP) There are n cities and 2 airline companies in a country. Between any two cities, there is exactly one two-way flight connecting them which is operated by one of the two companies. A mathematician plans a travel route, so that it starts and ends at the same city, passes through at least two other cities, and each city in the route is visited once. She finds out that wherever she starts and whatever route she chooses, she must take flights of both companies. Find the maximum value of n.*

Solution 12.2.3.2. *We claim that $n = 4$ is maximal. Observe that the value of n depends on whether K_n can be partitioned into two acyclic graphs. We will take as fact that acyclic graph on n vertices has no more than $n-1$ edges; from this, it follows that $2(n-1) \geq \frac{n(n-1)}{2} \implies n \leq 4$. For $n = 4$, label the vertices in clockwise order as A, B, C, D. Let the first airline operate the flights between A and B, A and C, and C and D. This works, so the proof is complete.*

Example 12.2.3.3. *(2002 IMO Shortlist C1) Let n be a positive integer. Each point (x, y) in the plane, where x and y are non-negative integers with $x + y < n$, is colored red or blue, subject to the following condition: if a point (x, y) is red, then so are all points (x, y) with $x \leq x$ and $y \leq y$. Let A be the number of ways to choose n blue points with distinct x-coordinates, and let B be the number of ways to choose n blue points with distinct y-coordinates. Prove that $A = B$.*

Solution 12.2.3.3. *Let the number of blue points with x-coordinate i be a_i, and let the number of blue points with y-coordinate i be b_i. We show that $a_0 a_1 a_2 \cdots a_{n-1} = b_0 b_1 b_2 \cdots b_{n-1}$. In turn, we must prove that $\{a_0, a_1, a_2, \ldots, a_{n-1}\}$ is in fact a permutation of $\{b_0, b_1, b_2, \ldots, b_{n-1}\}$.*

We induct upon n. Initially (after the trivial base case), consider the case when every point (x, y) s.t. $x + y = n + 1$ is blue. Ignoring these points, we have a configuration for $n - 1$, with blue columns of sizes and blue rows of sizes $a_0 - 1, a_1 - 1, \ldots, a_{n-2} - 1$. It follows that these form a permutation of $\{b_0 - 1, b_1 - 1, \ldots, b_{n-2} - 1\}$. As $a_{n-1} = b_{n-1} = 1$, we are done.

Henceforth, assume that the point $(k, n - k - 1)$ is colored red. Then its surrounding rectangle is entirely red, and by the induction hypothesis, in conjunction with $a_k = b_{n-k-1} = 0$, we are done.

12.2.4 Exercises

Exercise 109. *Show that a tree with n vertices has $n-1$ edges.*

Exercise 110. *Prove that a graph is bipartite iff it contains no cycles of odd length.*

Exercise 111. *(1997 South Africa TST P5) Six points are joined pairwise by red or blue segments. Must there exist a closed path consisting of four of the segments, all of the same color?*

Exercise 112. *(2010 IMO Shortlist C2) On some planet, there are $2N$ countries ($N \geq 4$). Each country has a flag N units wide and one unit high composed of N fields of size 1×1, each field being either yellow or blue. No two countries have the same flag. We say that a set of N flags is diverse if these flags can be arranged into an $N \times N$ square so that all N fields on its main diagonal will have the same color. Determine the smallest positive integer M such that among any M distinct flags, there exist N flags forming a diverse set.*

Exercise 113. *Consider six points in the plane, no three of which are collinear. Prove that there are two triangles with among the six points such that the longest side length of one of them is the shortest side length of the other.*

Extension Exercise 53. *(2014 USA TST) Let n be an even positive integer, and let G be an n-vertex graph with exactly $\frac{n^2}{4}$ edges, where there are no loops or multiple edges (each unordered pair of distinct vertices is joined by either 0 or 1 edge). An unordered pair of distinct vertices $\{x, y\}$ is said to be amicable if they have a common neighbor (there is a vertex z such that xz and yz are both edges). Prove that G has at least $2\binom{n/2}{2}$ pairs of vertices which are amicable.*

Extension Exercise 54. *There are n people in a room. Any group of m people ($m \geq 3$) in the room have a common friend, and moreover, this friend is unique for all such groups. Compute n in terms of m.*

12.3 Overview of Operations Research

One of the most central applications of discrete math in day-to-day operations is in optimization and improved decision-making. As we have already seen, probabilistic and statistical methods can be indispensable in solving problems in resource management, economics, and even reinforcement learning. These applications are the cornerstones of a subfield of discrete math known as *operations research* (OR): the study of problem-solving techniques that improve efficiency and lessen the amount of computing power or man-hours required to perform a task. By its very nature, OR ties in immensely with computer science (in determining the most efficient, least time- and resource-consuming algorithm to carry out a computation; for instance, using a logarithmic algorithm rather than a polynomial-time algorithm[45]).

OR encompasses a wide umbrella of sub-disciplines and areas of active research focus, such as information technology, financial engineering, policy modeling, transportation modeling, manufacturing, and revenue management. Furthermore, its ties with stochastic models and Markov chains are inseparable.

12.3.1 Markov Decision Processes

In the last section, we touched upon the memorylessness property of a random variable; that is, roughly speaking, its tendency to "forget" its history and behave with no regard for its past outcomes. For instance, a series of coin flips, with the corresponding random variable outcome being a single flip of heads, would be memoryless (as the chance of success at any given flip is unaffected by the outcomes of previous flips), but three heads in a row would *not* be memoryless (as it is much more likely to accomplish this after already having flipped two heads than after having flipped a tails). This property is also known as the Markov property. We can expand upon this, and the associated idea of a Markov process, into a more full-fledged conception of a *Markov decision process*.

A Markov decision process (MDP) models decision making in circumstances where outcomes have elements of both randomness and dependence on the decisions of a dedicated decision maker. Fundamentally, this harkens back to the states model of recursive probability, in that the process assumes some state s at any given time t, and the decision maker can choose any decision d available in state s. The process will then advance to the subsequent state s', and the corresponding reward will be a function of the old/new state and the action taken, namely $R_d(s, s')$.

> **Definition 12.10: Markov Decision Process**
>
> Formally, an MDP is an ordered 4-tuple $(S, D, P_d(s, s'), R_d(s, s'))$ (where $s' = s_{t+1}$, with t representing time) such that S is a finite set of states, D is a finite set of decisions (and D_s is then the set of decisions available *at state s*), $P_d(s, s') = P(s' \mid s_t) = s, d_t = d$ (where d_t is the decision made at time t), and $R_d(s, s')$ is defined as the reward from advancing from state s to state s'.

The probability that the process will advance to state s' is dependent on the decision d made by the decision maker. In essence, an MDP is an extension of a Markov chain, in that we also have the options of decision-making and rewards at our disposal.

> **Corollary 12.4: Reduction to a Markov Chain**
>
> If, in an MDP, there is only one action corresponding to each state, and all rewards are equal, then the MDP reduces to a Markov chain.

In working with MDPs, our core task is to determine a decision-making function π such that $\pi(s)$ uniquely determines the decision d_s to be made at state s that will provide the greatest benefit. Id est, the goal is to

---
[45]See §16, where we discuss the P/NP problem, and big O notation, in further detail.

maximize the cumulative reward, or the sum

$$\sum_{t=0}^{\infty} R_{d_t}(s_t, s_{t+1})$$

where $d_t = \pi(s_t)$ for all t. This function π is known as the *optimal policy*.

To determine the function π algorithmically, we can use a method known as *dynamic programming*. Consider the case where we know the functions P and R of an MDP. We require two arrays that are indexed by the state s: the *value* V and the *policy* π. Once the DP algorithm runs, the policy π will yield the solution and $V(s)$ will yield the (mean) sum of rewards from the actions taken at each state s.

The specifics of the DP algorithm consist of two primary steps: updating the value and the policy at each step s (hence the *dynamic* element of the DP algorithm; the optimal policy and the corresponding value are modified according to the previous outcomes of the algorithm from the chosen states). Indeed, we have that

$$\pi(s) := \arg\max_d P(s' \mid s, d) \left(R(s' \mid s, d) + \gamma \cdot V(s') \right)$$

$$V(s) := \sum_{s'} P_{\pi(s)}(s, s') \left(R_{\pi(s)}(s, s') + \gamma \cdot V(s') \right)$$

where $\arg\max_d$ denotes the value of d that maximizes $\pi(s)$ (i.e. $\arg\max_{x \in S \subset X} f(x) = \{x \mid \forall y \in S : f(y) \leq f(x)\}$), and where $\gamma \in [0, 1)$ denotes a *discount factor*: i.e. a constant that essentially amounts to the common ratio in a geometric series, which ensures that all sums of values converge to a finite real number.

12.3.2 Queueing Theory

In *queueing theory,* we turn to yet another application of OR in daily life: queues, better known as *long lines!* Whether they be at a fast food restaurant, at a busy convention, at the movie theater, or at the DMV, lines are everywhere - this is a fact of life. And yet, just as with everything else we have studied up to this point, they can be dissected mathematically in such a manner that we may even begin to *appreciate* them![46] Queueing theory mainly falls under the branch of OR for the versatile applications of its results in business decisions; namely, how to provide services in order to cut down on long lines and minimize wait times (and, therefore, improve customer experiences). The ideas motivating it have been used to great effect in telecommunications, traffic engineering, and, in particular, industrial engineering (as in factories, offices, and especially hospitals - where OR optimization can literally be a matter of life and death!)

At its simplest level, queueing theory can be considered in terms of a single *queueing node:* essentially just a black box that takes a task, or consumer, as input, and outputs the completed task or satisfied customer. However, said node is not a pure black box or oracle, as we do require some specific information about the inner workings of the box. We know that the node contains at least one server, with each server taking a certain amount of time to complete a job or task before being free to process another incoming task. This is akin to the role of a cashier; each cashier can only process one customer at a time. The number of cashiers is the number of servers; thus, a typical supermarket may have anywhere from 2 or 3 servers to 10 or more (assuming all checkout aisles are open[47]).

But say a supermarket (black box/queueing node) has only one cashier (server), who is currently serving Alice. Bob comes over to the checkout aisles, only to find that the only checkout aisle is already occupied, so he leaves the checkout area and hangs around waiting for the opening. This is an instance in which the server has no *buffer,* or area in which tasks pending processing by a server may idle while the server handles other tasks of higher priority (or which came in earlier). However, in a more realistic scenario, checkout aisles usually have a buffer of at least 2 customers, and in fact, multiple customers may place their goods onto

---

[46]Well, okay, maybe not, but hopefully after reading this, they'll be at least a little more tolerable. Seriously though, the DMV is notorious for a reason. Once I waited there for *seven hours* and still had unfinished business!

[47]As rare as this perfect scenario may be in real life.

the conveyor belt at the same time (separated by bars to indicate which items belongs to which customer). Thus, if Bob comes in while the cashier is scanning Alice's goods, he can still wait in the checkout aisle (or, more efficiently, place a divider bar after Alice's last good, then start placing his own goods onto the belt after the divider, as one would do in reality).

Each server in the black box also has an associated *buffer size*, which indicates how many tasks it can "hold" in reserve while another task is being processed. If this checkout aisle were to have buffer size 1, for instance, then if Eve were to try and check out, she would be unable to until Alice departs the checkout aisle, as the buffer of the server cannot accommodate another incoming task. She would, however, be able to start placing her goods after setting her divider bar if the aisle had buffer size 2. (Note that the number of available divider bars would also dictate the buffer size, since a customer cannot wait without placing his/her goods onto the belt, and s/he cannot place goods onto the belt unless s/he has first set up a divider bar separating his/her own goods from the next customer's goods. In this respect, the buffer size of a real-life supermarket aisle would be given by the capacity of the aisle length (in terms of the number of shopping carts that could physically fit within the aisle), as well as the number of available divider bars; namely, the minimum of the two.)

Furthermore, each queueing node can adapt a *service discipline* or *scheduling policy*, which can also differ from those of the other queueing nodes.

> **Definition 12.11: List of Scheduling Policies**
>
> This is not an all-inclusive list, but it will hopefully elucidate some of the different types of inner workings of a black box:
>
> - First come, first serve: Customers/tasks which have been waiting the longest (or come first) are served first (with service being done one at a time).
>
> - Last come, first serve: Customers which have been waiting the shortest (or come last) are served first.
>
> - Equal service distribution: The capacity of the server is split equally between all the customers, regardless of time of arrival/time remaining in the task/priority.
>
> - Priority-based scheduling: Those customers - or tasks - with higher priority are served first.
>
> - Shortest first: Customers/tasks with the work that takes the shortest amount of time to complete *overall* are served first.
>
> - Shortest *remaining* first: Customers/tasks with the work that takes the shortest amount of time *from the point of service* to complete are served first.

Should there be deficiencies in a chosen scheduling policy (in terms of leaving customers dissatisfied by virtue of waiting too long, or leaving critical tasks unfulfilled), there will be a *dropout* of customers/tasks when the customer decides to leave on his/her own volition, or switch to another server, which contributes to the server's *dropout rate*.

This dropout rate, along with a combination of other factors, can be used to compute the length of the queue:

> **Theorem 12.5: Queue Length**
>
> Given an arrival rate a, a dropout rate b, and a departure (finishing/servicing, non-dropout) rate c, the queue length L is given by $L = \dfrac{a-b}{c}$.

This is, in fact, a modification of *Little's Law,* which is an extremely simple statement with extremely deep layers behind it:

> **Theorem 12.6: Little's Law**
>
> The long-term average number of consumers l in a black box is equal to the product of the average arrival rate λ and the average time w that each consumer spends in the black box.

Example 12.3.2.1. *For example, if an average of 10 customers enter a computer repair store every hour, and each customer spends an average of 45 minutes in the shop, then the average number of consumers in the store at any given time is $10 \cdot 0.75 = 7.5$.*

Suppose the computer store advertises a discount, which boosts influx of consumers by half. Then there will be 15 customers per hour, each of whom must spend only 30 minutes in the shop as opposed to 45 in order for shop operation to remain the same. this can be accomplished by streamlining repair methods, improving employee training, or increasing the number of servers/repair employees (perhaps by a number close to 50%).

The sketch of the proof of Little's Law relies on integral calculus; specifically, the intuitive definition of the definite integral of a continuous, integrable function from one point to another as the area under the graph of that function in the xy-plane. Let the *total* number of arrivals in the time interval $[0, t]$ be n, so that the area under of the curve of the function n_t (the function representing the total number of arrivals up to time t) is given by

$$\int_0^t n_t \, dt = \sum_{i=1}^n t_i$$

where t_i is the number of arrivals at time i. The average time spent serving a consumer/performing a task is given by the limit

$$\lim_{n \to \infty} \frac{1}{n} \sum_{i=1}^n t_i = \bar{t}$$

The average number of consumers/tasks currently in the system (the black box) is given by

$$\lim_{t \to \infty} \frac{1}{t} \int_0^t n_t \, dt = \bar{n}$$

Finally, the average rate of arrival is the two-variable limit

$$\lim_{n,t \to \infty} \frac{n}{t} = \bar{\lambda}$$

Dividing both sides of the integral of n_t from 0 to t by t yields

$$\frac{1}{t} \int_0^t n_t \, dt = \frac{1}{t} \sum_{i=0}^n t_i = \frac{1}{n} \cdot \frac{n}{t} \sum_{i=0}^n t_i$$

Taking the limit of both sides as $t \to \infty$ (and, hence, $n \to \infty$) yields Little's Law.

12.3.3 Bayesian Search Theory

Bayesian statistics - the specific type of statistics which revolves around the Bayesian interpretation of probability (that probability expresses *belief* in an event) - and the field of statistics that uses such methods as Bayes' theorem and conditional probability, and Bayesian inference - is a cornerstone of *Bayesian search theory:* the statistical subfield that plays a major role in retrieving what has been lost, much like queue theory plays a crucial role in streamlining service operations. It has been used on several occasions to recover lost vessels at sea, as well as to locate the remains of aviation-related disasters (Malaysia Airlines Flight 370 being the most prominent example).

The general procedure in Bayesian search theory involves determining the probability distribution function for each plausible point at which an object may be located, and producing a corresponding probability density map, being sure to update the probabilities in the map continuously throughout the search. Indeed, it is quite reminiscent of a Monte Carlo simulation: we tend to formulate a large number of hypotheses (or spots where the object could lie) regarding the object; the larger the number of reasonable hypotheses, the greater the likelihood of eventually locating the object (since we only need one to actually be true in order to discover the object). In the same way, a Monte Carlo simulation seeks the largest possible number of data points so as to maximize the accuracy of its approximations.

In a typical Bayesian search, one typically begins conducting the search in the area that has the highest probability of holding the object. If the object is not there, one then proceeds to search the area which has the second-highest probability of holding the object, and so forth. This continues until there is no longer a sufficient probability that the object even remains at all under the formed hypotheses. As such, the growth rate of the probability of recovering the object slows over time, and in fact tends asymptotically to the probability that the object remains in the first place (if it is known for certain that the object *exists*, just that it has been lost, then this probability is 1).

Rigorously speaking, consider an array consisting of q grid squares (where q is a composite number), with each grid square equally likely to hold the object. Let p be the probability of successfully detecting the object in the grid square that contains it. If we search a square and find no (remains of) the object, then by Bayes' Theorem, the probability that the object actually does lie inside the square is

$$p' = \frac{\frac{1}{q} \cdot (1-p)}{\left(1 - \frac{1}{q}\right) + \left(\frac{1}{q}\right)(1-p)}$$

which we observe to be less than $\frac{1}{q}$. Hence, the probability of the object lying a square as we advance from hypothesis to hypothesis most to least likely) is decreasing, and eventually we will reach a point where it is deemed no longer worth searching for the object (due to resource expenditure that comes from the search operation). As for the other squares, say we had probability $\frac{1}{q}$ beforehand (as the probability is evenly distributed *a priori*.) After finding that one of the other squares does not contain the object, the probability of discovering the object in any of the other squares becomes $r \cdot \frac{1}{1 - \frac{p}{q}} > r$.

Chapter 12 Review Exercises

Exercise 114. *Show that any tree with at least two vertices is bipartite.*

Exercise 115. *(2019 AMC 12B) How many sequences of 0s and 1s of length 19 are there that begin with a 0, end with a 0, contain no two consecutive 0s, and contain no three consecutive 1s?*

Exercise 116. *There is a group of 9 people gathered at an airport waiting on the plane for an international business meeting. Each of them can speak at most three languages. Among any three of them, at least two can speak a common language. Prove that there are three people who speak a common language.*

Exercise 117. *A computer has an exponentially distributed lifespan. Explain why it would follow that its failure rate at any given time would remain constant.*

Exercise 118. *(2008 AIME I) Consider sequences that consist entirely of A's and B's and that have the property that every run of consecutive A's has even length, and every run of consecutive B's has odd length. Examples of such sequences are AA, B, and AABAA, while BBAB is not such a sequence. How many such sequences have length 14?*

Exercise 119. *(2001 AIME I) A mail carrier delivers mail to the nineteen houses on the east side of Elm Street. The carrier notices that no two adjacent houses ever get mail on the same day, but that there are never more than two houses in a row that get no mail on the same day. How many different patterns of mail delivery are possible?*

Exercise 120. *(1990 APMO) A graph G on n vertices satisfies all three of the following conditions:*

(i) No vertex has degree $n-1$.

(ii) For any 2 points A, B which are not adjacent, there exists exactly one C such that AC, BC are both edges.

(iii) There are no triangles.

Prove that all vertices have the same degree.

Exercise 121. *The pattern of sunny and rainy weather in Inconsistentown can be modeled by a Markov chain. If it is sunny one day, the probability that the next day will also be sunny is $\frac{4}{5}$. If it is rainy one day, the probability that the next day will also be rainy is $\frac{3}{5}$. What is the probability that it rains in Inconsistentown on Tuesday, given that it is sunny on Sunday?*

Exercise 122. *(2018 AIME I) Let $SP_1P_2P_3EP_4P_5$ be a heptagon. A frog starts jumping at vertex S. From any vertex of the heptagon except E, the frog may jump to either of the two adjacent vertices. When it reaches vertex E, the frog stops and stays there. Find the number of distinct sequences of jumps of no more than 12 jumps that end at E.*

Exercise 123. *(2002 USAMO P1) Let S be a set with 2002 elements, and let N be an integer with $0 \le N \le 2^{2002}$. Prove that it is possible to color every subset of S either blue or red so that the following conditions hold:*

1. *the union of any two red subsets is red;*

2. *the union of any two blue subsets is blue;*

3. *there are exactly N red subsets.*

Extension Exercise 55. *(2000 Putnam B2) Prove that*

$$\frac{\gcd(m,n)}{n}\binom{n}{m}$$

is an integer for all $n \ge m \ge 1$, $n, m \in \mathbb{Z}$.

Extension Exercise 56. *(1956 Putnam B5) Show that a graph with $2n$ nodes and $n^2 + 1$ edges necessarily contains a 3-cycle, but that we can find a graph with $2n$ points and n^2 edges without a 3-cycle.*

Extension Exercise 57. *(2018 HMIC P5) Let G be an undirected simple graph. Let $f(G)$ be the number of ways to orient all of the edges of G in one of the two possible directions so that the resulting directed graph has no directed cycles. Show that $f(G)$ is a multiple of 3 if and only if G has a cycle of odd length.*

Part V

Number Theory

Why Study Number Theory?

In the previous part, we explored the more pure, abstract structures of mathematics - discrete structures, which model non-continuous real-world phenomena. Number theory - best described as the study of integers - mirrors and closely replicates this focus. In the words of the great Carl Gauss, "Mathematics is the queen of the sciences, and number theory is the queen of mathematics."

In studying number theory, and its associated intricacies, such as those of the prime numbers, objects constructed from integers (such as rational numbers), and generalizations of the properties of integers to concepts such as algebraic and transcendental numbers, mathematicians seek to answer the underlying, fundamental questions of arithmetic. In fact, at an essential level, number theory *is* arithmetic: more precisely, the evolution of what was once the arithmetic of positive integers. In this way, we can learn a tremendous amount by analyzing the patterns of integers: seemingly the simplest possible mathematical constructs that end up revealing the deepest, most profound secrets of mathematics, and by extension, the universe.

Moreso, much like with other departments of math, we have much to gain from practicing number theory. Learning how to manipulate *numbers* - and their associated mathematical objects - with ease is a manifestation of the vital skills that belie mathematics as a whole. Furthermore, the reasons to study number theory go beyond what the subject itself has to offer in terms of content. Number theory is not only beautiful; it is perhaps the epitome of simplicity from complexity - the very essence of mathematics. Problems such as the Goldbach and Collatz conjectures, Fermat's Last Theorem, and the Twin Prime Conjecture have confounded mathematicians for centuries (and, in the case of all except Fermat's Last Theorem, continue to confound even the world's most brilliant mathematical minds), yet have their roots in the simplest of ideas and consist of the most elementary of formulations. Such is the true beauty and elegance of the subject, and such is hopefully motivating to the reader as we dive headfirst into the coming sections.

Chapter 13

Beginning Number Theory

13.1 Factors and Divisibility

One of the best ways to study numbers is to study how they are inter-related with *other* numbers. More specifically, what do certain numbers have in common with each other? An excellent way to categorize numbers according to their common characteristics is to analyze their *factors*, also known as *divisors*.

> **Definition 13.1: Factors of a Number**
>
> A *factor* of an integer N is any number n that can be multiplied by another integer k to produce N. That is, $N = n \cdot k$.

N is said to be a *multiple* of n in this case.

By convention, we usually abbreviate this idea of n being a factor of N (or n *dividing* N) as $n \mid N$ (read: "n divides N"). Though this may seem like an extremely simple concept, we can delve surprisingly deep into this to try and inter-relate the integers to each other.

13.1.1 Parity: Oddness and Evenness

Our study of factors begins with the idea of *parity*: simply put, the state of being even or odd. Otherwise stated, is 2 a factor of the number? If so, it is even; if not, it's odd. As in the first few chapters of the book, we can prove several facts about even and odd numbers. As a refresher, let's go with a basic example:

Example 13.1.1.1. *Show that the sum of any two even numbers is also even.*

Solution 13.1.1.1. *Proof.* Let the two even numbers be $2k$ and $2l$ for integers k and l. Then we have $2k + 2l = 2(k + l)$, which must be even since $k + l$ is always an integer whenever k and l are integers.[48] □

And a slightly more complicated one:

Example 13.1.1.2. *Show that the sum of any 2019 odd numbers is odd.*

Solution 13.1.1.2. *Don't let the 2019 intimidate you! All you need for this proof is that 2019 is an odd number.*

Proof. Recall that all odd numbers are of the form $2k + 1$ for some integer k. If we sum an odd number of odd numbers, then we will be adding 2019 1's, which sum to 2019 - an odd number. Factoring out the 2 will thus leave a remainder of 1, which is the very definition of an odd number; this completes the proof. □

---

[48] Indeed, the group of integers is *closed under addition*, meaning that we always obtain another integer when we add two integers. The group of integers is not, however, closed under division.

A more detailed one (but not any more complicated):

Example 13.1.1.3. *Show that the sum of the squares of two odd numbers and an even number is even.*

Solution 13.1.1.3. *Proof.* Let the two odd numbers be $2a+1$ and $2b+1$ for some integers a and b, and let the even number be $2c$ for some integer c. Then $(2a+1)^2 + (2b+1)^2 + (2c)^2 = (4a^2 + 4a + 1) + (4b^2 + 4b + 1) + 4c^2 = (4a^2 + 4a + 4b^2 + 4b + 4c^2) + 2 = 2(2a^2 + 2a + 2b^2 + 2b + 2c^2 + 1)$ which is even, since $2a^2 + 2a + 2b^2 + 2b + 2c^2 + 1$ is an integer for all integers a, b, c. $\square$

Note that parity only applies to integers, and not fractions, decimals, irrational numbers, or complex numbers. After all, it doesn't make much sense to describe the oddness or even-ness of $\frac{1}{2}$ or π^{e+3i-2}!

Let's try one conceptual example from an actual contest to solidify our problem-solving skills:

Example 13.1.1.4. *(2005 AMC 8) Suppose m and n are positive odd integers. Which of the following must also be an odd integer?*

(A) $m + 3n$

(B) $3m - n$

(C) $3m^2 + 3n^2$

(D) $(nm + 3)^2$

(E) $3mn$

Solution 13.1.1.4. *We evaluate each of the answer choices individually.*

Choice (A) is incorrect since $odd + 3 \cdot odd = odd + odd = even$. (Note that $3 \cdot odd = odd$ since the product of an odd number with another odd number is always odd.)

Choice (B) is also incorrect; we apply the same reasoning as we did to eliminate choice (A).

Choice (C) is incorrect, since $odd^2 = odd$, $3 \cdot odd = odd$, and finally, $odd + odd = even$.

Choice (D) is incorrect, because $mn = odd$, and $odd + 3 = even$. We then have $even^2 = even$.

This leaves choice $\boxed{(E)}$. *Indeed, $3mn = 3 \cdot odd \cdot odd = 3 \cdot odd = odd$.*

A way to fakesolve or "cheese" around this problem would be to set $m = n = 1$, from which $\boxed{(E)}$ is clearly the answer.

This may seem like a trivial concept, but as can be seen from the example above, parity is highly useful in proofs involving casework, or proofs by contradiction. Sometimes a simple idea can go a long way!

13.1.2 Divisibility Tests

It is trivially straightforward to identify the even/odd parity of any given integer just by looking at the units digit, but what about the "parity" with respect to some number other than 2, like 3, 5, 19, or 120? We can use a combination of divisibility tests to determine whether a given integer is a factor of a large number.

The following is a table of the most common divisibility tests for the positive integers from 2 to 11, inclusive.

n	Criterion
2	last digit is even
3	digit sum is a multiple of 3
4	last two digits form multiple of 4
5	units digit is 0 or 5
6	divisible by 2 and 3
7	see *
8	last three digits form multiple of 8
9	digit sum is a multiple of 9
10	units digit is 0
11	positive difference of alternating digit sums is a multiple of 11**

* Let $N = \overline{d_1 d_2 d_3 \ldots d_n}$. To determine whether 7 is a factor of N, take $2d_n$ and subtract it from the remainder of N (i.e. $\overline{d_1 d_2 d_3 \ldots d_{n-1}}$). Repeat until the resulting value is easily identifiable as a multiple, or not a multiple, of 7.

Example 13.1.2.1. *To determine whether 740,123 is a multiple of 7, we take 3, double it to get 6, and subtract it from 74,012 to get 74,006. We repeat this process: $74,006 \to 7,400 - 12 = 7,388 \to 738 - 16 = 722 = 700 + 21 + 1$ which is not a multiple of 7, so 740,123 is not a multiple of 7. (In fact, we can go a step further and say that 740,123 is exactly 1 more than a multiple of 7. The result of the divisibility test for 7 will not only determine divisibility, but also the remainder upon dividing by 7.)*

** To elaborate on the divisibility test for 11: take the sums of the digits in odd-numbered positions and in even-numbered positions, and then take their positive difference. Iff this difference is a multiple of 11, then the original number is a multiple of 11 as well.

Remark. For larger numbers, we can use the Chinese Remainder Theorem, based on their prime factorizations.

In particular, we should probably take a deeper look at the motivation behind the divisibility tests for 7 and 11, which upon first glance, seem quite random, even unrelated. The other tests are fairly straightforward, but why should the tests for 11, and especially 7, work at all?

Proving the divisibility rule for 7. Let N consist of two parts: A, which contains all digits aside from the units digit, and B, the units digit, so that the numerical value of N is $10A + B$. We want to show that, iff $A - 2B$ is a multiple of 7, then so is N. Setting $A - 2B = 7k$ for some positive integer k, we multiply both sides by 10 to obtain $10A - 20B = 70k$. Adding $21B$ to both sides yields $10A + B = 70k + 21 = 7(10k + 3)$, as desired (in this direction of the proof).

Next we tackle the opposite direction. If $10A + B = 7k$ for some positive integer k, then $10A - 20B = 7k - 21B$ (subtracting $21B$ from both sides as before) and $10(A - 2B) = 7(k - 3B)$. This implies that $A - 2B$ is a multiple of 7, since 10 is not divisible by 7. Hence, we have proven both directions, which completes the proof. $\square$

Next, we prove the divisibility rule for 11:

Proving the divisibility rule for 11. WLOG assume that N is an n-digit number, so that it can be written in the form $10^{n-1} \cdot d_1 + 10^{n-2} \cdot d_2 + \ldots + 10^1 \cdot d_{n-1} + 10^0 \cdot d_n$, where $d_1, d_2, d_3, \ldots, d_n$ are digits (in base 10; see §13.2). Note that $10 = 11^1 - 1$, $10^2 = 11 \cdot 9 + 1$, $10^3 = 11 \cdot 91 - 1$, $10^4 = 11 \cdot 909 + 1$, and so forth (with $10^n = 11k + 1$ whenever k is even and $10^n = 11k - 1$ whenever k is odd, for some positive integer k). Thus, each digit d_{n-a} for a even adds to the remainder when N is divided by 11, and each digit d_{n-a} for a odd subtracts from the remainder when N is divided by 11. It follows that we want the positive difference between the even- and odd-numbered digits to be a multiple of 11, which is what we set out to prove. $\square$

13.1.3 The Prime Factorization

Divisibility tests can only go so far. After all, what if we need to determine if a number is divisible by some larger prime, like, say, 43 or 97? There is no well-known, easy divisibility test for larger primes. If we want to determine whether or not any given number is a factor of some integer, then we take the *prime factorization* of the integer.

A prime factorization of a number N essentially breaks it down into its most basic components: its prime factors.

> **Definition 13.2: The Prime Factorization**
>
> The prime factorization of the positive integer N is given by $p_1^{k_1} \cdot p_2^{k_2} \cdot p_3^{k_3} \cdots p_n^{k_n}$, where $p_1, p_2, p_3, \ldots, p_n$ are the prime factors of n and $k_1, k_2, k_3, \ldots, k_n$ are exponents of the prime factors. (If a prime number is not present in the prime factorization of N, then its exponent is zero. This observation is key in establishing a vital theorem that will come shortly!)

To determine the prime factorization of a given integer, it is usually not too difficult to run through the first few primes, namely 2, 3, 5, and 7, and then divide the number by those primes that go evenly into the number (and repeat as necessary). However, there are cases in which the prime factors can be larger than expected, but thankfully, in a contest scenario, these can usually be seen immediately through an external observation.

Here are some elementary examples of the prime factorization in action:

Example 13.1.3.1. *Compute the prime factorizations of each of the following:*

(a) *45*

(b) *512*

(c) *1920*

(d) *10,152*

Solution 13.1.3.1.

(a) *45 is a multiple of 5, since its unit digit is 5. Thus, we can divide 45 by 5 to get 9, which we can identify easily as 3^2. Hence, the prime factorization of 45 is $\boxed{3^2 \cdot 5}$.*

(b) *Notice immediately that 512 is even, and that all subsequent quotients are also even. We can divide by 2 nine times, and thus $512 = \boxed{2^9}$. (Alternately, an astute reader might notice this immediately by inspection.)*

(c) *$1920 = 2000 - 80$, so we can pull out a factor of 80 to get $1920 = 80(25 - 1) = 80 \cdot 24$. We have $80 = 8 \cdot 10 = 2^3 \cdot (2 \cdot 5) = 2^4 \cdot 5$, and $24 = 2^3 \cdot 3$, so $1920 = (2^4 \cdot 5) \cdot (2^3 \cdot 3) = \boxed{2^7 \cdot 3 \cdot 5}$.*

(d) *Uh oh. This doesn't look like such an easy number to straight-up factor, does it? Well, we're in luck! By the difference of squares identity, note that $10,152 = 10,201 - 49 = 101^2 - 7^2 = (101+7)(101-7) = 108 \cdot 94$. Now write 108 as $3^3 \cdot 2^2$, and 94 as $47 \cdot 2$. The resulting prime factorization is $\boxed{3^3 \cdot 2^3 \cdot 47}$. (See what we mean about the large prime factors?)*

Remark. In the above example(s), we listed the prime factors in descending order of their exponents. It is equally acceptable to list the factors themselves in descending order (e.g. $5 \cdot 3^2$ instead of $3^2 \cdot 5$ for 45); the difference is purely cosmetic. However, we primarily list prime factorizations in descending exponent order throughout this book.

For extra practice, here is a seemingly impossible example that can be cracked easily with one astute observation:

Example 13.1.3.2. *What are the prime factors of 4,006,004,001? (The numbers 667 and 2,002,001 are both prime.)*

Solution 13.1.3.2. *Observe that* $4,006,004,001 = 1,004,006,004,001 - 1,000,000,000,000 = 1001^4 - 1000^4 = (1001^2)^2 - (1000^2)^2 = (1,002,001 + 1,000,000)(1,002,001 - 1,000,000) = (2,002,001)(2,001)$. *The number 2001 can be written as* $3 \cdot 667$ *by the divisibility test for 3. In addition, 667 and 2,002,001 are both prime, so the prime factorization of this enormous number is* $\boxed{2,002,001 \cdot 667 \cdot 3}$.

Needless to say, it's not always possible to be able to factor huge numbers this easily (in fact, it's extremely rare), but that's not to say that a number that isn't easily factorable is necessarily prime. That's precisely the whole reason the Great Prime Search still continues on to this day, and will likely bravely soldier on for decades, even centuries to come[49] - because we *don't* know if any given number is prime, and *can't* possibly know with our current methods. This is not to say that we should lose hope of ever nicely dealing with prime numbers, though, since there is a vast assortment of methods to handle them (see §13.4, "Studying Prime Numbers").

From the prime factorization of a number N, a very handy, very versatile, and extremely common trick follows:

Theorem 13.1: Number of Factors of an Integer

Let N be an integer with prime factorization $p_1^{k_1} p_2^{k_2} p_3^{k_3} \cdots p_n^{k_n}$. Then the number of integer factors of N (denoted $\tau(n)$) is given by $(k_1 + 1)(k_2 + 1)(k_3 + 1) \cdots (k_n + 1)$.

Proof. Each factor n of N has prime factors restricted to the set $\{p_1, p_2, p_3, \ldots, p_n\}$. For each prime factor p_i $(1 \leq i \leq n)$, we can choose what the exponent of p_i is in N: $0, 1, 2, 3, \ldots, k_i$. This gives us $k_i + 1$ choices, so there are $(k_1 + 1)(k_2 + 1)(k_3 + 1) \cdots (k_n + 1)$ choices in total by the multiplicative principle. □

We now present some examples.

Example 13.1.3.3. *Compute* $\tau(24)$, $\tau(450)$, *and* $\tau(2019)$.

Solution 13.1.3.3. $\tau(24) = \tau(2^3 \cdot 3) = (3+1)(1+1) = 4 \cdot 2 = \boxed{8}$. *We can verify that 24 has 8 factors by listing them out: 1, 2, 3, 4, 6, 8, 12, 24.*

Similarly, $\tau(450) = \tau(5^2 \cdot 3^2 \cdot 2) = (2+1)(2+1)(1+1) = 3 \cdot 3 \cdot 2 = \boxed{18}$.

We also have $\tau(2019) = \tau(3 \cdot 673) = (1+1)(1+1) = \boxed{4}$. *(This one is a little trickier, if only because it's not immediately obvious that 673 is prime. However, we can test all the primes up to* $\sqrt{673} < 26$, *namely all primes up to and including 23, and see that it is prime that way.)*

Here is a less straightforward example, but one that can nevertheless be approached directly with the formula:

Example 13.1.3.4. *Describe all positive integers N such that $2N$ has exactly twice as many factors as does N.*

Solution 13.1.3.4. *Let* $N = p_1^{k_1} p_2^{k_2} p_3^{k_3} \cdots p_n^{k_n}$ *for primes* $p_1, p_2, p_3, \ldots, p_n$ *and non-negative integers* $k_1, k_2, k_3, \ldots, k_n$. *If* $p_i = 2$ *for some* $1 \leq i \leq n$, *then* $2N = p_1^{k_1} p_2^{k_2} p_3^{k_3} \cdots p_i^{k_i+1} \cdots p_n^{k_n}$ *and* $\tau(2N) = \frac{k_i + 2}{k_i + 1} \cdot \tau(N)$. *In this case, we require* $\frac{k_i + 2}{k_i + 1} = 2 \implies k_i = 0$. *Hence, 2 must not be a prime factor of N. For all N without 2 as a prime factor, multiplying N by 2 introduces a new prime factor, and so the exponent of 2 goes from 0 to 1, multiplying the number of factors by* $\frac{1+1}{0+1} = 2$. *So our answer is* $\boxed{\text{all odd numbers.}}$

---

[49] Assuming that we don't resolve the P = NP question by then, anyway.

We can expand this trick to two more applications: finding the sum and product of the factors of a number.

> **Theorem 13.2: Sum and Product of Factors of an Integer**
>
> The sum of the factors of an integer $N = p_1^{k_1} p_2^{k_2} p_3^{k_3} \cdots p_n^{k_n}$ is equal to
>
> $$(p_1^{k_1} + p_1^{k_1-1} + p_1^{k_1-2} + \ldots + p_1 + 1) \cdots (p_2^{k_2} + p_2^{k_2-1} + p_2^{k_2-2} + \ldots + p_2 + 1)$$
>
> $$\cdots (p_n^{k_n} + p_n^{k_n-1} + p_n^{k_n-2} + \ldots + p_n + 1)$$
>
> $$= \prod_{i=1}^{n} \left(\sum_{j=0}^{k_i} p_i^j \right)$$
>
> and the product of its factors is equal to $N^{\frac{\tau(N)}{2}}$.

Proof. We first prove the sum of factors formula:

Proving the sum of factors formula. We proceed by induction; our base case is $i = 1$. This amounts to showing that the sum of the factors of p^k, for p a prime and k a non-negative integer, is $p^k + p^{k-1} + \ldots + p + 1$. The factors of p^k are just $1, p, p^2, \ldots, p^{k-1}, p^k$ (and since p^k has $k+1$ factors, these must be the only factors), so this proves the base case.

For the inductive hypothesis, assume the formula holds true for some positive integer i. For $i + 1$, the sum of the factors should be the sum of the factors of

$$p_1^{k_1} p_2^{k_2} p_3^{k_3} \cdots p_i^{k_i} p_{i+1}^{k_{i+1}} = \left(p_1^{k_1} p_2^{k_2} p_3^{k_3} \cdots p_i^{k_i} \right) \cdot p_{i+1}^{k_{i+1}}$$

By the inductive hypothesis, the sum of all factors of the first part of the expression (i.e. all factors that do not contain a prime factor of p_{i+1} is

$$\prod_{i=1}^{n} \left(\sum_{j=0}^{k_i} p_i^j \right)$$

which we will call P_i for brevity. If we include a factor of p_{i+1}, then we can go up to an exponent of k_{i+1}, so our desired sum is $P_i \cdot (1 + p_{i+1} + p_{i+1}^2 + \ldots + p_{i+1}^{k_{i+1}-1} + p_{i+1}^{k_{i+1}}) = P_{i+1}$ as desired. $\square$

To prove the product of factors formula, observe that, for each factor n of N, $\dfrac{N}{n}$ is also a factor of N, by the definition of a factor. There are $\dfrac{\tau(N)}{2}$ such pairs when $\tau(N)$ is even; if $\tau(N)$ is odd, then N is a perfect square[50] and $\sqrt{N}$ will be an integer. Thus, $\sqrt{N}$ contributes $N^{\frac{1}{2}}$ to the product of factors, covering all cases and completing the proof. $\square$

13.1.4 The Fundamental Theorem of Arithmetic

The *Fundamental Theorem of Arithmetic* is the unifying theorem of prime factorizations:

> **Theorem 13.3: Fundamental Theorem of Arithmetic**
>
> Each (positive) integer N has a unique prime factorization (up to the order in which the factors are

---

[50]If $\tau(N)$ is odd, then $(k_1 + 1)(k_2 + 1)(k_3 + 1) \cdots (k_n + 1)$ is odd, implying that all terms $(k_i + 1)$ are odd, and so that all k_i are even. If all k_i are even, they are divisible by 2, which implies that $\sqrt{N}$ (derived from halving all exponents of N) is an integer.

listed).

The theorem is two-fold: first, it says that every integer has a prime factorization, and second, it says that this prime factorization is unique to its corresponding integer. That is, we cannot represent one integer as the product of primes in two different ways.

Example 13.1.4.1. *$N = 12$ can be represented as $2^2 \cdot 3$ and only as $2^2 \cdot 3$; we cannot write it as, say, $4 \cdot 3$ or $2 \cdot 6$ (4 and 6 are not prime).*

Because we can represent certain composite integers in multiple different ways using non-prime factors (e.g. $18 = 2 \cdot 9$ and $18 = 3 \cdot 6$), we require the factors to be prime. In addition, since each integer has a unique factorization, this means we cannot tack on a bunch of 1's onto the prime factorization (keeping the value of the number the same), which shows that 1 is not a prime number (since if it were, we could tack them on freely with no constraint, producing a different integer each time we did so). This is where the traditional, but slightly mis-informed definition of a prime number as "having no factors other than 1 and itself" falls apart. Indeed, 1 meets this criterion, but fails the additional requisite of having *exactly* two factors, no more and no less, as it only has 1 factor: itself.

The formal proof of the theorem, as originally proposed by Euclid, proceeds as follows:

Proof. We first invoke Euclid's lemma:

> **Lemma 13.3: Euclid's Lemma**
>
> If a prime p divides ab, then it divides either a or b (or both), where a and b are natural numbers.

The proof of the Fundamental Theorem is two-fold. First, we need to prove *existence*: i.e. that every (positive) integer N is either prime or can be written as the product of primes, and second, we need to prove *uniqueness*: i.e. that each N has only one prime factorization.

To prove existence, we proceed by strong induction. For the base case, observe that 2 is prime, so we are done. For the inductive step, assume that the theorem holds for all positive integers between 1 and N, inclusive. We show that it holds for $N+1$ as well. If N is prime, then we are done since there is nothing to prove. On the other hand, if N is composite, then there exist integers a and b such that $N = ab$, and $1 < a \leq b \leq N$ (where we are assuming WLOG that $a \leq b$). By the induction hypothesis, we can write a and b themselves as products of primes, such that

$$a = p_1^{k_1} p_2^{k_2} p_3^{k_3} \cdots p_n^{k_n}$$
$$b = q_1^{l_1} q_2^{l_2} q_3^{l_3} \cdots q_n^{l_n}$$

and so
$$N = ab = p_1^{k_1} p_2^{k_2} p_3^{k_3} \cdots p_n^{k_n} q_1^{l_1} q_2^{l_2} q_3^{l_3} \cdots q_n^{l_n}$$

is a product of primes as well, which proves existence.

Henceforth, we show that the prime factorization of each integer N is unique. Assume the contrary: that N can be written as the product of prime numbers in two different ways, namely

$$N = p_1 p_2 p_3 \cdots p_m$$

$$N = q_1 q_2 q_3 \cdots q_n$$

(where we assume WLOG that all the exponents are 1)

We show that $m = n$, and furthermore that, up to order, $\{p_i : 1 \leq i \leq n\} = \{q_j : 1 \leq j \leq n\}$. Since $p_1 \mid N$, we have $p_1 \mid q_j$ for some $1 \leq j \leq n$ by Euclid's lemma. Re-labeling if necessary, say that $p_1 \mid q_1$. q_1 is prime, so its only divisors are 1 and itself, implying that either $p_1 = 1$ or $p_1 = q_1$. The former is impossible, however, since by definition, p_1 is prime. Hence, $p_1 = q_1$, and so

$$\frac{n}{p_1} = p_2 p_3 \cdots p_n = q_2 q_3 \cdots q_n$$

By the same reasoning as before, we have $p_2 = q_2$, $p_3 = q_3$, and in general, $p_i = q_i$ for all $1 \leq i \leq n$, which shows that $m \leq n$. Applying the proof in reverse shows that both $m \leq n$ and $n \leq m$, hence $m = n$ and the prime divisors are all equal to each other in some order. $\square$

13.1.5 Examples

Example 13.1.5.1. *(2014 MathCounts State Sprint) If n represents the number of seconds in a day, what is the largest prime factor of n?*

Solution 13.1.5.1. *It may be tempting to calculate n directly and then prime factorize it, but we actually don't need to do that - we can just pull out known factors of n. Namely, we know that $n = 24 \cdot 60 \cdot 60$ (24 hours $\cdot$ 60 minutes $\cdot$ 60 seconds), so $n = (2^3 \cdot 3) \cdot (2^2 \cdot 3 \cdot 5)^2 = (2^3 \cdot 3) \cdot (2^4 \cdot 3^2 \cdot 5^2) = 2^7 \cdot 3^3 \cdot 5^2$. The largest prime factor is thus $\boxed{5}$.*

Example 13.1.5.2. *What is the sum of the odd factors of 90?*

Solution 13.1.5.2. *The prime factorization of 90 is $5 \cdot 3^2 \cdot 2$, and an odd factor is any factor that does not contain a 2 in its prime factorization. This is essentially the sum of the factors of $5 \cdot 3^2$, which is $(5^1 + 5^0)(3^2 + 3^1 + 3^0) = 6 \cdot 13 = \boxed{78}$.*

Example 13.1.5.3. *(2014 Berkeley Math Tournament) Let m and n be integers such that $m + n$ and $m - n$ are prime numbers less than 100. Find the maximal possible value of mn.*

Solution 13.1.5.3. *Note that $m + n$ and $m - n$ differ by $2n$, which is always an even number for n an integer. Thus, $m + n$ and $m - n$ must be of the same parity. Since they are both prime numbers, they must both be odd (otherwise, they would both have to be equal to 2, forcing $m = 2$ and $n = 0$, which gives $mn = 0$ - clearly not optimal). To maximize mn, we want m and n as close together as possible (by AM-GM), and so they should both be as close to 50 as possible (so as to maximize the value of $m + n$). The largest prime number under 100 is 97, and indeed, $100 - 97 = 3$ is also prime, so $m = 50$ and $n = 47 \implies mn \leq \boxed{2350}$.*

13.1.6 Exercises

Exercise 124.

(a) How many factors does 2019 have?

(b) How many factors does 876 have?

(c) How many factors does $1^1 \cdot 2^2 \cdot 3^3 \cdots 10^{10}$ have?

Exercise 125. Show that the square of a prime has exactly 3 positive integral factors.

Exercise 126. Prime factorize each of the following:

(a) 54

(b) 97

(c) 1600

(d) 676

Exercise 127. What is the largest prime factor of $5904 = 10^4 - 8^4$?

Exercise 128. Compute the smallest positive integer with exactly 19 divisors.

Exercise 129. (2016 AMC 8) The number N is a two-digit number. When N is divided by 9, the remainder is 1. When N is divided by 10, the remainder is 3. What is the remainder when N is divided by 11?

Exercise 130. Describe the set of all positive integers with exactly 2019 positive integral factors.

Exercise 131. If the positive integer $6N$ has twice as many factors as does N, how many factors can N have?

Exercise 132. For $p \geq 3$ a prime number, show that the only positive integers with exactly p divisors are the perfect $(p-1)^{th}$ powers of a prime.

Exercise 133. Find the largest integer n for which $(3!)^n$ divides $2019!$.

Exercise 134. (2018 ARML Local Individual Round #5) Compute the number of triples of consecutive positive integers less than 50 whose product is both a multiple of 20 and 18.

Extension Exercise 58. (1986 AIME) Let S be the sum of the base 10 logarithms of all the proper divisors (all divisors of a number excluding itself) of 1000000. What is the integer nearest to S?

Extension Exercise 59. (2018 ARML Individual Round #1) Compute the greatest prime factor of N, where
$$N = 2018 \cdot 517 + 517 \cdot 2812 + 2812 \cdot 666 + 666 \cdot 2018$$

Extension Exercise 60. Let N be a positive integer with exactly 20 divisors. Does there exist a positive integer a such that $a \cdot N$ has exactly 19 divisors? If so, give at least one satisfactory ordered pair (N, a). If not, prove that no such ordered pair exists.

Extension Exercise 61. (2018 ARML Local Individual Round #9) Compute the number of positive integers N between 1 and 100 inclusive that have the property that there exist distinct divisors a and b of N such that $a + b$ is also a divisor of N.

Extension Exercise 62. (2013 Caltech-Harvey Mudd Math Competition) Determine all positive integers n whose digits (in decimal representation) add up to $\dfrac{n}{57}$.

Extension Exercise 63. (2016 ARML Individual Round #10) Compute the largest of the three prime divisors of $13^4 + 16^5 - 172^2$.

13.2 Bases Other than Base 10

How many fingers do you have? Seriously. It's a pretty easy question; you should be able to answer it. But **why** do you have the number of fingers that you do? A biologist might answer with the traditional evolutionary explanation - that we have 10 fingers and 10 toes because such a combination has proven optimal for survival and utility. It is the ideal amount for balance and grasping objects, and does not bear excess weight on our skeletal structure.

A mathematician, on the other hand, might be more interested in how we count using our fingers - and how other organisms having a number of fingers other than ten reflects how what we refer to as "counting" on a basic level is not universal. Far from it, in fact - we are among the few living organisms that do have 10 fingers and toes, which explains why we count by tens. But a survey of other animals reveals the importance of counting in other increments as well. Humans count in what is known as the **base-10** number system, but a frog or salamander might only have four digits on each hand, and certain species of birds may only have three fingers and four toes on each limb. Their number systems would likely be base-8 and base-6, respectively.

Different base number systems represent numerical quantities in different ways, with place values that depend on the specified base. For now, we will look at representations of the integers:

> **Definition 13.3: Base Number Systems**
>
> Let N be an integer (positive or negative). A representation of N in base b (where $b \geq 2$ is a positive integer) is a representation of N in the form
>
> $$b^n \cdot d_n + b^{n-1} \cdot d_{n-1} + \ldots + b^2 \cdot d_2 + b \cdot d_1 + d_0 + b^{-1} \cdot d_{-1} + b^{-2} \cdot d_{-2} + \ldots$$
>
> where d_{-n} denotes the nth digit to the right of the decimal point.

To avoid any ambiguity, we denote the original base-10 integer N with the notation N_{10} (or, if converting from a base $b \neq 10$, N_b).

Example 13.2.0.1. *Represent 2019_{10} as a base-5 integer.*

Solution 13.2.0.1. *The place values in base 5 (from right to left) are $5^0 = 1$, $5^1 = 5$, $5^2 = 25$, $5^3 = 125$, $5^4 = 625$, ... (and this is where we stop for this problem, since $5^5 = 3125 > 2019$, so it is unnecessary to use any digits past 5^4 to represent 2019_{10} in base 5). 625 goes into 2019 three times, so we fill in the 5^4 place value slot with a 3, then subtract off $625 \cdot 3 = 1875$ to get 144. 125 goes into 144 once, so the corresponding digit becomes a 1, leaving 19. The $5^2 = 25$ place value will have a zero, but the $5^1 = 5$ place value has a 3, and the $5^0 = 1$ units digit will be a 4. Hence, $2019_{10} = \boxed{31014_5}$.*

Example 13.2.0.2. *Represent 45_9 as a binary integer.*

Solution 13.2.0.2. *Recall that "binary" is another way of saying "base 2" (remember this by the prefix bi-, meaning "two," and also by its extreme versatility in computer science applications; computers use the digits 0 and 1 for practically any purpose you can think of). We first convert 45_9 to a base-10 integer, then convert that base-10 integer to base 2: $45_9 = 4 \cdot 9^1 + 5 \cdot 9^0 = 4 \cdot 9 + 5 = 36 + 5 = 41_{10}$. The largest power of 2 less than or equal to 41_{10} is $2^5 = 32$, so we fill in that place value with a 1. 2^4 has a 0, 2^3 has a 1, 2^2 and 2^1 both have 0's, and $2^0 = 1$ has a 1. The answer is $\boxed{101001_2}$.*

13.2.1 Bases Beyond Base 10

What if we wanted to go beyond base 10? For instance, what if we wanted to convert to and from base 16 (more commonly known as *hexadecimal* in its applications to computer science)? There are no numerical digits past 9, so what symbols could we possibly use to represent numerical quantities that are at least 10 within a single place value? The answer is intuitive enough: letters. More specifically, we conventionally use capital letters (in alphabetical order) to run through the values 10 through 35, then lowercase letters to go from 36 to 61, followed by the symbols + and / for 62 and 63 respectively, which is usually sufficient to cover $2^6 = 64$ values in a single slot (base 64 is almost always adequate for most real-world purposes). The actual base64 encoding system differs slightly in that it begins with the capital, then lowercase letters, and from 52-61, uses the standard base-10 digits 0 through 9, and finally uses + and / for 62 and 63 respectively.

Example 13.2.1.1. *Compute the hexadecimal (hex) encoding of* 4519_{10}.

Solution 13.2.1.1. *Since* $16^3 = 4096 < 4519$, *there will be a 1 in the* 16^3 *place value. Then* $4519 - 4096 = 423 = 256 + 16 \cdot 10 + 7$, *so in the* 16^2 *slot, there is a 1; in the* 16^1 *slot, there is the "digit" A; and in the* 16^0 *slot, there is a 7. The final representation of* 4519_{10} *in hex is* $\boxed{11A7_{16}}$.

Example 13.2.1.2. *What is the sum of all base-10 integers strictly less than 144 such that the base-12 representation of N has at least one alphabetical digit (A or B)? Express your answer as an integer in base 10.*

Solution 13.2.1.2. *We can actually solve this problem using PIE. Observe that this is the sum of the integers with "tens" (twelves?) digit A or B, plus the sum of the integers with units digit A or B, minus* $AA_{12}, AB_{12}, BA_{12},$ *and* BB_{12} *(the overlap between both cases). The sum of all integers beginning with A is* $(10 \cdot 12 + 0) + (10 \cdot 12 + 1) + (10 \cdot 12 + 2) + \ldots + (10 \cdot 12 + 11) = 12 \cdot 10 \cdot 12 + (0 + 1 + 2 + \ldots + 11) = 1440 + 66 = 1506$, *and likewise, the sum of all integers beginning with B is* $12 \cdot 11 \cdot 12 + 66 = 1650$. *Now we repeat for the integers with a units digit of A, to get* $10 + 22 + \ldots + 130 = 770$, *and a units digit of B, to get 782. Finally, we subtract* $AA_{12} = 130_{10}, AB_{12} = 131_{10}, BA_{12} = 142_{10},$ *and* $BB_{12} = 143_{10}$ *to obtain our final answer of* $\boxed{4162_{10}}$.

13.2.2 Fractional Bases

At first glance, the idea of counting in non-integer bases may seem patently absurd. How can we assign place values if we're not even working in the integers? But there is always a workaround. Enter fractional bases. The general idea might best be illustrated with a concrete example:

Example 13.2.2.1. *Convert* $\left(25\frac{1}{24}\right)_{10}$ *to base* $\frac{5}{2}$.

Solution 13.2.2.1. *Consider just the 25 for now. The place values are* $1, \frac{5}{2}, \frac{25}{4}, \frac{125}{8}$ *(since* $\frac{625}{16} > 25$). $25_{10} = \left(\frac{200}{8}\right)_{10}$, *so there is a 1 in the* $\frac{125}{8}$ *place, a 1 in the* $\frac{25}{4}$ *place (since we have* $\frac{75}{8}$ *left over, and* $\frac{25}{4} \cdot \frac{3}{2} = \frac{75}{8}$), *a 1 in the* $\frac{5}{2}$ *place (leaving us with* $\frac{5}{8}$), *and 0 in the units digit place. Adding back the* $\frac{1}{24}$ *yields a remainder of* $\frac{2}{3}$.

We now need to consider how to represent $\frac{2}{3}$ *using the place values beyond the decimal point, namely* $\frac{2}{5}, \frac{4}{25}, \frac{8}{125}, \ldots$. *Since* $\frac{\frac{2}{5}}{1 - \frac{2}{5}} = \frac{2}{3} \geq \frac{3}{4}$ *by the infinite geometric series formula, we can have all 1's after the decimal point, so our desired representation is* $\boxed{1110.\overline{1}_{\frac{5}{2}}}$.

This general solution procedure works for any rational base $b \in (-1, 1), b \neq 0$.

13.2.3 Negative Bases

Fundamentally, negative integer bases are no different from positive integer bases. The exponents dictating the place values simply represent more extreme positive or negative values, with the sign alternating from place value to place value.

Example 13.2.3.1. *Convert 77_{10} to base -7.*

Solution 13.2.3.1. *As with any ordinary positive integral base $b \geq 2$, the place values are no different, going $b^0 = 1, b^1 = b, b^2, b^3, \ldots$ right-to-left. The place values simply alternate in sign with a negative base, in this case $+1, -7, +49$ from right to left. (We need only go up to $+49$ in this example.) Since $49 < 77$, and the digits themselves are never negative, we will need to fill in the 7^2 place with a 2, leaving us to subtract 21 rather than add it. This is straightforward, however: simply fill in the -7 place with a 3, and the units place with a 0. This yields $77_{10} = \boxed{230_{-7}}$.*

At first, the notion of negative bases seems to defy our (correct) intuition up to this point - that each number should have exactly one representation in any given base. But this still holds true in any base, since if we go under the original integer, we are left with a negative place value that cannot possibly compensate, since digits cannot be negative. In the example above, if we let the $(-7)^2$ digit be 1 instead of 2, we could not cover the $+28$ difference with -7 and 1 place values. On the other hand, letting the $(-7)^2$ digit be a 3 would yield a -72 difference, which is too much for the -7 place value to accommodate for.

Even negative fractional bases still work out just fine under this method - in fact, any number that is easy to manipulate via exponentiation (and isn't a special case like ± 1 or 0) is a viable candidate for a base number system! This even works when the number to be converted is a non-integer, as we can see from the following example:

Example 13.2.3.2. *Convert $\dfrac{161}{144}$ from base 10 to base $-\dfrac{5}{4}$.*

Solution 13.2.3.2. *Observe that $\dfrac{161}{144} = \left(\dfrac{5}{4}\right)^2 - \dfrac{4}{9}$, and that $\dfrac{4}{9} = \dfrac{1}{1 - \left(-\dfrac{5}{4}\right)}$ by the infinite geometric series formula. Thus, $\left(\dfrac{161}{144}\right)_{10} = \boxed{100.\overline{1}_{-\frac{5}{4}}}$.*

13.2.4 Irrational Bases

Irrational numbers, by definition, cannot be represented as a quotient of two integers in simplest form. So how could we possibly work with them in terms of place values? Provided they are of the form $\sqrt[a]{r}$ for some positive integer a and rational number r, then it is not too difficult to raise this quantity to a certain exponent in order to turn it into a rational number, as in the following example:

Example 13.2.4.1. *What is the base $\sqrt{10}$ representation of 1110_{10}?*

Solution 13.2.4.1. *As always, we assign real numbers to each place value in base $\sqrt{10}$: $1, \sqrt{10}, 10, 10\sqrt{10}, 100, \ldots$ from right to left. $1110_{10} = (1000 + 100 + 10)_{10} = \left((\sqrt{10})^6 + (\sqrt{10})^4 + (\sqrt{10})^2\right)_{10} = \boxed{1010100_{\sqrt{10}}}$.*

However, certain irrational numbers (in particular, transcendental numbers, such as π or e) are vastly more complicated. Of course, it is still entirely possible to convert select base-10 quantities (such as, say, $e^2 + 3e + 4$ or $\pi + \dfrac{1}{\pi^4}$) to these bases. In particular, ϕ, the golden ratio, works especially well for this purpose:

Example 13.2.4.2. Convert 4_{10} to base $\phi = \dfrac{1+\sqrt{5}}{2}$.

Solution 13.2.4.2. Observe that ϕ is the positive solution to the quadratic equation $x^2 = x + 1$, so $\phi^2 = \dfrac{3+\sqrt{5}}{2}$ and $\dfrac{1}{\phi^2} = \dfrac{3-\sqrt{5}}{2}$. Thus, $4 = \phi^2 + 1 + \phi^{-2}$, and so $4_{10} = \boxed{101.01_\phi}$.

But for the vast majority of base-10 real numbers, such bases present a problem. Unless we resort to advanced methods such as the power series representation of e, we may not be able to represent rational quantities (or even almost all irrational quantities) in these exquisite irrational bases (and even those methods aren't guaranteed to work).

13.2.5 Complex-Base Systems

So far, we've been able to deal with any base that is a real number - with few exceptions. But could we possibly extend this any further - to, say, the complex numbers? Naturally, the answer is **yes**; but, yet again, with limitations. In a base such as base $i = \sqrt{-1}$, we cannot uniquely represent a complex quantity, since the powers of i cycle with length 4 (and so $i_{10} = 1_i$ becomes indistinguishable from 10000_i). However, it is still possible to perform uniquely-determined base conversions with some complex numbers:

Example 13.2.5.1. What is the base $2i$ representation of 11.75_{10}?

Solution 13.2.5.1. Our place values are, from right to left, $-0.25, 0.5i, 1, 2i, -4, -8i, 16$. From here, we can observe that $12_{10} = 16 - 4 = \boxed{10100.01_{2i}}$.

Of particular interest are base $2i$ and bases $-1 \pm i$, the former which is referred to as the *quater-imaginary* complex base. Note that complex bases, by convention, utilize the digits of base $|b|$, where b is the base and $|b|$ denotes the magnitude of b; i.e. $|b| = \sqrt{x^2 + y^2}$, where $b = x + iy$.

In base $2i$, the quater-imaginary base (originally proposed by Donald Knuth[51] in 1960) uniquely represents (almost) every complex number using the digits 0, 1, 2, and 3 (as $(2i)^2 = -4$). We can decompose a quater-imaginary number as follows:

$$\ldots + d_3 \cdot (2i)^3 + d_2 \cdot (2i)^2 + d_1 \cdot (2i) + d_0 + d_{-1} \cdot \frac{1}{2i} + d_{-2} \cdot \frac{1}{(2i)^2} + d_{-3} \cdot \frac{1}{(2i)^3} + \ldots$$

$$= d_3 \cdot (-8i) + d_2 \cdot (-4) + d_1 \cdot (2i) + d_0 + d_{-1} \cdot \left(-\frac{1}{2}i\right) + d_{-2} \cdot \left(-\frac{1}{4}\right) + d_{-3} \cdot \left(\frac{1}{8}i\right) + \ldots$$

$$= \left(\ldots + d_2(-4) + d_0 + d_{-2}\left(-\frac{1}{4}\right) + \ldots\right) + 2i\left(\ldots + d_1 \cdot (-4)^0 + d_{-1} \cdot (-4)^{-1} + \ldots\right)$$

Furthermore, note that $(-1 \pm i)^2 = \pm 2i$, so ultimately, these particular complex numbers produce the same results as the base $2i$, except with every place value horizontally "dilated" by a factor of 2. It is also of note that these bases, in particular, produce wonderfully fascinating *twindragon* fractals in the complex plane (the Argand diagram).[52]

Determining which complex numbers do, and do not, produce unique representations is left as a (challenging) exercise to the reader. (It may vastly help to consider the magnitude of the complex base b.)

13.2.6 Proving that Each Number has a Unique Representation

The title of this section may seem very misleading coming fresh off the statement from the very end of the last subsection - that certain complex numbers do, and do not, produce unique representations in complex bases. Lay those fears to rest - the unique representation theorem for bases *applies solely to positive integer bases*, as we have already seen that in non-integer bases, uniqueness is not guaranteed.

---

[51]Of fame for proposing *Knuth's up-arrow notation,* used to denote Graham's number.

[52]Unfortunately, we cannot depict these here due to computing limitations. For those interested, you may want to research the *Lindenmayer system.*

> **Theorem 13.4: Unique Base-b Representation Theorem**
>
> For each positive integer $b \geq 2$, every base-10 real number N_{10} has a unique representation in base b.

We can prove this by induction.

Proof. Let $N = d_0 + b \cdot d_1 + b^2 \cdot d_2 + \ldots + b^n \cdot d_n$ for some $n \in \mathbb{N}$, $0 \leq d_i < b$ for $0 \leq i \leq n$. We claim that the ordered $(n+1)$-tuple $(d_0, d_1, d_2, \ldots, d_n)$ is unique. Write $N = d_0 + qb$ for some $q \in \mathbb{N}$, and notice that d_0 is uniquely determined as the remainder when N is divided by b. Then consider $q = d_1 + b \cdot d_2 + b^2 \cdot d_3 + \ldots + b^{n-1} \cdot d_n$ and use induction to finish the proof. (The details are left as an exercise.) □

As an exercise, try to figure out why this would not necessarily apply to non-integer bases (and in particular, why irrational and complex bases are especially vulnerable).

13.2.7 Examples

Example 13.2.7.1. *Convert $\left(\dfrac{17}{6}\right)_{10}$ to base 2 (binary).*

Solution 13.2.7.1. *We have that $\dfrac{17}{6} = 2 + \dfrac{5}{6}$, so we clearly have 10 as the digits before the decimal point in binary. We now need to represent $\dfrac{5}{6}$ as a sum of reciprocals of powers of 2. Since $2^{-1} + 2^{-2} + 2^{-3} + \ldots = 1$, we need to find a sum of reciprocals of powers of 2 that sum to $\dfrac{1}{6}$, then subtract those off (i.e. turn their respective digits into 0's). Observe that $\dfrac{1}{6} = \dfrac{\frac{1}{8}}{1 - \left(\frac{1}{2}\right)^2}$, so we can begin with 2^{-3} and turn all successive 2^{-k} (where k is odd) into 0's. Thus, the resulting binary representation would be $\boxed{10.1\overline{10}_2}$.*

Example 13.2.7.2. *The base-b number 2019_b has 5 digits in base 10. What is the smallest possible positive integer value of b? Explain.*

Solution 13.2.7.2. *Converting 2019_b to base 10, we obtain $2019_b = (2b^3 + b + 9)_{10}$. In order for this to have 5 digits, we require $2b^3 + b + 9 \geq 10^4$, or $b(2b^2 + 1) \geq 10^4 - 9$. We can make reasonable estimates, beginning with $b = 20$. This is excessive, so we reduce our estimate, then maybe raise it a little bit until we reach our desired answer of $b = \boxed{18}$.*

Example 13.2.7.3. *How many of the first 2019 binary positive integers have a digit of 1 in the 2^5 place?*

Solution 13.2.7.3. *The 2^5 place will have a 1 whenever the base-10 value of the integer leaves a remainder of between 2^5 and $2^6 - 1$ when divided by 2^6 (i.e. 32 numbers for every "block" of 64). This covers half of the numbers up to $64 \cdot 31 = 1983$, or 992 of them. The base-10 integers from 1984 to 2015, inclusive, do not have a 1 in the 2^5 place, but all of them from 2016 to 2019 do, thereby adding 4 to our count - the final count being $\boxed{996}$ integers.*

Example 13.2.7.4. *(1967 AHSME) Let the product $(12)(15)(16)$, each factor written in base b, equals 3146 in base b. Let $s = 12 + 15 + 16$, each term expressed in base b. Then s, in base b, is*

Solution 13.2.7.4. *From the first part of the problem, we have $(b+2)(b+5)(b+6) = b^3 + b^2 + 4b + 6$ (first converting everything to base 10). Expanding and then canceling like terms yields $6b^2 + 24b + 27 = b^3$, which has solution $b = 9$ by inspection. Then $s_9 = 12_9 + 15_9 + 16_9 = 3 \cdot 9 + (2 + 5 + 6) = 40_{10} = \boxed{44_9}$.*

13.2.8 Exercises

Exercise 135. *Convert each of the following to base 10:*

(a) 99_{11}

(b) 101011101_2

(c) 2019_{16}

(d) 345_{-6}

(e) $20321_{\sqrt{10}}$

(f) 10010.01001_i

Exercise 136. *In base b, the positive integer N has d digits. Describe the interval of possible values for N in base 10.*

Exercise 137. *(2003 AMC 10A) A base-10 three digit number n is selected at random. What is the probability that the base-9 representation and the base-11 representation of n are both three-digit numerals?*

Exercise 138. *(2015 AMC 10A) Hexadecimal (base-16) numbers are written using numeric digits 0 through 9 as well as the letters A through F to represent 10 through 15. Among the first 1000 positive integers, there are n whose hexadecimal representation contains only numeric digits. What is the sum of the digits of n?*

Exercise 139. *(2013 AMC 10A) In base 10, the number 2013 ends in the digit 3. In base 9, on the other hand, the same number is written as $(2676)_9$ and ends in the digit 6. For how many positive integers b does the base-b-representation of 2013 end in the digit 3?*

Exercise 140. *Explain how we can determine a unique base-b representation for any $0 < b < 1$, $b \in \mathbb{R}$.*

Exercise 141. *In base $-2i$, what is the representation of 63_{10}? (Hint: First consider the representation of -63_{10}.)*

Exercise 142. *Compute the number of base-10 positive integers that have the same number of digits in base 3 and base 5.*

Exercise 143. *What is the smallest positive integer b such that 2019 has 2 or fewer digits in base b?*

Exercise 144. *Describe a formula for the number of trailing zeros of a positive integer N in base b, where $b \in \mathbb{N}$.*

Exercise 145. *(2015 ARML Team Round #3) A positive integer has the Kelly Property if it contains a zero in its base-17 representation. Compute the number of positive integers less than 1000 (base 10) that have the Kelly Property.*

Exercise 146. *Find the base-10 three-digit number whose base-4 representation is the reverse of its base-5 representation.*

Extension Exercise 64. *Call a base-10 positive integer (m,n)-palindromic if it is a palindrome in both base 4 and base 5. What is the set of $(4,5)$-palindromic base-10 positive integers?*

Extension Exercise 65. *(2018 AIME I) The number n can be written in base 14 as $\underline{a}\,\underline{b}\,\underline{c}$, can be written in base 15 as $\underline{a}\,\underline{c}\,\underline{b}$, and can be written in base 6 as $\underline{a}\,\underline{c}\,\underline{a}\,\underline{c}$, where $a > 0$. Find the base-10 representation of n.*

Extension Exercise 66. *(2017 AIME I) A rational number written in base eight is $\underline{ab}.\underline{cd}$, where all digits are nonzero. The same number in base twelve is $\underline{bb}.\underline{ba}$. Find the base-ten number $\underline{abc}$.*

Extension Exercise 67. *(2016 HMMT November Guts Round) On the blackboard, Amy writes 2017 in base a to get 133201_a. Betsy notices she can erase a digit from Amy's number and change the base to base b such that the value of the number remains the same. Catherine then notices she can erase a digit from Betsy's number and change the base to base c such that the value still remains the same. Compute, in decimal, $a + b + c$.*

Extension Exercise 68. *(2013 USAMO P5) Given postive integers m and n, prove that there is a positive integer c such that the numbers cm and cn have the same number of occurrences of each non-zero digit when written in base ten.*

13.3 Modular Arithmetic: Part 1

Divisibility is nice and all, but what happens when one number *isn't* evenly divisible by another? If we want to take the quotient of these numbers, we will necessarily end up with a remainder. This remainder is also known as the *modulus* with respect to the divisor, hence the name *modular arithmetic* of the study of remainders and divisors.

In modular arithmetic, we define one crucial operator around which we base the rest of our analysis: the *modulo*, or *mod*, operator.

> **Definition 13.4: Modulo Operator**
>
> Let non-zero real numbers a, b, and c satisfy the equation $\frac{a}{b} = q \text{ R } c$ (where the R denotes remainder). Then we say that $a \pmod{b} = c$, or $a \equiv c \pmod{b}$ (read "a modulo b is equal to c," and "a is equivalent to c modulo b," respectively). Note that $0 \leq a \pmod{b} < b$.

Example 13.3.0.1. *What are* $20 \pmod{19}$, $19 \pmod{20}$, $45 \pmod{6}$, *and* $2019.45 \pmod{45}$?

Solution 13.3.0.1. *We have* $20 \pmod{19} = 1$, *since the remainder upon dividing 20 by 19 is equal to 1. Since 19 leaves a remainder of 19 upon being divided by 20,* $19 \pmod{20} = 19$. *45 leaves a remainder of 3 when divided by 6, so* $45 \pmod{6} = 3$. *Finally, 2019.45 leaves a remainder of 39.45 when divided by 45 (with quotient 44), so* $2019.45 \equiv 39.45 \pmod{45}$.

We can also equate certain modular equations with each other, as in the following subsection.

13.3.1 Modular Congruences

We can derive *modular equivalences*, or *modular congruences*, from the equality of two remainders upon being divided by the same divisor. The most common - and apt - analogy for modular equivalence is a 12-hour clock, which operates in modulo 12. Every 12 hours, the time returns to its time 12 hours ago, so 1:00 remains 1:00 (even though 12 hours have passed), and 3:00 becomes 5:00 after 14 hours. In this way, we can say that 12:00 is "equivalent" to 0:00.

> **Definition 13.5: Modular Congruence**
>
> In general, we may write $a = kc + b$ for some $k \in \mathbb{Z}$ if $a \equiv b \pmod{c}$. We may also write a and b individually as $a = pn + r$ and $b = qn + r$, where r is the remainder when both a and b are divided by c, and $p, q \in \mathbb{Z}$. Subtracting the two, we get $a - b = n(p - q)$.

Example 13.3.1.1. *Evaluate* $20192019 \pmod{2018}$.

Solution 13.3.1.1. *Note that 20182018 is a multiple of 2018, so* $20192019 \equiv 20192019 - 20182018 = 10001 \pmod{2018} = 10001 - 5(2018) \pmod{2018} = -89 \pmod{2018} = -89 + 2018 \pmod{2019} = \boxed{1929} \pmod{2018}$.

An important property of the modulo relation/operator is that it is an equivalence relation; that is, it has the properties of reflexivity, symmetry, and transitivity.

Proof.

- Reflexivity: For all a, we have $a \equiv a \pmod{c}$. This is trivial (since $a - a = 0 \equiv 0 \pmod{c}$).

- Symmetry: We have $a \equiv b \pmod{c}$ if and only if $b \equiv a \pmod{c}$. This, too, is trivial.

- Transitivity: For all a, b, c, d, if we have $a \equiv b \pmod{d}$ and $b \equiv c \pmod{d}$, then we also have $a \equiv c \pmod{d}$. Let $a = kd + b$, $b = ld + c$, so that $a = d(k + l) + c \equiv c \pmod{d}$, as desired.

This shows that the modulo operator is an equivalence relation. □

Furthermore, if we let $a \equiv b \pmod{c}$, then this equivalence has quite a few levels of compatibility:

> **Theorem 13.5: Compatibilities of Modular Congruence**
>
> If $a \equiv b \pmod{c}$, then it follows that
>
> - $a + k \equiv b + k \pmod{c}$.
> - $ka \equiv kb \pmod{c}$.
> - $a^k \equiv b^k \pmod{c}$. It is not, however, true that $k^a \equiv k^b \pmod{c}$ in general!
> - $p(a) \equiv p(b) \pmod{c}$, where $p(x)$ is a polynomial in terms of x with integer coefficients.
>
> In addition, if we have $a_1 \equiv b_1 \pmod{c}$ and $a_2 \equiv b_2 \pmod{c}$, then
>
> - $a_1 \pm a_2 \equiv b_1 \pm b_2 \pmod{c}$ (with the respective signs the same).
> - $a_1 a_2 \equiv b_1 b_2 \pmod{c}$.
>
> for any $k \in \mathbb{Z}$.

Proof. For the following, let $a = pc + r$ and $b = qc + r$, where $p, q \in \mathbb{Z}$ and $0 \leq r < c$.

- We have $a + k = pc + r + k$ and $b + k = qc + r + k$, and $pc \equiv qc \pmod{c}$ (which we have assumed with proof). The proof follows.

- We have $ka = pck + rk$ and $kb = qck + rk$, so then $pck \equiv qck \pmod{c}$ (using the fact that $a + k \equiv b + k \pmod{c}$), which is true since both are multiples of c.

- This implies $a^k - b^k \equiv 0 \pmod{c}$, and $(a - b) \mid (a^k - b^k)$ for all $k \in \mathbb{Z}$, so $(a - b) \mid c$ as desired.

- Define
$$p(x) = c_n x^n + c_{n-1} x^{n-1} + \ldots + c_2 x^2 + c_1 x + c_0$$
Then $p(a) = c_n a^n + c_{n-1} a^{n-1} + \ldots + c_2 a^2 + c_1 a + c_0$. If we substitute $x = b$, then we obtain $p(b) = c_n b^n + c_{n-1} b^{n-1} + \ldots + c_2 b^2 + c_1 b + c_0$. It follows that $p(b) - p(a) = c_n(a^n - b^n) + c_{n-1}(a^{n-1} - b^{n-1}) + \ldots + c_2(a^2 - b^2) + c_1(a - b)$. Since $(a - b) \mid c$, and $(a - b)$ is a factor of each term of $p(b) - p(a)$, it follows that $p(a) - p(b) \mid c$, and thus that $p(a) \equiv p(b) \pmod{c}$.

- Let $a_1 = pc + b_1$ and $a_2 = qc + b_2$, with $p, q \in \mathbb{Z}$. Then $a_1 + a_2 = c(p + q) + (b_1 + b_2) \equiv b_1 + b_2 \pmod{c}$. (The proof for subtraction is identical.)

- Continuing from the above, we have $a_1 a_2 = c^2 \cdot pq + pcb_2 + qcb_1 + b_1 b_2 = c(c \cdot pq + pb_2 + qb_1) + b_1 b_2 \equiv b_1 b_2 \pmod{c}$.

These are the basic properties of the mod operator that will be used to prove other properties in later chapters. □

There are several additional properties of the modulo operator, but these are the ones appropriate for this section and most commonly used in a competition setting. As for real-world applications, modular arithmetic relies primarily on these basic tenets, since they ultimately provide a defining structure. More advanced properties of mods, such as Fermat's Little Theorem, Wilson's Theorem, or quadratic residues, will be covered in future sections. For now, we will take a look at special cases where we are taking the modulo operator with respect to an odd number.

13.3.2 Symmetry of the Mod Operator

Let's examine a specific case: when the modulus is an odd number. For example, let's say that the modulus is 7. In the modular equation $a^2 \equiv b \pmod{7}$, we can run through the squares of each possible value of a that would produce a distinct value for b ($0 \leq a \leq 6$):

a	$a^2 \pmod{7}$
0	0
1	1
2	4
3	2
4	2
5	4
6	1

Figure 13.1: A table depicting symmetry of squares $\pmod{7}$.

Notice that the values of $a^2 \pmod{7}$ are symmetric about $a = \frac{7}{2}$, with $3^2 \equiv 4^2 \pmod{7}$. (This is because $4^2 - 3^2$ is a multiple of 7; indeed, with any odd number p, $\left(\frac{p-1}{2}\right)^2 \equiv \left(\frac{p+1}{2}\right)^2 \pmod{p}$, not just prime p.) This is a key observation, because it carries so much potential for casting great insight into the symmetry of number theory as a whole! Indeed, as we shall see in studying modular inverses, symmetry manifests in a wide variety of places, many unexpected.

But can we extend this symmetry to higher powers? In particular, let's take a look at even-numbered powers first. If we want to examine the fourth powers, all we have to do is square the second powers (in Figure 13.1 above) to obtain a new table, as follows:

a	$a^4 \pmod{7}$
0	0
1	1
2	2
3	4
4	4
5	2
6	1

Figure 13.2: A revised table, depicting symmetry of fourth powers $\pmod{7}$.

So it seems like the symmetry is still there. And, indeed, it remains with any even power (proving this is left as an exercise). But what about odd powers? If we substitute a with $p - a$ and then raise it to the same power, the Binomial Theorem will still guarantee that the result is the same with respect to a given modulus p. Thus, nothing has changed - the symmetry is a permanent part of raising quantities in modular equations to powers! (Taking the cubes of each integer from 0 to 6, inclusive, and then taking the remainder upon division by 7, reveals that the symmetry is still there.)

But the question still remains: is this symmetry only a feature of odd moduli? What if we try an even modulus, like mod 6? If we do this, we produce the following table:

a	a^2 (mod 6)
0	0
1	1
2	4
3	3
4	4
5	1

Figure 13.3: Yet another table, depicting symmetry of second powers (mod 6).

Sure looks like it! We can rigorously prove this, in fact:

Proof. We claim that, for *any* positive integers p and $n \geq 2$, we have $a^n \equiv (p-a)^n \pmod{p}$.

Expanding $(p-a)^n$ by the Binomial Theorem yields

$$\binom{n}{0}p^n - \binom{n}{1}p^{n-1}a + \binom{n}{2}p^{n-2}a^2 - \ldots - \binom{n}{n-1}pa^{n-1} + \binom{n}{n}a^n$$

and subtracting a^n yields an expression which contains p in all of its terms. Hence, $p \mid (a^n - (p-a)^n) \implies a^n - (p-a)^n \equiv 0 \pmod{p}$, and so $a^n \equiv (p-a)^n \pmod{p}$, as desired. □

So yes, the symmetry is still there! It is constant for all moduli and all powers - it is an intrinsic part of the modulo operation. And with that, we venture into territory that may not be as neatly symmetric and orderly, but it nonetheless just as important.

13.3.3 Modular Inverses

We have well established how to perform the arithmetic operations of addition, subtraction, and multiplication under modulo p. But how do we *divide*? The answer remains fundamentally the same as in our standard number system - we multiply by the inverse.

> **Definition 13.6: Modular Inverse**
>
> In modulo p, the inverse of a, denoted a^{-1}, is the smallest positive integer b such that $a^{-1} \cdot b \equiv 1 \pmod{p}$.

Here are a few examples to illustrate the concept:

Example 13.3.3.1. *Compute 5^{-1} (mod 6), 19^{-1} (mod 21), and 4^{-1} (mod 2019).*

Solution 13.3.3.1. *In all computations, we want to find integers b_i such that $5b_1 \equiv 1 \pmod{6}$, $19b_2 \equiv 1 \pmod{21}$, and $4b_3 \equiv 1 \pmod{2019}$, respectively. That is, $5b_1 = 6k_1 + 1$, $19b_2 = 21k_2 + 1$, and $4b_3 = 2019k_3 + 1$, respectively. Since $6 \equiv 1 \pmod{5}$, $5 \mid 6k_1 + 1$ when $k_1 \equiv 4 \pmod 5$, or just $k_1 = 4$. Thus, $5b_1 = 25$, or $b_1 = \boxed{5}$. Similarly, we obtain $b_2 = \boxed{10}$ and $b_3 = \boxed{505}$ (taking the difference between 19 and 21 as an aid in the second example).*

By extension, we can compute quotients directly:

Example 13.3.3.2. *What is the value of $\dfrac{4}{5}$ (mod 9)?*

Solution 13.3.3.2. *This is equal to $4 \cdot 5^{-1}$ (mod 9) $= 4 \cdot 2 = \boxed{8}$ (mod 9). (We can also obtain 8 by multiplying by the reciprocal of $\dfrac{4}{5}$, $\dfrac{5}{4}$, to obtain 10, which is equivalent to 1 (mod 9).)*

We make a key observation here regarding the existence of an inverse modulo p:

Theorem 13.6: Existence of a Modular Inverse

With respect to $(\bmod\ p)$, the inverse of a, a^{-1} (for $0 \leq a < p$) exists if and only if $\gcd(a,p) = 1$ (i.e. a and p are relatively prime).

Proof. Assume the contrary; that there exists an inverse of an a such that $\gcd(a,p) \neq 1$. Then there would exist a b such that $ab \equiv 1 \pmod{p}$, or $ab = 1 + kp$ for some $k \in \mathbb{Z}$. Let $\gcd(a,p) = c$; then $a = 0, c, 2c, 3c, \ldots, pc, (p+1)c, \ldots$. At each value of a, the product ab is always a multiple of c, which 1 is not. Therefore, no such inverse exists. $\square$

The actual process of determining modular inverses can be carried out via inspection for small values of p, but as p grows larger, it becomes considerably more difficult to run through the list of potential values. Instead, we can use the Euclidean Algorithm to single out the smallest value whose product with a produces $1 \pmod{p}$:

Theorem 13.7: Existence from the Euclidean Algorithm

Suppose we wish to compute $a^{-1} \pmod{p}$, where $0 \leq a < p$ and $\gcd(a,p) = 1$. From the Euclidean Algorithm, we know that there exist integers x and y such that $ax + py = 1$ (which actually serves as an alternate, briefer proof of Theorem 13.6).

Example 13.3.3.3. *Using the Euclidean algorithm, determine whether $35^{-1} \pmod{192}$ exists.*

Solution 13.3.3.3. *The answer is* $\boxed{\text{yes.}}$ *Since* $\gcd(35, 192) = 1$ *by the Euclidean Algorithm, an inverse exists. (The specifics: $192 - 5 \cdot 35 = 17$, and $35 - 2 \cdot 17 = 1$, from which it is clear that $\gcd(17, 1) = 1$ and so $\gcd(35, 192) = 1$ as well. (Indeed, it is not too hard to verify that $35^{-1} \pmod{192} = 11$ by inspection.)*

13.3.4 Examples

Example 13.3.4.1. *Compute the remainder when $12345\ldots19$ is divided by 11.*

Solution 13.3.4.1. *The divisibility rule for 11 can not only be used to determine whether or not a number is divisible by 11, but also to compute the exact remainder upon division by 11. (Proving this is left as an exercise.) There are $9 + 10 \cdot 2 = 29$ digits in the number in total. Of these 29 digits, $1, 3, 5, 7, 9, 0, 1, 2, 3, \ldots 9$ (or 15) digits are in the odd-numbered positions, while $2, 4, 6, 8, 1, 1, 1, \ldots, 1$ (with 14 digits, 10 of those being 1's) are in the even-numbered positions. Summing both individually and then taking their positive difference yields 40, which is equivalent to $\boxed{7} \pmod{11}$.*

Example 13.3.4.2. *For how many positive integers n with $1 \leq n < 2019$ does the modular inverse $n^{-1} \pmod{2019}$ exist?*

Solution 13.3.4.2. *We want to compute the number of positive integers n in the specified range such that n is relatively prime to 2019; i.e. $\gcd(n, 2019) = 1$. Since 2019 factors as $3 \cdot 673$, we exclude all multiples of 3 and all multiples of 673 (we need not worry about overlap, since 2019 is excluded). There are 672 multiples of 3 and 2 multiples of 673, so the number of valid n is $2018 - 674 = \boxed{1344}$. (After going through §13.6, one can instantly see that this is just $\phi(2019)$.)*

Example 13.3.4.3. *What is the remainder when $3^{3^{3^3}}$ is divided by 28?*

Solution 13.3.4.3. *Consider just the expression 3^n, and list the remainders when it is divided by 28 as n ranges through the first few positive integers: $3^1 = 3$, $3^2 = 9$, $3^3 = 27 = 281$, $3^4 = 81 = 28 \cdot 3 - 3$, $3^5 = 28 \cdot 9 - 9$, $3^6 = 28 \cdot 2727 = 28 \cdot 26 + 1$. At this point, we notice that the cycle of remainders when 3^n is divided by 28 repeats whenever n increases by 6. Thus, we have reduced the problem to computing the remainder when $3^{3^3} = 3^{27}$ is divided by 6. But this is the easy part: $6 = 3 \cdot 2$, and 3^{27} is odd, so it must leave a remainder of 3 upon being divided by 6. Hence, $3^{3^{3^3}} \equiv 3^3 = \boxed{27} \pmod{28}$.*

13.3.5 Exercises

Exercise 147. *(2016 Berkeley Math Tournament)* What is the sum of all positive integers less than 30 divisible by 2, 3, or 5?

Exercise 148. For how many positive integers $n \in [1, 100]$ does 7 divide $11^n \pm 1$?

Exercise 149. Compute the remainder when
$$\sum_{k=1}^{2019} k!$$
is divided by 7.

Exercise 150. Compute the smallest positive integer x such that x^2 leaves a remainder of one when divided by seven.

Exercise 151. *(1964 IMO P1)*

(a) Find all positive integers n for which $2^n - 1$ is divisible by 7.

(b) Prove that there is no positive integer n for which $2^n + 1$ is divisible by 7.

Exercise 152.

(a) What is the units digit of 9^{19}?

(b) What is the units digit of 7^{2019}?

(c) What is the **tens** digit of 7^{2019}?

(d) What is the units digit of $1^2 + 2^3 + 3^4 + \ldots + 9^{10}$?

Exercise 153. *(2000 AMC 12)* Mrs. Walter gave an exam in a mathematics class of five students. She entered the scores in random order into a spreadsheet, which recalculated the class average after each score was entered. Mrs. Walter noticed that after each score was entered, the average was always an integer. The scores (listed in ascending order) were 71, 76, 80, 82, and 91. What was the last score Mrs. Walters entered?

Exercise 154. *(2016 Berkeley Math Tournament)* Let $g_0 = 1$, $g_1 = 2$, $g_2 = 3$, and $g_n = g_{n-1} + 2g_{n-2} + 3g_{n-3}$. For how many $0 \leq i \leq 100$ is it that g_i is divisible by 5?

Exercise 155. For a positive integer n, let s_n be the sum of the first n positive integers, and let S_n be the sum of the first n perfect squares. Find the remainder when
$$\sum_{k=1}^{2018} S_k - s_k$$
is divided by 1000.

Exercise 156. *(2011 AMC 10B)* What is the hundreds digit of 2011^{2011}?

Exercise 157. *(2016 Berkeley Math Tournament)* What are the last two digits of $9^{8^{7^2}}$?

Exercise 158. *(2017 AIME I)* For a positive integer n, let d_n be the units digit of $1 + 2 + \cdots + n$. Find the remainder when
$$\sum_{n=1}^{2017} d_n$$
is divided by 1000.

Exercise 159. *(2011 AIME I) Let R be the set of all possible remainders when a number of the form 2^n, n a non-negative integer, is divided by 1000. Let S be the sum of the elements in R. Find the remainder when S is divided by 1000.*

Exercise 160. *For what positive integral values of n is $2^n + 3^n + 6^n$ a multiple of 7?*

Extension Exercise 69. *What is the rightmost nonzero digit of 20^{18} in base 18?*

Extension Exercise 70. *(2018 AIME I) Find the least positive integer n such that when 3^n is written in base 143, its two right-most digits in base 143 are 01.*

Extension Exercise 71. *What is the smallest positive integer n such that $2 \cdot 5^n$ is 1 less than a multiple of 49?*

Extension Exercise 72. *What is the remainder when $1 \cdot 10^1 + 2 \cdot 10^2 + 3 \cdot 10^3 + \ldots + 2019 \cdot 10^{2019}$ is divided by 11?*

Extension Exercise 73. *(1989 AIME) One of Euler's conjectures was disproved in the 1960s by three American mathematicians when they showed there was a positive integer n such that $133^5 + 110^5 + 84^5 + 27^5 = n^5$. Find the value of n.*

13.4 Studying Prime Numbers

So far, we have studied divisibility in a comfortable amount of detail. But what about prime numbers - those positive integers which are not multiples of any integer other than 1 and themselves - the loners of the number system? Turns out, they are far from loners, for they have a slew of intrinsically interesting properties that bind them together and make them just as important to study, if not more so, than composite numbers.

13.4.1 The Sieve of Eratosthenes

In order to study prime numbers, we must first know what they are. Naturally, doing so is a crucial part of studying them to begin with, and in fact, plays a major role in shaping the field of number theory to this day. Perhaps the most logical starting point would be the *Sieve of Eratosthenes*, an algorithm which can be used to compute all primes p up to a certain positive integer n. (In the table below, we have used $n = 100$, the most traditional example).

	2	3	4	5	6	7	8	9	~~10~~
11	~~12~~	13	~~14~~	~~15~~	~~16~~	17	~~18~~	19	~~20~~
~~21~~	~~22~~	23	~~24~~	~~25~~	~~26~~	~~27~~	~~28~~	29	~~30~~
31	~~32~~	~~33~~	~~34~~	~~35~~	~~36~~	37	~~38~~	~~39~~	~~40~~
41	~~42~~	43	~~44~~	~~45~~	~~46~~	47	~~48~~	~~49~~	~~50~~
~~51~~	~~52~~	53	~~54~~	~~55~~	~~56~~	~~57~~	~~58~~	59	~~60~~
61	~~62~~	~~63~~	~~64~~	~~65~~	~~66~~	67	~~68~~	~~69~~	~~70~~
71	~~72~~	73	~~74~~	~~75~~	~~76~~	~~77~~	~~78~~	79	~~80~~
~~81~~	~~82~~	83	~~84~~	~~85~~	~~86~~	~~87~~	~~88~~	89	~~90~~
~~91~~	~~92~~	93	~~94~~	~~95~~	~~96~~	97	~~98~~	~~99~~	~~100~~

Figure 13.4: Using the Sieve of Eratosthenes to determine the primes up to 100.

The algorithm for the Sieve of Eratosthenes proceeds as follows:

1. Begin by setting $p = 2$.

2. Then, strike out all multiples of p (not including p itself) up to n.

3. Increment p up to the next prime number that has not been stricken out. If no such number exists, stop. Otherwise, repeat step 2 until the algorithm terminates.

4. When the algorithm terminates, the numbers that remain are all the primes less than or equal to n.

In this way, we can demonstrate that we need only check divisibility for primes up to $\sqrt{n}$, since the algorithm will not reach multiples of primes that have not already been stricken out (with $11 \cdot 13$, for instance, the product is $143 > 100$). Note that some numbers have been crossed out more than once (such as 30 and 42). (As an exercise: can you convert this algorithm to pseudocode?)

13.4.2 Divisibility and Prime Number Equations

We now take a look at the solvability of equations involving prime numbers in terms of divisibility. Since primes are so intertwined with multiples, factors, and divisibility, a natural place to start our discussion would be with algebraic identities that involve factors. For instance, consider the unassuming difference of squares identity:

$$a^2 - b^2 = (a + b)(a - b)$$

for all real a, b. The decomposition into factors $(a + b)(a - b)$ is naturally conductive to studying prime numbers. Let us suppose that a and b are both odd prime numbers. We can segue directly from the observation that a and b are both odd to the implication that $a + b$ and $a - b$ are both even, and so the product $(a + b)(a - b) = a^2 - b^2$ must be divisible by $2^2 = 4$. This is a *prime* example (see what we did there?) of how we can use primes to determine divisibility in seemingly unrelated contexts.

Next, let's take a step toward simplicity and examine the parity associated with prime numbers. Though it may seem that parity is a trivial idea, it is also a very significant one, as we can use it to prove crucial facts about primes. For instance, consider the following:

Example 13.4.2.1. *Show that no consecutive four odd numbers can all be prime.*

Solution 13.4.2.1. *Let these numbers be a, $a+2$, $a+4$, $a+6$. If a is prime, then it is not a multiple of 3; this leads us to consider the numbers modulo 3. Either $a \equiv 1$ (mod 3) or $a \equiv 2$ (mod 3). In the former case, we have $a + 2 \equiv 0$ (mod 3), so it cannot be prime. Similarly, in the latter case, it follows that $a + 4 \equiv 0$ (mod 3), so it, too, is not prime. Thus, we have shown that no four consecutive positive odd prime integers exist.*

Let's work through another example, this one related to parity in a less direct manner:

Example 13.4.2.2. *Prove that the product of any two consecutive primes has exactly four positive divisors.*

Solution 13.4.2.2. *Since both numbers are primes, their product will consist of two prime factors. Thus, it will have $(1+1)(1+1) = 4$ factors, as desired.*

This is also an opportune time to introduce *twin primes:* primes that are consecutive odd numbers.

> **Definition 13.7: Twin Primes**
>
> An ordered pair (m, n) with m, n positive integers is an ordered pair of *twin primes* if we can write $m = p_i$, $n = p_{i+1}$, where p_n denotes the nth prime number (with $p_1 = 2$, $p_2 = 3$, ...).

Example 13.4.2.3. *Describe the product of twin primes (mod 4). Show that no two twin primes have a product divisible by 4.*

Solution 13.4.2.3. *All prime numbers greater than 2 are odd, so they must be equivalent to either 1 (mod 4) or 3 (mod 4). In an ordered pair of twin primes, one prime is 1 (mod 4) and the other is 3 (mod 4). When the twin primes are 2 (mod 4) apart, their product is 3 (mod 4); when the twin primes are a multiple of 4 apart, their product is 1 (mod 4) (either way, since $1^2 \equiv 1$ (mod 4), and $3^2 \equiv 1$ (mod 4) as well). $(2,3)$ is an edge case, since 2 is the only even prime; then $2 \cdot 3 = 6 \equiv 2$ (mod 4). This covers all cases; hence, there is no pair of twin primes whose product is divisible by 4.*

Since antiquity, we have known that there are an infinite number of primes. However, the *Twin Prime Conjecture*, which posits that there are an infinite number of *twin* primes, is an open problem.

13.4.3 Fermat and Mersenne Primes

A *Fermat number*, named after Pierre de Fermat (of Fermat's Last Theorem fame, and Fermat's Little Thorem somewhat-fame, too), has several interesting properties that we should analyze before we begin to discuss the real meat of the section: *Fermat primes*.

> **Definition 13.8: Fermat Numbers**
>
> A *Fermat number* is a positive integer of the form $F_n = 2^{2^n} + 1$, where n is a non-negative integer.

A *Fermat prime* is simply any Fermat number that happens to be prime. But what is a Fermat number, *really?* We can derive some basic properties of the Fermat numbers, first by inspection, and then prove them via induction. The sequence $(F_n)_{n=1}^{\infty}$ of the Fermat numbers begins $F_0 = 3$, $F_1 = 5$, $F_2 = 17$, $F_3 = 257$, $F_4 = 65,537$, ... By looking at the first few terms and searching for a pattern (which is, admittedly, somewhat arduous), we find that $F_n = \left(\prod_{k=1}^{n-1} F_k\right) + 2$ for all $n \geq 1$. Proving this can be done by induction and

145

is left as an exercise to the reader.

In addition, relating to the recursive product of Fermat numbers, we have the formula

$$F_n = (F_{n-1} - 1)^2 + 1 \tag{13.1}$$

which can be proven by observing that $F_{n-1} = 2^{2^{n-1}} + 1$, and that squaring this minus 1 yields $(F_{n-1} - 1)^2 = 2^{2^n}$, such that $F_n = 2^{2^n} + 1$, as desired *a priori*.

Furthermore, we have the following recursive equations, which are more complicated to derive (but still fall to proofs by induction):

$$F_n = F_{n-1} + 2^{2^{n-1}} \cdot \prod_{k=0}^{n-2} F_k \tag{13.2}$$

$$F_n = F_{n-1}^2 - 2(F_{n-2} - 1)^2 \tag{13.3}$$

for all $n \geq 1$. From the fact that $F_n = \left(\prod_{k=0}^{n-1} F_k\right) + 2$ in particular, we obtain a direct corollary, known as *Goldbach's Theorem* (not to be confused with the unrelated open problem *Goldbach's Conjecture*):

> **Theorem 13.8: Goldbach's Theorem**
>
> All Fermat numbers are relatively prime to each other.

Proof. Let i and j be non-negative integers with $0 \leq i \leq j$, and assume to the contrary that $\gcd(F_i, F_j) = k > 1$. Then k divides F_j, as well as the product $F_0 F_1 F_2 \cdots F_{j-1}$. Thus, k also must divide their difference, which is 2. However, this forces $k = 2$, which leads to the desired contradiction, as all Fermat numbers are odd. □

One of the cornerstone proofs revolving around Fermat numbers is the proof that, if $2^n + 1$ is an odd prime, then n must be a power of 2 (such that the only odd primes of the form $2^n + 1$ are necessarily Fermat primes).

Proof. Let n be a positive integer that is not a power of 2. Then it will have an odd prime factor $p \geq 3$, such that $n = pq$, where $1 \leq q < n$. Using the identity

$$a^n - b^n = (a - b)\left(\sum_{k=0}^{n-1} a^k b^{n-k-1}\right)$$

we can substitute $a = 2^q$, $b = -1$, and $c = p$. Since p is odd, $2^q + 1 \mid 2^{cq} + 1$, or $2^q + 1 \mid 2^n + 1$. Because $q < n$, $2^n + 1$ cannot be prime, which forces n to be a power of 2. □

We can also use modular arithmetic to prove some significant results regarding the Fermat numbers. For instance, save for $F_0 = 3$ and $F_1 = 5$, the units digit of a Fermat number is always 7:

Proof. This follows immediately from the recursive formula $F_n = (F_{n-1} - 1)^2 + 1$, as well as from the definition of F_n; note that $2^{2^n} \equiv 6 \pmod{10}$ for all $n \geq 2$, as $2^n \equiv 0 \pmod 4$ for all $n \geq 2$). □

We now turn our attention to the Fermat primes. The sequence of Fermat primes is interesting in that hardly anything is known about it; as of 2019, the only known Fermat primes are $F_0 = 2^1 + 1 = 3$ through $F_4 = 2^{16} + 1$. Fermat himself had originally conjectured that all Fermat numbers were also prime; however, Euler had disproven this by showing that the prime factorization of $F_5 = 2^{32} + 1$ was given by $641 \cdot 6,700,417$. Euler's proof roughly proceeded as follows:

Proof of Euler's counterexample, F_5. We have that $641 = 2^7 \cdot 5 + 1 = 5^4 + 2^4$, so $2^7 \cdot 5 \equiv -1 \pmod{641}$ and $5^4 \equiv -2^4 \pmod{641}$. Raising the first equivalence to the fourth power yields $2^{28} \cdot 5^4 \equiv 1 \pmod{641}$, which in conjunction with the second equivalence, yields $2^{32} + 1 \equiv 0 \pmod{641}$. □

Many of the natural questions about Fermat numbers are open problems - is there a Fermat number (let alone prime) with no perfect square in its prime factorization? Is F_n composite for all $n \geq 5$? There may be infinitely many primes, but are there infinitely many *Fermat* primes? Furthermore, are there even infinitely many composite Fermat numbers? More advanced methods do exist that may aid in at least partially resolving these pressing questions, but for now, it is perhaps better to leave the mystery of the Fermat numbers intact. This way it will hopefully make for inspiration and marvel on the part of the reader!

Henceforth, let's discuss the nature of *Mersenne primes*, named for 1600s French scholar Marin Mersenne. This variant of prime number is quite similar to the Fermat prime, in that it is any prime that is expressible in the form $M_n = 2^n - 1$ for some positive integer n (where the numbers that are merely numbers of the form $2^n - 1$ are *Mersenne numbers*). Like with the Fermat primes, it is still unknown whether the Mersenne primes are finite or infinite, but there are considerably more connections that have been made with other disciplines of number theory (one in particular which will be the subject of our attention in this very section).

In studying Mersenne primes, we first turn to an important theorem regarding their primality:

> **Theorem 13.9: The Primality of Mersenne Numbers**
>
> A necessary (but not sufficient) condition for M_n to be prime is that n must be prime.

Proof. Suppose otherwise. Then we can write $n = ab$, where a and b are positive integers not equal to 1. We have

$$2^{ab} - 1 = (2^a - 1)\left(1 + 2^a + 2^{2a} + \ldots + 2^{(a(b-1))}\right)$$
$$= (2^b - 1)\left(1 + 2^b + 2^{2b} + \ldots + 2^{b(a-1)}\right)$$

which is a contradiction. $\square$

Note that a counterexample to the false statement that all M_n with n prime are themselves prime is relatively easy to find by inspection; namely $M_{11} = 2^{11} - 1 = 2047 = 89 \cdot 23$. In addition, we can observe that all Mersenne primes are congruent to 3 (mod 4) (with the first Mersenne prime being $M_2 = 3$, not $M_1 = 1$). This is an immediate consequence of the definition of the nth Mersenne number, M_n.

These are properties that can be gleaned in a relatively straightforward manner from the definition of the Mersenne numbers, but what is less apparent is the strength of the link between the Mersenne primes and the perfect numbers. As a refresher, a *perfect number* is any positive integer such that the sum of its positive proper divisors is equal to itself (where a *proper divisor* is any divisor of a number other than the number itself). As early as the 4th century BCE, Euclid had proven that, if $M_p = 2^p - 1$ is a Mersenne prime, then $(2^p - 1)2^{p-1} = \frac{1}{2}M_p(M_p + 1)$ is a perfect number. Written in this form, this is evidently the M_p^{th} triangular number. In the 1700s, Euler showed that all even perfect numbers assume this form (which, combined with Euclid's proof, comprises the Euclid-Euler Theorem). Neither Euclid's nor Euler's proof is not exceptionally inaccessible; indeed, we can list both proofs here without too strongly deviating from our intended scope:

Proving the Euclid-Euler Theorem. We will tackle Euclid's assertion first. One way to reframe the assertion is to consider a finite geometric series A with first term 1, common ratio 2, and sum $S = 2^p - 1$ (and thus p terms). Multiplying $2^p - 1$ by $T = 2^{p-1}$, the final term in the series, yields $(2^p - 1)(2^{p-1})$, which Euclid claims is a perfect number for all prime p. Observe that the geometric series B with common ratio 2 and first term S has all terms in A multiplied by T, and so its sum is the sum of the terms in A multiplied by T, or $S(2T - 1) = 2ST - S$. Adding back S, the sum of the terms in A, yields $2ST = 2(2^p - 1)(2^{p-1})$. Since A and B do not have any terms in common, they contain all divisors of S, in turn because S is prime. Thus, ST has factors that sum to $2ST$, or proper divisors that sum to ST, proving that it is perfect.

The second part of the theorem follows from the multiplicativity[53] of the sum of divisors function (which we will denote by $\sigma(n)$). In the context of Euler's proof, $\sigma(n)$ counts *all* positive divisors of n, including n itself (not just the proper divisors). It follows that n is perfect iff $\sigma(n) = 2n$.

In proving the first direction of Euler's statement (i.e. that all even perfect numbers can be written as $(2^p - 1)2^{p-1}$ for some prime number p), we can write $\sigma((2^p - 1)2^{p-1}) = \sigma(2^p - 1)\sigma(2^{p-1})$ due to the multiplicativity of σ. Since the sum of the factors of 2^{p-1} is $2^p - 1$, and the sum of the factors of $2^p - 1$ is 2^p, it follows that $\sigma((2^p-1)2^{p-1}) = 2^p(2^p-1) = 2\cdot 2^{p-1}(2^p-1)$, which implies that $2^{p-1}(2^p-1)$ is a perfect number.

In the other direction, start with an even perfect number $2^k \cdot n$, for k a non-negative integer and n an odd positive integer. For $2^k \cdot n$ to be a perfect number, we require $\sigma(2^k \cdot n) = 2^{k+1} \cdot n$, or $2^{k+1} \cdot n = (2^{k+1}-1)\cdot \sigma(n)$, since $\sigma(2^k \cdot n) = \sigma(2^k) \cdot \sigma(n)$ and $\sigma(2^k) = 2^{k+1} - 1$. The factor $2^{k+1} - 1$ on the RHS is at least 3, and is a factor of n (which is the only odd factor on the LHS). Thus, $m = \dfrac{n}{2^{k+1} - 1}$ is a proper divisor of n. Dividing the equation $2^{k+1} \cdot n = \sigma(2^k \cdot n) = (2^{k+1} - 1) \cdot \sigma(n)$ through by $2^{k+1} - 1$ yields $2^{k+1} \cdot m = \sigma(n)$. Note that $\sigma(n) = n + m + S$, where S is the sum of the other divisors of n. It follows that $n + m + S = 2^{k+1} \cdot m + S$, from which we obtain $m = 1$ and $n = 2^{k+1} - 1$, with n prime. Hence, we can attain the desired form $2^k(2^{k+1} - 1)$. □

In the search for Mersenne primes, we have progressed somewhat further than we have with Fermat primes, but our understanding of them is still a far cry from 100% comprehensive understanding (and it likely will be for decades, perhaps centuries, to come, as with the vast majority of open problems. Assuming such a thing as "100% understanding" even exists to begin with. Such is a fact of mathematics!) We have developed algorithms with polynomial time[54] or greater efficiency to determine Mersenne primes. The most famous and widely-used of these is the Lucas-Lehmer primality test:

> **Theorem 13.10: Lucas-Lehmer Primality Test**
>
> For prime $p > 2$, M_p is a Mersenne prime iff $M_p \mid S_{p-2}$, where $S_0 = 4$ and $S_n = (S_{n-1})^2 - 2$ for all $n \geq 1$.

These algorithms, of course, only came about with the advent of the Digital Revolution in the latter half of the 20th century; manual computation predominated before then. The first four Mersenne primes (M_2, M_3, M_5, M_7) were known since antiquity, likely before 1000 AD. In 1772, Euler proved that $M_{31} = 2^{31}-1$ was prime. Curiously, the next Mersenne number shown to be prime was M_{127} in 1876, but it was followed by M_{61} seven years later, and then by M_{89} and M_{107} in 1911 and 1914 respectively.

In 1952, Alan Turing's work on the Manchester Mark 1 - one of the earliest computers in history - led to the discovery of M_{521} as a Mersenne prime, followed almost immediately by M_{607}. In the same year, M_{1279}, M_{2203}, and M_{2281} were subsequently found to be prime. As computing power and technology rapidly progressed, so did the efficiency with which Mersenne numbers could be tested for primality. The first *megaprime* - a prime number with at least one million digits - was $M_{6,972,593}$, discovered in 1999.

At the turn of the 21st century, the distributed prime computing project GIMPS (Great Internet Mersenne Prime Search) began to gain a great amount of footing (though it originated in 1996). In September 2008, mathematicians at the University of California, Los Angeles, won $100,000 for discovering a Mersenne prime with 13 million digits (and the prize was for a Mersenne prime with at least 10 million digits). Fast-forwarding to the present day, as of December 7, 2018 (with the finding officially verified two weeks later, on December 21, 2018), the largest known (and 51st discovered) Mersenne prime is $M_{82,589,933}$, with nearly 25 million digits in decimal; this is also the largest known prime number. As of 2019, it is known that the eight largest known primes are also Mersenne primes.

---

[53] A *multiplicative* function $f : \mathbb{N} \mapsto \mathbb{C}$ is any function with $f(1) = 1$ such that, for all coprime positive integers m, n in the domain of f, $f(m)f(n) = f(mn)$. Furthermore, it is *completely multiplicative* if $f(m)f(n) = mn$ for *any* positive integers m, n, not necessarily coprime.

[54] See §16 for an introduction to big-O notation, where we invoke the concept of "polynomial time" and related algorithms.

13.4.4 Examples

Example 13.4.4.1. *(2009 ARML Individual Round #2) Define a reverse prime to be a positive integer N such that when the digits of N are read in reverse order, the resulting number is a prime. For example, the numbers 5, 16, and 110 are all reverse primes. Compute the largest two-digit integer N such that the numbers N, $4N$, and $5N$ are all reverse primes.*

Solution 13.4.4.1. Firstly, note that $N < 100$. Any reverse prime must have its first digit odd (unless it is 2). Thus, we can check N in the 90s. However, this forces $5N \in [450, 495]$, which will not work. We now check values of $N \in [70, 79]$; indeed, $N = \boxed{79}$, $4N = 316$, and $5N = 395$ are all reverse primes (since 97, 613, and 593 are all prime).

Example 13.4.4.2. *If a and p are positive integers such that $a^p - 1$ is prime, prove that $a = 2$ or $p = 1$ (but not both).*

Solution 13.4.4.2. *Proof.* Observe that $a \equiv 1 \pmod{a-1}$, so $a^p \equiv 1 \pmod{a-1}$ as well. Then $a^p - 1 \equiv 0 \pmod{a-1}$, so $a - 1 \mid (a^p - 1)$. Since $a^p - 1$ is prime, either $a - 1 = a^p - 1 \implies a = a^p \implies a^{p-1} = 1$ or $p = 1$ (but $a^{p-1} = 1$ is a contradiction, since 1 is not prime) or $a - 1 = 1, -1$. The latter implies $a = 2, 0$, but we cannot have $a = 0$, since a must be a positive integer. Therefore, either $p = 1$ or $a = 2$. $\square$

Example 13.4.4.3. *In terms of n, how many digits does F_n, the nth Fermat number, have?*

Solution 13.4.4.3. In general, the number of digits $D(n)$ of a positive integer n is given by the largest integer less than or equal to $\log_{10}(n) + 1$ (which can be seen intuitively by testing edge cases like $n = 9$ or $n = 10$).[55] Since $F_n = 2^{2^n} + 1$, it follows that $\boxed{D(F_n) = \lfloor \log_{10}(2^{2^n} + 1) + 1 \rfloor = 1 + \lfloor \log_{10}(2^{2^n} + 1) \rfloor}$. We can closely approximate this as $1 + \lfloor 2^n \log_{10}(2) \rfloor$ by logarithm properties.

---

[55] More concisely, we can write $D(n) = \lfloor \log_{10}(n) + 1 \rfloor$, where the brackets represent *floor notation* (see section 6 of this chapter for a definition of this notation).

13.4.5 Exercises

Exercise 161. *If the digits of a positive integer sum to 69, show that the number cannot be prime.*

Exercise 162. *Show that 2 and 3 are the only primes that differ by 1.*

Exercise 163. *If p and q are both prime numbers, and $p > q > 2$, show that $p^2 - q^2$ is a multiple of 8.*

Exercise 164. *Let p be a prime number. When is $p^2 + p + 1$ equal to the square of an integer?*

Exercise 165. *(2000 AMC 12) Two different prime numbers between 4 and 18 are chosen. When their sum is subtracted from their product, which of the following numbers could be obtained?*

- *21*
- *60*
- *119*
- *180*
- *231*

Exercise 166. *Can the number $2^{2017} - 1$ be prime? What about $2^{2019} - 1$?*

Exercise 167. *(2015 AIME I) There is a prime number p such that $16p + 1$ is the cube of a positive integer. Find p.*

Exercise 168. *If $\gcd(a, b) = p$, where p is a prime number, what are the possible values of $\gcd(a^3, b)$?*

Exercise 169. *(1999 AIME) Find the smallest prime that is the fifth term of an increasing arithmetic sequence, all four preceding terms also being prime.*

Exercise 170. *Show that $19 \nmid (4n^2 + 4)$ for all $n \in \mathbb{N}$.*

Exercise 171. *(2005 AMC 12A) Call a number prime-looking if it is composite but not divisible by $2, 3$, or 5. The three smallest prime-looking numbers are $49, 77$, and 91. There are 168 prime numbers less than 1000. How many prime-looking numbers are there less than 1000?*

Exercise 172. *Show that $5 \mid (7^{4k+2} + 11^{3k+1})$ for all $k \in \mathbb{N}$.*

Extension Exercise 74. *Prove that there are infinitely many primes of the form $4k + 1$, $k \in \mathbb{N}$.*

Extension Exercise 75. *Show that, for all positive integers $n \geq 2$, $n^4 + 4$ is composite.*

Extension Exercise 76. *Show that if p is prime and is congruent to 3 (mod 4), then*

$$\left(\frac{p-1}{2}\right)! \equiv \pm 1 \pmod{p}$$

(Hint: Use Wilson's Theorem.)

Extension Exercise 77. *(2002 AMC 12A) Several sets of prime numbers, such as $\{7, 83, 421, 659\}$ use each of the nine nonzero digits exactly once. What is the smallest possible sum such a set of primes could have?*

Extension Exercise 78. *Prove that there always exists at least one prime number between n and $1 + n!$, for $n \geq 2$ a positive integer. (You may use Bertrand's Postulate, which guarantees the existence of a prime number between k and $2k$ for all natural numbers $k \geq 2$.)*

13.5 Introduction to Diophantine Equations

We can extend our asset of algebra tools, as well as what we have learned in number theory thus far, to begin our foray into the field of *Diophantine analysis*. Essentially, Diophantine analysis is the study of *Diophantine equations*. Hailing from Diophantus of Alexandria circa the 3rd century BCE, Diophantine equations are those in which we are only concerned with integer solutions (i.e. solutions for which every variable is an integer). In a system of Diophantine equations, there are fewer equations than there are unknowns. Despite the setup of these equations being remarkably simple on the surface, underneath their exterior lies an endless amount of mystery and intrigue that characterizes many other areas of mathematics.

13.5.1 Linear Diophantine Equations

The simplest type of Diophantine equation is a linear equation: that is, any equation of the form $ax + by = c$ where a, b, and c are integer constants. Solving this equation is only possible if $\gcd(a, b) \mid c$; the proof of this can be found in §4.?, where we discuss the Euclidean Algorithm, as well as in §5.?, where we discuss Bézout's Identity. To briefly recap, Bézout's Identity guarantees the existence of integers e and f such that $ae + bf = d$, where $d = \gcd(a,b)$. Using the fact that $d \mid ax + by$, we can show that $d \mid c$, and that other solutions assume the form $(x+kv, y-ku)$, where k is an arbitrary integer, and u and v are equal to $\dfrac{a}{d}$ and $\dfrac{b}{d}$ respectively.

But what is the actual solution method for the equation $ax + by = c$, provided a solution exists? For the purposes of this explanation, we will assume, without loss of generality, that $c = 1$ (since we can just divide through by c if $c \neq 1$). Consider the resulting equation (mod ab). Then $ax + by \equiv 1 \pmod{ab}$, which implies $ax \equiv 1 - by \pmod{ab}$ or $ax \equiv ab - by + 1 = b(a - y) + 1 \pmod{ab}$. Then $x \equiv \dfrac{b(a-y)+1}{a}$ (mod ab), which immediately yields a solution in y.

Example 13.5.1.1. *Solve the equation $2x + 3y = 7$ over the integers.*

Solution 13.5.1.1. *Consider the equation (mod 6). We have $2x + 3y \equiv 1 \pmod 6$, which implies $2x \equiv 1 - 3y \equiv 7 - 3y \pmod 6$, or $x \equiv \dfrac{7-3y}{2} \pmod 6$. Then whenever y is odd, there will exist an integral solution (x,y) (since $\dfrac{7-3y}{2}$ is an integer iff $7 - 3y$ is even). More specifically, if $y \equiv 1 \pmod 6$, then $2x \equiv 4 \pmod 6$, and either $x \equiv 2 \pmod 6$ or $x \equiv 5 \pmod 6$. We can apply similar logic to the cases $y \equiv 3, 5 \pmod 6$. Our solution set is* $\boxed{\{(x,y)\} = \left\{\left(x, \dfrac{7-2x}{3}\right)\right\}}$ *whenever $x \equiv 2 \pmod 3$ (finding the solution sets for the other two cases is left as a straightforward exercise to the reader).*

In general, we can use modular arithmetic to our advantage when solving these types of equations. Working in a certain modulus effectively serves as a "sieve" for absurd solutions, eliminating those that clearly would not work until we arrive at feasible solutions. For instance, consider the following example of a common word problem:

Example 13.5.1.2. *Ryan the coin collector has nickels and rare, 1870s 3-cent coins. If he has exactly $20.19 in his piggy bank, how many possible combinations of coins could he have?*

Solution 13.5.1.2. *Let x be the number of nickels, and let y be the number of 3-cent coins. We know that $2019 \equiv 0 \pmod 3$ and $2019 \equiv 4 \pmod 5$, so $673 \cdot 3 + 0 \cdot 5 \implies (x,y) = (673, 0)$ is a solution. Subtracting 3 increases the modulo 5 value by 3 each time, so it will repeat in the cycle $\mathbf{0} \to 3 \to 1 \to 4 \to 2 \to \mathbf{0} \to \ldots$ Thus, $(x,y) = (668, 3)$ is also a solution, as is $(663, 6)$, or any ordered pair of the form $(673 - 5k, 3k)$. This forces $5k \leq 673$, or $k \leq 134$. Hence, $k \in [0, 134]$, and so there are $134 - 0 + 1 = \boxed{135}$ possible coin combinations.*

We also present a very slightly changed, but significantly more complicated version of the same example.

Example 13.5.1.3. *Ryan the coin collector has nickels and rare, 1870s 3-cent coins. If he has up to $20.19 in his piggy bank, how many possible combinations of coins could he have?*

This variant is left as an exercise to the reader. (Hint: Consider the resulting Diophantine equation wrt (mod 15). What are the possible starting values (mod 3) and (mod 5) for nickels and 3-cent coins?)

13.5.2 Homogeneous Diophatine Equations

A *homogeneous* Diophantine equation is one whose nonzero terms all have the same degree; for instance, $x^2 + y^2 + z^2 = 3$ or $x^2 + xy + y^2 = 7$. Geometrically, these types of equations boil down to finding all lattice points that lie on a hypersurface (a surface in greater than 3 dimensions). Solving the generalized homogeneous Diophantine equation in any number of variables is an extremely difficult and inaccessible problem, to the point where we have more unanswered questions regarding them than we do answered questions. The prototypical example is Fermat's Last Theorem, which invokes the special case $x^d + y^d = z^d$ for $d \geq 3$. This is far from the only type of homogeneous Diophantine equation (especially considering it is only in 3 variables, whereas a homogeneous Diophantine equation may have *any* number of variables), but this alone took over 300 years to fully prove.

With a homogeneous Diophantine equation, we can find solutions modulo p, where p is some positive integer. By reducing any given homogeneous equation to mod p, we can use an argument that considers the possible values of each variable with respect to mod p.

Example 13.5.2.1. *Solve the Diophantine equation $x^2 + y^2 + z^2 = 4w^2$.*

Solution 13.5.2.1. *Consider the equation (mod 4). Since $n^2 \equiv 0, 1$ (mod 4) for all $n \in \mathbb{Z}$, we have $x^2 + y^2 + z^2 \equiv 0, 1, 2, 3$ (mod 4), and $4w^2 \equiv 0$ (mod 4) for all $w \in \mathbb{Z}$, so the only possible solutions arise from when the LHS is equivalent to 0 (mod 4). Thus, x, y, z are all even.*

13.5.3 Simon's Favorite Factoring Trick

A favorite factoring method of renowned Art of Problem Solving user Dr. Simon Rubinstein-Salzedo, and a classical competition math technique, the aptly-named *Simon's Favorite Factoring Trick* (or SFFT) is a form of "completing the rectangle," much like the technique of completing the square that you have probably learned in middle or high school algebra class. You may have also used it in a simplified form - namely factoring by grouping. However, SFFT is more generalized, and covers all cases where we have an equation of the form $xy + a \cdot x + b \cdot y = c$, where a, b, c are constants.

> **Theorem 13.11: Simon's Favorite Factoring Trick**
>
> For all constants $a, b, c \in \mathbb{R}$, we can factor the expression $xy + ax + by$ as $(x + b)(y + a) - ab$.

It immediately follows that the equation $xy + ax + by = c$ can be re-written as $(x+b)(y+a) = c+ab$, thereby becoming more conducive to solution as a Diophantine equation. Undoubtedly, in a competition setting, the two most common special cases are $(x+1)(y+1) = xy + x + y + 1$ and $(x-1)(y-1) = xy - x - y + 1$.

Example 13.5.3.1. *Find all ordered pairs of integers (x, y) satisfying $xy + 2x + y = 19$.*

Solution 13.5.3.1. *Using SFFT, we have $xy + 2x + y = (x+1)(y+2) - 2 \implies (x+1)(y+2) = 21$. 21 has factors $\pm 1, \pm 3, \pm 7, \pm 21$. 21 has 8 factors, so the total number of satisfactory integer ordered pairs is $\boxed{8}$.*

Let's try an example where we must exclude choice cases:

Example 13.5.3.2. *Find all pairs of positive integers (x, y) such that $xy + x + y = 24$.*

Solution 13.5.3.2. *This time, we are seeking only positive integer solutions, so $(x+1)(y+1) = 25 \implies x+1, y+1 = \{\pm 1, \pm 5, \pm 25\}$. But the negative solutions should be discarded, so we are left with $x + 1 = 1, 5, 25$. We cannot have $x + 1 = 1$, nor can we have $x + 1 = 25$, since $x = 24$, which would imply that $y + 1 = 1 \implies y = 0$, so we are left with $\boxed{(4, 4)}$ as our only solution in positive integers.*

And one more for good measure, which combines both the driving ideas behind the previous two examples:

Example 13.5.3.3. *Find all ordered pairs of integers (x, y) such that $x^2y^2 + 2xy^2 + 3x^2y + 6xy = 5$.*

Solution 13.5.3.3. *We have $(xy+3x)(xy+2y)-2 = 5 \implies (x(y+3))(y(x+2)) = 7 \implies xy(y+3)(x+2) = 7$. 7 is prime, so either $xy = \pm 1$ or $xy = \pm 7$ (in turn, forcing $(y+3)(x+2) = \pm 7$ or $(y+3)(x+2) = \pm 1$ respectively). But this implies $x, y = \pm 1$ and $x, y \in \{\pm 1, \pm 7\}$ respectively, none of which satisfy the original equation. Hence, the equation has no integer solutions in x, y.*

Can we extend SFFT to two-variable expressions with a leading coefficient other than 1? What about expressions of more than two variables, or higher degree than 2? These are the types of questions that we leave to the reader to consider, discuss, and reflect upon for this section.

13.5.4 Examples

Example 13.5.4.1. *A son and his father have ages whose digits are the reverse of each other (e.g. 25 and 52, not that these are actually their ages). We know that the father's age is equal to twice the son's age plus seven. What are their ages?*

Solution 13.5.4.1. *Let the father's age be $\overline{AB} = 10A + B$, and let the son's age be $\overline{BA} = 10B + A$. We have the Diophantine equation $10A + B = 2(10B + A) + 7$ (since A, B must be integers - they are digits), which simplifies to $8A = 19B + 7$. We can either determine the solution by inspection, or by taking the equation (mod 8). Noting that $19B \equiv 1$ (mod 8), and that $19 \equiv 3$ (mod 8), with 3^{-1} (mod 8) = 3, we obtain $A = 8$, $B = 3$, so the father's age is $\boxed{83}$ and the son's age is $\boxed{38}$. (The second solution involves a technique discussed in subsection 3 of this section.)*

Example 13.5.4.2. *Compute the number of ordered pairs of positive integers (x, y) such that $xy + 20x + 19y = 2020$.*

Solution 13.5.4.2. *By SFFT, we can factor the equation as $(x + 20)(y + 19) = 2020 + 20 \cdot 19 = 2400$. Then any factor of 2400 that is strictly greater than 20 may be the value of 20 (and, similarly, any factor that is strictly greater than 19 may be the value of $y + 19$). The prime factorization of 2400 is $2^5 \cdot 5^2 \cdot 3$, so it has $(5 + 1)(2 + 1)(1 + 1) = 36$ factors. Of these 36 factors, 12 of them are less than or equal to 20, so there are 24 possible values of $x + 20$. However, we must consider $y + 20$ as well, which requires that $x + 20 < \frac{2400}{20} = 120$. By symmetry, there are $12 - 1 = 11$ factors of 2400 greater than or equal to 120, so there are $36 - 12 - 11 = \boxed{13}$ solutions to the equation in positive integers.*

Example 13.5.4.3. *(2015 AMC 10A) Consider the set of all fractions $\frac{x}{y}$, where x and y are relatively prime positive integers. How many of these fractions have the property that if both numerator and denominator are increased by 1, the value of the fraction is increased by 10%?*

Solution 13.5.4.3. *We can set up the equation $\frac{x+1}{y+1} = \frac{x}{y} \cdot \frac{11}{10} = \frac{11x}{10y}$, so $10y(x+1) = 11x(y+1) \implies 10xy + 10y = 11xy + 11x \implies 10y = xy + 11x \implies y(10 - x) = 11x$ by cross-multiplication and subsequent simplification. 11 is prime, and $10 - x < 11$, so y must be a multiple of 11. If $y = 11$, then $x = 5$. If $y = 11k$ for some positive integer $k \geq 2$, then $10 - x = \frac{x}{k} \implies 10 = x \cdot \left(\frac{k+1}{k}\right)$. Clearly, $x < 5$ does not work; for $x > 5$, $x = 8, 9$ are both solutions (with $k = 4, 9$ respectively), so $y = 44, 99$ respectively. However, $\gcd(8, 44) \neq 1$ and $\gcd(9, 99) \neq 1$, so these cases do not count. The only solution (x, y) with x and y relatively prime is $(5, 11)$, and indeed $\frac{5}{11} \cdot \frac{11}{10} = \frac{6}{12}$; the answer is $\boxed{1}$.*

Example 13.5.4.4. *Find all pairs of base-10 digits x, y such that $\frac{x}{y} = \overline{xy}\%$, where $\overline{xy}$ is a two-digit number (and x can be 0).*

Solution 13.5.4.4. *We require $\frac{x}{y} = \frac{10x + y}{100}$, or $100x = 10xy + y^2 \implies 100x - 10xy = y^2 \implies 10x(10 - y) = y^2 \implies 10x = \frac{y^2}{10 - y}$. This implies $10 \mid y^2$, or $10 \mid y$, but this forces $y = 0$ or $y \geq 10$ - contradiction.*

Hence, $\boxed{\text{there exist no such ordered pairs }(x,y)}$ *(It is worth noting that* $\dfrac{3}{8} = 37.5\%$ *and* $\dfrac{8}{9} = 88.\overline{8}\%$*, which round to 38% and 89% respectively. However, the problem concerns exact values.)*

13.5.5 Exercises

Exercise 173. *Solve the Diophantine equation $20x - 19y = 45$.*

Exercise 174.

(a) *Find all ordered pairs of positive integers (x, y) such that $x^2 - y^2 = 20$.*

(b) *Find all ordered pairs of positive integers (x, y) such that $x^2 - y^2 = 2019$.*

Exercise 175. *Find the sum of all solutions for x in the system of equations $xy = 2$, $x + y = 3$.*

Exercise 176. *(2015 Berkeley Mini Math Tournament) Given integers a, b, c satisfying $abc + a + c = 12$, $bc + ac = 8$, $b - ac = -2$, what is the value of a?*

Exercise 177. *(2016 AMC 10A) How many ways are there to write 2016 as the sum of twos and threes, ignoring order? (For example, $1008 \cdot 2 + 0 \cdot 3$ and $402 \cdot 2 + 404 \cdot 3$ are two such ways.)*

Exercise 178. *(2011 AMC 10B) In multiplying two positive integers a and b, Ron reversed the digits of the two-digit number a. His erroneous product was 161. What is the correct value of the product of a and b?*

Exercise 179. *Compute the prime number p such that $n^2 = 19p + 1$, where n is a positive integer.*

Exercise 180. *Find all ordered pairs (a, b) of positive integers such that $a^2 - a + b - b^2 = 36$.*

Exercise 181. *Find all ordered pairs of integers (x, y) for which $4^y - x^2 = 615$.*

Exercise 182. *Find all rational ordered pairs (x, y) that satisfy the equation $5x^2 + 2y^2 = 1$.*

Exercise 183. *Find all positive integers n such that $n \pm 76$ are both perfect cubes.*

Exercise 184. *How many ordered pairs of positive integers (m, n) are there such that $mn + 2m + 2n = 176$?*

Exercise 185. *Solve the equation $\dfrac{3}{x} - \dfrac{7}{y} = \dfrac{1}{5}$ over $\mathbb{Z}$.*

Exercise 186. *(2015 AMC 10A) The zeroes of the function $f(x) = x^2 - ax + 2a$ are integers. What is the sum of the possible values of a?*

Exercise 187. *(2008 AMC 12B) A rectangular floor measures a by b feet, where a and b are positive integers with $b > a$. An artist paints a rectangle on the floor with the sides of the rectangle parallel to the sides of the floor. The unpainted part of the floor forms a border of width 1 foot around the painted rectangle and occupies half of the area of the entire floor. How many possibilities are there for the ordered pair (a, b)?*

Extension Exercise 79. *(1996 AIME) The harmonic mean of two positive integers is the reciprocal of the arithmetic mean of their reciprocals. For how many ordered pairs of positive integers (x, y) with $x < y$ is the harmonic mean of x and y equal to 6^{20}?*

Extension Exercise 80. *Find all pairs of positive integers (a, b) with $a > b$ and $a^2 + b^2 = 40,501$.*

Extension Exercise 81. *(2008 AIME I) There exist unique positive integers x and y that satisfy the equation $x^2 + 84x + 2008 = y^2$. Find $x + y$.*

Extension Exercise 82. *(1984 AHSME) The number of distinct pairs of integers (x, y) such that $0 < x < y$ and $\sqrt{1984} = \sqrt{x} + \sqrt{y}$ is*

Extension Exercise 83. *(2005 AMC 12A) How many ordered triples of integers (a, b, c), with $a \geq 2$, $b \geq 1$, and $c \geq 0$, satisfy both $\log_a b = c^{2005}$ and $a + b + c = 2005$?*

Extension Exercise 84. *(2011 AIME I) For some integer m, the polynomial $x^3 - 2011x + m$ has the three integer roots a, b, and c. Find $|a| + |b| + |c|$.*

13.6 Special Functions in Number Theory

In number theory, we often come across special functions that are not traditionally covered in a middle school, or even high school, math curriculum. These functions are instrumental in providing for slick, succinct solutions to number theoretic problems, yet they are at times tragically glossed over in the curriculum. Here, we define, describe, and use to the best effect several examples of the most prominent and significant functions in the discipline, beginning with two that may seem intuitively familiar on a basic level: the *floor* and *ceiling* functions.

13.6.1 The Floor and Ceiling Functions

The floor and ceiling functions, mercifully, are extremely simple to define on the real numbers:

> **Definition 13.9: Floor and Ceiling Functions**
>
> Consider some real number r; then the *floor* of r, denoted $\lfloor r \rfloor$, is the largest integer less than or equal to r. Similarly, the *ceiling* of r, denoted $\lceil r \rceil$, is the smallest integer greater than or equal to r.

To demonstrate, $\lfloor 1 \rfloor = 1$, $\lfloor 2.5 \rfloor = 2$, $\lceil \pi \rceil = 4$, and $\lfloor -e \rfloor = -3$.

We can also define the *round* function, which is self-explanatory, and the *fix* function, which rounds to the nearest integer *in the direction of zero*:

> **Definition 13.10: Round and Fix Functions**
>
> The *round* function, denoted round(r), rounds r to the nearest integer (using the standard convention of rounding 0.5 up, regardless of the sign of r). The *fix* function, denoted fix(r), rounds r up if it has a value *strictly greater* than 0.5 after the decimal point, and down otherwise, *if it is positive*. If $r < 0$, then fix rounds r up if it has a value less than or equal to -0.5 after the decimal point, and rounds r down otherwise.

Example 13.6.1.1. *Examples of the round and fix functions in action:*

$round(2.7) = 3$, $round(2.4) = 2$, $round(-4.5) = -4$, $fix(0.9) = 1$, $fix(3.5) = 3$, $fix(-4.5) = -4$.

Note that the only difference between the round and fix functions lies at r such that $r < 0$ and $r = k.5$ for some non-negative integer k. In such cases, while round(r) = $k + 1$, fix(r) = k. That is, for all $r < 0$, round(r) = fix(r).

13.6.2 The Fractional Part Function

The *fractional part* function, similarly, has a (fairly) straightforward definition on $\mathbb{R}$:

> **Definition 13.11: Fractional Part Function**
>
> The *fractional part* of $r \in \mathbb{R}$, denoted $\{r\}$, is equal to $r - \lfloor r \rfloor$. $\{r\}$ gives the value after the decimal point of r (for non-negative r), and 1 minus the value after the decimal point of r (for negative r), such that $0 \leq \{r\} < 1$ for all r.

Example 13.6.2.1. *For positive r, examples of the fractional part include $\{0.6\} = 0.6$, $\{1.89\} = 0.89$, and $\{201,945\} = 0$. For negative r, examples include $\{-0.25\} = 0.75$, $\{-1\} = 0$, and $\{-2\pi\} = 7 - 2\pi$.*

It should be mentioned here that the definition of $\{r\}$ for negative r is surrounded by controversy. Some authors extend the definition for positive r to negative r, as we do here. Other authors, however, prefer to define $\{r\} = |r| - \lfloor |r| \rfloor$ for $r < 0$, using the logic that the fractional part should be interpreted literally as that which comes after the decimal point (so, for instance, $\{-0.8\} = 0.8$, not 0.2). Others still define the fractional part over $\mathbb{R}$ via a piecewise function:

$$\{r\} = \begin{cases} r - \lfloor r \rfloor & r \geq 0 \\ r - \lceil r \rceil & r < 0 \end{cases}$$

Using this third definition, we obtain negative values for $\{r\}$ if r itself is negative. Under this definition, we can also write $\{r\} = |r| - \text{sgn}(r) \cdot \lfloor |r| \rfloor$, where $\text{sgn}(r)$, the *signum* function, is defined such that

$$\text{sgn}(r) = \begin{cases} 1 & r > 0 \\ 0 & r = 0 \\ -1 & r < 0 \end{cases}$$

In this way, it is possible to obtain up to *three* different values for $\{r\}$! An example is $\{-\pi\}$; under the first definition (and the one that will be used in this book), $\{-\pi\} = 4 - \pi$, under the second definition, $\{-\pi\} = \pi - 3$, and under the third definition, $\{-\pi\} = 3 - \pi < 0$. This is a rather interesting example of why it is of utmost importance for mathematicians to clarify what they mean when they use certain notation! Though all of these definitions are equally correct, and each has its strong and weak points, we will be going with the first definition (namely, $\{r\} = r - \lfloor r \rfloor$), due to its simplicity and applicability to all real numbers r.

13.6.3 Euler's Totient Function

Though we have already discussed Euler's totient function (also known as the *phi function*, due to how it is customarily denoted) in some detail in prior sections, we will delve deeper into it here. To recap, here is the formal definition of the function:

> **Definition 13.12: Euler's Totient Function**
>
> *Euler's totient function*, denoted $\phi(n)$ for n a positive integer, returns the number of positive integers $k \leq n$ such that $\gcd(k, n) = 1$.

Example 13.6.3.1. *Compute $\phi(19)$, $\phi(45)$, and $\phi(200,000)$.*

Solution 13.6.3.1. Since 19 is prime, all positive integers under 19 are relatively prime with it, so $\phi(19) = \boxed{18}$. $45 = 3^2 \cdot 5$, so $\phi(45) = 45 \cdot \frac{2}{3} \cdot \frac{4}{5} = \boxed{24}$. Finally, $\phi(200,000) = \phi(2 \cdot 10^5) = \phi(2^6 \cdot 5^5) = 200,000 \cdot \frac{1}{2} \cdot \frac{4}{5} = \boxed{80,000}$.

An important property of the phi function is its *multiplicativity*:

> **Definition 13.13: Multiplicativity of a Function**
>
> A function $f : \mathbb{N} \mapsto \mathbb{C}$ is *multiplicative* if $f(1) = 1$ and, for all coprime positive integers m and n, $f(m)f(n) = f(mn)$.

For now, we use this property without proof. This will factor into our next point of analysis: an explicit formula for $\phi(n)$.

> **Theorem 13.12: An Explicit Formula for $\phi(n)$**
>
> Let n have prime factorization $p_1^{k_1} p_2^{k_2} p_3^{k_3} \cdots p_j^{k_j}$, where the p_i ($1 \leq i \leq j$) are the prime factors of n

and the k_i are non-negative integers. Then

$$\phi(n) = n \cdot \prod_{i=1}^{j}\left(1 - \frac{1}{p_i}\right)$$

Proof. We first introduce a key lemma:

Lemma 13.12: Value at a Prime Power

If p is a prime number and $k \geq 1$ is an integer, then $\phi(p^k) = p^k\left(1 - \frac{1}{p}\right)$.

Proof. Because p is prime, the only divisors of p^k are also the multiples of p (with these being mutually inclusive). There are p^{k-1} multiples of p less than or equal to p^k; hence, $\phi(p^k) = p^k - p^{k-1} = p^{k-1} \cdot (p-1) = p^k \cdot \left(1 - \frac{1}{p}\right)$. $\square$

By the Fundamental Theorem of Arithmetic, assuming $n \geq 2$, we have a unique decomposition of n into the powers of its prime factors. Using the multiplicative property of ϕ, we obtain

$$\phi(n) = \phi(p_1^{k_1}) \cdot \phi(p_2^{k_2}) \cdot \phi(p_3^{k_3}) \cdots \phi(p_j^{k_j})$$

$$= p_1^{k_1}\left(1 - \frac{1}{p_1}\right) \cdot p_2^{k_2}\left(1 - \frac{1}{p_2}\right) \cdot p_3^{k_3}\left(1 - \frac{1}{p_3}\right) \cdots p_j^{k_j}\left(1 - \frac{1}{p_j}\right)$$

$$= (p_1^{k_1} p_2^{k_2} p_3^{k_3} \cdots p_j^{k_j}) \cdot \prod_{i=1}^{j}\left(1 - \frac{1}{p_i}\right) = n \cdot \prod_{i=1}^{j}\left(1 - \frac{1}{p_i}\right)$$

which is the formula for $\phi(n)$. $\square$

Since we have proven the explicit formula for $\phi(n)$, we can now directly prove that $\phi(n)$ is a multiplicative function as follows:

Proof. Let $A = \prod_{a=1}^{b} p_a^{k_a}$ and $B = \prod_{a=b+1}^{j} p_a^{k_a}$ Note that $n = AB$, and

$$\phi(A) = A \prod_{a=1}^{b}\left(1 - \frac{1}{p_a}\right)$$

$$\phi(b) = B \prod_{a=b+1}^{j}\left(1 - \frac{1}{p_a}\right)$$

$$\phi(a)\phi(b) = AB \cdot \prod_{a=1}^{j}\left(1 - \frac{1}{p_a}\right)$$

$$= n \cdot \prod_{a=1}^{j}\left(1 - \frac{1}{p_a}\right) = \phi(n) = \phi(ab)$$

as desired. $\square$

Example 13.6.3.2. *Find $\phi(102)$, $\phi(3^{24})$, and $\phi(lcm(2^2, 3^3, 5^5, 7^7, \ldots, 19^{19}))$.*

Solution 13.6.3.2. *We have $\phi(102) = \phi(2) \cdot \phi(3) \cdot \phi(17)$ (using the fact that ϕ is multiplicative), or $1 \cdot 2 \cdot 16 =$ $\boxed{32}$. In addition, $\phi(3^{24}) = 3^{24} \cdot \left(1 - \dfrac{1}{3}\right) = \boxed{2 \cdot 3^{23}}$. Finally, $lcm(2^2, 3^3, 5^5, 7^7, \ldots, 19^{19}) = 2^2 3^3 5^5 7^7 \cdots 19^{19}$, the totient of which is*

$$\left(1 - \frac{1}{2}\right)\left(1 - \frac{1}{3}\right)\left(1 - \frac{1}{5}\right)\cdots\left(1 - \frac{1}{19}\right) = \frac{1}{2} \cdot \frac{2}{3} \cdot \frac{4}{5} \cdots \frac{18}{19}$$

(and you can do the rest).

13.6.4 The Divisor Functions

We can also define several other functions related to the divisors of a positive integer n:

> **Definition 13.14: Divisor Functions**
>
> The divisor function, $\tau(n)$, counts the number of positive integer factors of n. Id est, if $n = p_1^{k_1} p_2^{k_2} p_3^{k_3} \cdots p_j^{k_j}$, then $\tau(n) = \displaystyle\prod_{i=1}^{j}(k_i + 1)$.
>
> The sum of divisors function, $\sigma(n)$, can also be generalized to the sum of the pth powers of the divisors function, $\sigma_p(n) = \displaystyle\sum_{d \mid n} d^p$.

We have already proven the formula for $\tau(n)$ in §13.1.3, "The Prime Factorization." Note that $\sigma_0(n)$ is the sum of the divisors of n, or

$$\prod_{i=1}^{j}\left(\sum_{m=0}^{k_i} p_1^m\right).$$

Observe that, for prime p, we have the following identities:

$$\sigma_0(p) = 2, \sigma_0(p^n) = n + 1, \sigma_1(p) = p + 1. \tag{13.4}$$

In addition, $\sigma_p(n)$ for any p is multiplicative, but not *completely* so. That is, it is only multiplicative in relatively prime a, b, such that $\sigma_p(a) \cdot \sigma_p(b) = \sigma_p(ab) \iff \gcd(a, b) = 1$. Proving this is left as an exercise to the reader.

13.6.5 Examples

Example 13.6.5.1. *Solve the equation $2x^2 + 3y^2 + 5xy = 760$ given that $x + y = 19$.*

Solution 13.6.5.1. *By SFFT, the given equation simplifies to $(x+y)(2x+3y) = 760$, or $(x+y)(2(x+y)+y) = 760$. Now we substitute in $x + y = 19$ to obtain $y = 2$, or $x = 17$, yielding the solution $\boxed{(x, y) = (17, 2)}$. (If you had trouble finding the factorization, try working with the fact that $2 + 3 = 5$.)*

Example 13.6.5.2. *Show that for all odd n, $\phi(n) = \phi(2n)$.*

Solution 13.6.5.2. *Recall that $\phi(n) = n \cdot \left(1 - \dfrac{1}{p_1}\right)\left(1 - \dfrac{1}{p_2}\right)\cdots\left(1 - \dfrac{1}{p_k}\right)$. Since n is odd, it does not have 2 as a prime factor. In the formula, we are multiplying n by 2, but we are also introducing a new prime factor of 2, which multiplies the value of $\phi(n)$ by $1 - \dfrac{1}{2} = \dfrac{1}{2}$. This cancels out the 2; hence, $\phi(n) = \phi(2n)$ for all odd n.*

Example 13.6.5.3. *For what non-negative real values of n is $\lfloor n^2 + n \rfloor < \lceil n^2 \rceil$?*

Solution 13.6.5.3. *We will examine the functions $f(n) = \lfloor n^2 + n \rfloor$ and $g(n) = \lceil n^2 \rceil$ separately. For $0 < n \leq 1$, $g(n) = 1$, and for $0 \leq n < \dfrac{-1 + \sqrt{5}}{2}$, $f(n) = 0$. Thus, an interval of solutions is $n \in \left(0, \dfrac{\sqrt{5} - 1}{2}\right)$. For $1 < n \leq \sqrt{2}$, $g(n) = 2$, and $f(n) < 2$ whenever $n^2 + n < 2$, or $n^2 + n - 2 < 0 \implies n < 1$. But when $n < 1$, $g(n) < 2$ as well, so no more solutions can exist (as $g(n)$ grows faster than does $f(n)$, which can be verified by graphing both functions). Hence the interval of solutions is* $\boxed{\left(0, \dfrac{\sqrt{5} - 1}{2}\right)}$.

13.6.6 Exercises

Exercise 188. *What is the value of* $\left\lceil \dfrac{2+2^3}{3+2^2} \right\rceil$?

Exercise 189. *Either prove or disprove that* $\lfloor -x \rfloor = \lfloor x \rfloor$ *for all* $x \in \mathbb{R}$.

Exercise 190. *Prove or disprove that* $\lfloor nx \rfloor = n\lfloor x \rfloor$ *for all positive integers* $n > 2$.

Exercise 191. *Show that, for all* $x \in \mathbb{R}$, $\lfloor 2x \rfloor = \lfloor x \rfloor + \left\lfloor x + \dfrac{1}{2} \right\rfloor$.

Exercise 192. *(2003 AMC 10B) Compute*
$$\lfloor \sqrt{1} \rfloor + \lfloor \sqrt{2} \rfloor + \lfloor \sqrt{3} \rfloor + \cdots + \lfloor \sqrt{16} \rfloor.$$

Exercise 193. *Show that* $\lfloor \sqrt{n^2 + n} \rfloor = n$ *for all* $n \in \mathbb{N}$.

Exercise 194. *Evaluate the sum*
$$\sum_{k=1}^{\infty} \left\lfloor \dfrac{30}{k} \right\rfloor.$$

Exercise 195. *In how many trailing zeros does 2019!!! end? (n!!! denotes the triple factorial; i.e.* $n \cdot (n-3) \cdot (n-6) \cdots$).

Exercise 196. *(2014 Berkeley Math Tournament) For a positive integer* n, *let* $\phi(n)$ *denote the number of positive integers between 1 and* n, *inclusive, which are relatively prime to* n. *We say that a positive integer* k *is total if* $k = \dfrac{n}{\phi(n)}$, *for some positive integer* n. *Find all total numbers.*

Exercise 197. *For a real number* x, *let* $\lfloor x \rfloor$ *be the largest integer less than or equal to* x. *For example,* $\lfloor 1 \rfloor = 1$ *and* $\lfloor 2.99 \rfloor = 2$. *The area under the graph of* $y = \lfloor x^2 - 2x \rfloor$ *between* $x = 0$ *and* $x = 5$ *can be written in the form* $N - R$, *where* N *is a positive integer and* R *is a sum of radicals. Find* N.

Exercise 198. *(2018 AMC 10A) What is the greatest integer less than or equal to*
$$\dfrac{3^{100} + 2^{100}}{3^{96} + 2^{96}}?$$

Exercise 199. *(1983 AIME) Let* a_n *equal* $6^n + 8^n$. *Determine the remainder upon dividing* a_{83} *by* 49.

Exercise 200. *(2015 AMC 10B/12B) Cozy the Cat and Dash the Dog are going up a staircase with a certain number of steps. However, instead of walking up the steps one at a time, both Cozy and Dash jump. Cozy goes two steps up with each jump (though if necessary, he will just jump the last step). Dash goes five steps up with each jump (though if necessary, he will just jump the last steps if there are fewer than 5 steps left). Suppose the Dash takes 19 fewer jumps than Cozy to reach the top of the staircase. Let s denote the sum of all possible numbers of steps this staircase can have. What is the sum of the digits of s?*

Exercise 201. *What fraction of the positive integers up to* 20192019 *are relatively prime with it? (Note:* $10{,}001 = 137 \cdot 73$.)

Extension Exercise 85. *Let* $f(x) = \left\lfloor \dfrac{x}{7} \cdot \left\lfloor \dfrac{37}{x} \right\rfloor \right\rfloor$, *where* $1 \leq x \leq 45$ *is an integer. How many distinct values does* $f(x)$ *assume?*

Extension Exercise 86. *(2011 AMC 12B) For every* m *and* k *integers with* k *odd, denote by* $\left[\dfrac{m}{k}\right]$ *the integer closest to* $\dfrac{m}{k}$. *For every odd integer* k, *let* $P(k)$ *be the probability that*
$$\left[\dfrac{n}{k}\right] + \left[\dfrac{100-n}{k}\right] = \left[\dfrac{100}{k}\right]$$
for an integer n *randomly chosen from the interval* $1 \leq n \leq 99!$. *What is the minimum possible value of* $P(k)$ *over the odd integers* k *in the interval* $1 \leq k \leq 99$?

Extension Exercise 87. *(2010 AIME I) For each positive integer n, let $f(n) = \sum_{k=1}^{100} \lfloor \log_{10}(kn) \rfloor$. Find the largest value of n for which $f(n) \leq 300$.*

Extension Exercise 88. *(2015 AIME I) For each integer $n \geq 2$, let $A(n)$ be the area of the region in the coordinate plane defined by the inequalities $1 \leq x \leq n$ and $0 \leq y \leq x \lfloor \sqrt{x} \rfloor$, where $\lfloor \sqrt{x} \rfloor$ is the greatest integer not exceeding $\sqrt{x}$. Find the number of values of n with $2 \leq n \leq 1000$ for which $A(n)$ is an integer.*

Extension Exercise 89. *Prove Hermite's Identity:*

Theorem 13.13: Hermite's Identity

$$\lfloor na \rfloor = \lfloor a \rfloor + \left\lfloor a + \frac{1}{n} \right\rfloor + \ldots + \left\lfloor a + \frac{n-1}{n} \right\rfloor$$

Chapter 13 Review Exercises

Exercise 202. What is the remainder when 2018^2 is divided by 2019?

Exercise 203. This book became available for pre-order on March 14, 2019 (π Day!), a Thursday. What day of the week is the day that falls exactly 314 days after π Day 2019?

Exercise 204. Let m and n be two odd numbers with $m > n$. What is the largest integer that is a factor of all positive integers of the form $m^2 - n^2$?

Exercise 205. Let a_n be the sum $1 + 2 + 4 + \ldots + 2^n$. Compute the remainder when $a_1 + a_2 + a_3 + \ldots + a_{2019}$ is divided by 100.

Exercise 206. Compute the units digit of $\sum_{i=1}^{100}(i!)^2$.

Exercise 207. *(1997 AIME)* Sarah intended to multiply a two-digit number and a three-digit number, but she left out the multiplication sign and simply placed the two-digit number to the left of the three-digit number, thereby forming a five-digit number. This number is exactly nine times the product Sarah should have obtained. What is the sum of the two-digit number and the three-digit number?

Exercise 208. *(2009 AIME I)* Call a 3-digit number geometric if it has 3 distinct digits which, when read from left to right, form a geometric sequence. Find the difference between the largest and smallest geometric numbers.

Exercise 209. *(2016 ARML Local Individual Round #9)* The integer $Z = 104,060,001$ is the product of three distinct prime numbers. Compute the largest prime factor of Z.

Exercise 210. Determine the number of distinct ordered pairs of positive integer solutions (m, n) to the equation $20m + 19n = 2019$.

Exercise 211. *(2014 AIME II)* The repeating decimals $0.\overline{abab}ab$ and $0.\overline{abcabc}abc$ satisfy

$$0.\overline{abab}ab + 0.\overline{abcabc}abc = \frac{33}{37},$$

where a, b, and c are (not necessarily distinct) digits. Find the three digit number abc.

Exercise 212. Compute the remainder when 7^{2018} is divided by 650.

Exercise 213. *(2000 AIME I)* Find the least positive integer n such that no matter how 10^n is expressed as the product of any two positive integers, at least one of these two integers contains the digit 0.

Exercise 214. *(1994 AIME)* The increasing sequence $3, 15, 24, 48, \ldots$ consists of those positive multiples of 3 that are one less than a perfect square. What is the remainder when the 1994th term of the sequence is divided by 1000?

Exercise 215. *(2013 ARML Individual Round #9)* Compute the smallest positive integer base b for which 16_b is prime and 97_b is a perfect square.

Exercise 216. *(2016 AMC 12B)* The sequence (a_n) is defined recursively by $a_0 = 1$, $a_1 = \sqrt[19]{2}$, and $a_n = a_{n-1} a_{n-2}^2$ for $n \geq 2$. What is the smallest positive integer k such that the product $a_1 a_2 \cdots a_k$ is an integer?

Exercise 217. Let a_n be the tens digit of 7^n. What is the remainder when $\sum_{k=1}^{2018} a_k$ is divided by 1000?

Extension Exercise 90.

(a) *(2016 AMC 12A)* How many ordered triples (x, y, z) of positive integers satisfy $\text{lcm}(x, y) = 72$, $\text{lcm}(x, z) = 600$ and $\text{lcm}(y, z) = 900$?

(b) (2016 AMC 12B) There are exactly $77,000$ ordered quadruplets (a, b, c, d) such that $\gcd(a, b, c, d) = 77$ and $\operatorname{lcm}(a, b, c, d) = n$. What is the smallest possible value for n?

Extension Exercise 91. (2016 AMC 12B) For a certain positive integer n less than 1000, the decimal equivalent of $\frac{1}{n}$ is $0.\overline{abcdef}$, a repeating decimal of period of 6, and the decimal equivalent of $\frac{1}{n+6}$ is $0.\overline{wxyz}$, a repeating decimal of period 4. What is n?

Extension Exercise 92. (1968 IMO P2) Find all natural numbers x such that the product of their digits (in decimal notation) is equal to $x^2 - 10x - 22$.

Extension Exercise 93. (1975 IMO P4) When 4444^{4444} is written in decimal notation, the sum of its digits is A. Let B be the sum of the digits of A. Find the sum of the digits of B. (A and B are written in decimal notation.)

Extension Exercise 94. Show that $\lim_{n \to \infty} \frac{\tau(n)}{\phi(n)} = 0$.

Chapter 14

Intermediate Number Theory

14.1 Modular Arithmetic: Part 2

We continue our study of modular arithmetic with a crucial cornerstone of Euclid's analysis of remainders and divisibility: none other than the almighty *Chinese Remainder Theorem* (CRT, for short).

14.1.1 Chinese Remainder Theorem

The CRT, in particular, is useful in the construction of computational methods with very large integers that would ordinarily be impossible to work with by hand. A very early formulation of the problem that eventually led to the modern, full-fledged statement of the CRT was roughly as follows:

> "There is a collection of things whose number is unknown. If we count them by threes, we have two left over; by fives, we have three left over; and by sevens, two are left over. How many things are there?"

This formulation is widely credited to the Chinese mathematician Sunzi, from circa the 3rd century AD. Later formulations of the problem invoked Gauss' notions of modular congruence, and Gauss himself applied the motivating ideas to the problem of determining when the Gregorian calendar repeats (where Gauss used the term "period"). Though we may be able to solve this problem specifically using the methods already at our disposal, the CRT serves to provide a rigorous foundation for them by proving existence for all such cases:

> **Theorem 14.1: Chinese Remainder Theorem**
>
> There are two equivalent ways to state the theorem.
>
> Let $n_1, n_2, n_3, \ldots, n_k \geq 2$ be integers, and let $N = \prod_{i=1}^{k} n_i$. The CRT says that, if the n_i are pairwise relatively prime (i.e. for all $1 \leq i < j \leq k$, $\gcd(n_i, n_j) = 1$), and if $0 \leq a_1, a_2, a_3, \ldots, a_k$ are integers such that $a_i \leq k_i$ for all $i \in [1, k]$, then there exists *exactly* one integer $x \in [0, N-1]$ for which the remainder of division of x by n_i is a_i.
>
> We may also state the CRT in terms of modular congruences (as Gauss probably would have preferred). If the n_i are pairwise relatively prime, and if $a_1, a_2, a_3, \ldots, a_k$ are integers such that $a_i < n_i$ for all $i \in [1, k]$ (as in the first formulation), then there exists a *unique* non-negative integer $x \leq N-1$ such that
>
> $$x \equiv a_1 \pmod{n}$$
> $$x \equiv a_2 \pmod{n}$$
> $$\vdots$$

$$x \equiv a_k \pmod{n}$$

As always, we must prove both existence and uniqueness. With the CRT, however, it is possible to prove them independently from each other, meaning that we do not need to rely on the proof of existence for the proof of uniqueness. There is an existence proof which is dependent on the uniqueness proof, but we will save this for last. For now, we will tackle the first existence proof, then the uniqueness proof, and close with the second existence proof.

First existence proof. This proof is *constructive*; i.e. we can construct x explicitly. We use an induction argument, where we begin with the base case of two modular congruences, and then extend this inductively to any number of congruences.

In the case of two moduli, namely the congruences

$$x \equiv a_1 \pmod{n_1}$$

$$x \equiv a_2 \pmod{n_2}$$

with $\gcd(n_1, n_2) = 1$, Bézout's Identity guarantees the existence of integers m_1, m_2 such that $m_1 n_1 + m_2 n_2 = 1$, given that $\gcd(n_1, n_2) = 1$. (m_1 and m_2 themselves can be found using the Euclidean Algorithm.) We have that $x = a_1 m_2 n_2 + a_2 m_1 n_1 = a_1(1 - m_1 n_1) + a_2(m_1 n_1) = a_1 + m_1 n_1 (a_2 - a_1)$, which shows that $x \equiv a_1 \pmod{n_1}$. The proof that $x \equiv a_2 \pmod{n_2}$ is identical.

Now consider the generalized system of congruences

$$x \equiv a_1 \pmod{n_1}$$

$$x \equiv a_2 \pmod{n_2}$$

$$\vdots$$

$$x \equiv a_k \pmod{n_k}$$

The first two congruences have a solution as extrapolated from the base case, thereby reducing the problem to a solution of $k-1$ modular congruences. We can repeat this until we get down to 2 congruences, which we can solve in the same manner as before - thereby proving existence. $\square$

Uniqueness proof. Assume that there exist two solutions to the system of congruences: x and y. Since x and y yield the same remainder upon being divided by n_i for all $i \in [1, k]$, $x - y \mid n_i$, $\forall i \in [1, k]$. Then $N = n_1 n_2 n_3 \cdots n_k$ is also a multiple of $x - y$, so $x \equiv y \pmod{N}$. But this may only be the case whenever $x = y$. $\square$

Second existence proof. Consider the map

$$x \mapsto (x \pmod{n_1}, x \pmod{n_2}, x \pmod{n_3}, \ldots, x \pmod{n_k})$$

By the proof of uniqueness, this map is an injective map, and is also surjective, since the domain and codomain have the same number of elements. Hence, the map is a bijection, which proves existence. $\square$

Example 14.1.1.1. *Solve the system of modular congruences*

$$x \equiv 0 \pmod{5}$$

$$x \equiv 1 \pmod{7}$$

$$x \equiv 3 \pmod{8}$$

Solution 14.1.1.1. *Since the moduli (5, 7, and 8) are pairwise coprime, there necessarily exists an x satisfying the congruences with $0 \leq x < lcm(5,7,8) = 280$. Since $x \equiv 0 \pmod 5$, $x = 5k_1$ for some $k_1 \in \mathbb{Z}$. In addition, $x \equiv 1 \pmod 7 \implies 7k_2 + 1, k_2 \in \mathbb{Z}$, and $x \equiv 3 \pmod 8 \implies 8k_3 + 3, k_3 \in \mathbb{Z}$. Thus $5k_1 = 7k_2 + 1 = 8k_3 + 3 \implies 7k_2 = 8k_3 + 2 \implies 8k_3 \equiv 5 \pmod 7$. Then $k_3 \equiv 5 \pmod 7$ (since $8 \equiv 1 \pmod 7$). Assume WLOG that $k_3 = 5$, so that $x \equiv 8(5) + 3 = 43 \pmod{56}$. By similar logic, $5k_1 = 56k_4 + 43 \implies 0 \equiv k_4 + 3 \pmod 5$ after reducing the equation to $\pmod 5$; this implies $k_4 \equiv 2 \pmod 5$. Hence, $k_4 = 2$ (since $0 \leq k_4 < 5$), so $x = 56(2) + 43 = \boxed{155} \pmod{280}$.*

Essentially, the CRT guarantees existence; using the Euclidean algorithm and/or divisibility rules, we can actually extend this to a computational solution. Note that x is "unique," in the sense that it can only assume one value *with respect* to a certain modulus. Needless to say, there are an infinite number of x that satisfy the congruences over the integers, but over the group $\mathbb{Z}/280$ (i.e. the group of integers modulo 280) in this example, x is uniquely determined.

Even if the moduli are not pairwise relatively prime, some systems of congruences can still have solutions. The following theorem states exactly when any given system of congruences is solvable:

Theorem 14.2: Solvability w.r.t. Non-Relatively Prime Moduli

The system of simultaneous congruences

$$x \equiv a_1 \pmod{n_1}$$
$$x \equiv a_2 \pmod{n_2}$$

is solvable in $x \in \mathbb{Z}/\text{lcm}(n_1 n_2)$ if and only if $a_1 \equiv a_2 \pmod{\gcd(n_1, n_2)}$.

Corollary 14.2: Existence for Coprime Moduli

If n_1, n_2 are coprime, then $\gcd(n_1, n_2) = 1$, and by definition, $n_1 \equiv n_2 \pmod 1$ for all $n_1, n_2 \in \mathbb{Z}$.

Proof. We can use Euclid's Algorithm to show that there exist integers s, t such that $tm_1 + sm_2 = \gcd(m_1, m_2)$. Since $n_2 - n_1 \equiv 0 \pmod{\gcd(m_1, m_2)}$, we have $tm_1 + sm_2 = n_2 - n_1$ (making a change of variables (t', s') for t and s if necessary). Thus, $x = tm_1 + n_1 = -sm_2 + n_2$, which proves existence.

To prove uniqueness, assume there exist two integers x, y such that both satisfy the system of congruences. We can observe that $x - y \equiv 0 \pmod{n_1}$, $x - y \equiv 0 \pmod{n_2}$ which together yield $x - y \equiv 0 \pmod{\text{lcm}(n_1, n_2)}$. $\square$

Similarly, we can show that the system of simultaneous congruences

$$x \equiv a_1 \pmod{n_1}$$
$$x \equiv a_2 \pmod{n_2}$$
$$x \equiv a_3 \pmod{n_3}$$
$$\vdots$$
$$x \equiv a_k \pmod{n_k}$$

is solvable in x iff all a_i, a_j are pairwise equivalent $\pmod{\gcd(n_i, n_j)}$ for $1 \leq i < j \leq k$. In addition, the solution is unique $\pmod{\text{lcm}(n_1, n_2, n_3, \ldots, n_k)}$ (which we have already shown). The first part of this proof will be left as an exercise for the reader.

But the previous exposition has admittedly been quite conceptual. How do we apply the CRT to more real-life, or at least competition math, situations? A typical tactic is to consider a given modulus and split it into two or more of its factors, as optimal. For instance, consider the following seemingly-straightforward modular arithmetic example that falls to CRT:

Example 14.1.1.2. *Find the last two digits of* 49^{2019}.

Solution 14.1.1.2. *This is equivalent to computing* 49^{2019} (mod 100)*. We have that* $100 = 25 \cdot 4$*, so it suffices to compute* 49^{2019} (mod 25) *and* 49^{2019} (mod 4)*. We have* $49^{2019} \equiv (-1)^{2019}$ (mod 25) $= -1$*, and* $49^{2019} \equiv 1^{2019} = 1$ (mod 4)*. Hence, we seek* x *(*$0 \leq x < 100$*) such that* $x \equiv -1$ (mod 25) *and* 1 (mod 4)*. By inspection,* $x \equiv \boxed{49}$ (mod 100)*.*

Though it is definitely possible to exploit the fact that $49^2 = 2401 \equiv 1$ (mod 100), this is not the aim of this section. Instead, we have opted for the more "immediate" CRT solution, though this is by no means any more or less correct than a "bash" solution, or a solution using, say, Euler's totient function.

We close out this section by making one more equivalent statement of the CRT, which first requires another definition:

> **Definition 14.1: Residue Classes**
>
> Let $f(x) : \mathbb{Z} \mapsto \mathbb{Z}$ be a function from the integers to the integers. Then the residue class of f modulo n is the set of all possible values of f (mod n).

Example 14.1.1.3. *The residue class of* $f : \mathbb{Z} \mapsto \mathbb{Z}$*,* $f(x) = x^2$ *modulo 5 is* $\{0, 1, 4\}$ *(since* $0^2 \equiv 0$ (mod 5)*,* $1^2, 4^2 \equiv 1$ (mod 5)*,* $2^2, 3^2 \equiv 4$ (mod 5)*).*

In modulo n, there are exactly n possible values (mod n), and so the residue class for (mod n) has n values, namely the set $\{0, 1, 2, 3, \ldots, n-2, n-1\}$.

Similarly, we can define a *complete residue system* of f:

> **Definition 14.2: Complete Residue System**
>
> A set of numbers $\{a_0, a_1, a_2, \ldots, a_{n-1}, a_n\}$ constitutes a *complete residue system* for f modulo n if, for each $0 \leq i \leq n$, $a_i \equiv i$ (mod n) (with duplicate values, of course, discarded).

Example 14.1.1.4. *The complete residue system of* $f : \mathbb{Z} \mapsto \mathbb{Z}$*,* $f(x) = x^2$ *modulo 5 is* $\{0, 1, 4\}$*.*

Using the notion of residue classes, we may restate the CRT as follows:

> **Theorem 14.3: Restatement of CRT in Terms of Residue Classes**
>
> If m and n are coprime positive integers greater than 1, then every pair of residue classes (mod m), (mod n) corresponds to a residue class (mod mn).

The proof of Theorem 14.3 is left as an exercise to the reader. (Hint: try to establish a bijection.)

14.1.2 Working with Periodic Sequences

Just as we can apply modular arithmetic to systems of congruences and singular expressions containing exponents, so we can apply them to sequences of numbers. In this section we will analyze some of the crucial centerpieces of number theory that revolve around both finite and infinite *periodic sequences,* and many of the beautiful, profound patterns vested within them.

Say that we want to compute the sum of the first n terms of a sequence a_n. For smaller values of n, this can be done easily by hand. But for larger values of n, it would take an eternity to list out all the terms up to a_n (not to mention, this method is extremely vulnerable to simple calculation errors). Instead, a much more viable approach would be to look for blatant patterns in the terms of the sequence. More specifically, do the terms ever repeat after a certain point (or even from the very beginning)? If so, how often do they repeat? If a sequence "loops" in this sense, we say that it is a *periodic* sequence:

> **Definition 14.3: Periodic Sequences**
>
> A sequence (a_n) is *periodic* with least period (or just period) p if, for all $n \geq 1$, $a_{n+p} = a_n$.

Example 14.1.2.1. *The sequence $\{1, 2, 3, 2, 1, 2, 3, \ldots\}$ is periodic (with $p = 3$), as is $a_n = \sin\left(\dfrac{\pi n}{2019}\right)$ (with $p = 4038$, since $\sin(x + 2\pi k) = \sin(x)$ for all $k \in \mathbb{Z}$; i.e. $\sin$ itself is periodic with $p = 2\pi$).*

A computational example taking advantage of the periodic nature of a sequence might proceed like the following:

Example 14.1.2.2. *For positive integers n, let the sequence (a_n) be defined as follows: $a_1 = 2$, $a_2 = 3$, and for all $n \geq 3$, a_n is the remainder when a_{n-1}^{n-2} is divided by 10. Compute the remainder when $a_1 + a_2 + a_3 + \ldots + a_{2019}$ is divided by 1000.*

Solution 14.1.2.1. *Lets list out the first few terms of this sequence: $2, 3, 8, 1, 8, 1, 8, 1, \ldots$. We notice that the 8s and 1s will repeat indefinitely, as $8^1 \equiv 8 \pmod{10}$ and $1^8 \equiv 1 \pmod{10}$. Since a_3 is the first 8 and a_4 is the first 1, there will be $\dfrac{2018 - 4}{2} + 1 = 1008$ blocks of 8s and 1s, plus another 8 to account for a_{2019}. The sum of the 2 and 3 is 5, the sum of the 1008 blocks of 8 and 1 is $1008 \cdot 9 = 8072$, and the final 8 brings the total up to $8085 \implies \boxed{85}$.*

A considerably more difficult, more abstract example plays with the idea of having a periodic function nested within another periodic function:

Example 14.1.2.3. *(2017 AMC 12B) The functions $\sin(x)$ and $\cos(x)$ are periodic with least period 2π. What is the least period of the function $\cos(\sin(x))$?*

Solution 14.1.2.2. *Note that $-1 \leq \sin(x) \leq 1$ for all $x \in \mathbb{R}$, and so $\cos(1) \leq \cos(\sin(x)) \leq 1$ for all $x \in \mathbb{R}$ (because $\cos$ is an even function; i.e. $\cos(-x) = \cos(x)$ for all $x \in \mathbb{R}$). Furthermore, every time $\sin(x) = 0$, $\cos(\sin(x)) = 1$ (which can be seen by considering the unit circle, with $\sin x$ as a parameter). The smallest value of $x > 0$ for which $\cos(\sin(x)) = 1$ is the smallest $x > 0$ for which $\sin(x) = 0$, which is $x = \boxed{\pi}$.*

Let's delve a little deeper into periodicity now. From this basic definition, we can extend the idea of a constantly repeating sequence to several generalizations, beginning with some that may seem intuitively clear and going through others that may be significantly less obvious:

> **Theorem 14.4: Periodicity of Roots of Unity**
>
> The sequence of powers of a root of unity is periodic.

Proof. Let this root of unity be z (such that $z^n = 1$ for some positive integer n), and let $(a_n) = \{z^1, z^2, z^3, \ldots\}$ be the sequence of powers of z. Then $z^n = 1$, and $z^{n+1} = z = z^1$, so $z^1 = z^{n+1}$. Similarly, $z^k = z^{n+k}$, and so $a_k = a_{n+k}$ for all n, k, which is the definition of periodicity. □

A related corollary generalizes the theorem to specific groups:

> **Corollary 14.4: Periodicity of Groups of Finite Order**
>
> Theorem 14.4 holds true for any group with finite order. That is, where G is a group, for any element

$g \in G$, the sequence $(a_n) := a_n = g^n$ is periodic.

We can apply this fact to a similar property of any function $f : X \mapsto X$ (with $x \in X$):

Definition 14.4: Periodic Points of a Function

A *periodic point* of a function $f : X \mapsto X$ is a point $x \in X$ such that the sequence

$$x, f(x), f^2(x), f^3(x), \ldots$$

(the *orbit* of f) is periodic (where $f^2(x) = f(f(x))$, $f^3(x) = f(f(f(x)))$, and in general, $f^n(x)$ denotes the result of applying f to itself n times)..

It is also important to make the distinction between a periodic sequence and a sequence that is *eventually* periodic. Whereas a periodic sequence with period p has $a_{n+p} = a_n$ for *all* $n \in \mathbb{N}$, this does not necessarily hold for eventually periodic sequences:

Definition 14.5: Eventually Periodic Sequences

A sequence (a_n) is *eventually periodic* (or EP) with period p if there exists $n \in \mathbb{N}$ for which, for all $m \geq n$, $a_{m+p} = a_m$.

Informally, we can "lop off" the first few terms of an eventually periodic sequence to obtain a periodic sequence (namely, the first $n-1$ terms).

Using this definition, we can make a crucial statement regarding the decimal expansion of any rational number:

Theorem 14.5: Periodicity of a Rational Decimal Expansion

The base-10 decimal expansion of any rational number is eventually periodic.

Proof. Let this number be $r = \dfrac{a}{b}$ for $a, b \in \mathbb{Z}$, $\gcd(a,b) = 1$. Suppose that r has a base-10 decimal expansion given by

$$r = \frac{a}{b} = c.d_1 d_2 d_3 \ldots$$

where $0 \leq d_i \leq 9$ for all $i \in \mathbb{N}$. Then $10^k \cdot r = c_k.d_{k+1}d_{k+2}d_{k+3}\ldots$ where c_k is such that $a \cdot 10^k = bc_k + r_k$, and $0 \leq r_k \leq b-1$ is the remainder upon dividing $a \cdot 10^k$ by b.

It follows that $\dfrac{r_k}{b} = 0.d_{k+1}d_{k+2}d_{k+3}\ldots$ By the Pigeonhole Principle, the remainder and decimal expansion will eventually repeat, such that $r_k = r_{k+p}$ for some $p \in \mathbb{N}$. Therefore, we arrive at $0.d_{k+1}d_{k+2}d_{k+3}\ldots = 0.d_{p+1}d_{p+2}d_{p+3}\ldots$ and the decimal expansion of r is periodic with period $p-k$. □

There is also another type of periodicity, *asymptotic periodicity*:

> **Definition 14.6: Asymptotically Periodic Sequences**
>
> A sequence (a_n) is *asymptotically periodic* (or AP) if there exists a periodic sequence (b_n) such that $\lim_{n \to \infty}(a_n - b_n) = 0$ (where a_n and b_n denote the nth terms of the sequences (a_n) and (b_n) respectively, *not* the sequences themselves).

Example 14.1.2.4. *An example of an AP sequence is*

$$\left\{\frac{1}{2}, \frac{1}{2}, \frac{3}{2}, \frac{5}{2}, \frac{1}{4}, \frac{3}{4}, \frac{7}{4}, \frac{11}{4}, \frac{1}{8}, \frac{7}{8}, \frac{15}{8}, \frac{23}{8}, \ldots\right\}$$

with respect to the periodic sequence $\{0, 1, 2, 3, 0, 1, 2, 3, \ldots\}$.

Loosely speaking, each term of an asymptotically periodic sequence approaches the corresponding term of a periodic sequence.

14.1.3 Examples

Example 14.1.3.1. *(2018 ARML Local Individual Round #4) For each real number x, let $f(x) = \sin\left(\frac{\pi x}{3}\right) + \cos\left(\frac{\pi x}{2}\right)$. Compute the least $p > 0$ such that $f(x + p) = f(x)$ for all real x.*

Solution 14.1.3.1. As $\sin\left(\frac{\pi x}{3}\right)$ has period 6, and $\cos\left(\frac{\pi x}{2}\right)$ has period 4, $f(x)$ has period $lcm(6, 4) = \boxed{12}$.

Example 14.1.3.2. *Show that the sequence $(a_k)_{k=1}^{\infty}$, $a_k = k^k$, is periodic modulo 10, and compute the period of the sequence.*

Solution 14.1.3.2. *Since k is itself periodic mod 10, so will a_k. This is a well-defined property of sequences; the period of the index will influence the period of the sequence (which is 10; in this case it remains constant, with each $k \in [0, 9]$ being mapped to a distinct mod 10 value).*

14.1.4 Exercises

Exercise 218. *For all positive integers $1 \leq n \leq 2019$, what is the largest value of $2019 \pmod{n}$?*

Exercise 219. *(2016 A-Star Math Tournament) Compute the remainder when $\underbrace{56666 \ldots 6666}_{2016 \text{ sixes}}$ is divided by 17.*

Exercise 220. *What is the remainder when 149162536496481100121 (the perfect squares from 1^2 to 99^2 concatenated together) is divided by 99?*

Exercise 221.

(a) *(2017 AMC 10B/12B) Let $N = 123456789101112\ldots 4344$ be the 79-digit number that is formed by writing the integers from 1 to 44 in order, one after the other. What is the remainder when N is divided by 45?*

(b) *If we write all of the positive integers from 1 to 19, inclusive, in consecutive order, and call the resulting number N, what is the remainder when N is divided by 330?*

(c) *(2017 AIME I) Let $a_{10} = 10$, and for each integer $n > 10$ let $a_n = 100a_{n-1} + n$. Find the least $n > 10$ such that a_n is a multiple of 99.*

Exercise 222. *(2018 Berkeley Math Tournament) Find*

$$\sum_{i=1}^{2016} i(i+1)(i+2) \bmod 2018$$

Exercise 223. *(2018 Berkeley Math Tournament) What is the remainder when $201820182018\ldots$ [2018 times] is divided by 15?*

Exercise 224. *For how many positive integers $x \leq 100$ is $3^x - x^2$ divisible by 5?*

Exercise 225. *Show that $111\ldots 111$ with 91 ones is composite.*

Exercise 226. *Compute the remainder when $1 \cdot 1! + 2 \cdot 2! + 3 \cdot 3! + \ldots + 2015 \cdot 2015!$ is divided by 2017. (Recall that 2017 is prime.)*

Exercise 227. *(2012 Stanford Math Tournament) Define a number to be boring if all the digits of the number are the same. How many positive integers less than 10,000 are both prime and boring?*

Exercise 228. *(2014 HMMT February Guts Round) Find the number of ordered quadruples of positive integers (a, b, c, d) such that a, b, c, and d are all (not necessarily distinct) factors of 30 and $abcd > 900$.*

Exercise 229. *(2014 HMMT February Team Round) Compute*

$$\sum_{k=0}^{100} \left\lfloor \frac{2^{100}}{2^{50} + 2^k} \right\rfloor.$$

Exercise 230. *(2004 AIME II) Let S be the set of integers between 1 and 2^{40} whose binary expansions have exactly two 1's. If a number is chosen at random from S, the probability that it is divisible by 9 is $\frac{p}{q}$, where p and q are relatively prime positive integers. Find $p + q$.*

Exercise 231. *Let $S(n)$ be the sum of the digits of n. Prove that $S(n) \equiv n \pmod{9}$.*

Exercise 232. *Find all ordered pairs of integers (m, n) such that $m^4 + n! = 2016$.*

Extension Exercise 95. *(2005 USAMO P2) Prove that the system*

$$x^6 + x^3 + x^3y + y = 147^{157}$$
$$x^3 + x^3y + y^2 + y + z^9 = 157^{147}$$

has no solutions in integers x, y, and z.

Extension Exercise 96. *Prove that if $n > 4$ is composite, then $(n-1)!$ is a multiple of n.*

Extension Exercise 97. *(2000 IMO Shortlist N4) Find all triplets of positive integers (a, m, n) such that $a^m + 1 \mid (a+1)^n$.*

Extension Exercise 98. *Find all ordered pairs of integers (m, n) such that $m^4 - n! = 2016$.*

Extension Exercise 99. *(1976 USAMO P3) Determine all integral solutions of $a^2 + b^2 + c^2 = a^2b^2$.*

14.2 Primality Theorems

We continue our analysis of prime numbers by expanding our territory past the numbers themselves and into some of the applications in which they are the most useful. The set of prime numbers, in and of itself, is already quite interesting, but when we take it our study of primes to a new level by placing them within other polynomials, functions, and expressions including exponents, we can truly begin to witness, and understand, the astonishing elegance of these building blocks of our number system.

14.2.1 Fermat's Little Theorem

Seemingly the best place to begin is *Fermat's Little Theorem*, commonly abbreviated FLT (and not to be confused with Fermat's *Last* Theorem, though the given context should almost always be enough to unambiguously differentiate the two). This handy tool in our ever-growing number-theoretic toolbox provides an impactful, yet very brief, statement that fits the pieces together in the jigsaw puzzle of prime moduli:

> **Theorem 14.6: Fermat's Little Theorem**
>
> Given relatively prime integers a, p, with p prime,
> $$a^{p-1} \equiv 1 \pmod{p}.$$

An equivalent form states that $a^p \equiv a \pmod{p}$; to prove this, multiply both sides of the congruence by a.

Proof. We use the FLT Rearrangement Lemma:

> **Lemma 14.6: FLT Rearrangement Lemma**
>
> The list $\{a, 2a, 3a, \ldots, a(p-1)\}$ is a rearrangement of $\{1, 2, 3, \ldots, p-1\}$ modulo p (that is, the second list is a permutation of the original list when its elements are reduced modulo p).

Proof. For contradiction, assume there exist two m, n such that $am \equiv an \pmod{p}$. But $\gcd(a, p) = 1$, and this violates the properties of modular arithmetic.

To elaborate, no term in the first list can be congruent to $0 \pmod{p}$, as a is coprime with p; this follows from Euclid's Lemma. Thus, each number in the first list must be found somewhere in the second list. To show that the two lists are one and the same $\pmod{p}$, we can invoke the contradiction argument above. The implication would be that $m \equiv n \pmod{p}$ by the cancellation law, but since both m and n lie between 0 and $p-1$, they must be equal. Hence, $a, 2a, 3a, \ldots, a(p-1)$ must be distinct $\pmod{p}$, completing the proof. $\square$

Then we note that
$$a^{n-1}(p-1)! \equiv (p-1)! \mod p$$
which implies
$$a^{n-1} \equiv 1 \pmod{p}$$
(We divide both sides by $(p-1)!$ which is relatively prime to p.) $\square$

We can also prove FLT using a combination of induction and the Binomial Theorem, as follows:

FLT induction proof using the Binomial Theorem. The base case is $a = 1$, but this is trivial (since every number divides zero). Suppose as the induction hypothesis that $p \mid (a^p - a)$. We then examine $(a+1)^p - (a+1) = ((a+1)^p - a^p - 1) + (a^p - a)$.

By the Binomial Theorem,

$$(a+1)^p = a^p \cdot \binom{p}{0} + a^{p-1} \cdot \binom{p}{1} + a^{p-2} \cdot \binom{p}{2} + \ldots + a \cdot \binom{p}{p-1} + 1 \cdot \binom{p}{p},$$

so $(a+1)^p - a^p - 1 = a^{p-1} \cdot \binom{p}{1} + a^{p-2} \cdot \binom{p}{2} + \ldots + a \cdot \binom{p}{p-1}$. Then observe that p is a factor of the RHS of the second equation (as $p \mid \binom{p}{k}$ for all $1 \leq k \leq p-1$), so $p \mid ((a+1)^p - a^p - 1)$, or $p \mid ((a+1)^p - (a+1))$ by the induction hypothesis. $\square$

It should be noted that the converse of FLT is not true; notable exceptions are the Carmichael numbers (see the following section, "Fermat Psuedoprimes"). But if we strengthen the false statement by adding an extra condition, then we end up with a true form, known as *Lehmer's theorem*:

> **Theorem 14.7: Lehmer's Theorem**
>
> If there exists an integer a such that $a^p \equiv a \pmod{p}$, and for all primes $q \mid (p-1)$, $a^{\frac{p-1}{q}} \not\equiv 1 \pmod{p}$, then p is prime.

We will not be proving this theorem here, but we will note that it is instrumental in a primality test known as the *Lucas-Lehmer primality test*, which is extremely useful in determining whether Mersenne numbers[56] are prime.

Fermat Pseudoprimes

Going back to the statement that the converse of FLT is false, p need not be prime if $a^p \equiv a \pmod{p}$. The set of counterexamples is the set of *Fermat pseudoprimes p in base a*. If a positive integer p is a Fermat pseudoprimes in base a for all a relatively prime with p, then p is a *Carmichael number*. For some sense of scale, the smallest Fermat pseudoprime in base 2 is 341, and the smallest Carmichael number is 561 (both of which are multiples of 11, interestingly. Coincidence? We'll leave it to you to decide!)

14.2.2 Euler's Totient Theorem

In a similar vein as Fermat's Little Theorem, *Euler's Totient Theorem* provides a majorly helpful statement regarding the power of an integer modulo m:

> **Theorem 14.8: Euler's Totient Theorem**
>
> Denote by $\phi(n)$ Euler's totient function (i.e. the number of positive integers less than or equal to n that are relatively prime to n). For a an integer and m a positive integer with $\gcd(a, m) = 1$, we have $a^{\phi(m)} \equiv 1 \pmod{m}$.

Proof. Let A be the set of positive integers relatively prime to m, namely

$$A = \{n_1, n_2, n_3, \ldots, n_{\phi(m)}\}$$

We claim that the set

$$B = \{an_1, an_2, an_3, \ldots, an_{\phi(m)}\}$$

is a permutation (rearrangement) of A modulo m. Since all elements in B are relatively prime with m, they are all distinct modulo m, and so B has the same elements as does A. It follows that

$$\prod_{k=1}^{\phi(m)} n_k \equiv a^{\phi(m)} \cdot \prod_{k=1}^{\phi(m)} n_k$$

---

[56] A Mersenne number is a positive integer of the form $2^n - 1$ for $n \in \mathbb{N}$.

$$\implies a^{\phi(m)} \equiv 1 \pmod{m}$$

as desired. □

> **Corollary 14.8: Fermat's Little Theorem**
>
> Fermat's Little Theorem is, indeed, a special case of ETT, where m is prime (and $\phi(m) = m - 1$ as a result).

> **Corollary 14.8: Modular Exponentiation**
>
> This second corollary of ETT can also be a corollary of FLT if n is prime.
>
> For every positive integer n, if a is an integer such that $\gcd(a, n) = 1$, then
> $$x \equiv y \pmod{\phi(n)} \implies a^x \equiv a^y \pmod{n}$$

Proof of Corollary 14.8. If $x \equiv y \pmod{\phi(n)}$, then $x = y + k \cdot \phi(n)$ for some integer k. It thus follows that
$$a^x = a^{y+k\cdot\phi(n)} = a^y \cdot (a^{\phi(n)})^k \equiv a^y \cdot 1^k = a^y \pmod{n}$$
as desired. □

The core ideas motivating the proof of Corollary 14.8 also play a pivotal role in the Lifting the Exponent Lemma (see §15).[57]

Example 14.2.2.1. *Compute the remainder when 2017^{2018} is divided by 2019.*

Solution 14.2.2.1. *We first note that $\gcd(2017, 2019) = 1$ (since not only is 2017 itself prime, it does not have a factor of 3 or 673, and is not a multiple of 2019). Thus, $2017^{\phi(2019)} \equiv 1 \pmod{2019} \implies (-2)^{1344} \equiv 1 \pmod{2019}$.*

To finish off the problem, observe that $2^{672} \equiv 1 \pmod{2019}$. Using the Chinese Remainder Theorem, we claim that $2^{672} \equiv 1 \pmod{3}$ and $2^{672} \equiv 1 \pmod{673}$ (where $\gcd(3, 673) = 1$). The first congruence is clear from observing that $2^2 = 4 \equiv 1 \pmod{3}$. To prove the second congruence, we note that $\gcd(2, 673) = 1$, and so $2^{\phi(673)} = 2^{672} \equiv 1 \pmod{673}$. Thus, $2^{672} \equiv 1 \pmod{2019}$ as well. This, together with $2^{1334} \equiv 1 \pmod{2019}$, implies $2^{2016} \equiv 1 \pmod{2019}$, or $2^{2018} \equiv 2017^{2018} \equiv \boxed{4} \pmod{2019}$.

14.2.3 Wilson's Theorem

Yet another one of the simply-stated yet deeply profound statements of number theory, *Wilson's Theorem* states the following about any prime p:

> **Theorem 14.9: Wilson's Theorem**
>
> For any prime number p, $(p-1)! \equiv -1 \pmod{p}$.

Wilson's Theorem, unlike FLT, serves as a primality test, since p is prime iff $(p-1)! \equiv -1 \pmod{p}$. In addition, for $n > 4$ composite, $(n-1)! \equiv 0 \pmod{n}$.[58]

---

[57] LTE

[58] To prove this, observe that n has (at least) two prime factors, p_1 and p_2. Since $p_1, p_2 < n$ strictly, both p_1 and p_2 are included in the prime factorization of $(n-1)! = (n-1) \cdot (n-2) \cdots 2 \cdot 1$.

Proof. Assume $p > 3$ (the cases $p = 2$ and $p = 3$ are trivial). Each integer between 1 and $p - 1$, inclusive, is relatively prime to p, so for each of these integers a_{p-1}, there exists an integer b such that $a_{p-1}b \equiv 1 \pmod{p}$. This b is uniquely determined, and furthermore, $a = b$ iff $a = 1$ or $a = p - 1$. The other integers can be grouped into pairs with product equivalent to 1 $\pmod{p}$, so $2 \cdot 3 \cdot 4 \cdots (p-2) \equiv 1 \pmod{p}$, or $(p-1)! \equiv -1 \pmod{p}$ after multiplying both sides by $p - 1$. $\square$

Alternate proof using FLT. Suppose $p \geq 3$ is an odd prime. Consider the polynomial

$$f(x) = \prod_{k=1}^{p-1}(x-k)$$

with degree $p - 1$. $f(x)$ has a leading term of x^{p-1}, constant term $(p-1)!$, and roots of $1, 2, 3, \ldots, p-1$.

Next, consider the polynomial $g(x) = x^{p-1} - 1$, which has degree $p - 1$ and leading term x^{p-1}. By FLT, $g(x)$ has the same roots as does $f(x)$ modulo p.

Finally, consider $h(x) = f(x) - g(x)$. With respect to mod p, h yet again has the same roots as f, and $\deg h \leq p - 2$. By Lagrange's Theorem (see the following section), $h \equiv 0 \pmod{p}$ cannot have more than $p - 2$ incongruent solutions $\pmod{p}$, so $h \equiv 0 \pmod{p}$ and its constant term is $(p-1)! + 1 \equiv 0 \pmod{p}$, which can be rearranged to obtain Wilson's Theorem. $\square$

> **Corollary 14.9: Wilson's Theorem for $p \equiv 1 \pmod{4}$**
>
> Iff $p \equiv 1 \pmod{4}$ is a prime number, with $p = 4k + 1$ for some $k \in \mathbb{N}$, then $(2k)!^2 \equiv -1 \pmod{p}$.

Proof. Using Wilson's Theorem for any odd prime $p = 2m + 1$, begin with the congruence

$$1 \cdot 2 \cdot 3 \cdots (p-1) \equiv -1 \pmod{p}$$

and rearrange it so that it becomes

$$1 \cdot (p-1) \cdot 2 \cdot (p-2) \cdots m \cdot (p-m) \equiv -1 \pmod{p}$$

Then it follows that

$$\prod_{i=1}^{m} i^2 \equiv (-1)^{m+1} \pmod{p}$$

$$\implies m!^2 \equiv (-1)^{m+1} \pmod{p}$$

Now let $p = 4k + 1$ for some integer k, and let $m = 2k$. Then $(2k!)^2 \equiv -1^{2k+1} \pmod{p}$, or $(2k!)^2 \equiv -1 \pmod{p}$, as desired (since $2k + 1$ is odd, and -1 raised to any odd power is still -1). $\square$

This is equivalent to the statement that -1 is a quadratic residue $\pmod{p}$ (see the following section, §14.3, "Symmetry and Quadratic Reciprocity").

In principle, Wilson's theorem can be used as a primality test, since it is both a necessary and sufficient condition for a number p to be prime. However, in principle, it is horribly inefficient, as it requires computing $(p-1)! \pmod{p}$, which is a computationally complex problem.[59]

Gauss later generalized Wilson's Theorem to a more rigorous form, which considers moduli that are prime powers, and not just prime numbers themselves:

---

[59] A problem's *computationally complexity* is a measure of its dependence on computational resources. The more resources (e.g. time, processing power, memory) a problem requires, the more complex it is said to be.

> **Theorem 14.10: Gauss' Generalization of Wilson's Theorem**
>
> Let $P(n)$ denote the product of the positive integers less than or equal to n that are relatively prime to n. Then
> $$P(n) \equiv \begin{cases} -1 \pmod{n} & n = 4, p^a, 2 \cdot p^a \\ 1 \pmod{n} & \text{otherwise} \end{cases}$$
> where p is a prime number and a is an arbitrary positive integer.

Before we begin to prove this compelling expansion of an already bottomlessly fascinating theorem, we should first rigorously answer the natural question upon which the statement relies: why should $P(n)$ always be equivalent to $\pm 1 \pmod{n}$?

Proof. For $n = 1, 2$, this is trivial, so assume that $n > 2$.

Let S_n be the set of positive integers less than or equal to n that are coprime with n, namely
$$S_n = \{a_1, a_2, a_3, \ldots, a_{\phi(n)}\}$$

For every integer $a_i \in S_n$, there exists a unique integer a_j with $1 \leq j \leq \phi(n)$ such that $a_i a_j \equiv 1 \pmod{n}$. If $a_i \neq a_j$, then by definition, $a_i a_j \equiv 1 \pmod{n}$, and there is nothing to prove.

If, however, $a_i = a_j$ (i.e. $i = j$), then note that if $a_i^2 \equiv 1 \pmod{n}$, then we must also have $(n - a_i)^2 \equiv 1 \pmod{n}$. Furthermore, since $n > 2$, $a_i \not\equiv -a_i \pmod{n}$, as a_i is relatively prime to n, which requires that n be odd. Hence, it follows that $x(n - x) \equiv -1 \pmod{n}$. Thus, the product $P(n) = \prod_{i=1}^{\phi(n)} a_i \equiv \pm 1 \pmod{n}$, which is what we originally claimed. $\square$

At last, we can proceed to prove the Gaussian generalization:

Proof. Note: The following proof is adapted from a 2009 paper of R. Andrew Ohana, "A Generalization of Wilson's Theorem."[60]

For $n = 1, 2$, the result is trivial, so assume $n \geq 3$. Because a_i is relatively prime with n, as before, there exists an a_j such that $a_i a_j = 1$.

Using similar logic as in the immediately preceding proof, we can "pair up" a_i terms such that an end term is paired up with a beginning term, and so on down the line. Consider $f(x) = x^2 - 1$ with roots $x = \pm 1$. $i = j$ iff one of a_i or a_j is a root of f. If a_i is a root of f, then so must be $n - a_i$, as $n > 2$.

Let P_1 denote the product of the roots of $f(x) = x^2 - 1$. We claim that, with the exception of $n = 4, p^a, 2 \cdot p^a$, the number of roots of the corresponding polynomial is a multiple of 4. Consider $n = 2^a$; if $n = 4$, then there are two roots. If $a \geq 3$, then if one of the factors of f is a multiple of 2, then so is the other, but only one may also be a multiple of 4 (or some other larger power of 2). If $2 \mid (x + 1)$, but $4 \nmid (x + 1)$, then $f(2) \equiv 0$ with respect to the quotient ring $\mathbb{Z}/(2^{a-1})$. This yields roots of $x = 1$ and $x = 2^{a-1} - 1$. Similarly, if $2 \mid (x - 1)$ but $4 \nmid (x - 1)$, then $x = -1$ and $= -1 - 2^{a-1}$, so all four roots are clearly distinct.

Now consider the general form, $n = 2^a \cdot p_1^{k_1} p_2^{k_2} p_3^{k_3} \cdots p_j^{k_j}$ where the p_i ($1 \leq i \leq j$) are the prime factors of n. We seek the roots of $f(x)$ over $\mathbb{Z}/(2^a)$ and $\mathbb{Z}/(p_i^{k_i})$. Using CRT, we can conclude that the roots of f in $\mathbb{Z}/n$ come from selecting a root from each of the $p_i^{k_i}$ terms and computing the roots over $\mathbb{Z}/(p_i^{k_i})$. If n is odd ($a = 0$), then each of the equations will have two solutions, so there will be 2^j roots in total. If $a = 1$, then there will still be 2^j roots, as $f(x)$ will have a single root over $\mathbb{Z}/2$ (i.e. $-1 \equiv 1 \pmod{2}$). If $a = 2$, there will be two roots over $\mathbb{Z}/4$, and so the total number of roots is instead 2^{j+1}. Finally, if $a \geq 3$,

---
[60] Ohana, 2009.

then there will always be four roots in $\mathbb{Z}/2^a$, making the total number of roots 2^{j+2}. All of these quantities are divisible by 4, so unless $n = 4, p^a, 2 \cdot p^a$, the number of roots will be a multiple of four.

To conclude, observe that if $f(a_i) = 0$, then $a_i(n - a_i) = -1$. If f has a number of roots that is divisible by 4, then $Q_1 = 1$; otherwise, $Q_1 = -1$. Let Q_2 be the product of the values of a_i that are *not* roots of f (and if there are none, then $Q_2 = 1$ by convention). If such a_i do exist, however, then necessarily $Q_2 = 1$ (since $a_i a_j = 1$). So $Q_2 = 1$, and since $P(n) = Q_1 Q_2$, we obtain the desired generalization, namely

$$P(n) \equiv \begin{cases} -1 \pmod{n} & n = 4, p^a, 2 \cdot p^a \\ 1 \pmod{n} & \text{otherwise} \end{cases}$$

thereby completing the proof. □

There is one last point to address before we will have covered all our bases for Wilson's Theorem. A natural extension of the problem of finding primes lies in determining which primes p have their squares divide $(p-1)! + 1$. Such primes p are known as *Wilson primes*. The only known Wilson primes are 5, 13, and 563, the first being trivial by inspection and the second also not impossible to verify by hand (and the third, similarly, not too hard to verify using a computer program such as Wolfram Alpha or Mathematica). However, if there were to exist any more Wilson primes, they would need to exceed $2 \cdot 10^{13}$ (20 trillion).[61]

14.2.4 Lagrange Theorem

Named for Joseph-Louis Lagrange, an Enlightenment-era highly prolific mathematician, physicist, and astronomer, the *Lagrange Theorem* reflects a certain simplicity that, while somewhat in stark contrast to his most well-known and important contributions in the fields of physics and mathematical analysis, is nonetheless just as much of a stepping stone in the world of number theory:

> **Theorem 14.11: Lagrange Theorem**
>
> If p is a prime number, and $f(x)$ is a polynomial with integer coefficients, then either
>
> - every coefficient of $f(x)$ is a multiple of p, or
> - $f(x) \equiv 0 \pmod{p}$ has at most $\deg f$ incongruent solutions wrt $\pmod{p}$. (Two solutions are *incongruent* wrt $\pmod{p}$ if they are not equivalent $\pmod{p}$.)

Proof. Let $g(x)$ be the polynomial obtained from taking the coefficients of $f(x)$ modulo p. We have that $p \mid g(x)$ iff $g(x) = 0$, and furthermore that $g(x)$ has a number of roots less than or equal to $\deg g(x)$.

More specifically, observe that $g(x) = 0$ iff all coefficients of f are multiples of p. Assume that $g(x) \neq 0$; then $\deg g(x)$ is well-defined, and additionally $\deg g(x) \leq \deg f(x)$. To prove that $p \mid g(x) \iff g(x) = 0$, observe that $g(x) = 0 \iff f(x) \equiv 0 \pmod{p}$ from substitution and division over $\mathbb{Z}/p$, i.e. $p \mid f(x)$. To prove that the number of roots of $g(x)$ is not greater than $\deg g(x)$, note that $\mathbb{Z}/p$ is a field (which we will not rigorously prove here, due to group theory being outside the scope of this book), in turn because it is a finite integral domain, which immediately finishes the proof. Alternatively, we can apply Euclid's division algorithm (or the Fundamental Theorem of Algebra) to conclude that a non-zero polynomial can only have as many roots as its degree.

To finish off, note that two solutions $f(x_1), f(x_2) \equiv 0 \pmod{p}$ are incongruent iff $x_1 \not\equiv x_2 \pmod{p}$. The number of incongruent solutions is then equal to the number of roots of $g(x)$, which itself is less than or equal to $\deg g(x) \leq \deg f(x)$. □

---

[61] As of 2012; originally 500 million in 2009, according to a result cited in Ohana (2009), "A Search for Wierferich and Wilson Primes," by R. Crandall, K. Dilcher, and C. Pomerance.

It is at this point that we take a bit of a digression to briefly revisit Fermat's Little Theorem, equipped with our newly-acquired tool of Lagrange's Theorem. Using elementary group theory, along with Lagrange's Theorem, we can also prove Fermat's Little Theorem:

Alternate FLT proof using Lagrange's Theorem. First observe that the set $G = \{1, 2, 3, \ldots, p-1\}$ is in fact a group, equipped with the operation of multiplication (mod p). We show that G obeys all axioms of a group, namely closure under multiplication (mod p) (that the product of any two elements in G modulo p is also in G), associativity (that for all $a, b, c \in G$, $(a \cdot b) \cdot c = a \cdot (b \cdot c)$, where $\cdot$ is the operation with which the group G is equipped, in this case multiplication modulo p), the existence of an identity element (an element $i \in G$ such that, for all $a \in G$, $a \cdot i = i \cdot a = a$), and the existence of an inverse element (an element $b = a^{-1} \in G$ such that, for all $a \in G$, $a \cdot b = b \cdot a = i$, where i denotes the identity element as above).

Since $0 \notin G$, for any two elements $a, b \in G$, $a \cdot b \in G$ as well (as all possible values of the operation (mod p), except for 0, are in G); this proves closure. Since multiplication over $\mathbb{R}$ is associative, so is multiplication over $\mathbb{Z}/p$. (Proving this is left as an exercise.) In addition, the identity element is 1, since for any $a \in G$, $a \cdot 1 = 1 \cdot a = a$. Finally, we prove the existence of an inverse element $b \in G$ (this will take a bit more doing).

First note that $\gcd(b, p) = 1$, so we can apply Bezout's identity to conclude that there must exist integers x, y such that $bx + py = 1$. Reducing this (mod p) yields $bx \equiv 1$ (mod p), or $b = x^{-1}$ over $\mathbb{Z}/p$ (with $x \neq 0$). Hence, G is a group, as it satisfies all the axioms of a group.

Using the fact that G is a group, let $a \in [1, p-1]$ be an integer, $a \in G$. Let k be the smallest positive integer such that $a^k \equiv 1$ (mod p) (the *order* of a modulo p; see §15.3, "Order and Cyclotomic Polynomials," for an expanded definition). Then the list $\{1, a, a^2, \ldots, a^{k-1}\}$, when reduced modulo p, is a subgroup[62] of G with order k, and so by Lagrange's Theorem, $k \mid \operatorname{ord}_p(G) = p - 1$ (the order of G modulo p). Then $p - 1 = km$ for some $m \in \mathbb{N}$, from which it follows that

$$a^{p-1} \equiv a^{km} = (a^k)^m \equiv 1^m \equiv 1 \pmod{p}$$

as desired. $\square$

Indeed, Euler himself found himself inspired and motivated by the above proof, and constructed the following third proof of the theorem, building on the ideas from the second proof:

Euler's FLT proof. Let A be the set $\{1, a, a^2, \ldots, a^{k-1}\}$ reduced modulo p. If $A = G$, then we are done since $k = p - 1$ and $k \mid (p-1)$ as a result. If $A \neq G$, there exists $b_1 \in G \backslash A$.

Let A_1 be the set $\{b_1, ab_1, a^2 b_1, \ldots, a^{k-1} b_1\}$, again reduced modulo p. We have $|A_1| = k$ (to prove this, assume the contrary; then there would exist $m, n \in A_1$ such that $a^m b_1 \equiv a^n b_1$ (mod p), but this would imply $a^m \equiv a^n$ (mod p), which is impossible). In addition, by similar logic, no element in A_1 can also be in A.

Now consider the set $B = A \cup A_1$, where $|B| = 2k$. If $B = G$, then $2k = p - 1$, and so $k \mid (p-1)$. If $B \neq G$, then there exists $b_2 \in G \backslash B$, and we can repeat the process for sets $A_2, A_3, A_4, \ldots$ But since G is finite, the process must terminate eventually, which shows that $k \mid (p-1)$. $\square$

14.2.5 Examples

Example 14.2.5.1. *Show that $\dfrac{18! + 1}{437}$ is an integer.*

Solution 14.2.5.1. *Note that $437 = 23 \cdot 19$, so it suffices to show $19 \mid (18! + 1)$ and $23 \mid (18! + 1)$. The first of these follows from Wilson's Theorem with $p = 19$. The latter follows again from Wilson's Theorem, only we need to backtrack from $22! \equiv -1 \equiv 22$ (mod 23), or $21! \equiv 1$ (mod 23). Hence, by inverses, $20! \equiv 11$ (mod 23), $19! \equiv 4$ (mod 23), and $18! \equiv -1$ (mod 23), as desired.*

---

[62] A subset $H \subseteq G$ that is itself a group.

Example 14.2.5.2. *Compute* $(2016! - 2015!) \pmod{2017}$.

Solution 14.2.5.2. $2016! - 2015! = 2015! \cdot 2015$. By Wilson's Theorem, $2016! \equiv 2016 \pmod{2017} \implies 2015! \equiv 1 \pmod{2017}$, so $2016! - 2015! \equiv \boxed{2015} \pmod{2017}$.

Example 14.2.5.3. *(2015 Berkeley Math Tournament) Suppose $k > 3$ is a divisor of $2p + 1$, where p is prime. Prove that $k \geq 2p + 1$.*

Solution 14.2.5.3. *WLOG assume k is also prime. Then $2^p \equiv -1 \pmod{k}$, or $2^{2p} \equiv 1 \pmod{k}$. By FLT, $2^{k-1} \equiv 1 \pmod{k}$, so $k \geq 2p + 1$.*

14.2.6 Exercises

Exercise 233. *(2016 ARML Local Individual Round #6) Compute the remainder when $100!$ is divided by 103.*

Exercise 234. *What is the remainder when 3^{2019} is divided by 19?*

Exercise 235. *Compute $2^{20} + 3^{30} + 4^{40} + 5^{50} + 6^{60} \pmod{7}$.*

Exercise 236. *(2008 AMC 12A) Let $k = 2008^2 + 2^{2008}$. What is the units digit of $k^2 + 2^k$?*

Exercise 237. *(2002 AMC 12B) The positive integers $A, B, A - B,$ and $A + B$ are all prime numbers. What is their sum?*

Exercise 238. *Let p be a prime number, and let k be a positive integer. Prove that the decimal representation of $\dfrac{k}{p}$ consists of $d \mid (p-1)$ repeating digits.*

Exercise 239. *Show that the converse of Fermat's Little Theorem is false.*

Exercise 240. *(2017 HMMT February Guts Round) At a recent math contest, Evan was asked to find 2^{2016} $\pmod{p}$ for a given prime number p with $100 < p < 500$. Evan has forgotten what the prime p was, but still remembers how he solved it:*

- *Evan first tried taking 2016 modulo $p - 1$, but got a value e larger than 100.*

- *However, Evan noted that $e - \dfrac{1}{2}(p-1) = 21$, and then realized the answer was $-2^{21} \pmod{p}$.*

What was the prime p?

Extension Exercise 100. *(1989 IMO P5) Prove that for each positive integer n, we can find a set of n consecutive integers such that none of them is a power of a prime number.*

Extension Exercise 101. *(2003 IMO Shortlist N6) Let p be a prime number. Prove that there exists a prime number q such that for every integer n, the number $n^p - p$ is not divisible by q.*

Extension Exercise 102. *(2016 HMMT February Algebra Test) Define $\phi^!(n)$ as the product of all positive integers less than or equal to n and relatively prime to n. Compute the number of integers $2 \leq n \leq 50$ such that n divides $\phi^!(n) + 1$.*

14.3 Symmetry and Quadratic Residues/Reciprocity

] When dealing with modular congruences with respect to a prime modulus, the beautiful symmetry vested within modular arithmetic begins to show its true colors. The study of *quadratic reciprocity* allows us to compute what values that a perfect square may assume modulo p, and thus to seamlessly work with operations such as the *Legendre symbol*, and to prove weighty statements such as the *law of quadratic reciprocity*. Furthermore, quadratic residues have a veritable slew of applications in the field of cryptography, where they serve as aids in finding the square root of a large number, and in constructing the Rabin and Goldwasser-Micali cryptosystems.

Before we study quadratic reciprocity, we must go through one prerequisite definition: that of *quadratic residues*.

> **Definition 14.7: Quadratic Residue** (mod p)
>
> Let p be an odd prime number. If there exists a non-zero integer a such that $a^2 \equiv q \pmod{p}$, then q is said to be a *quadratic residue modulo p*. (Traditionally, we exclude $q = 0$ from the set of quadratic residues modulo p.) Otherwise, q is a *quadratic non-residue modulo p*.

Example 14.3.0.1. *The set of quadratic residues* (mod 7) *is* $\{1, 2, 4\}$, *since* $1^2 \equiv 6^2 \equiv 1 \pmod{7}$, $2^2 \equiv 5^2 \equiv 4 \pmod{7}$, *and* $3^2 \equiv 4^2 \equiv 2 \pmod{7}$ *(note that* $a^2 \equiv (7-a)^2 \pmod{7}$; *this result has already been proven, but we can also demonstrate it by expanding both sides to obtain* $a^2 \equiv 49 - 14a + a^2 \pmod{7} \implies 0 \equiv 49 - 14a \pmod{7} \implies 7(7 - 2a) \equiv 0 \pmod{7}$, *which is clearly true. The same goes for any congruence of the form* $a^2 \equiv (n-a)^2 \pmod{n}$).

Remark. Because of the symmetry $a^2 \equiv (n-a)^2 \pmod{n}$, the values of a to check in order to determine the set of quadratic residues are limited to the interval $a \in \left[1, \lfloor \frac{p}{2} \rfloor \right]$.

If p is an odd prime, there are $\dfrac{p+1}{2}$ quadratic residues (mod p), by Euler's Criterion (see §14.3.3). If q is a quadratic residue modulo p, then $q^{-1} \pmod{p}$ is also a residue, and vice versa. If q is a non-residue, then so is q^{-1}. Thus, it follows that there are an equal number of (non-zero) residues and non-residues modulo p. Furthermore, the product of two residues, or two non-residues, is a residue, while the product of a residue and a non-residue is a non-residue.

If the modulus p is a prime power, namely p^n for some positive integer $n \geq 2$, then the properties of the quadratic residues (mod p^n) differ between $p = 2$ and p odd. In the former case, note that all odd squares are equivalent to 1 (mod 4), so q can be a residue modulo p iff $q \equiv 1 \pmod{8}$. Thus, a nonzero number is a residue (mod p^n) if it can be expressed in the form $4^k(8n+1)$.

In the latter case, modulo an odd prime power p^n, residues and non-residues obey the same properties as they do modulo p, provided they are relatively prime to p.

14.3.1 The Legendre Symbol

Building upon our fundamental idea of quadratic residues and non-residues, we can now expand this to a crucial definition that provides the backbone for much of our future study of modular arithmetic:

> **Definition 14.8: The Legendre Symbol**
>
> Let p be an odd prime number. Then the *Legendre symbol*, which takes an integer a as input, is defined

as

$$\left(\frac{a}{p}\right) = \begin{cases} 1 & a \text{ is a QR modulo } p,\ a \not\equiv 0 \pmod{p} \\ -1 & a \text{ is a QNR modulo } p \\ 0 & a \equiv 0 \pmod{p} \end{cases}$$

The original definition provided by Legendre was a condensed version of the above:

$$\left(\frac{a}{p}\right) = a^{\frac{p-1}{2}} \pmod{p}, \left(\frac{a}{p}\right) \in \{-1, 0, 1\}$$

Remark. By Euler's Criterion, these two definitions are equivalent.

We now list and prove some crucial properties of the Legendre symbol:

Theorem 14.12: Properties of the Legendre Symbol

- The Legendre symbol is *periodic* in its first (top) argument; i.e. if $a \equiv b \pmod{p}$, then $\left(\frac{a}{p}\right) = \left(\frac{b}{p}\right)$ (that is, either a and b are both QRs modulo p or neither of them are).

- In its top argument, the Legendre symbol is completely multiplicative; i.e. $\left(\frac{a}{p}\right) \cdot \left(\frac{b}{p}\right) = \left(\frac{ab}{p}\right)$ for all integers a and b, even if a and b are not coprime.

- If p is an odd prime, the product of two QRs is a QR, as is the product of two QNRs (modulo p). Similarly, the product of a QR and QNR is a QNR mod p. A special case is $\left(\frac{x^2}{p}\right)$, which is 1 if $x \nmid p$, and 0 if $x \mid p$. If p is not an odd prime, it is still guaranteed that the product of two QRs is a QR, and that the product of a QR and QNR is a QNR, but it is not certain that the product of two QNRs is a QR.

- The first supplement to the law of quadratic reciprocity: $\left(\frac{-1}{p}\right) = (-1)^{\frac{p-1}{2}}$ which is equal to 1 if $p \equiv 1 \pmod{4}$ and -1 if $p \equiv -1 \pmod{4}$.

- The second supplement to the law of quadratic reciprocity: $\left(\frac{2}{p}\right) = (-1)^{\frac{p^2-1}{8}}$ which is equal to 1 if $p \equiv 1, 7 \pmod{8}$ or -1 if $p \equiv 3, 5 \pmod{8}$.

Proof.

- This follows directly from symmetry modulo p (namely, the property $a^2 \equiv (p-a)^2 \pmod{p}$).

- This is an Exercise.

- To show that the product of two QRs is itself a QR, take $x \equiv a^2 \pmod{p}$, $y \equiv b^2 \pmod{p}$, so that $xy \equiv (ab)^2 \pmod{p}$.

- This follows immediately from Euler's Criterion (§14.3.3) with $a = -1$. The rest directly follows from proving the second property.

- Consider the set $S = \{2, 4, 6, 8, \ldots, p-1\}$ of even positive integers up to, and including, $p - 1$. By Gauss' Lemma (Lemma 14.13 in the following section), we have $\left(\frac{2}{p}\right) = (-1)^n$, where n is the number of elements in S with least positive residue[63] greater than $\frac{p}{2}$ modulo p. Since all elements of S are

---

[63]The smallest positive integer in the residue class modulo p.

less than p, it suffices to count how many of them are less than (or equal to) $\frac{p}{2}$. This is just $\lfloor \frac{p}{4} \rfloor$, so $n = \frac{p-1}{2} - \lfloor \frac{p}{4} \rfloor$. If we consider each residue class modulo 8 separately, n is even iff $p \equiv 1, 7 \pmod{8}$, and odd iff $p \equiv 3, 5 \pmod{8}$. A final application of Gauss' Lemma finishes the proof.

These properties will all be put to extensive use in proofs of the law of quadratic reciprocity, as well as in secondary lemmas that play essential roles in deriving that result. □

We can define alongside the traditional Legendre symbol, a generalization to all odd numbers in the lower argument: the *Jacobi symbol*.

> **Definition 14.9: Jacobi Symbol**
>
> For any integer a and positive odd integer n, the *Jacobi symbol* $\left(\frac{a}{n}\right)$ (with the same notation as the Legendre symbol, the two being differentiated based on context) is the product of the Legendre symbols with lower arguments consisting of the prime factors of n, such that
> $$\left(\frac{a}{n}\right) = \prod_{i=1}^{k} \left(\frac{a}{p_i}\right)$$
> where $p_1, p_2, p_3, \ldots, p_k$ are the prime factors of n.

> **Corollary 14.12: For p an Odd Prime**
>
> When p is an odd prime, the Jacobi and Legendre symbols assume the same value.

Much like the Legendre symbol, the Jacobi symbol has a wide berth of practical applications, such as in computer science, cryptography, factorization, and especially primality testing. In addition, the two symbols share most of their intrinsic properties, such as complete multiplicativity in each argument (if the other is fixed), modular equivalence, and the law of quadratic reciprocity and its two supplements (see the following section).

However, **there is one key difference:** if $\left(\frac{a}{n}\right) = 1$, whether a is a quadratic residue mod n cannot be determined. In order for a to be a QR mod n, it must be a QR modulo every prime factor of n. However, the Jacobi symbol can assume the value of 1 if a is a non-residue modulo an even number of prime factors of n.

14.3.2 The Law of Quadratic Reciprocity

Initially conjectured by Legendre along with Euler, and proven by Gauss, the *law of quadratic reciprocity* was collectively known to them as the "golden theorem" (and for good reason), and to Gauss in particular as the "fundamental theorem," presumably of all of number theory. To this day, the theorem's generalizations to other number-theoretic disciplines, modern algebra, and algebraic geometry, are of exceptional mathematical interest.

> **Theorem 14.13: Law of Quadratic Reciprocity**
>
> Let p and q be *distinct* odd primes. Then
> $$\left(\frac{p}{q}\right)\left(\frac{q}{p}\right) = (-1)^{\frac{p-1}{2} \cdot \frac{q-1}{2}}$$
> where $\left(\frac{p}{q}\right)$ denotes the Legendre symbol.

Using this theorem, we can easily calculate a Legendre symbol based on the output of the product of $\left(\frac{p}{q}\right)$ and $\left(\frac{q}{p}\right)$; that is, we can determine whether or not q is a QR modulo p (or, equivalently, whether p is a QR modulo q, as necessary). However, we cannot determine what the actual solutions are to the resulting congruence; the law of quadratic reciprocity is *non-constructive* in this sense.

Much like the famous Pythagorean Theorem, the (admittedly less famous) law of quadratic reciprocity has a wealth of proofs, numbering well into the hundreds! We will present two of the more accessible proofs that, in the authors' view, most perfectly embody the brilliant elegance characteristic of the theorem itself.

First proof: Using Gauss' Lemma.

Note: The following is adapted from William Stein, 2007.[64]

For this proof, we will be using Gauss' Lemma, which derives a very simple closed-form expression for $\left(\frac{q}{p}\right)$ in terms of a reduction of a set modulo p:

> **Lemma 14.13: Gauss' Lemma**
>
> Let p be an odd prime, and let a be an integer not divisible by p. Consider the set
> $$S = \left\{a, 2a, 3a, \ldots, \frac{a(p-1)}{2}\right\}$$
> and reduce it modulo p, such that for each element of the set we can find a number in the interval $\left(-\frac{p}{2}, \frac{p}{2}\right)$ that is congruent to the original element modulo p. Let v be the number of negative numbers in the reduced set. Then
> $$\left(\frac{q}{p}\right) = (-1)^v.$$

Proof of Gauss' Lemma. Reducing S to another set T, we obtain
$$S = \left\{a, 2a, 3a, \ldots, \frac{a(p-1)}{2}\right\} \equiv \left\{1, -1, 2, -2, \ldots, \frac{p-1}{2}, \frac{1-p}{2}\right\}$$

Note that no element appears more than once; if there were to exist a duplicate element, then either two elements of S would be congruent modulo p, or two distinct elements of S would sum to 0 (mod p). Thus, T must be of the form
$$T = \left\{\pm 1, \pm 2, \ldots, \pm \frac{p-1}{2}\right\}$$

Multiplying S and T together yields
$$a \cdot 2a \cdot 3a \cdots \frac{(p-1)a}{2} \equiv (\pm 1) \cdot (\pm 2) \cdot (\pm 3) \cdots \left(\pm \frac{p-1}{2}\right) \pmod{p}$$

---
[64]Source: https://wstein.org/edu/2007/spring/ent/ent-html/node52.html#lem:even

Hence, $a^{\frac{p-1}{2}} \equiv \pm 1 \pmod{p}$, from which the desired result follows. □

We can now apply a second lemma derived by Euler:

> **Lemma 14.13: Euler's Lemma**
>
> Let p be an odd prime, and let a be a positive integer with $p \nmid a$. If $q \equiv \pm p \pmod{4a}$ is a prime number, then $\left(\frac{a}{p}\right) = \left(\frac{a}{q}\right)$.

Proof. Let $S = \{a, 2a, 3a, \ldots, \frac{a(p-1)}{2}\}$. Define a union of intervals

$$I = \left(\frac{p}{2}, p\right) \cup \left(\frac{3p}{2}, 2p\right) \cup \left(\frac{5p}{2}, 3p\right) \cup \ldots \cup \left(\left(b - \frac{1}{2}\right)p, bp\right)$$

where $b = \frac{a}{2}$ or $\frac{a-1}{2}$, whichever is an integer.

By Gauss' Lemma, $\left(\frac{a}{p}\right) = (-1)^{|S \cap I|}$, since we are considering the number of negative elements in S. To actually compute $|S \cap I|$, note that

$$|S \cap I| = \left|\frac{1}{a} \cdot S \cap \frac{1}{a} \cdot I\right|$$

where a is a non-zero constant, and as such,

$$\frac{1}{a} \cdot S = \left\{1, 2, 3, \ldots, \frac{p-1}{2}\right\}$$

$$\frac{1}{a} \cdot I = \left(\frac{p}{2a}, \frac{p}{a}\right) \cup \left(\frac{3p}{2a}, \frac{2p}{a}\right) \cup \ldots \cup \left(\frac{p(2b-1)}{2a}, \frac{bp}{a}\right)$$

If we express p in the form $4ac + r$, then we have another union of intervals

$$J = \left(\frac{r}{2a}, \frac{r}{a}\right) \cup \left(\frac{3r}{2a}, \frac{2r}{a}\right) \cup \ldots \cup \left(\frac{p(2b-1)}{2a}, \frac{bp}{a}\right)$$

Note that $\frac{1}{a} \cdot I$ and J differ only by an even integer. We now invoke the following lemma:

> **Lemma 14.13: Integers in an Interval**
>
> Let a and b be rational numbers, and let $(a, b) \cap \mathbb{Z} = \{x \in \mathbb{Z} \mid a \leq x \leq b\}$. Then for any integer n,
>
> $$|(a, b) \cap \mathbb{Z}| = |(a, b + 2n) \cap \mathbb{Z}| \pmod{2}$$
>
> $$|(a, b) \cap \mathbb{Z}| = |(a - 2n, b) \cap \mathbb{Z}|$$
>
> assuming that each interval is non-empty.

Proof. Since n is positive, we have $(a, b + 2n) = (a, b) \cup [b, b + 2n)$. There are $2n$ integers in this interval, namely $\lceil b \rceil, \lceil b \rceil + 1, \lceil b \rceil + 2, \ldots, \lceil b \rceil + 2n - 1$ in the interval $[b, b + 2n)$, so the first congruence is true. Similarly, $(a, b - 2n) = (a, b) \setminus [b - 2n, b)$, so the proof is identical and we are done. □

which implies that $v = \left|\frac{1}{a} \cdot S \cap J\right| \pmod 2$. Therefore, v is dependent only on $p \pmod{4a}$, and so $q \equiv p \pmod{4a} \implies \left(\frac{a}{p}\right) = \left(\frac{a}{q}\right)$.

If instead $q \equiv -p \pmod{4a}$, then the proof remains identical with the exception of changing the sign in the rest of the congruence accordingly (i.e. r becomes $4a - r$). Similarly, applying Lemma 14.13 (Integers in an Interval) again, we obtain $\left(\frac{a}{p}\right) = \left(\frac{a}{q}\right)$, which completes the proof of Euler's Lemma. □

From here, we can perform algebraic manipulations directly to obtain the law of quadratic reciprocity.

Suppose that $p \equiv q \pmod 4$. Assume WLOG that $p > q$; then $p - q = 4a$ and $p = 4a + q$. Therefore,

$$\left(\frac{p}{q}\right) = \left(\frac{4a+q}{q}\right) = \left(\frac{4a}{q}\right) = \left(\frac{4}{q}\right)\left(\frac{a}{q}\right) = \left(\frac{a}{q}\right)$$

$$\left(\frac{q}{p}\right) = \left(\frac{p-4a}{p}\right) = \left(\frac{-4a}{p}\right) = \left(\frac{-1}{p}\right)\left(\frac{a}{p}\right)$$

By Lemma 14.13 (Euler's Lemma), $\left(\frac{a}{p}\right) = \left(\frac{a}{q}\right)$; it follows that

$$\left(\frac{p}{q}\right) \cdot \left(\frac{q}{p}\right) = \left(\frac{-1}{p}\right) = (-1)^{\frac{p-1}{2}} = (-1)^{\frac{p-1}{2} \cdot \frac{q-1}{2}}$$

since $\dfrac{p-1}{2}$ is even iff $\dfrac{q-1}{2}$ is also even.

Now suppose that $p \not\equiv q \pmod 4$ (more specifically, $p \equiv -q \pmod 4$, so that $p + q = 4a$ and

$$\left(\frac{p}{q}\right) = \left(\frac{4a-q}{q}\right) = \left(\frac{a}{q}\right)$$

$$\left(\frac{q}{p}\right) = \left(\frac{4a-p}{p}\right) = \left(\frac{a}{p}\right)$$

Applying Euler's Lemma one more time, $\left(\frac{a}{p}\right) = \left(\frac{a}{q}\right)$. Since $(-1)^{\frac{p-1}{2} \cdot \frac{q-1}{2}}$, this completes the proof of quadratic reciprocity. □

Second proof: Eisenstein's geometric proof. First, we must introduce and prove *Eisenstein's Lemma:*

Lemma 14.13: Eisenstein's Lemma

For distinct odd primes p and q,

$$\left(\frac{q}{p}\right) = (-1)^{\sum_{u \text{ even}} \lfloor \frac{qu}{p} \rfloor}$$

Proof. Let u be an *even* integer with $2 \leq u \leq p - 1$. Define $r(u)$ as the least positive residue of $qu \pmod p$. If we examine the set of all possible values of $(-1)^{r(u)} \cdot r(u) \pmod p$, we discover that they are all even, as well as distinct (which can be proven by contradiction - doing so is left as an exercise). Since there are exactly $\dfrac{p-1}{2}$ of them, and they are distinct mod p, they are a permutation of the even integers from 2 to $p - 1$ inclusive. As such, we can take their product to obtain

$$(2q \cdot (-1)^{r(2)}) \cdot (4q \cdot (-1)^{r(4)}) \cdots (q(p-1) \cdot (-1)^{r(p-1)}) \equiv 2 \cdot 4 \cdot 6 \cdots (p-1) \pmod p,$$

so $q^{\frac{p-1}{2}} \equiv (-1)^{r(2)+r(4)+r(6)+\ldots+r(p-1)} \pmod p$. We can simultaneously write $\dfrac{qu}{p} = \left\lfloor \dfrac{qu}{p} \right\rfloor + \dfrac{r(u)}{p}$, and notice that $\left\lfloor \dfrac{qu}{p} \right\rfloor \equiv r(u) \pmod 2$. Finishing off the proof is the subsequent observation that $q^{\frac{p-1}{2}} \equiv (-1)^{\sum_{u \text{ even}} \left\lfloor \frac{qu}{p} \right\rfloor} \pmod p$ (since the LHS is just equal to $\left(\frac{q}{p}\right)$ by Euler's criterion). □

To prove the law itself, the crux of Eisenstein's argument consists of geometrically representing the sum $\sum_{u \text{ even}} \left\lfloor \frac{qu}{p} \right\rfloor$ as a sum of the number of lattice points (points with integer coordinates in the plane) with even x-coordinate strictly within a triangle $\triangle ABC$. Since q is odd, $q-1$, the total number of lattice points in each column of Eisenstein's grid representation, will always be even, and so the number of lattice points in each column and within the interior of $\triangle ABC$, as well as the complement of $\triangle ABC$, will be congruent modulo 2. It follows that $\left(\frac{p}{q}\right) = (-1)^\alpha$ and $\left(\frac{q}{p}\right) = (-1)^\beta$, where α and β are the total number of lattice points in the triangles bounded by the vertices $(0,0)$, $\left(\frac{p}{2}, 0\right)$, and $\left(\frac{p}{2}, \frac{q}{2}\right)$; and $(0,0)$, $\left(\frac{p}{2}, \frac{q}{2}\right)$, and $\left(0, \frac{q}{2}\right)$. Line $\overline{AY}$ has no lattice points (because $\gcd(p,q) = 1$); in addition, the number of lattice points in the rectangle with vertices at $(0,0)$, $\left(\frac{p}{2}, 0\right)$, $\left(\frac{p}{2}, \frac{q}{2}\right)$, and $\left(\frac{q}{2}, 0\right)$. We finally obtain the desired result:

$$\left(\frac{p}{q}\right)\left(\frac{q}{p}\right) = (-1)^{\alpha+\beta} = (-1)^{\frac{p-1}{2} \cdot \frac{q-1}{2}}$$

which completes the proof. $\square$

The length and depth of these two proofs alone should hopefully give some perspective into the sheer complexity - and yet beauty - of the law of quadratic reciprocity! Though this is a theorem that does require pre-requisite definitions, these definitions, once committed to memory, are nicely intuitive and can be manipulated with great flexibility.

14.3.3 Euler's Criterion

Euler's Criterion, initially proposed by Euler in 1748, concisely summarizes and condenses perhaps the most critical of the Legendre symbol's properties: its multiplicativity, closed form, and constriction to ± 1 or 0.

> **Theorem 14.14: Euler's Criterion**
>
> Let p be an odd prime, and let a be an integer with $\gcd(a, p) = 1$. Then
>
> $$\left(\frac{a}{p}\right) = a^{\frac{p-1}{2}}, \left(\frac{a}{p}\right) = 0, \pm 1$$

This definition is in fact a reformulation of the definition of the Legendre symbol, but in this section in particular, we will focus on proving the criterion rather than merely stating it as a corollary.

Proof. We use the fact that the set of residue classes modulo p (for p a prime) is a field. Since p is an odd prime, we can apply Lagrange's Theorem to conclude that a polynoomial with degree k can have at most k roots. Id est, $x^2 \equiv a \pmod{p}$ can have up to 2 solutions for each a. Therefore, there are at least $\frac{p-1}{2}$ distinct QRs mod p.

Because $\gcd(a, p) = 1$, by FLT, $a^{p-1} \equiv 1 \pmod{p}$, or

$$\left(a^{\frac{p-1}{2}} - 1\right)\left(a^{\frac{p-1}{2}} + 1\right) \equiv 0 \pmod{p}$$

Using the fact that the integers modulo p form a field, one of the factors must be zero.

Furthermore, if a is a QR mod p, namely $a \equiv x^2 \pmod{p}$, then $a^{\frac{p-1}{2}} \equiv x^{p-1} \equiv 1 \pmod{p}$ by another application of FLT. One final application of Lagrange's Theorem allows us to complete the proof, since no more than $\frac{p-1}{2}$ values of a can make the first factor equal to zero, but there are also at least $\frac{p-1}{2}$ distinct QRs modulo p (excluding 0). The QNRs make the second factor zero; this is Euler's Criterion. $\square$

With all of the fundamental concepts covered, we can now begin to delve into more computational and practical territory. Consider the following somewhat contest-like example:

Example 14.3.3.1.

(a) What are the QRs modulo 19?

(b) For which prime numbers p is 19 a QR modulo p?

Solution 14.3.3.1.

(a) Observe that a is a QR mod 19 iff $\left(\frac{a}{19}\right) = 1$, i.e. $a^9 \equiv 1 \pmod{19}$ by Euler's Criterion. We can then test each value from 1 through 9, inclusive, to determine if it is a QR (but this is nigh-impossible without a computational aid). Instead, a much more efficient way is to simply square each number in that interval and see which remainders they produce mod 19. The set of QRs turns out to be $\boxed{\{1, 4, 5, 6, 7, 9, 11, 16, 17\}}$.

(b) We require $19^{\frac{p-1}{2}} \equiv 1 \pmod{p}$; testing primes p can be done in this manner via Euler's Criterion. We will not provide the answer here due to the tedium (and lack of illustrative benefit) of computation, but this is the general solution path.

14.3.4 Lagrange's Four-Square Theorem

Yet another of the innocuous and innocent-looking, yet simultaneously complex and endlessly-inspiring theorems of modern number theory, *Lagrange's Four-Square Theorem*, also known in some circles as *Bachet's Conjecture*, presents a statement regarding the positive integers that could not be much simpler on its surface:

> **Theorem 14.15: Lagrange's Four-Square Theorem**
>
> Every positive integer can be written as the sum of four (not necessarily distinct) perfect squares.

Proof. We invoke Euler's Four-Square Theorem, a simplified, weaker version of Lagrange's counterpart:

> **Lemma 14.15: Euler's Four-Square Theorem**
>
> If positive integers $x = x_1^2 + x_2^2 + x_3^2 + x_4^2$ and $y = y_1^2 + y_2^2 + y_3^2 + y_4^2$ can be written as the sums of four perfect squares, then so can their product xy.

The proof of this lemma is straightforward polynomial expansion, the details of which will be left as an exercise to the reader. Specifically, the end result should be that

$$xy = (x_1y_1 + x_2y_2 + x_3y_3 + x_4y_4)^2 + (x_1y_2 - x_2y_1 + x_3y_4 - x_4y_3)^2$$
$$+ (x_1y_3 - x_2y_4 - x_3y_1 + x_4y_2)^2 + (x_1y_4 + x_2y_3 - x_3y_2 - x_4y_1)^2$$

In addition, the theorem trivially holds for the base cases 1 and 2, so we can proceed with the remainder of the proof. It is sufficient to prove the theorem for all odd primes p. First note that the QRs of $a^2 \pmod{p}$ are distinct whenever $0 \leq a \leq \frac{p-1}{2}$, and so a is a root of $x^2 - c$ over the finite field $\mathbb{Z}/p\mathbb{Z}$, where $c = a^2 \pmod{p}$. $p - a$ is also a root of $x^2 - c$; by Lagrange's Theorem, these must be the only roots. Similarly, for $0 \leq b \leq \frac{p-1}{2}$, $-b^2 - 1$ is distinct, so by Pigeonhole, there exist a, b such that $a^2 \equiv -b^2 - 1 \pmod{p}$; i.e. $a^2 + b^2 + 1^2 + 0^2 = np$ for some $n \in \mathbb{N}$.

More generally, let m be the smallest positive integer such that $mp = a^2 + b^2 + c^2 + d^2$ is expressible as the sum of four squares, with $m \leq n$, and also $m < p$. We claim that $m = 1$; the proof is by contradiction. Namely, we show that there exists an $r \in \mathbb{N}$, $r < m$, for which rp is also the sum of four squares.

For each of a, b, c, d, consider the corresponding x, y, z, w in the same modulo m residue class and between $-\frac{m+1}{2}$ and $\frac{m}{2}$, inclusive. It follows that $x^2 + y^2 + z^2 + w^2 = mr$.

Applying Lemma 14.15 again, we obtain $m^2 pr = \alpha^2 + \beta^2 + \gamma^2 + \delta^2$ for some $\alpha, \beta, \gamma, \delta \in \mathbb{Z}$. But each of these is divisible by m, since a, b, c, d and x, y, z, w are congruent modulo m in some order. It follows that, if we define $z_1 m = \alpha, z_2 m = \beta, z_3 m = \gamma, z_4 m = \delta$, then $z_1^2 + z_2^2 + z_3^2 + z_4^2 = rp$, which is a contradiction. $\square$

14.3.5 Examples

Example 14.3.5.1. *Evaluate* $\left(\frac{3}{5}\right)$, $\left(\frac{20}{19}\right)$, $\left(\frac{59}{19}\right)$, *and* $\left(\frac{2019}{1009}\right)$.

Solution 14.3.5.1. $\left(\frac{3}{5}\right) = -1$, $\left(\frac{20}{19}\right) = \left(\frac{5}{19}\right) \cdot \left(\frac{4}{19}\right) = 1 \cdot 1 = 1$, $\left(\frac{59}{19}\right) = \left(\frac{2}{19}\right) = -1$, $\left(\frac{2019}{1009}\right) = \left(\frac{1}{1009}\right) = 1$.

Example 14.3.5.2. *Show that if $a \equiv b \pmod{n}$, $\left(\frac{a}{n}\right) = \left(\frac{b}{n}\right)$ (where $\left(\frac{n}{k}\right)$ denotes the Jacobi symbol).*

Solution 14.3.5.2. *This is an immediate consequence of the fact that the quadratic residue is invariant with respect to modulo n. That is, whether a number is the solution of a modular congruence involving x^2 is independent of its actual value, only what its value is modulo n.*

14.3.6 Exercises

Exercise 241. Let $\left(\frac{a}{n}\right) = 1$ (Jacobi symbol). Does it follow that a is a quadratic residue $\pmod{n}$?

Exercise 242. Calculate $\left(\frac{1001}{2019}\right)$.

Exercise 243. Show that $\left(\frac{a}{p}\right)$ (Legendre symbol) is a multiplicative function.

Exercise 244. Explain how we can test whether a number is prime using the Jacobi symbol.

Exercise 245. Find all solutions to the modular congruence $x^2 + 20x - 19 \equiv 0 \pmod{2017}$, or show that none exist.

Exercise 246. Find all primitive roots $\pmod{7}$.

Extension Exercise 103. Let g and h be primitive roots of p. Can gh be a primitive root of p?

Extension Exercise 104. *(2018 Berkeley Math Tournament)* Evaluate

$$\prod_{j=1}^{50}\left(2\cos\left(\frac{4\pi j}{101}\right) + 1\right).$$

Chapter 14 Review Exercises

Exercise 247. *For which prime numbers p is $\left(\dfrac{2^p}{p}\right) = 1$?*

Exercise 248. *(2002 AIME I) Harold, Tanya, and Ulysses paint a very long picket fence. Harold starts with the first picket and paints every h^{th} picket; Tanya starts with the second picket and paints every t^{th} picket; and Ulysses starts with the third picket and paints every u^{th} picket. Call the positive integer $100h + 10t + u$ paintable when the triple (h,t,u) of positive integers results in every picket being painted exactly once. Find the sum of all the paintable integers.*

Exercise 249. *(2012 Stanford Math Tournament) Find the sum of all integers x, $x \geq 3$, such that 201020112012_x (that is, 201020112012 interpreted as a base x number) is divisible by $x - 1$.*

Exercise 250. *(2012 Winter OMO) Find the largest prime number p such that when $2012!$ is written in base p, it has at least p trailing zeroes.*

Exercise 251. *(2015 HMMT February Guts Round) Find the least positive integer $N > 1$ satisfying the following two properties:*

- *There exists a positive integer a such that $N = a(2a1)$.*
- *The sum $1 + 2 + \ldots + (N-1)$ is divisible by k for every integer $1 \leq k \leq 10$.*

Exercise 252. *Show that an integer n has $\left(\dfrac{n}{p}\right) = 1$ for all prime numbers p iff n is a perfect square.*

Exercise 253. *Show that there exists $n \in \mathbb{N}$ such that $n < \sqrt{p} + 1$ is a quadratic nonresidue modulo p.*

Exercise 254. *(2014 Fall OMO) Find the sum of all positive integers n such that $\tau(n)^2 = 2n$, where $\tau(n)$ is the number of positive integers dividing n.*

Exercise 255. *(2008 AIME II) Find the largest integer n satisfying the following conditions:*

(i) n^2 can be expressed as the difference of two consecutive cubes;

(ii) $2n + 79$ is a perfect square.

Exercise 256. *(2015 Fall OMO) Toner Drum and Celery Hilton are both running for president. A total of 2015 people cast their vote, giving 60% to Toner Drum. Let N be the number of "representative sets of the 2015 voters that could have been polled to correctly predict the winner of the election (i.e. more people in the set voted for Drum than Hilton). Compute the remainder when N is divided by 2017.*

Extension Exercise 105. *(1998 IMO Shortlist N2) Determine all pairs (a,b) of real numbers such that $a\lfloor bn \rfloor = b\lfloor an \rfloor$ for all positive integers n.*

Extension Exercise 106. *(1998 IMO Shortlist N6) For any positive integer n, let $\tau(n)$ denote the number of its positive divisors (including 1 and itself). Determine all positive integers m for which there exists a positive integer n such that $\tau(n^2)\tau(n) = m$.*

Extension Exercise 107. *(1994 IMO P4) Find all ordered pairs (m,n) where m and n are positive integers such that $\dfrac{n^3 + 1}{mn - 1}$ is an integer.*

Extension Exercise 108. *(2004 IMO Shortlist N2) The function ψ from the set $\mathbf{N}$ of positive integers to itself is defined by the equality*

$$\psi(n) = \sum_{k=1}^{n}(k,n), n \in \mathbb{N}$$

where (k,n) denotes the greatest common divisor of k and n.

(a) Prove that $\psi(mn) = \psi(m)\psi(n)$ for every two relatively prime $m, n \in \mathbf{N}$.

(b) Prove that for each $a \in \mathbf{N}$ the equation $\psi(x) = ax$ has a solution.

(c) Find all $a \in \mathbf{N}$ such that the equation $\psi(x) = ax$ has a unique solution.

Chapter 15

Advanced Number Theory

15.1 Lifting the Exponent (LTE)

We will open this concluding chapter with one of the most central, graceful, and striking results of modern number theory: the *lifting the exponent lemma,* or *LTE lemma* as it's commonly known. Just as its name suggests, we "lift the exponent" by establishing a modular equivalence between two expressions, one a sum of two powers that are derived from the other expression, in the spirit of the "freshman's dream": the typically-incorrect equation $(x+y)^p = x^p + y^p$, which actually does hold over a commutative ring of characteristic p.[65]

Before we introduce the LTE lemma itself, we must define a key piece of notation: the *p-adic valuation.*

> **Definition 15.1: p-adic Valuation**
>
> For a given prime number p, the *p-adic valuation* of n, denoted $v_p(n)$, is the largest power of p that evenly divides n. (We can easily compute this using Legendre's formula.)

Remark. By convention, we define $v_p(0) = \infty$ for all p.

Example 15.1.0.1. *For example, $v_2(19) = 0$, $v_3(81) = 4$, and $v_5(1250) = v_5(5^4 \cdot 2) = 4$.*

The p-adic valuation has a sort of "linearity" in that it mirrors the addition/subtraction property of logarithms. In addition, it has three noteworthy special cases:

> **Corollary 15.0: Valuation of Rational Numbers**
>
> Regarding $v_p\left(\dfrac{n}{d}\right)$, if $p \mid n$, then $v_p\left(\dfrac{n}{d}\right) = v_p(n)$, and if $p \mid d$, then $v_p\left(\dfrac{n}{d}\right) = -v_p(d)$. If p divides neither n nor d, then $v_p\left(\dfrac{n}{d}\right) = 0$. In general, $v_p\left(\dfrac{n}{d}\right) = v_p(n) - v_p(d)$. (By extension, $v_p(ab) = v_p(a) + v_p(b)$ for all $a, b \in \mathbb{Q}$.)

Remark. This also implies that $v_p(a+b) \geq \min\{v_p(a), v_p(b)\}$ (with equality when $v_p(a) \neq v_p(b)$).

Second remark. Clearly p cannot divide both n and d, since $\gcd(n,d) \geq p$, which is a contradiction (as $\dfrac{n}{d}$ must be rational). This shows that the theorem statement is exhaustive.

We can state one more important property of the p-adic valuation, namely the *p-adic absolute value (norm):*

---

[65]Over any such ring, by the Binomial Theorem, all coefficients except for the first and last ones reduce to zero, since p is a factor of all of them.

Definition 15.2: The p-adic Norm

The *p-adic absolute value*, or *p-adic norm*, of a rational number n is given by $|n|_p = p^{-v_p(n)}$ and $|n|_p = 0$ if $n = 0$.

With all of these valuable tools and ideas in hand, we can now proceed to state and prove the LTE lemma:[66]

Theorem 15.1: Lifting the Exponent Lemma

Let $a, b \in \mathbb{Z}$, $n \in \mathbb{N}$, and p a prime number with $p \nmid ab$.

For $p = 2$: We split this up into two separate cases: n even and n odd (where $p \mid (a-b)$ in the first case and $p \mid (a+b)$ in the second case). For n even, $v_2(a^n - b^n) = v_2(a-b) + v_2(a+b) + v_2(n) - 1$. For n odd, $v_2(a^n + b^n) = v_p(a+b) + v_p(n)$ (from substituting $-b$ in place of b).

For odd primes p: If $p \mid (a-b)$, then $v_p(a^n - b^n) = v_p(a-b) + v_p(n)$. If $p \mid (a+b)$, then $v_p(a^n + b^n) = v_p(a+b) + v_p(n)$.

Before proving the LTE theorem itself, we will introduce two key (sub-) lemmas:

Lemma 15.1: When n and p are Relatively Prime (Lemma 15.1A)

Let a, b be integers, let n be a positive integer, and let p be any prime. If $\gcd(n, p) = 1$, $p \mid (a-b)$, but $p \nmid ab$, then $v_p(a^n - b^n) = v_p(a-b)$. (If $p \mid (a+b)$ and $p \nmid ab$, then $v_p(a^n + b^n) = v_p(a+b)$.)

Proof of Lemma 15.1A. Factoring $a^n - b^n$ as $(a-b)(a^{n-1} + a^{n-2}b + a^{n-3}b^2 + \ldots + ab^{n-2} + b^{n-1})$, we claim that $p \nmid a^{n-1} + a^{n-2}b + a^{n-3}b^2 + \ldots + ab^{n-2} + b^{n-1}$. Given that $p \mid a - b$, it follows that $a - b \equiv 0 \pmod{p}$, or that $a \equiv b \pmod{p}$. Then

$$a^{n-1} + a^{n-2}b + a^{n-3}b^2 + \ldots + ab^{n-2} + b^{n-1} \equiv a^{n-1} \cdot n \neq 0 \pmod{p}$$

This completes the proof. (The proof for the case $a^n + b^n$ with $p \mid a + b$ is similar.) $\square$

Lemma 15.1: Generalization of Lemma 15.1 (Lemma 15.1B)

For *any* integers $a, b, n \in \mathbb{N}$, p prime, if $p \mid (a-b)$ and $p \nmid ab$, then $v_p(a^n - b^n) = v_p(a-b) + 1$.

Proof of Lemma 15.1B. Let $k = v_p(a-b)$, so that $a = mp^k + b$, with $p \nmid m$. We have

$$a^p - b^p = (mp^k + b)^p - b^p$$

$$= \sum_{i=0}^{p} \left(\binom{p}{i} (mp^k)^{p-i} \cdot b^i \right) - b^p$$

$$= \sum_{i=0}^{p-2} \left(\binom{p}{i} (mp^k)^{p-i} \cdot b^i \right) + \binom{p}{p-1} \cdot (mp^k) b^{p-1}$$

$$= \sum_{i=0}^{p-2} \left(\binom{p}{i} (mp^k)^{p-i} \cdot b^i \right) + mp^{k+1} b^{p-1}$$

---

[66] The common term in the mathematical literature is "lemma," but we will be referring to it as a "theorem." Just because, I guess.

This implies that $p^{k+1} \mid a^p - b^p$.

Furthermore, observe that all terms in the expression for $a^p - b^p$ have a factor of p^{k+2}, but since $p \nmid m, b$, $p^{k+2} \nmid mp^{k+1}b^{k-1}$. Therefore $p^{k+2} \nmid a^p - b^p$, and so $v_p(a^p - b^p) = k + 1 = v_p(a - b) + 1$, as desired. □

Now we can proceed to prove the LTE theorem itself:

Proving the LTE Lemma. We first prove LTE for the odd prime p, $p \mid (a - b)$ case. If we let $k = v_p(n)$, then we have $n = mp^k$, where $p \nmid m$. Then $v_p(a^n - b^n) = v_p(a^{p^k} - b^{p^k})$. Repeatedly applying Lemma 15.1.B yields the desired result, specifically $v_p(a^n - b^n) = v_p(a - b) + k$. To prove LTE for the $p \mid (a + b)$ subcase is similar.

If $p = 2$, on the other hand, then it suffices to show that $v_2(a^{2^n} - b^{2^n}) = v_2(ab) + n$ for the n odd case. Factoring the LHS of this equation yields

$$(a^{2^{n-1}} + b^{2^{n-1}})(a^{2^{n-2}} + b^{2^{n-2}}) \cdots (a^2 + b^2)(a + b)(a - b).$$

Since $a, b \equiv 1, 3 \pmod 4$, it follows that $a^{2^n}, b^{2^n} \equiv 1 \pmod 4$, or $a^{2^n} + b^{2^n} \equiv 2 \pmod 4$. We are done (at least in this case), since the power of 2 in all factors (except $a - b$) is 1.

Finally, if $p = 2$ and n is even, then consider an arbitrary odd integer $q = 2k + 1$ for some $k \in \mathbb{Z}$. We know that $q^2 \equiv 1 \pmod 4$, so $4 \mid (a^2 - b^2)$ for a, b odd. Let m be another odd integer, and let l be a positive integer with $n = m \cdot 2^l$. Then we have

$$v_2(a^n - b^n) = v_2(a^{m \cdot 2^l} - b^{m \cdot 2^l})$$
$$= v_2((a^2)^{2^l} - (b^2)^{2^l})$$
$$\vdots$$
$$= v_2(a^2 - b^2) + l - 1$$
$$= v_2(a - b) + v_2(a + b) + v_2(n) - 1$$

as desired. □

So how can we put this to good use? As it turns out, the LTE theorem is extremely handy in not only simplifying the Legendre formula for numbers with absurdly large exponents (as well as sums or differences of such quantities) and providing a sort of recursive method, but also in solving certain types of Diophantine equations and rigorously proving that some set of solutions is exhaustive.

In the following section, we will take a look at a few applications of the LTE theorem to computational examples that may or may not be doable without the lemma, but that are always clean and elegant with the lemma.

15.1.1 Computational Examples of LTE in Action

Perhaps LTE's single greatest use is in extending Legendre's formula for $v_p(n)$ to circumstances in which the number n is either very large or a sum of exponential quantities that cannot be further simplified.

Example 15.1.1.1. *Compute $v_3(7^{2019} + 2^{2019})$.*

Solution 15.1.1.1. *In this case, $p = 3$ is an odd prime, $a = 7$, $b = 2$, and $n = 2019$ (with $3 \mid 7 + 2 = 9$), with an expression of the form $v_p(a^n + b^n)$, so this yields $v_3(7^{2019} + 2^{2019}) = v_3(7 + 2) + v_3(2019) = v_3(9) + 1 = \boxed{3}$.*

Essentially, LTE expedites the process of computing $v_3(7^{2019})$ and $v_3(2^{2019})$ individually using Legendre's formula by bypassing the traditional process completely in favor of a single lemma.

For the following more complicated example, we will leave the intricacies to the reader (we don't want to rob the joy of finally solving a brutally difficult problem from you, after all!) but we will give a slight hint - a nudge in the right direction, if you will - in the solution area.

Example 15.1.1.2. Denote
$$f(x) = \frac{x^{x^{x^x}}}{x^{x^x}} - \frac{x^x}{x}$$

and

$$P = f(2)f(3)f(4)\cdots f(100).$$

Find the largest integer n such that $2^n \mid P$.

Solution 15.1.1.2. First simplify the given expression to

$$f(x) = x^{\left(x^{x^x} - x^x\right)} - x^{x-1}$$

then apply LTE repeatedly to obtain an expression for

$$v_2\left(x^{\left(x^{x^x} - x^x\right)}\right)$$

This is a general solution method that will hopefully provide the first spark of inspiration to solve the rest of the problem.[67]

But application of a given formula is far from the sole use of LTE; we can also apply it to proving the existence and uniqueness of immediate (and, in some cases, not-so-immediate) solutions to an equation.

15.1.2 Using LTE in Solving Diophantine Equations

LTE is quite versatile in solving Diophantine equations, namely showing that the (almost always) trivial solutions are the only ones. Though it may seem completely obvious in a short-answer context to just search for solutions to a Diophantine equation by inspection and record that as the answer, in a proof-based contest, this will not fly. However, LTE is an equally effective strategy that *will* fly!

Example 15.1.2.1. Find all solutions to the equation $(n-1)! + 1 = n^m$ over the positive integers.

Solution 15.1.2.1. First note that either $n = 1$ or n is prime (as if there exists a divisor $d \geq 2$ of n, it will divide both $(n-1)!$ and n^m). By inspection, $(n,m) = (2,1)$ is a solution, and $n = 1$ produces no solutions. Thus, it suffices to consider the cases where n is an odd prime.

If m is odd, and $n > 3$, we have $v_2(n^m - 1) = v_2(n-1) < v_2((n-1)!)$. (If $n = 3$, then there is no solution to the equation.) Thus, we have ruled out the possibility of m being odd.

We have $v_2((n-1)!) = v_2(m) + v_2(n+1) + v_2(n-1) - 1$, which implies $v_2((n-2)!) = v_2(n+1) + v_2(m) - 1$. Note that $v_2((n-2)!) \geq \dfrac{n-3}{2}$ (consider the prime factorization of $(n-2)!$), so $m \geq \dfrac{1}{n+1} \cdot 2^{\frac{n-1}{2}} \geq n$ when $n \geq 18$ (by inspection). Finally, observe that, for $n > 1$, $n^n > (n-1)! + 1$, so we have effectively reduced the problem to trial and error. We have another solution $(n,m) = (5,2)$, which is indeed the only other one than $(n,m) = (2,1)$.

And one more for good measure:

Example 15.1.2.2. *(Brilliant.org)* Let n be a square-free integer. Show that there is no pair of positive integers (x,y) with $\gcd(x,y) = 1$ such that

$$(x+y)^3 \mid x^n + y^n$$

Solution 15.1.2.2. Assume the contrary. First, suppose that n is even; this implies that, if there is an odd prime $p \mid x+y$, then $x^n + y^n \equiv x^n + (-x)^n \equiv 2x^n \pmod{p}$, which in turn implies $p \mid x \implies p \mid y$, but this is a contradiction.

---

[67] For those interested, the answer is 5,206.

Next, suppose that n is odd. If there is an odd prime $p \mid x+y$, then we get that $3v_p(x+y) \leq v_p(x^n + y^n)$. Applying LTE on the RHS yields

$$3v_p(x+y) \leq v_p(x^n + y^n) = v_p(x+y) + v_p(n) \leq v_p(x+y) + 1 \implies v_p(x+y) \leq \frac{1}{2}$$

but this, too, is a contradiction since we have assumed that $p \mid x+y$ (implying that $v_p(x+y) \geq 1$).

Now the only case left to consider is when $x+y$ is a power of 2. But $v_2(x^n + y^n) = v_2(x+y)$ iff n is odd, since we can write

$$\frac{x^n + y^n}{x+y} = x^{n-1} - x^{n-2}y + \ldots - xy^{n-2} + y^{n-1},$$

which contains an odd number of odd terms. It is thus impossible that $(x+y)^3 \mid (x^n + y^n)$, since the power of 2 on the LHS is greater than that on the RHS - contradiction.

Hence the proof is complete.

Evidently, LTE is not just a computational tool: it is, indeed, a way to simplify the complicated and prove one's intuition correct. It is the ultimate way to circumvent one's natural frustrations of knowing the answer, but being unable to prove it - with the LTE lemma in hand, there are few Diophantine equations left whose solutions remain a mystery!

15.1.3 Examples

Example 15.1.3.1. Using the LTE, *compute* $v_3(5^{2019} - 2^{2019})$.

Solution 15.1.3.1. *We have* $v_3(5^{2019} - 2^{2019}) = v_3(5-2) + v_3(2019) = 1 + 1 = \boxed{2}$.

Example 15.1.3.2. *Determine all ordered triples* (x, y, p) *of positive integers* x, y *and positive prime numbers p such that $p^x - y^p = 1$.*

Solution 15.1.3.2. *Note that $p > 1$, as it is prime. We can split the problem into two cases depending on x: $x = 1, x > 1$.*

If $x = 1$, $p = y^p + 1$. If p is odd, then $y^p + 1 = (y+1)(y^{p-1} - y^{p-2} + \ldots - y + 1)$, so $y^p + 1$ is not prime. Hence $p = 2$, and so $y = 1$, leading to the solution $(p, x, y) = (2, 1, 1)$.

If $x > 1$, p^x and x^y are both perfect powers. But the only pair of perfect powers that differ by 1 is $(2^3, 3^2)$. Hence $(p, x, y) = (3, 2, 2)$.

These are our only two solutions.

15.1.4 Exercises

Exercise 257. Show that $v_p(mn) = v_p(m) + v_p(n)$ for all primes p and positive integers m, n.

Exercise 258. Show that $v_p(m + n) = \min\{v_p(m), v_p(n)\}$ for all prime p, positive integers m, n, and whenever $v_p(m) \neq v_p(n)$.

Exercise 259. Evaluate $v_2(2020^{2019} + 20^{2019})$.

Exercise 260. Let n be a squarefree integer. Show that there does not exist a pair of positive integers (x, y) with $\gcd(x, y) = 1$ such that $(x + y)^3$ is a factor of $x^n + y^n$.

Exercise 261. Prove that 2 is a primitive root of 3^k for all $k \in \mathbb{N}$.

Exercise 262.

(a) Prove Kummer's Theorem:

> **Theorem 15.2: Kummer's Theorem**
>
> Given integers $n \geq m \geq 0$ and prime p,
> $$\nu_p\left(\binom{n}{m}\right)$$
> equals the number of carries when m is added to $n - m$ in base p.

(b) Generalize Kummer's Theorem to the multinomial coefficient. You do not need to prove this generalized form.

Extension Exercise 109. *(2000 IMO P5)* Does there exist a positive integer n such that n has exactly 2000 prime divisors and n divides $2^n + 1$?

Extension Exercise 110. If x and y are positive reals such that $x - y$, $x^2 - y^2$, $x^3 - y^3$, ... are all positive integers, prove that x and y are themselves positive integers.

Extension Exercise 111. *(1990 IMO P3)* Determine all integers $n > 1$ such that $\dfrac{2^n + 1}{n^2}$ is an integer.

Extension Exercise 112. *(1991 IMO Shortlist P18)* Find the highest degree k of 1991 for which 1991^k divides the number $1990^{1991^{1992}} + 1992^{1991^{1990}}$.

15.2 Continued Fraction Representations and the Pell Equation

For this section (which will be somewhat more computationally-focused than usual, albeit still fundamentally driven by rigorous proof), we will delve into two number theory topics that, despite seemingly having zero relation to each other on the surface, are actually inseparably interlinked. In the study of *continued fraction representations,* we analyze those fractions which do not terminate; i.e. fractions which cannot be represented as a finite sum of integer terms and reciprocals of integer terms. These special types of fractions have remarkable properties that relate not only to the subject of our second topic in this section, but also to areas such as the Euclidean algorithm and analytic theory.

In analyzing the *Pell equation* - a general form of a specific subset of Diophantine equations that is usually infeasible to solve hand - we often use methods that include continued fraction representations! Studied since antiquity but still enshrouded in mathematical mystique to this day, the almighty Pell equation is one of the finest testaments to the idea of "simple-looking does not mean simple to solve."

15.2.1 The Continued Fraction Representation

Much like decimals can be classified as *terminating* or *repeating* based on the nature of the digits in their decimal representations, so can fractions be classified as terminating or, in this case, *continued fractions.* A terminating fraction is the type we are all familiar with - rational numbers; namely, quotients of relatively prime integers. A continued fraction, like a repeating decimal, is derived from an infinite number of sums of a fraction and the reciprocal of another integer term, whereas a terminating fraction is the finite sum of a fraction and reciprocals of integers.

Example 15.2.1.1. *An example of a terminating (finite) fraction is $\dfrac{1}{3}$ or $\dfrac{1}{2 + \dfrac{3}{4 + \dfrac{5}{6}}}$. An example of a continued fraction would be $\dfrac{1}{1 + \dfrac{2}{3 + \dfrac{5}{8 + \ldots}}}$.*

In addition, we can define the notion of a *generalized continued fraction,* which consists of functions or variables in place of integers. However, we will be focusing on integer continued fraction representations throughout this section, as they are by far more prevalent in the context of competition math (and the general ideas and concepts belying integer continued fraction representations also carry over to their generalized forms, so there is little need to cover them in detail twice).

To be completely rigorous, we define a continued fraction as follows:

> **Definition 15.3: Continued Fraction Representation**
>
> The[a] *continued fraction representation* of a real number r is an expression of the form
>
> $$a_0 + \dfrac{b_0}{a_1 + \dfrac{b_1}{a_2 + \dfrac{b_2}{a_3 + \ldots}}}$$
>
> where the a_i and b_i can be real or complex numbers. If, for all i, $b_i = 1$, then the expression is known as a *simple continued fraction.*
>
> [a] Note that the continued fraction representation (CFR) of any given real number is unique. We will prove this shortly.

Remark. A continued fraction may have $a_0 = 0$; in this case, the CFR is still perfectly valid. a_0 may also be negative, or the sole part of the CFR (in which case we say the CFR is *degenerate*).

Before proceeding further with our analysis of CFRs, this would perhaps be an opportune time to briefly discuss the motivation behind them. If we consider any given rational number $\frac{a}{b}$, then chances are that $a > b$. In such cases, it seems natural to split this number into two parts: its fractional part $\{\cdot\} \in [0, 1)$, and its integer part $\lfloor \cdot \rfloor$. Then the remainder (the fractional part) can be written as the reciprocal of a rational number that is itself greater than 1; the process repeats indefinitely (finitely in the case of terminating representations).

Computing CFRs

Consider some real number r. As above, define $f = \lfloor r \rfloor$ and $F = r - f$ (the fractional part of r).

> **Definition 15.4: CFR Notation**
>
> The CFR of the real number r given f and F is denoted $[f; a_1, a_2, \ldots]$ where $[a_1, a_2, a_3, \ldots]$ is the CFR of $\frac{1}{F}$, or alternatively,
>
> $$a_0 + \mathbb{K}_{i=1}^{n} \frac{1}{a_i}$$

We can define an algorithm to compute the CFR of any given real number, finite or infinite. This algorithm effectively ties in with the Euclidean Algorithm for divisibility by reducing each fraction to its simplest components:

> **Theorem 15.3: An Algorithm for CFRs**
>
> Begin with $F = r - f$, and set $a_0 = f$. Then consider $r - a_0$; if this difference is 0, then halt. If it is non-zero, consider $\frac{1}{r - a_0}$, and repeat the process until $r = a_0$.

> **Corollary 15.3: Termination of the CFR**
>
> The algorithm described above will eventually terminate if and only if r is rational.

Proof. If r is rational, then it can be written as $\frac{a}{b}$. Applying Euclid's Algorithm, we can write $a = a_1 q + r_1$, $b = a_2 r_1 + r_2$, where $0 \leq r_1 < q$, $0 \leq r_2 < r_1$. We can also write

$$r_1 = a_3 r_2 + r_3, 0 \leq r_3 < r_2$$

$$r_2 = a_4 r_3 + r_4, 0 \leq r_4 < r_3$$

$$\vdots$$

$$r_{n-3} = a_{n-1} r_{n-2} + r_{n-1}, 0 \leq r_{n-1} < r_{n-2}$$

$$r_{n-2} = a_n r_{n-1}$$

Note that $r_1, r_2, r_3, \ldots, r_{n-1}$ has $r_i < r_{i-1}$ for all $i \in [2, n-1]$, with each r_i a non-negative integer. In addition, the sequence converges to zero in a finite number of steps, so there are at most n instances of a_i's.

Beginning with the original number $r = \frac{a}{b} \in \mathbb{Q}$, write

$$\frac{a}{b} = a_1 + \frac{1}{\left(\frac{q}{r_1}\right)}$$

$$\frac{q}{r_1} = a_2 + \cfrac{1}{\left(\cfrac{r_1}{r_2}\right)}$$

$$\frac{r_1}{r_2} = a_3 + \cfrac{1}{\left(\cfrac{r_2}{r_3}\right)}$$

$$\vdots$$

$$\frac{r_{n-3}}{r_{n-2}} = a_{n-1} + \cfrac{1}{\left(\cfrac{r_{n-2}}{r_{n-1}}\right)}$$

$$\frac{r_{n-2}}{r_{n-1}} = a_n$$

Each of these equations can then be substituted into the previous equation to obtain the desired result, namely

$$r = \frac{a}{b} = a_1 + \cfrac{1}{a_2 + \cfrac{1}{a_3 + \cfrac{1}{a_4 + \ldots}}}$$

which is a finite simple CFR. $\square$

Let's work through two concrete examples: one finite CFR, and one infinite.

Example 15.2.1.2. *Determine the CFR of $\frac{13}{5}$.*

Solution 15.2.1.1. *Since $2 < \frac{13}{5} < 3$, $a_0 = 2$, and the remaining portion is $\frac{3}{5}$. Taking the reciprocal of this to get $\frac{5}{3}$ and subtracting 1, we get $\frac{2}{3}$, whose reciprocal is $\frac{3}{2}$ (and whose difference with 1 is then $\frac{1}{2}$). However, $\frac{1}{2}$ has a reciprocal of 2, and subtracting 1 from 2 yields 1, so we stop here. Our CFR is*

$$2 + \cfrac{1}{1 + \cfrac{1}{1 + \cfrac{1}{2}}}$$

represented in bracket notation as $\boxed{[2; 1, 1, 2]}$.

Example 15.2.1.3. *Determine the CFR of $\sqrt{2}$.*

Solution 15.2.1.2. *Since $1 < \sqrt{2} < 2$, we have $a_0 = 1$, and then $\frac{1}{\sqrt{2} - a_0} = \sqrt{2} + 1$. Then we can set $a_1 = 1$, which yields $\sqrt{2}$ once again. We can see from here that the CFR will be infinitely repeating, since we obtain the same term that we started with. More specifically, the CFR evaluates to*

$$1 + \cfrac{1}{2 + \cfrac{1}{2 + \cfrac{1}{2 + \ldots}}}$$

by the CFR algorithm (or $\boxed{[1; \overline{2}]}$ in bracket notation).

And now for an example in reverse:

Example 15.2.1.4. *What real number corresponds to the representation $[1; \overline{1}]$? (Here the bar over the 1 represents that it repeats indefinitely, much like with a bar over a digit in the decimal representation of a real number, as in $0.\overline{45} = \frac{5}{11}$.)*

Solution 15.2.1.3. *The bracket notation lists the floor of the desired value r, followed by the set of denominators in the rest of the CFR. That is, $\lfloor r \rfloor = 1$, and all subsequent denominators are 1, so the long form of the CFR is*

$$1 + \cfrac{1}{1 + \cfrac{1}{1 + \cfrac{1}{1 + \ldots}}}$$

Note that the denominator of the very first fraction is just a duplicated version of the entire expression, so we can substitute $x = 1 + \cfrac{1}{1 + \cfrac{1}{1 + \ldots}}$ to obtain the equation $x = 1 + \frac{1}{x}$, which simplifies into the quadratic $x^2 = x + 1$ that can be solved to obtain $x = r = \boxed{\dfrac{1 + \sqrt{5}}{2}}$.

Each and every real number also has a series of rational approximations of its actual value, which are called *convergents*.

Definition 15.5: Convergents of a Real Number

The sequence of *convergents* of $r \in \mathbb{R}$ is given by $\{a_0, a_0 + \frac{1}{a_1}, a_0 + \cfrac{1}{a_1 + \cfrac{1}{a_2}}, \ldots\}$; i.e. the set of portions of the continued fraction representation of r up to a_i for any given non-negative integer i.

The larger the term of the convergent, the closer it will approximate the true value of r. Even-numbered convergents are smaller than r, while odd-numbered convergents are larger than r. (Why should this be the case? For instance, why can't odd-numbered convergents be smaller and even-numbered ones be larger?) Furthermore, note that $\phi = \dfrac{1 + \sqrt{5}}{2} = 1 + \cfrac{1}{1 + \cfrac{1}{1 + \ldots}}$ has all 1's in its infinite CFR, and so it is the irrational number whose convergents approximate it the "worst" out of all irrationals. "Better" approximations in early convergents manifest in the CFRs or irrational numbers such as e and π, which contain large terms in the first few instances of a_i.

Proving Existence and Uniqueness

By the CFR algorithm, it seems intuitive that every rational number should have *at least* one representation as a continued fraction, as well as *no more than* one (since we are repeatedly taking the reciprocal and subtracting the floor, neither of which are multi-valued functions).[68]

Proving existence. Let $r = \dfrac{a}{b}$, so that $a_0 = \left\lfloor \dfrac{a}{b} \right\rfloor$. Then by the CFR algorithm, there exists an a_1 such that $[a_1, a_2, a_3, \ldots] = \dfrac{1}{r - a_0} \in \mathbb{Q}$, given that $r \in \mathbb{Q}$. Subsequently, there exists an a_2 such that $[a_2, a_3, \ldots] \in \mathbb{Q}$ whenever $a_1 \in \mathbb{Q}$. By strong induction, we can show that, if r and all of $a_1, a_2, a_3, \ldots, a_i$ are rational numbers, then a_{i+1} is also rational, which shows that the CFR given by $r = [a_0; a_1, a_2, a_3, \ldots]$ must exist for all $r \in \mathbb{Q}$. □

---

[68] However, if we were dealing with complex numbers, this would be a different story. This is why we are dealing strictly in the reals for the purposes of this section.

Proving uniqueness. Note: If we impose no restrictions on the coefficients a_i, then the CFR is in fact *not unique!* Consider the two CFRs of $\frac{1}{2}$, namely $0 + \frac{1}{2}$ and $0 + \cfrac{1}{1 + \cfrac{1}{1}}$. However, if we require that $a_n > 1$, where a_n is the last element in the finite CFR, then uniqueness does indeed hold.

We prove this by induction. Suppose that $r = [a_0; a_1, a_2, a_3, \ldots, a_n]$ where $a_n > 1$. Then denote $P = [a_1, a_2, a_3, \ldots, a_n]$, and it follows that $\|P\| > 1$. Thus, $a_0 \leq x < a_0 + 1$, which implies $a_0 = \lfloor x \rfloor$. Note that a_0 is uniquely determined by x, which in turn implies $x - a_0 = \lfloor P \rfloor$. By induction, P is uniquely determined by the value of $x - a_0$, and so the entire CFR for $r \in \mathbb{Q}$ is uniquely determined as well. □

Special CFRs

In this subsection we list several of the "special" CFRs; i.e. the CFRs of constants that are particularly worth noting.

> **Theorem 15.4: "Special" CFRs**
>
> The (infinite) CFR of $e \approx 2.71828$ can be written succinctly as
>
> $$[2; 1, 2, 1, 1, 4, 1, 1, 6, 1, 1, 8, \ldots]$$
>
> where each "block" of three a_i's after the first 1 consists of the even number $2i$ immediately followed by two 1's.
>
> The CFR of π can be written (quite clumsily and arbitrarily) as
>
> $$[3; 7, 15, 1, 292, 1, 1, 1, 2, 1, 3, 1, \ldots]$$
>
> but to date, no pattern has been discovered that represents all terms in this representation. Instead, mathematicians have opted to express π in terms of its *generalized* CFR, namely
>
> $$\pi = \cfrac{4}{1 + \cfrac{1^2}{2 + \cfrac{2^2}{2 + \cfrac{3^2}{2 + \ldots}}}} = \cfrac{2}{1 + \cfrac{1^2}{3 + \cfrac{2^2}{5 + \cfrac{3^2}{7 + \ldots}}}}$$

As a dual exercise to the reader: firstly, can you prove the listed CFRs for e and π? Secondly, can you determine the (generalized) CFR of e^n and π^n for any given positive integer n? (Hint: Using the CFR algorithm boils down to extended computation; is there any way to shortcut this process?)

15.2.2 The Pell Equation

In this section, alongside the theory of continued fraction representations, we will also examine the *Pell equation*, a specific sub-category of Diophantine equation named for mathematician John Pell.[69]

> **Definition 15.6: The Pell Equation**
>
> A *Pell equation* is any Diophantine equation of the form $x^2 - ny^2 = \pm 1$, where n is a positive integer that is not a perfect square.
>
> A *Pell-like equation* is any Diophantine equation of the form $x^2 - ny^2 = k$, where k can be any integer.

The general graph of the Pell equation in $\mathbb{R}^2$ is that of a hyperbola (recalling that the parent function of a hyperbola is of the form $\frac{x^2}{a^2} - \frac{y^2}{b^2} = 1$). If this hyperbola passes through a lattice point $P = (x,y)$, then P is a solution to the Pell equation.

Joseph-Louis Lagrange - of Lagrange's Theorem fame - had proven a landmark result regarding the generalized Pell equation:

> **Theorem 15.5: Infinitude of Distinct Integer Solutions**
>
> If n is *not* a perfect square, then $x^2 - ny^2 = \pm 1$ has infinitely many distinct integer solutions (x,y).

Proof.

Note: The case where n is a perfect square is trivial (simply rewrite $x^2 - ny^2$ as $(x + y\sqrt{n})(x - y\sqrt{n})$ by the difference of squares identity).

If n is not a perfect square, let $c_1 \geq 2$ be an integer. We claim that there exist integers a_1, b_1 such that $a_1 - b_1\sqrt{n} < \frac{1}{c_1}$ and $b_1 \leq c_1$. Let (s_k) be the sequence defined by $s_k = \lfloor k\sqrt{n} + 1 \rfloor$; then $0 \leq s_k - k\sqrt{n} \leq 1$, for all $k \in [0, c_1]$. By Pigeonhole, there exist integers i, j, p such that $i < j$, $0 \leq i, j, p \leq c_1$, and

$$\frac{p-1}{c_1} < s_i - i\sqrt{n} \leq \frac{p}{c_1}$$

$$\frac{p-1}{c_1} < s_j - j\sqrt{n} \leq \frac{p}{c_1}$$

which together imply $a_1 = s_i - s_j, b_1 = j - i$.

It follows that

$$a_1 - b_1\sqrt{n} < \frac{1}{c_1}$$

$$a_1 + b_1\sqrt{n} < \frac{1}{c_1} + 2\sqrt{n}$$

$$a_1^2 - nb_1^2 < \frac{2b_1}{c_1}\sqrt{n} + \frac{1}{c_1} < 2\sqrt{n} + 1$$

In general, we can construct sequences of integers $(a_k), (b_k), (c_k)$ such that $a_k - b_k\sqrt{n} < \frac{1}{c_k}$, $a_k^2 - nb_k^2 < 2\sqrt{n} + 1$, and $a_k - b_k\sqrt{n} > \frac{1}{c_{k+1}}$. Then the t_k are pairwise distinct for all $k \in \mathbb{N}$. However, by Pigeonhole,

---

[69] Although due to a mistaken attribution to him by Euler; the equation had actually been studied extensively by Brahmagupta before him. This is far from the only case of a mathematical misnomer!

there must exist another pair of infinite sequences a_{y_k}, b_{y_k} and integers f, g such that

$$a_{y_f}^2 - n \cdot b_{y_f}^2 = a_{y_g}^2 - n \cdot b_{y_g}^2 = H$$

$$a_{y_f} \equiv a_{y_g} \pmod{H}, b_{y_f} \equiv b_{y_g} \pmod{H}$$

and, finally,

$$\frac{a_{y_f}}{b_{y_f}} \neq \frac{a_{y_g}}{b_{y_g}}$$

The last step is to define variables that can be manipulated to obtain the direct form of Pell's equation, as follows.

Let

$$\alpha = a_{y_f} \cdot a_{y_g} - n \cdot b_{y_f} \cdot b_{y_g}$$

$$\beta = a_{y_f} \cdot b_{y_g} - a_{y_g} \cdot b_{y_f}$$

Observe that $\alpha^2 - n \cdot \beta^2 = H^2$, and furthermore,

$$\alpha = a_{y_f} \cdot a_{y_g} - n \cdot b_{y_f} \cdot b_{y_g} \implies H \mid \alpha$$

This implies $H \mid \beta$ as well, so $\left(\dfrac{\alpha}{H}, \dfrac{\beta}{H}\right)$ is a nontrivial solution to the generalized Pell equation, completing the proof of existence of *at least one solution*.

Furthermore, to prove the infinitude of solutions under the requisite conditions, if we suppose that some (x_0, y_0) is the minimal solution to the Pell equation $x^2 - ny^2 = 1$.[70] Then, to determine the family of other solutions, note that if (a, b) and (c, d) are also solutions, then $(ac + nbd, bc + ad)$ is another solution (hint: multiply the resulting equations from substituting each ordered pair into the Pell equation). We can then obtain all solutions of the form (x_n, y_n) from $(x_0^2 - ny_0^2)^{n+1}$, which can be verified experimentally.[71]

For completeness, we will show that this solution family is the only such set of solutions. For contradiction, assume the existence of another, non-minimal, solution (r, s). Then there exists an integer m such that

$$x_m + \sqrt{n} \cdot y_m < r + q\sqrt{n} < x_{m+1} + \sqrt{n} \cdot y_{m+1}$$

Multiplying through by $x_m - \sqrt{n}y_m$ yields

$$x_0 + y_0\sqrt{n} > (rx_m - n \cdot sy_m) + \sqrt{n}(sx_m - ry_m) > 1$$

However, we observe that

$$(rx_m - sy_m)^2 - n \cdot (sx_m - ry_m)^2 = (r + s\sqrt{n})(r - s\sqrt{n})(x_m + y_m\sqrt{n})(x_m - y_m\sqrt{n}) = 1$$

which is a contradiction, as $(rx_m - n \cdot sy_m, sx_m - ry_m)$ would be a solution smaller than the minimal solution. □

Most importantly, however, in studying the *Pell equation,* we inevitably run into our good friend, the continued fraction. But *why?* What possible connection could these two seemingly-unrelated constructs have with each other? To develop an answer, we can consider the "families" of solutions to the Pell equation.

The Pell equation, unlike many other classes of Diophantine equations, is not one whose solutions merely consist of the obvious trivial solutions with small integers (such as, say, $(0, 1)$ or $(2, 3)$). Here, we temporarily

---

[70] Here we have made WLOG the *a priori* assumption that $x^2 - ny^2 = 1$, as opposed to -1.

[71] Regarding the derivation of this expression, this is a kind of "hand-wave" that nevertheless works out just fine. Naturally, the reader may be questioning the rigor of the proof with this step, but we can still prove this just as rigorously as any other fact (and in fact, we leave it as an exercise to do so). The only difference is that we *can* "guess" the desired form, but such guesswork is by no means absolutely necessary.

eschew the idea that a Diophantine equation must have a small, or even finite, family of solutions in favor of a bigger-picture perspective. As is evident from the proof of the infinitude of solutions to the Pell equation for n a non-square positive integer, there is in fact an entire family of solutions for all such n, which segues into the natural question of whether similar "families" of the values of the denominators in continued fraction representations lend themselves to infinitude as well. And indeed they do - in fact, we can sum up the relationship between Pell's equation and CFRs in one fell swoop:

> **Theorem 15.6: Solutions to the Pell Equation from CFR Convergents**
>
> For n a non-square positive integer, if p is the period[a] of the CFR of $\sqrt{n}$, and $C_k = \dfrac{P_k}{Q_k}$ is the kth convergent of that CFR, then all solutions to the Pell equation $x^2 - ny^2 = 1$ are of the form (P_{ai}, Q_{ai}) for i a positive integer.
>
> More explicitly, from the minimal solution (x_1, y_1), we can construct the equation
>
> $$x_k + y_k \sqrt{n} = (x_1 + y_1 \sqrt{n})^k$$
>
> as well as the recurrence relations
>
> $$x_{k+1} = x_1 x_k + n y_1 y_k$$
>
> $$y_{k+1} = x_1 y_k + y_1 x_k$$
>
> ---
> [a]The smallest positive integer p for which $a_i = a_{i+p}$ for all i.

Proof. Let (x, y) be a solution in general. Then

$$\left| \sqrt{n} - \frac{x}{y} \right| = \left| \frac{y\sqrt{n} - x}{y} \right|$$

$$= \left| \frac{ny^2 - x^2}{y(y\sqrt{n} + x)} \right| = \left| \frac{1}{y^2 \left(\sqrt{n} + \frac{x}{y} \right)} \right| < \frac{1}{2y^2}$$

We claim that this implies that (x, y) is a convergent to $\sqrt{n}$ in its CFR representation; to prove this, we invoke Lemma 15.6:

> **Lemma 15.6: The Pell Equation Convergent Inequality**
>
> Any rational number $\dfrac{p}{q}$ satisfying the equation
>
> $$\left| r - \frac{p}{q} \right| < \frac{1}{2q^2}$$
>
> is a convergent of the real number r.

Proof. Given some satisfactory $\dfrac{p}{q}$, it suffices to prove that $\dfrac{p}{q}$ also satisfies the *law of best approximation*:

> **Theorem 15.7: Law of Best Approximation**
>
> Let $r \in \mathbb{R}$ have convergent $\dfrac{p_n}{q_n}$ with $n \geq 2$ a positive integer. If p and q are integers such that $0 < q \leq q_n$, $\dfrac{p}{q} \neq \dfrac{p_n}{q_n}$, then $|q_n \cdot r - p_n| < |qr - p|$.

Proof. Suppose $q = q_n$. Then we have $p \neq p_n$, which implies that

$$\left| \frac{p}{q} - \frac{p_n}{q_n} \right| = \frac{|p - p_n|}{q_n} \geq \frac{1}{q_n}.$$

Furthermore,

$$\left| r - \frac{p_n}{q_n} \right| < \frac{1}{q_n \cdot q_{n+1}} < \frac{1}{2q_n}$$

since q_n is strictly increasing and $q_0 = 1$. Applying the Triangle Inequality, we obtain

$$\left| r - \frac{p}{q} \right| > \left| \frac{p}{q} - \frac{p_n}{q_n} \right| - \left| r - \frac{p_n}{q_n} \right| > \frac{1}{q_n} - \frac{1}{2q_n} = \frac{1}{2q_n} > \left| r - \frac{p_n}{q_n} \right|.$$

Finally, multiplying both sides by q_n yields the desired result (in this case).

In general, suppose $0 < q < q_n$. Then define X, Y such that

$$Xp_n + Yp_{n-1} = p$$
$$Xq_n + Yq_{n-1} = q.$$

We then have

$$X = \frac{pq_{n-1} - qp_{n-1}}{p_n q_{n-1} - p_{n-1} q_n}$$
$$Y = \frac{pq_n - qp_n}{p_n q_{n-1} - p_{n-1} q_n}.$$

The denominators both reduce to ± 1, so we have $X = \pm(pq_{n-1} - qp_{n-1})$ and $Y = \pm(pq_n - qp_n)$. Thus, $x, y \neq 0$. Since $q_n > q$, one of x and y is positive while the other is negative. Then $r - \dfrac{p_n}{q_n}$ and $r - \dfrac{p_{n+1}}{q_{n+1}}$ have opposite signs as well, which in turn implies that $rq_n - p_n$ and $rq_{n-1} - p_{n-1}$ have opposite signs. Therefore, $x(rq_n - p_n)$ and $y(rq_{n-1} - p_{n-1})$ are of the same sign. From here, we finish by noting that

$$rq - p = x(rq_n - p_n) + y(rq_{n-1} - p_{n-1})$$
$$\implies |rq - p| = |x(rq_n - p_n)| + |y(rq_{n-1} - p_{n-1})|$$
$$\implies |rq - p| > |rq_{n-1} - p_{n-1}| > |rq_n - p_n|$$

as desired. $\square$

We claim $\frac{p}{q}$ is the best approximation of r. Let $\frac{P}{Q} \neq \frac{p}{q}$ such that

$$|Qr - P| \leq |qr - p| = q\left|r - \frac{p}{q}\right| < \frac{1}{2q}$$

It follows that

$$\frac{1}{q \cdot Q} \leq \frac{|p \cdot Q - p \cdot q|}{Q \cdot Q} = \left|\frac{p}{q} - \frac{P}{Q}\right|$$

$$\leq \left|r - \frac{p}{q}\right| + \left|\frac{P}{Q} - r\right| < \frac{1}{2q^2} + \frac{1}{2Q \cdot q}$$

$$= \frac{q + Q}{2q^2 \cdot Q}$$

This implies that $\frac{q+Q}{2q} > 1$, or $Q > q$. Thus, $\frac{p}{q}$ is the best approximation of r, as desired. $\square$

By Lemma 15.6, the proof is complete. $\square$

15.2.3 Examples

Example 15.2.3.1. *Solve the equation $x^2 - 3y^2 = 1$ over the integers.*

Solution 15.2.3.1. *Observe that, if (x,y) is a solution, then so is $(2x+3y, x+2y)$. This can be seen by approximating $\sqrt{3}$ via its continued fraction convergents, after observing that the equation takes on the form of a Pell equation. (This shows that the equation has infinitely many solutions over $x, y \in \mathbb{Z}$.)*

Example 15.2.3.2. *Convert each of the following between a continued fraction and a real number, or evaluate if specified to do so:*

(a) $[2;1;3]$

(b) $\frac{2019}{19}$

(c) *Evaluate* $[2;1;1;\ldots]$

(d) e

(e) $\sqrt{19}$

Solution 15.2.3.2.

(a) $2 + \cfrac{1}{1 + \frac{1}{3}} = \frac{11}{4}$

(b) $106 + \cfrac{1}{3 + \cfrac{1}{1+\frac{1}{4}}}$

(c) *This is* $\lim_{n \to \infty} \frac{F_{n+2}}{F_n} = \frac{3 + \sqrt{5}}{2}$

(d) $[2; 1, 2, 1, 1, 4, 1, 1, 6, 1, 1, 8, \ldots]$

(e) $[4; \overline{2, 1, 3, 1, 2, 8}]$

15.2.4 Exercises

Exercise 263. Find all integer solutions to the equation $x^2 - 7y^2 = 2$.

Exercise 264. Let n be a nonzero integer. Show that if $x^2 - dy^2 = n$ has at least one solution (x, y), then it has infinitely many solutions.

Exercise 265. What are the first three convergents of

(a) e^2, and

(b) $\sqrt{5}$?

Exercise 266. *(2017 HMMT February Guts Round)* Find the smallest possible value of $x + y$ where $x, y \geq 1$ and x and y are integers that satisfy $x^2 - 29y^2 = 1$.

Exercise 267. Show that x is a Fibonacci number iff either $5x^2 + 4$ or $5x^2 - 4$ is a perfect square.

Exercise 268. *(1991 AIME)* Find A^2, where A is the sum of the absolute values of all roots of the following equation:

$$x = \sqrt{19} + \cfrac{91}{\sqrt{19} + \cfrac{91}{\sqrt{19} + \cfrac{91}{\sqrt{19} + \cfrac{91}{\sqrt{19} + \cfrac{91}{x}}}}}.$$

Exercise 269. Determine linear transformations onto the trivial solution $(3, 2)$ of the Pell equation $x^2 - 2y^2 = 1$ that will produce more solutions for x and y.

Extension Exercise 113. *(1999 IMO Shortlist P3)* Prove that there exist two strictly increasing sequences (a_n) and (b_n) such that $a_n(a_n + 1)$ divides $b_n^2 + 1$ for every natural n.

Extension Exercise 114. *(2000 Putnam A2)* Prove that there exist infinitely many integers n such that $n, n+1, n+2$ are each the sum of the squares of two integers.

Extension Exercise 115. Describe an algorithm to find the nth triangular number that is also a perfect square.

Extension Exercise 116. Prove that the solutions to the Pell equation $x^2 - dy^2 = 1$ form a group under multiplication.

Extension Exercise 117. *(2004 IMO Shortlist A3)* Does there exist a function $s : \mathbb{Q} \mapsto \{-1, 1\}$ such that if x and y are distinct rational numbers satisfying $xy = 1$ or $x + y \in \{0, 1\}$, then $s(x)s(y) = -1$? Justify your answer.

15.3 Order and Cyclotomic Polynomials

For the very last section in our foray into the great, big expanse of number theory, we will introduce the idea of *order* and *cyclotomic polynomials* - two closely related mathematical objects that share the fundamental building blocks of the prime numbers. The idea of cyclotomic polynomials is in fact an extension of order, which we shall see as we delve into some of their most curious and wonderfully striking properties.

15.3.1 Order

A key element in the construction of cyclotomic polynomials is the notion of *order* of a prime, as cyclotomic polynomials revolve around the roots of unity and theorems pertaining to exponentiation that follows a periodic cycle (e.g. as an elementary example, 2^k (mod 5), which follows the cycle $(2, 4, 3, 1, 2, \ldots)$).

> **Definition 15.7: Multiplicative Order Modulo a Prime**
>
> Let p be a prime, and consider an integer $a \not\equiv 0$ (mod p). Then the *multiplicative order* (or just *order*)[a] of a modulo p is the smallest positive integer m such that $a^m \equiv 1$ (mod p). We denote this by $\mathrm{ord}_p a$ (alternatively, ord a (mod p)).
>
> [a] Also known as the *Haupt-exponent*.

> **Corollary 15.7: Existence and Finitude of Order**
>
> The order $\mathrm{ord}_p a$, for any prime p and $a \in \mathbb{Z}, a \neq 0$, exists, and furthermore is finite.

Proof. To show that the order exists, we note that the powers of a can only assume a finite number of values (mod n) by the Pigeonhole Principle. Assume that there exist two powers of a, namely a^s and a^t (where we assume WLOG that $s > t$), such that $a^s \equiv a^t$ (mod n). Because $\gcd(a, n) = 1$, a has an inverse modulo n, and multiplying both sides of the initial congruence by a^{-t} yields $a^{s-t} \equiv 1$ (mod n). This proves the existence of the order for all prime p.[72]

The finitude of the order follows from Fermat's Little Theorem, which guarantees that $a^{p-1} \equiv 1$ (mod p); in particular, $\mathrm{ord}_p a \leq p - 1$ for all p, a. □

Example 15.3.1.1. *In the example of 2^k (mod 5), we have $a = 2$, $p = 5$; then $\mathrm{ord}_5(2) = \boxed{4}$, as $2^4 = 16 \equiv 1$ (mod 5), and 4 is clearly minimal.*

We can now introduce the *fundamental theorem of orders*:

> **Theorem 15.8: Fundamental Theorem of Orders**
>
> Suppose that $a^n \equiv 1$ (mod p). Then $\mathrm{ord}_p(a) \mid n$.

Proof. Let $n = k \cdot \mathrm{ord}_p(a) + l$ for some integers k, l, $1 \leq l \leq \mathrm{ord}_p(a)$ (such that n is *not* an even multiple of $\mathrm{ord}_p(a)$). Then $a^n = a^{k \cdot \mathrm{ord}_p(a) + l} \equiv 1^k \cdot a^l \equiv a^l$ (mod p) $\neq 1$ by definition. Note that $a^l \not\equiv 1$ (mod p) since $l < \mathrm{ord}_p(a)$, and the order is the *minimal* integer for which $a^{\mathrm{ord}_p(a)} \equiv 1$ (mod p). □

The idea of order, though it may seem unrelated with cyclotomic polynomials at this point, is directly related with the idea of *primitive roots*, which indeed play a pivotal role in defining cyclotomic polynomials:

---

[72]This can actually be made easier by using the fact that we are working in the multiplicative group of integers modulo n, but this is not necessary to complete the proof.

> **Theorem 15.9: Primitive Roots**
>
> Let p be a prime. Then there exists a *primitive root* r such that $\text{ord}_p(r) = p - 1$.

Proof. This is Fermat's Little Theorem (proven in §14.2). □

Henceforth, consider the following corollary of Lagrange's Theorem:

> **Theorem 15.10: Order Divides $\phi(p)$**
>
> For all prime p, $\text{ord}_p(a) \mid \phi(p)$.

Proof. Note that $\phi(p) = p - 1$, and $a^{p-1} \equiv 1 \pmod{p}$. It immediately follows that $\text{ord}_p(a)$ must divide $\phi(p) = p - 1$. □

(Exercise: How does this relate to Lagrange's Theorem? Hint: Try to generalize this theorem to all n, not necessarily prime, in place of prime p.)

> **Corollary 15.10: Sufficient Condition for a Primitive Root**
>
> If $\text{ord}_p(a) = \phi(p)$, then a is a primitive root $\pmod{p}$.

Example 15.3.1.2. *What are the primitive roots $\pmod{11}$ and $\pmod{13}$?*

Solution 15.3.1.1. *We require $\text{ord}_{11}(a) = \phi(11) = 10$ for $p = 11$ (and, similarly, $\text{ord}_{13}(a) = \phi(13) = 12$. Then $a^{10} \equiv 1 \pmod{11}$ and $a^{12} \equiv 1 \pmod{13}$ respectively, or $a = 2$ in both cases by inspection (with $a = 9$, $a = 11$ also being primitive roots in the respective cases $p = 11, 13$).*

Note that primitive roots *do not always exist!* For instance, if we take $n = 12$ (working in modulo 12), then no integer satisfies the resulting congruence, so no primitive root exists. Indeed, 12 is far from the only integer that has no primitive root $\pmod{n}$. We can actually prove that only certain integers have primitive roots:

> **Theorem 15.11: Necessary Condition for Primitive Roots mod Composites**
>
> For n a composite integer, a primitive root $\pmod{n}$ exists iff $n = 2, 4, p^k, 2p^k$, where p is an odd prime and k is a positive integer.

Proof. For $n = 2$, $n = 4$, the theorem holds by inspection. For the more general cases $n = p^k, 2p^k$, write n as $2^a \cdot p_1^{e_1} \cdot p_2^{e_2} \cdot p_3^{e_3} \cdots p_k^{e_k}$. Further define $\psi(2^a) = \phi(2^a)$ if $0 \leq a \leq 2$ and $\frac{1}{2}\phi(2^a)$ if $a \geq 3$. Then define

$$\lambda(n) = \text{lcm}(\psi(2^a), \phi(p_1^{e_1}), \phi(p_2^{e_2}), \phi(p_3^{e_3}), \ldots, \phi(p_k^{e_k}))$$

If $\gcd(a, n) = 1$, then $a^{\lambda(n)} \equiv 1 \pmod{n}$,[73] so $\lambda(n) \mid \phi(n)$. If $\lambda(n) < \phi(n)$, then there are no primitive roots modulo n. We can easily check that $\lambda(n) = \phi(n)$ for the specified values of n, which completes the proof. □

We present one last intriguing property of note regarding order:

> **Theorem 15.12: Decimal Expansion and Order**
>
> For a positive integer n such that $\gcd(n, 10) = 1$, we have that $\text{ord}_n(10) = p$, where p is the period of

---

[73] Proof: By Euler's totient theorem, $a^{\phi(n)} \equiv 1 \pmod{n}$, and $\lambda(n) \equiv \phi(n) \pmod{n}$ since $\lambda(n)$ is the LCM of all terms in the closed form of $\phi(n)$, namely the prime factors of n.

the decimal expansion of $\frac{1}{n}$.

Example 15.3.1.3. *As an illustrative example, let $n = 7$; then $\frac{1}{7} = 0.\overline{142857}$ has period 6, and indeed, $ord_7(10) = 6$.*

This result was proven by Glaisher in 1878, and refined by Lehmer in 1941.[74]

15.3.2 Cyclotomic Polynomials

The *cyclotomic polynomials* are a specific class of polynomials that share properties that are intrinsically connected to the idea of order with respect to primes:

> **Definition 15.8: Cyclotomic Polynomial**
>
> A *cyclotomic polynomial* is a monic polynomial (i.e. a polynomial with leading coefficient 1) in a specifically-defined sequence. Namely, the nth cyclotomic polynomial, denoted $\psi_n(x)$, is the monic polynomial with integer coefficients that so happens to be the minimal polynomial[a] over the field of rationals having all primitive n^{th} roots of unity as roots. That is,
>
> $$\psi_n(x) = \prod_{\substack{1 \leq k \leq n \\ \gcd(k,n)=1}} \left(x - e^{2i\pi \cdot \frac{k}{n}}\right)$$
>
> (since $e^{2i\pi} = 1$ by Euler's formula).
>
> ---
> [a]The polynomial with smallest degree having a specified value as a root.

We now state a particularly crucial and useful theorem uniting divisors with the theory of cyclotomic polynomials:

> **Theorem 15.13: Product of Cyclotomic Polynomials over Divisors**
>
> For any positive integer n,
>
> $$\prod_{d \mid n} \psi_d(x) = x^n - 1$$
>
> That is, each nth root of unity is also a primitive d^{th} root of unity for a unique $d \mid n$.

Proof. We have

$$x^n - 1 = \prod_{k=1}^{n} \left(x - e^{\frac{2i\pi \cdot k}{n}}\right)$$

$$= \prod_{d \mid n} \prod_{\substack{k=1 \\ \gcd(k,n)=d}} \left(x - e^{\frac{2i\pi \cdot k}{n}}\right)$$

$$= \prod_{d \mid n} \phi_{\frac{n}{d}} = \prod_{d \mid n} \phi_d$$

where the final step follows from the one-to-one correspondence between $d \mapsto \frac{n}{d}$ over all divisors $d \mid n$. $\square$

Subsequently, we can immediately derive some examples of cyclotomic polynomials for certain integers n:

---

[74]For the interested reader, we provide the following works of theirs: "Periods of Reciprocals of Integers Prime to 10" and "Guide to Tables in the Theory of Numbers," respectively."

> **Theorem 15.14: Elementary Examples of Cyclotomic Polynomials**
>
> If n is prime,
> $$\psi_n(x) = \frac{x^n - 1}{x - 1} = 1 + x + x^2 + \ldots + x^{n-1}$$
> If $n = 2p$ for p a prime,
> $$\psi_n(x) = 1 - x + x^2 - \ldots + x^{p-1}$$
> If $n \geq 3$ is an odd integer, then we have
> $$\psi_{2n}(x) = \psi_n(-x)$$
> If n is a prime power (i.e. of the form p^k for some prime p, $k \in \mathbb{N}$), then
> $$\psi_n(x) = \psi_p\left(x^{p^{k-1}}\right) = \sum_{i=0}^{p-1} x^{i \cdot p^{k-1}}$$
> More generally, if $n = p^k \cdot m$ with $\gcd(m, p) = 1$, then
> $$\psi_n(x) = \psi_{pm}\left(x^{p^{k-1}}\right)$$
> From these formulas, we can then derive the equation
> $$\psi_n(x) = \psi_q\left(\sqrt[q]{x}\right)$$

Of particular interest is $\psi_{105}(x)$, which is the first cyclotomic polynomial to have a coefficient other than 0 or ± 1 (the anomalous coefficient is -2, which corresponds to the x^7 term). (Why do you think this is the case? Hint: 105 is the product of the three smallest odd prime numbers.)

Henceforth, we analyze the values that a cyclotomic polynomial may assume. Over all real values of x (not just integral x), $\psi_n(x) > 0$ for all $n \geq 3$ (as the roots of a cyclotomic polynomial are all non-real). For $n = 1, 2$, we simply have $\psi_n(x) = x - 1$ and $\psi_n(x) = x + 1$ respectively. For $n \geq 2$, $\psi_n(0) = 1$ (i.e. the constant term is always 1), $\psi_n(1) = 1$ if n is not a prime power, and $\psi_n(1) = p$ if $n = p^k$ for prime p and $k \geq 1$ an integer.

Outside of these special cases, the order with respect to modulo p, for p a prime, is heavily intertwined with the study of cyclotomic polynomials, for we can generalize the main results of our analysis in the previous section to all primes p. If p is a prime and b is an integer coprime with p, then $\text{ord}_p(b) = n$, where n is the smallest positive integer such that $p \mid (x^n - 1)$. A direct consequence of this - in conjunction with the definition of order - is that, if $\text{ord}_p(b) = n$, then $p \mid \psi_n(b)$. (Note, however, that the converse is false. One can derive a counterexample by inspection.)

Proof. By definition, $p \mid b^n - 1$. If $p \nmid \psi_n(b)$, then there must exist another factor $\psi_k(b)$ of $b^n - 1$ that is a multiple of p, and would divide $b^k - 1$. Hence, n would not be the order of $b \pmod{p}$. □

Using Möbius Inversion

In order to expedite the process of determining $\psi_n(x)$ for sufficiently large n, we might consider using a tactic known as *Möbius inversion*, in which we take advantage of the properties of the *Möbius function* to derive the cyclotomic polynomials. First, of course, we must define this function:

Definition 15.9: Möbius Function

Denoted by $\mu(n)$ for a positive integer n, the *Möbius function* is the sum of the primitive n^{th} roots of unity. More explicitly:

$$\mu(n) = \begin{cases} 1 & n = 1 \\ 0 & \exists a > 1 \text{ such that } a^2 \mid n \\ (-1)^k & n \text{ is the product of } k \text{ distinct primes} \end{cases}$$

In particular, for all prime p, $\mu(p) = -1$.

Corollary 15.14: Multiplicativity of $\mu(n)$

$\mu(n)$ is a multiplicative function.

Proof. Observe that $\mu(1) = 1$. For any $m, n \in \mathbb{N}$, consider three cases: one case where m and n are both prime, another case where one of m or n is prime (WLOG assume m) and the other has exactly k distinct prime factors ($k \geq 2$), and a final case where m and n have a and b prime factors, respectively. The first case immediately yields $(-1)(-1) = 1$, so we are done. In the second case, observe that $\mu(m) = -1$ and $\mu(n) = (-1)^k$. If m is *not* a prime factor of n, then $\mu(mn) = (-1)^{k+1} = (-1) \cdot (-1)^k$. If m is a prime factor of n, then $mn = m^2 \cdot p_1^{e_1} \cdot p_2^{e_2} \cdot p_3^{e_3} \cdots p_{k-1}^{e_{k-1}}$ and $\mu(mn) = (-1)^k$ still, since no new prime factors have been introduced to the prime factorization of mn as opposed to that of n. Finally, in the third case, if m has a distinct prime factors and n has b distinct prime factors, mn will have $a+b$ distinct prime factors, such that $\mu(m) = (-1)^a$, $\mu(n) = (-1)^b$, and $\mu(mn) = (-1)^{a+b} = (-1)^a \cdot (-1)^b$ (since m and n must be coprime, by the definition of a multiplicative function). This exhausts all cases; hence, the proof is complete. $\square$

The following is another quite useful property of the Möbius function:

Theorem 15.15: Sum over Divisors

$$\sum_{d \mid n} \mu(d) = \left\lfloor \frac{1}{n} \right\rfloor$$

This theorem is instrumental in laying down the framework for the Möbius inversion formula, which in turn plays a critical role in shaping our understanding of cyclotomic polynomials.

Proof. For $n = 1$, the theorem clearly holds. For $n > 1$, we write $n = p_1^{e_1} \cdot p_2^{e_2} \cdot p_3^{e_3} \cdots p_k^{e_k}$. In the sum, the only terms other than 0 are attained at $d = 1$ and those divisors d which are the products of distinct primes (using the definition of the Möbius function). Hence,

$$\sum_{d \mid n} \mu(d) = \mu(1) + \sum_{i=1}^{k} \mu(p_i) + \sum_{1 \leq i < j \leq k} \mu(p_i p_j) + \ldots + \mu(p_1 p_2 p_3 \cdots p_k)$$

$$= \binom{k}{0} + \binom{k}{1} \cdot (-1) + \binom{k}{2} \cdot (-1)^2 + \ldots + \binom{k}{k} \cdot (-1)^k = 0$$

as desired. $\square$

An extension of Theorem 15.16 states the following:

Theorem 15.16: Extending the Sum Over Divisors

Let n be a positive integer; then

$$\sum_{d \mid n} \mu(d) \cdot \frac{n}{d} = \phi(n)$$

Proof. Define the constant function $1(k) = 1$ for all k. Then write

$$\phi(n) = \sum_{\gcd(d,n)=1} 1(d)$$

so that the original sum becomes

$$\sum_{k=1}^{n} \left\lfloor \frac{1}{\gcd(n,k)} \right\rfloor$$

By Theorem 15.16, we obtain

$$\phi(n) = \sum_{k=1}^{n} \left(\sum_{d \mid \gcd(n,k)} \mu(d) \right)$$

$$= \sum_{k=1}^{n} \sum_{\substack{d \mid n \\ d \mid k}} \mu(d)$$

Essentially, for $d \mid n$, we are summing over all $k \in [1, n]$ that are multiples of d. Setting $k = qd$ for $q \in \mathbb{Z}$, we have that $1 \leq k \leq n \iff 1 \leq q \leq \frac{n}{d}$. Therefore, we can write

$$\phi(n) = \sum_{d \mid n} \left(\sum_{q=1}^{\frac{n}{d}} \mu(d) \right)$$

$$= \sum_{d \mid n} \mu(d) \cdot \sum_{q=1}^{\frac{n}{d}} 1(q)$$

$$= \sum_{d \mid n} \mu(d) \cdot \frac{n}{d}$$

which completes the proof. $\square$

At last, with the Möbius function defined, and hopefully clarified, we can proceed to discuss the technique of Möbius inversion. The motivating idea behind this technique is that functions can be expressed as sums of other functions (i.e. as *linear combinations* of those functions), namely over divisors of the argument. Such equations involving these sums are of the general form $f(n) = \sum_{d\mid n} g(N)$. Here, $g(n)$ can be solved in terms of f by way of Möbius inversion.

> **Theorem 15.17: Möbius Inversion**
>
> Let f and g be arithmetic functions[a] such that $f(n) = \sum_{d \mid n} g(d)$ for all n. Then
>
> $$g(n) = \sum_{d \mid n} \mu(d) \cdot f\left(\frac{n}{d}\right) = \sum_{d \mid n} \mu\left(\frac{n}{d}\right) \cdot f(d)$$
>
> [a] An *arithmetic function* $f(n)$ is any function with domain $\mathbb{N}$ and range $R \subset \mathbb{C}$, equipped with the properties of additivity and multiplicativity.

Surprisingly, unlike the other proofs we have conquered up to this point, the proof of Möbius inversion is short and sweet (relatively speaking, at least):

Proof. Consider the sum

$$\sum_{d|n} f\left(\frac{n}{d}\right) \cdot \mu(d) = \sum_{d|n} \left(\sum_{r|\frac{n}{d}} g(r) \cdot \mu(d)\right)$$

$$= \sum_{rd|n} g(r) \cdot \mu(d)$$

$$= \sum_{r|n} \left(\sum_{d|\frac{n}{r}} \mu(d)\right) \cdot g(r)$$

Note that

$$\sum_{d|\frac{n}{r}} \mu(d)$$

is equal to 0 if $r \neq n$, and 1 if $r = n$, so the entire expression simplifies to $g(n)$. $\square$

As an exercise to the reader, try to prove the following theorem:

> **Theorem 15.18: Multiplicativity of Möbius Inversion**
>
> If f, g are arithmetic functions with $f(n) = \prod_{d|n} g(d)$, then
>
> $$g(n) = \prod_{d|n} f\left(\frac{n}{d}\right)^{\mu(d)}$$
>
> $$= \prod_{d|n} f(d)^{\mu\left(\frac{n}{d}\right)}$$

(The solution will not be provided here; this is purely to spur your interest, and for your own edification!)

As an application of this theorem that will segue into our next topic, consider the general cyclotomic polynomial $\psi_n(x)$:

$$x^n - 1 = \prod_{d|n} \psi_d(x)$$

Applying multiplicative Möbius inversion yields

$$\psi_n(x) = \prod_{d|n} (x^d - 1)^{\mu\left(\frac{n}{d}\right)}$$

which can be used as a more explicit method of computing $\psi_n(x)$ for large n (as opposed to having to expand using the definition, which grows to be wholly impractical - even altogether infeasible - for any n over around 10 or so).

Example 15.3.2.1. *Compute $\psi_{24}(2)$.*

Solution 15.3.2.1. *Using multiplicative Möbius inversion,*

$$\psi_{24}(x) = \prod_{d|24} (x^d - 1)^{\mu\left(\frac{24}{d}\right)}$$

$$= (x - 1)^{\mu(24)} \cdot (x^2 - 1)^{\mu(12)} \cdot (x^3 - 1)^{\mu(8)} \cdots (x^{24} - 1)^{\mu(1)}$$

$$= \frac{(x^4 - 1)(x^{24} - 1)}{(x^8 - 1)(x^{12} - 1)} = \frac{x^{12} + 1}{x^4 + 1} = x^8 - x^4 + 1$$

$$\implies \psi_{24}(2) = 2^8 - 2^4 + 1 = \boxed{241}$$

Irreducibility of the Cyclotomic Polynomials

A crucial result of the cyclotomic polynomials $\psi_n(x)$ is their *irreducibility* over the integers for all n; i.e. that they cannot be factored, or "split," into the product of polynomials with integer coefficients. (For instance, the polynomial $x^2 + x - 6$ is reducible over $\mathbb{Z}$ as $(x+3)(x-2)$, but not over $\mathbb{N}$, since the constant terms, 3 and -2, are in $\mathbb{Z}$ but not both in $\mathbb{N}$.)

> **Theorem 15.19: Irreducibility of Cyclotomic Polynomials over $\mathbb{Z}$**
>
> The cyclotomic polynomials $\psi_n(x)$ are all irreducible over $\mathbb{Z}$.

Before we can prove this central theorem of irreducibility, we must first introduce a few key lemmas:

> **Lemma 15.19: Multiple Zeroes over a Field (15.20A)**
>
> Over a field $\mathbb{F}$, a polynomial $p(x)$ has a multiple zero in some extension of $\mathbb{F}$ iff both p and p' have a common factor with positive degree in $\mathbb{F}[x]$.

Proof. The proof of this involves techniques beyond the scope of this book, so we will take it as fact. However, for those interested, the proof can be found in *Contemporary Abstract Algebra* (Gallian, 2010). □

> **Lemma 15.19: Monic Polynomials over the Integers (15.20B)**
>
> Let $g(x), h(x) \in \mathbb{Z}[x]$ with $h(x)$ monic. If, over $\mathbb{Q}[x]$, $h(x) \mid g(x)$ for all x, then $h(x) \mid g(x)$ over $\mathbb{Z}[x]$.

Proof. We induct upon the degree of $g(x)$, starting with the base case $\deg g(x) = m, \deg h(x) = n$ with $m = n$. Then we have $g(x) = a \cdot h(x)$ for some constant $a \in \mathbb{Z}$. Because a is an integer, $h(x) \mid g(x)$ over $\mathbb{Z}[x]$.

Assume that $m > n$; for all d such that $n \leq d < m$, we define a polynomial $f(x) \in \mathbb{Z}[x]$ with $\deg f = d$ such that $h(x) \mid f(x)$ over $\mathbb{Q}[x]$. Then define another polynomial $k(x) = g(x) - ah(x) \cdot x^{m-n}$; observe that $h(x) \mid k(x)$ over $\mathbb{Q}[x]$. Furthermore, $\deg k < m$, and $k(x)$ has integer coefficients. If k is the zero polynomial, then $h(x)$ clearly divides $g(x)$ over $\mathbb{Z}[x]$. Assume $k(x) \neq 0$; then since $h(x)$ divides $k(x)$, we must have $\deg k > n$. It follows that $h(x) \mid k(x)$ over $\mathbb{Z}[x]$, and so $h(x) \mid (k(x) + ah(x) \cdot x^{m-n}) \implies h(x) \mid g(x)$ over $\mathbb{Z}[x]$. □

And one more lemma for good measure:

> **Lemma 15.19: Associativity of the Exponent (15.20C)**
>
> Let p be a prime number. If $g(x) \in \mathbb{Z}_p[x]$, then $g(x)^p = g(x^p)$.

Proof. Consider the function $\gamma : \mathbb{Z}_p[x] \mapsto \mathbb{Z}_p[x]$, $\gamma(f(x)) = (f(x))^p$. Also, suppose that $f(x), g(x) \in \mathbb{Z}_p[x]$. We have that $\gamma(f(x)g(x)) = f(x)^p \cdot g(x)^p$. Recall that, if p is prime, then for any integer j such that $0 < j < p$, the binomial coefficient $\binom{p}{j}$ is a multiple of p.[75] It follows from the Binomial Theorem that

$$\gamma(f(x) + g(x)) = (f(x) + g(x))^p$$

$$= \sum_{j=0}^{p} \binom{p}{j} \cdot f(x)^p \cdot g(x)^p$$

---

[75]There will always be a factor of p in the expansion of the binomial coefficient, since p is prime.

$$= f(x)^p + g(x)^p + \sum_{j=1}^{p-1} \binom{p}{j} f(x)^p \cdot g(x)^p$$

$$\equiv \gamma(f(x)) + \gamma(g(x))$$

where the last step follows from the fact that $\mathbb{Z}_p[x]$ has characteristic p. This shows that the function γ is operation-preserving.

Now take some arbitrary constant polynomial $a \in \mathbb{Z}_p[x]$. If $a \neq 0$, then $\gcd(a, p) = 1$; by Fermat's Little Theorem, $a^p \equiv a \pmod{p}$. Thus, $\gamma(a) = a^p = a$. (For $a = 0$, the desired result is immediate.) If $g(x) \in \mathbb{Z}_p[x]$, then we have $g(x) = \sum_{i=0}^{n} a_i x^i$. It follows that

$$g(x)^p = \gamma(g(x)) = \gamma\left(\sum_{i=0}^{n} a_i x^i\right)$$

$$= \sum_{i=0}^{n} \gamma(a_i) \cdot \gamma(x^i)$$

$$= \sum_{i=0}^{n} \gamma a_i x^{p \cdot i} = g(x^p)$$

This completes the proof. $\square$

Finally, we have all the tools necessary to prove that all cyclotomic polynomials are irreducible over the integers:

Proof. This proof is perhaps the most straightforward and elegant (in that it does not involve any techniques beyond those used to prove the lemmas) of those discovered by Gauss, as well as other mathematicians following in his wake.

Let $f(x) \in \mathbb{Z}[x]$ be a monic, irreducible polynomial that is a factor of $\psi_n(x)$. We claim that every root of $\psi_n(x)$ is also a root of $f(x)$.

Note that $\psi_n(x) \mid (x^n - 1)$ over $\mathbb{Z}[x]$, and so there exists a polynomial $g(x) \in \mathbb{Z}[x]$ such that $f(x)g(x) = x^n - 1$. Then let ω be such that $f(\omega) = 0$ and ω is a primitive n^{th} root of unity. Let p be a prime such that $\gcd(n, p) = 1$, from which it follows that ω^p is also a primitive n^{th} root of unity. To prove this, we invoke the following additional lemma:

> **Lemma 15.19: Condition for a Primitive Root of Unity (15.20D)**
>
> Let $n, k \in \mathbb{N}$, and let ξ be a primitive n^{th} root of unity (such that $\xi^n = 1$ and n is minimal). Then ξ^k is a primitive n^{th} root of unity iff k and n are coprime.

Proof. Let d be such that $\xi^{kd} = 1$. It follows that $n \mid kd$. If $\gcd(k, n) = 1$, then $n \mid d$. But $d \mid n$ as well, so $d = n$, and so ξ^k is primitive. Otherwise, if $\gcd(k, n) \neq 1$, then $\xi^{\frac{kn}{\gcd(k,n)}} = 1$, so $d < n$, which implies that ξ^k is not primitive. $\square$

It follows that $(\omega^p)^n - 1 = f(\omega^p) \cdot g(\omega^p) = 0$, and so ω^p must be a root of either $f(x)$ or $g(x)$. Suppose $f(\omega^p) \neq 0$; then $g(\omega^p) = 0$, which implies that ω is a root of $g(x^p)$. Moreover, $f(x)$ has ω as a root, so $f(x)$ is the minimal polynomial for ω over $\mathbb{Q}$. Thus, $f(x) \mid g(x^p)$ over $\mathbb{Q}[x]$, and by extension, $f(x) \mid g(x^p)$ over $\mathbb{Z}[x]$ as well (because f is monic).

Define another polynomial $h(x) \in \mathbb{Z}[x]$ such that $f(x)h(x) = g(x^p)$. Henceforth, define three polynomials $\hat{f}(x), \hat{g}(x), \hat{h}(x)$ whose coefficients are those of $f(x), g(x),$ and $h(x)$ reduced $\pmod{p}$. By Lemma 15.20C, it follows that $\hat{g}(x)^p = \hat{g}(x^p) = \hat{f}(x) \cdot \hat{h}(x)$. Subsequently observe that $\hat{f}(x)$ and $\hat{g}(x)$ share an irreducible factor $k(x)$ such that there exist polynomials $l(x), m(x) \in \mathbb{Z}_p(x)$ with $k(x)l(x) = \hat{f}(x)$ and $k(x)m(x) = \hat{g}(x)$.

This yields
$$x^n - 1 = \hat{f}(x) \cdot \hat{g}(x) = k(x)^2 \cdot l(x)m(x)$$
Hence, in some extension field of $\mathbb{Z}_p$, $x^n - 1$ has a multiple root. By Lemma 15.20A, $x^n - 1$ and $(x^n - 1)' = nx^{n-1}$ share a non-constant polynomial common factor. Note that all non-constant factors of nx^{n-1} are of the form Cx^j, where $C \mid n$ and $1 \leq j \leq n-1$. Any factor of this form cannot divide $x^n - 1$, so $x^n - 1$ and nx^{n-1} must be coprime. This is our desired contradiction, which yields $f(\omega^p) = 0$. Thus, if ω is a primitive n^{th} root of unity such that $f(\omega) = 0$ and p is a prime with $p \nmid n$, then $f(\omega^p) = 0$.

Let ξ be a primitive root of unity, not equal to ω. Then there exists an integer k such that $\omega^k = \xi$. By Lemma 15.20D, $\gcd(k, n) = 1$, so it follows that $k = p_1 p_2 p_3 \cdots p_r$, where $p_i \nmid n$ for all $1 \leq i \leq r$. Then $\omega, \omega^{p_1}, \omega^{p_1 p_2}, \ldots, \omega^k$ are all roots of f that are also primitive n^{th} roots of unity. In particular, $f(\xi) = 0$, which implies every primitive n^{th} root of unity is a root of f. Therefore, $f(x)$ and $\psi_n(x)$ share the same roots. At this point, using the fact that $\psi_n(x)$ is monic (which will not be proven here; this is a section exercise) allows us to conclude that $\psi_n(x)$ and $f(x)$ are in fact equal, and so $\psi_n(x)$ is irreducible over $\mathbb{Z}$; this completes the proof.[76] □

15.3.3 Examples

Example 15.3.3.1. *Let p be an odd prime. Use the fact that there exists a primitive root $\pmod{p}$ to prove Wilson's Theorem.*

Solution 15.3.3.1. *Let r be a primitive root $\pmod{p}$. Then the integers $1, 2, 3, \ldots, p-1$ are congruent to $r, r^2, r^3, \ldots, r^{p-1}$ in some order. Thus,*
$$(p-1)! \equiv r \cdot r^2 \cdot r^3 \cdots r^{p-1} \pmod{p}$$
$$\equiv r^{\frac{p(p-1)}{2}} \pmod{p}$$
$$\equiv (-1)^p \pmod{p} \equiv -1 \pmod{p}$$
which is the statement of Wilson's Theorem.

Example 15.3.3.2. *Evaluate $\mathrm{ord}_7 13$, $\mathrm{ord}_5 2019$, and $\mathrm{ord}_p 2^p$ for any prime p.*

Solution 15.3.3.2. *$\mathrm{ord}_7 13 = 2$, since $13^2 \equiv 1 \pmod{7}$; $\mathrm{ord}_5 2019 = 2$, $\mathrm{ord}_p 2^p = p - 1$ unless $p + 1 = 2^k$ for some k, in which case $\mathrm{ord}_p 2^p = k$*

---

[76] For further reference, the interested reader may want to consult the paper "Several Proofs of the Irreducibility of the Cyclotomic Polynomials" (Weintraub), at https://www.lehigh.edu/~shw2/c-poly/several_proofs.pdf, from which this proof was adapted.

15.3.4 Exercises

Exercise 270. *Evaluate $\text{ord}_7 13$, $\text{ord}_5 2019$, and $\text{ord}_p 2^p$ for any prime p.*

Exercise 271. *Show that $\deg(\psi_n) = \phi(n)$.*

Exercise 272. *Prove that n is a divisor of $\phi(a^n - 1)$ for all positive integers a, n.*

Exercise 273. *Prove that the nth roots of unity form a group under multiplication.*

Exercise 274. *Let $n > 1$ be an odd integer. Show that $\psi_{2n}(x) = \psi_n(-x)$ for all x.*

Exercise 275. *Describe the relationship between the Euler totient function $\phi(n)$ and the primitive n^{th} roots of unity for any positive integer n.*

Exercise 276. *Express $\psi_n(x)$ in terms of the Mobius function, $\mu(n)$.*

Extension Exercise 118. *(2006 IMO Shortlist N5, modified) Find all integer solutions (x, y) to the equation $x + x^2 + x^3 + x^4 + x^5 + x^6 - y^5 = -2$.*

Extension Exercise 119. *Prove that all cyclotomic polynomials have integer coefficients.*

Extension Exercise 120. *For each $n \in \mathbb{N}$, prove that there are infinitely many primes $p \equiv 1 \pmod{n}$.*

Extension Exercise 121. *(2003 IMO P6) Let p be a prime number. Prove that there exists a prime number q such that for every integer n, the number $n^p - p$ is not divisible by q.*

Extension Exercise 122. *Prove Wedderburn's Theorem:*

> **Theorem 15.20: Wedderburn's Theorem**
>
> *A finite division ring is a field.*

Chapter 15 Review Exercises

Exercise 277. *Compute $v_3(2019^{4^5} - 1920^{5^4})$.*

Exercise 278. *Prove that, for any $n \in \mathbb{N}$, we have*
$$\sum_{d \mid n} \phi(d) = n$$

Exercise 279. *(2005 Canadian MO P2) Let (a, b, c) be a Pythagorean triple, i.e., a triplet of positive integers with $a^2 + b^2 = c^2$.*

(a) Prove that $\left(\dfrac{c}{a} + \dfrac{c}{b}\right)^2 > 8$.

(b) Prove that there does not exist any integer n for which we can find a Pythagorean triple (a, b, c) satisfying $\left(\dfrac{c}{a} + \dfrac{c}{b}\right)^2 = n$.

Exercise 280. *Describe the positive integers n such that*
$$\sum_{i=1}^{n} \sum_{j=1}^{n} (i-j)^2$$
is a perfect square.

Exercise 281. *Define the root mean square of positive integers $a_1, a_2, a_3, \ldots, a_n$ to be*
$$rms(a_1, a_2, a_3, \ldots, a_n) = \sqrt{\dfrac{a_1^2 + a_2^2 + a_3^2 + \ldots + a_n^2}{n}}$$
Compute the smallest positive integer n such that $rms(1, 2, 3, \ldots, n)$ is also an integer.

Exercise 282. *(2001 IMO Shortlist N5) Let $a > b > c > d$ be positive integers and suppose that $ac + bd = (b + d + a - c)(b + d - a + c)$. Prove that $ab + cd$ is not prime.*

Exercise 283. *(2017 USAMO P1) Prove that there are infinitely many distinct pairs (a, b) of relatively prime positive integers $a > 1$ and $b > 1$ such that $a^b + b^a$ is divisible by $a + b$.*

Extension Exercise 123. *(2016 IMO Shortlist N3) Define $P(n) = n^2 + n + 1$. For any positive integers a and b, the set*
$$\{P(a), P(a+1), P(a+2), \ldots, P(a+b)\}$$
is said to be fragrant if none of its elements is relatively prime to the product of the other elements. Determine the smallest size of a fragrant set.

Extension Exercise 124. *(2017 IMO Shortlist N5) Find all pairs (p, q) of prime numbers with $p > q$ for which the number*
$$\dfrac{(p+q)^{p+q}(p-q)^{p-q} - 1}{(p+q)^{p-q}(p-q)^{p+q} - 1}$$
is an integer.

Extension Exercise 125. *Prove that all elements of the infinite sequence*
$$10001, 100010001, 1000100010001, \ldots$$
are composite.

Extension Exercise 126. *Using the properties of the Jacobi symbol, show that 561 is the smallest Carmichael number.*

Extension Exercise 127. *If a and n are integers, then $\left(\dfrac{a}{n}\right) = 1$ does not necessarily guarantee that a is a quadratic residue modulo n. Approximately how often does the Jacobi symbol incorrectly identify quadratic residues?*

Part VI

Open Problems in Mathematics and Fun Proofs!

Chapter 16

Through the Looking Glass into Some Open Problems

16.1 A Brief Introduction to the Millennium Prize Problems

The world-famous Millennium Prize Problems: seven of the most notoriously difficult problems in mathematics, each worth a whopping $1,000,000 if solved. Only one of these - the Poincaré Conjecture - has actually been solved,[77] by Grigori Perelman in 2003.[78] The other six remain as captivatingly enigmatic as they were when they were first proposed.

We won't actually attempt to make progress on any of these in this chapter, but we will discuss the background behind each one, the obstacles that lie on the path to a solution, and the mathematical repercussions of finally discovering a solution.

16.1.1 Poincaré Conjecture

To date, this is the only solved Millennium Prize problem (which is why we're discussing it first). Conjectured by Henri Poincaré, the conjecture-turned-theorem concerns a 4-dimensional hypersphere (henceforth *3-sphere* for brevity) and its homeomorphism to 3-manifolds:

> **Theorem 16.1: Poincare Conjecture**
>
> Every simply connected, closed 3-manifold is homeomorphic to the 3-sphere.

To even begin to dissect this deceptively complicated statement, we must first define all of the terminology, beginning with the concept of a *topological space*:

> **Definition 16.2: Topological Space**
>
> A topological space is a set of points, accompanied by a set of neighborhoods for each point, that satisfies topological axioms. In order to define these axioms, first we must define a *neighborhood*:

---

[77] As of April 2019.
[78] The proof was confirmed to be correct in 2006, however.

> **Definition 16.2: Topological Neighborhood**
>
> Let X be a set of points (X can be empty), and let $N : x \mapsto N(x)$ (where $x \in X$) be a function mapping a point $x \in X$ to a *neighborhood topology* $N(x)$ consisting of subsets of X. Then the elements of $N(x)$ are neighborhoods of x with respect to N. N is a neighborhood topology iff all of the axioms below are satisfied, and X is hence a *topological space*:
>
> - If N is a neighborhood of x, then $x \in N$. That is, each point belongs to all of its neighborhoods.
> - If $N \subseteq X$, and includes a neighborhood of x, then N is a neighborhood of x.
> - The intersection of two neighborhoods of x is itself a neighborhood of x.
> - For all neighborhoods N of x, there exists a neighborhood M of x with $M \subseteq N$ such that N is a neighborhood of m for all points $m \in M$.

We then define the concept of a topological *homeomorphism*:

> **Definition 16.3: Homeomorphism**
>
> A homeomorphism is a continuous function between two topological spaces that has a continuous inverse function; it is a mapping between two topological spaces that preserves all topological properties of the first space.

Example 16.1.1.1. *For example, consider the open interval (a, b), where $a, b \in \mathbb{R}$, $a < b$. This interval is homeomorphic to $\mathbb{R}$, as (a, b) is homeomorphic to (c, d) where (c, d) is another open interval. Namely, let $(c, d) = \left(-\frac{\pi}{2}, \frac{\pi}{2}\right)$, and let $f(x) = \tan x$, which is both continuous on (c, d) for any $c, d \in \mathbb{R}$, $c < d$, and has an inverse function continuous on (c, d) (specifically, $f^{-1}(x) = \arctan x$).*

Next, we define the generalized *manifold*:

> **Definition 16.4: Definition of a Manifold**
>
> In the topological context, a *manifold* is a topological space that resembles Euclidean space at each of its points. Id est, each point of an n-dimensional manifold (or n-manifold for short) has a neighborhood that is homeomorphic to $\mathbb{R}^n$.
>
> In particular, a *3-manifold* is a manifold for $n = 3$, for which each point $x \in X$ has a neighborhood N that is homeomorphic to $\mathbb{R}^3$.

Finally, we define the terms "simply connected" and "closed" for a 3-manifold:

> **Definition 16.5: Connectedness in Topological Spaces**
>
> First we must define a few notions of connected-ness in a topological space.
>
> A *path* in a topological space X from point x to point y (where $x, y \in X$) is a continuous function $f : [0, 1] \mapsto X$ such that $f(0) = x$ and $f(1) = y$. Intuitively, a curve between x and y in a topological space X is said to be a path if we can draw a straight line from x to y entirely within X.[a]
>
> A *path-connected* topological space X is one such that, for any two points $x, y \in X$, there is a path between x and y.

A *simply connected* topological space X is a path-connected space such that every path between two points in X can be continuously transformed (i.e. transmuted while staying entirely within X) into another path in X while preserving the two endpoints.

---

[a]This differs slightly from the definition of X as a *convex set*, i.e. a set of points such that, for any two points $x, y \in X$, and a parameter $t \in [0, 1]$, $xt + y(1-t) \in X$. For convex sets, the straight line must be of minimal distance, "as the crow flies," representing a linear combination of the coordinates of x and y as imagined in $\mathbb{R}^2$ as dictated by the parameter $t \in [0, 1]$.

Example 16.1.1.2. *Any circle (or, more generally, ellipse; even more generally, any n-sphere) is both path-connected and simply-connected, but an annulus (a circle with a smaller concentric circular region in the center removed) is path-connected but not simply-connected (as we cannot transmute the path from two points on the boundary of the inner circle any further while still preserving the endpoints; this would require that the path extend into the cut-out circle, which defies the path-connected condition).*

We require one more definition before proceeding with our discussion of what makes this particular conjecture memorable and significant:

Definition 16.6: Closure of Manifolds

We say a manifold is *closed* if it is compact (i.e. closed and bounded) and without boundary. In a sense, it is "finite."

We now have all of the necessary tools to study the conjecture further. But *why* have mathematicians gone to such painstaking lengths to put this particular conjecture to rest? At its core, the Poincaré Conjecture had sought out an answer to the then-pressing question of when two topological spaces could be considered the "same." The importance? Five words: the shape of the universe. The collective work of scientists and mathematicians Alexander Friedmann, Georges Lemaître, Howard P. Robertson, and Arthur Geoffrey Walker, the FLRW model of the universe, upon which most cosmologists currently agree, depends on a thorough solution of the Poincaré Conjecture (Nieto 8).According to this model, the radius of the universe, which has the properties of a 3-manifold, expands according to the density of vacuum energy (Solà 1). More so, our entire understanding of what three-dimensional shapes truly *are* relies on the conjecture, since 3-manifolds are, in essence, just three-dimensional objects, which pervade our everyday lives and the fabric of our very existence.

The second, natural question: *how? How did the proof come about, and what exactly did its solution mean for the future of topology?* The answer this time is much more long-winded. Hang on tight and keep your hands and feet in the vehicle; it's going to be a bumpy ride!

First, the monumental task of proving the long-standing conjecture needed to begin with being able to *visualize higher dimensions than what we, as humans, are naturally accustomed to*, but dimensions nevertheless low enough to still be interesting to topologists. Among the lowest dimensions (where "lowest" here means 1- or 2-dimensional), topology is, and was, easily understood, as much so as in dimensions 5 or higher (where topologists had ready access to an invaluable tool, known as the *h-cobordism theorem*, that allowed them to prove the generalized form of the conjecture).

However, in 3 and 4 dimensions, the particular manifestation of the Generalized Poincaré Conjecture that ensues is what we know today as just the "Poincaré Conjecture." For 3 and 4 dimensions, a crucial condition for the *h*-cobordism theorem to work fails (but this was not yet known in 1904, when the conjecture was originally proposed).

Thus emerged in 1900 the roots of modern topology in higher dimensions - and what better way for the field to be born than at the hands of Henri Poincaré himself! Poincaré had claimed that *homology* - roughly speaking, a method of associating algebraic objects to topological spaces - was in and of itself sufficient to show what would eventually evolve into the conjecture bearing his namesake. Four years later, however, he would come to discover a counterexample to his very own claim in the form of the *Poincaré homology sphere*, a 3-manifold with the homology of a 3-sphere and a finite fundamental group[79] (which showed that homology alone was not enough to precisely state the conjecture). This led him to wonder: did all 3-manifolds with the homology of the 3-sphere, and a finite fundamental group, have to *be* 3-spheres? Using this solitary question as a starting point, Poincaré managed to condense his curiosities down to the Poincaré Conjecture the same year, hence laying down the first building blocks of mathematicians' efforts to crack the then-toughest and most interesting problem in topology thus far.

For nearly thirty years following Poincaré's initial statement of the conjecture, no one had seemed to make any considerable progress. The problem lie untouched, with nary a dent made in its seemingly-impenetrable armor. In the 1930s, however, mathematician and co-founder of homotopy theory J. H. C. Whitehead proposed a proof, but then redacted it. This was far from an exercise in futility, though, as many of his insights paved the way for further developments in the field as a whole. A specific type of manifold, the *Whitehead manifold,* had drawn topologists' attention as a result of Whitehead's study of 3-manifolds that were simply connected, yet not homeomorphic to $\mathbb{R}^3$.

Throughout the mid-20th century, several other mathematicians took their cracks at proving the conjecture, only to tragically run into more brick walls. It was in 1958, however, that the mathematical community finally had its first breakthrough in the path to solving the Poincaré Conjecture. R. H. Bing, who himself had attempted, and failed, to prove the "strong" conjecture, managed to prove a weaker version of it, while at the same time imposing conditions that, when fulfilled, would guarantee the solution of the strong conjecture. The following is Bing's "weak" Poincaré Conjecture:[80]

> **Theorem 16.2: The "Weak" Poincare Conjecture**
>
> If every simple closed curve of a compact 3-manifold is contained within a 3-sphere, then the 3-manifold is homeomorphic to the 3-sphere.

Seven years later, in 1965, Bing proceeded to establish with Polish topologist Karol Borsuk the Bing-Borsuk conjecture (and prove it for the base cases $n = 1$ and $n = 2$), which would prove instrumental in ultimately laying the Poincaré Conjecture to rest. In 1978, Włodzimierz Jakobshe demonstrated that proving the Bing-Borsuk conjecture for $n = 3$ would imply the Poincaré Conjecture. But this mathematical odyssey was still far from over.

It is here where our story takes a somewhat devious turn. In the years to come, the Poincaré Conjecture became notorious for its difficulty, and the mathematical community began to regard announced proofs with heightened skepticism, inasmuch as many mathematicians began to harbor reservations about their own work.[81] Indeed, the book *Poincaré's Prize* by George Szpiro chronicles the numerous failed attempts at resolving the conjecture.

In 1982, two major events opened even more doors for those eager to prove the conjecture. Michael Freedman had proven the conjecture in four dimensions, which definitively defied the conventional wisdom that the generalized conjecture was false.[82] (Earlier, in 1961, Stephen Smale had proven the generalized conjecture in all dimensions 5 and higher.) The same year, differential geometer Richard S. Hamilton had applied the Ricci flow to manifolds in his paper "Three-manifolds with positive Ricci curvature" (Hamilton 255-306), thereby illustrating the general framework for proving special cases of the conjecture. However, he could not actually prove the main conjecture itself - this he left to the flow of time.

---

[79]In very loose terms, a fundamental group, in reference to a topological space, is a property that relates to the shape of the space (namely if it has any holes, and if so, where). It allows one to determine when two paths within the space can be transmuted into each other while leaving their shared base point unchanged.

[80]The proof of this modified theorem can be found in Bing's paper in the *Annals of Mathematics*, Volume 68, Issue 1 (Bing 17-37).

[81]A resurgence of this phenomenon took place with the Riemann Hypothesis in 2018, when renowned Fields Medalist and Abel Prize winner Sir Michael Atiyah announced a proof that was almost immediately shown to be critically flawed.

[82]In addition, his work here had paved the way for yet another offshoot of the conjecture, the *smooth* Poincaré conjecture, which as of the publication of this book, remains an open problem.

Twenty long years after the combined contributions of Freedman and Hamilton, it appeared that there was finally a solution! Grigori Perelman, who would eventually complete the final proof, began in 2002 to sketch the proof and complete the framework laid out by Hamilton. The year after, he had finally dealt the finishing blow: the proof of the Poincaré Conjecture was officially complete!

...Almost. It took another three years for Perelman's proof to undergo comprehensive review; during this time the papers were subject to intense mathematical scrutiny and skepticism which had almost reached a critical mass. Despite all of this, his proof survived review and was confirmed to be correct in the summer of 2006, earning the triumphant mathematician the Fields Medal (which he refused on the grounds that he viewed his contributions to the final proof as no greater than Hamilton's), a spot on the cover of *Science* magazine with his proof as the Breakthrough of the Year, and a permanent place in history as the first person to ever solve a Millennium Prize Problem!

This particular conjecture also has an interesting manner of enticing the brave, valorous souls before Perelman into climbing the mountain but never actually reaching the summit. Like the other Millennium Prize Problems, the Poincaré Conjecture had stumped the world round for a full century before finally having a proper resolution. However, the Poincaré Conjecture in particular stands as a powerful testament to the sheer determination and passionate willpower of mathematicians in both the collaborative and independent spheres. The conjecture has befuddled many a professional mathematician for their entire lifetime, insofar as topologists' obsession with it has been given the affectionate nickname of "Poincaritis" ("The Poincaré Conjecture: Its Past, Present, and Future"). Those who have grappled with it had found little solace, or even a clue as to which general direction they should have proceeded in. Countless valiant mathematicians have either surrendered or succumbed to their own mortalities, but only a few have made the breakthroughs that ultimately led to the proof of the conjecture.

16.1.2 P vs NP

Arguably the most famous of the unsolved Millennium Prize Problems,[83] the P vs. NP problem, in its most simplified form, asks whether or not any problem whose solution can be quickly and easily verified can also be solved quickly and easily. The technicalities of the problem statement are much more complicated, but the problem ultimately boils down to this one seemingly simple question. Can we solve problems forwards that we know to be reverse-engineerable?

First we must introduce a concept from computer science: *computational complexity theory*, namely *time complexity*. At the most fundamental level, computational complexity theory (CCT) studies and classifies computational problems according to their inherent difficulty to solve, or *complexity*. These problems, which are by definition solvable by a series of mechanically-feasible steps as part of an algorithm readable and executable by a computer, are part of *complexity classes*:

> **Definition 16.7: Complexity Classes**
>
> A *complexity class* is a set of problems that demand a similar amount of resources to solve (i.e. have similar complexities). That is, these problems can be solved by some machine/computer C with $O(f(n))$ of some resource R, where n is the input size.

Here O (for *order*, as in the *order* of a function) denotes *big-O notation*, which describes the asymptotically limiting behavior of a function. More specifically, we can derive the following relationship between two functions under this notation:

> **Theorem 16.3: Big-O Notation Relationships**

---

[83]This or the Riemann Hypothesis; it depends on who you ask.

> Consider two functions $f(x)$ and $g(x)$, with $f(x)$ either real- or complex-valued and $g(x)$ real-valued. Iff there exist constants $c > 0$ and $x_0 > 0$ such that $f(x) \leq c \cdot g(x)$ whenever $x \geq x_0$, then $f(x) = O(g(x))$.
>
> Furthermore, we say $f(x) = O(g(x))$ as $x \to a$ for some $a \in \mathbb{R}$ iff there exist $\delta > 0$, $c > 0$ such that $|f(x)| \leq c \cdot g(x)$ whenever $0 < |x - a| < \delta$.

Example 16.1.2.1. *For polynomials $p(x)$ with $\deg p(x) = n$, $O(p(x)) = x^n$, since the limit $\lim_{x \to a} p(x)$ is dictated by the highest exponent of p.*

Proof. Let $p(x) = a_n x^n + a_{n-1} x^{n-1} + \ldots + a_1 x + a_0$. To prove the statement that $O(p(x)) = x^n$, we show that there exist $x_0, c \in \mathbb{R}$ such that for all $x > x_0$, we have $|f(x)| \leq c \cdot x^n$. Let $x_0 = 1$ and $c = \sum_{i=0}^{n} a_i$. Then for all $x > x_0$:

$$\left| a_n x^n + a_{n-1} x^{n-1} + \ldots + a_1 x + a_0 \right| \leq a_n x^n + a_{n-1} x^n + \ldots + a_1 x^n + a_0 x^n = x^n \cdot \left(\sum_{i=0}^{n} a_i \right)$$

Hence, $|p(x)| \leq cx^n$ as desired. □

Big-O notation also has the following properties of linearity:

> **Theorem 16.4: Properties of Big-O Notation**
>
> (a) If $f_1 = O(g_1)$ and $f_2 = O(g_2)$, then $f_1 + f_2 = O(g_1 + g_2)$.
>
> (b) If $f_1 = O(g_1)$ and $f_2 = O(g_2)$, then $f_1 f_2 = O(g_1 g_2)$.
>
> (c) $f \cdot O(g) = O(fg)$.

Proving these is left as an exercise to the reader.

Going back to the complexity class definition, we say that if the growth rate of the amount of resources required by a problem p_1 is $f(p_1)$, and the growth rate of the amount of resources required by problem p_2 is $f(p_2)$, and $f(p_1) > f(p_2)$, then p_1 and p_2 are in different complexity classes. Complexity classes indicate an *approximation* of the amount of some resource (whether it be time in seconds, memory in bytes, or a number of algorithmic operations); they do not provide an exact measurement (and cannot, due to variations in the implementation of two algorithms, even ones with the same order).

As well as classifying problems, we can also classify the algorithms used to solve those problems by their time complexity, or the amount of time required for a solution given an input function. The time complexity of an algorithm n, abbreviated $T(n)$, depends on the nature of the function in big-O notation.

Example 16.1.2.2. *Common examples of time complexity classes include* constant time *($O(c)$ for some constant c, used to find the median in a sorted list, for instance)*, linear time *($O(n)$, used to find a small/large item in an array)*, logarithmic time *($O(\log n)$, used for binary search[84])*, quasi-linear time *($O(n \log n)$)*, factorial time *($O(n!)$, used to solve the traveling salesman problem)*, and polynomial time *($O(n^k)$ for some positive integer k)*.

The last of these, polynomial time, is of particular interest in solving the P vs. NP problem. The problem asks whether any problem that can be verified *in polynomial time* can also be solved *in polynomial time*. Indeed, the abbreviations "P" and "NP" stand for "polynomial time" and "non-deterministic polynomial time." P is the class of problems/tasks that can actually be *solved* in polynomial time, while NP is the class of problem/tasks that can only be *verified* in polynomial time.

---

[84] An algorithm in which we repeatedly divide a list in half and search for the desired item in both halves, until the possible locations of the item are narrowed down to just one.

There are numerous examples of real-world problems that can definitively be classified as P or NP. For instance, obtaining the prime factorizations of *sufficiently small* positive integers is a problem in P, whereas solving a Sudoku puzzle (with enough blank spaces) or graph coloring are problems in NP. However, the vast majority of problems cannot be strictly classified as either P or NP; rather, to classify these, we must invoke the idea of *NP-completeness*.

A problem is *NP-complete*, or in the *NPC class*,[85] if it belongs to both the NP and *NP-hard* classes. First we must define the NP-hard class:

> **Definition 16.9: NP-Hardness**
>
> A problem P is *NP-hard* if, for every problem $Q \in$ NP, Q can be *reduced* to P in polynomial time.
>
> > **Definition 16.9: Reduction**
> >
> > A *reduction* is an algorithm that transforms a problem into an equally difficult, or more difficult, one. More specifically, if A and B are two subsets of the natural numbers, and F is a set of functions from $\mathbb{N}$ to $\mathbb{N}$, then we say A is *reducible* to B iff there exists a function $f \in F$, such that for all $x \in \mathbb{N}$, $x \in A \iff f(x) \in B$. We denote this by $A \leq_F B$. (This definition may seem counter-intuitive, but this apparent discrepancy may be resolved with an example. If we program a machine that only knows how to do addition, subtraction, multiplication, and division, then programming it to multiply x and y by computing $\frac{(x+y)^2-(x^2+y^2)}{2}$ will necessitate at least as many operations, and so squaring is at least as hard as multiplication in the sense that it is just as demanding.)
>
> Informally speaking, P is at least as difficult to solve (requiring at least as many resources) as the hardest problems in NP.

To elaborate on the NP-completeness definition, we say that NP-complete problems are more difficult (more resource-consuming) than most other NP problems, as every NP problem must be reducible to an NP-complete problem. The earliest example[86] of an NP-complete problem (not just NP or NP-hard problems) is the *SAT problem*,[87] which concerns the conditions necessary to satisfy a Boolean formula.[88] In other words, can a Boolean formula always have its variables replaced with "true" or "false" such that the formula evaluates to "true"? Equivalently phrased, do there exist Boolean formulas that always evaluate to "false," regardless of the values of their input variables?

Other famous examples of NP-complete problems include the *set-splitting problem*, which concerns the family of subsets of a set, and the partitions of that set, the *longest path problem*, which seeks to find a simple path (a path which visits each vertex at most once) of maximal length in a graph, and the intricate mechanics of games such as FreeCell, Minesweeper, Light Up, Battleship, the Rubik's Cube,[89] puzzles of Japanese origin such as Nurikabe, Kakuro, and Sudoku, and even video games such as Tetris and Super Mario Bros.[90]

To date, the P vs. NP problem, despite its infamy in the mathematical and computer science communities, remains unsolved. But why? What are the obstacles that belie the path to laying the seemingly-intuitive and intimidating symbol-free problem to rest, and why have they proven so uncooperative with mathematicians

---

[85] No, unfortunately, NPC doesn't stand for "non-player character"; this isn't an MMORPG. (But it would be awesome to see one based on taking down the seven Millennium Prize dragons of the Clay Institute!)
[86] In that this problem was the first to be proven NP-complete, by the Cook-Levin theorem.
[87] SAT is an abbreviation for "satisfiability."
[88] A formula in terms of variables that have the binary values of "true" and "false."
[89] Demaine, Eisenstat & Rudoy 2017.
[90] That *Super Mario Bros. 1-3/Lost Levels/World, Donkey Kong Country 1-3*, and all games in *The Legend of Zelda, Metroid,* and *Pokémon* series, are all NP-complete was proven in the 2015 paper "Classic Nintendo Games are (Computationally) Hard" (Aloupis, Demaine, Guo & Viglietta 2015).

and computer scientists?

As it turns out, the issue lies not with the problem itself, but rather with the progress that has already been made. There are several previously-undiscovered barriers on a hypothetical solution path, namely the limitations of *relativization, natural proofs,* and *algebrization*. Additionally, methods previously used to attempt to attack the problem have shown that existing proof methods are not powerful enough to actually resolve whether P = NP; the implication is that a new set altogether of techniques or understanding is required.

Relativization revolves around the idea of an *oracle,* or "black box" of sorts that takes a question in constant time as input and outputs an answer. Every computational algorithm can make a query to this black box, but some boxes operate differently than others with respect to the P = NP problem in particular. The black boxes are *Turing machines,* which are nested within each other, with machine M_1 at the "center" (much like concentric circles) and machine M_i depending on the output of machine M_{i-1} for all $i \geq 2$. The technique of proving a theorem or statement using these nested black boxes does not change even if the language used to process it changes, hence why the technique is known as relativization. In 1975, Theodore Baker, John Gill, and Robert Solovay showed that there exist two oracles A and B such that P = NP with respect to A, but P $\neq$ NP with respect to B. As relativization proofs only work for statements that are true with respect to all potential oracles, this showed that the P = NP problem was unsolvable using only relativization techniques.

The class of proof techniques known as *natural proofs* was devised in 1994 by mathematicians and theoretical computer scientists, Alexander Razborov and Steven Rudich. A natural proof, in computational complexity theory, is a type of proof that shows that certain complexity classes differ. Specifically, attempted natural proofs of P = NP tend to deal with Boolean functions, namely by placing and proving lower bounds on their complexity. In this case, natural proofs have limitations that prevent them from being a viable method of proving the P vs. NP problem. Razborov and Rudich, treating P and NP as complexity classes, showed that, based on the existence of pseudo-random functions, P and NP could not be separated by natural proofs. First, they proposed the idea of "discrepancy" of a Boolean function, which would differ between the ideal of polynomial time oracles and the reality of functions in NP. Then, they proceeded to show that only functions with low discrepancy could be computed, and finally, that it would suffice to show that *any* NP problem (not necessarily NP-complete) has a high discrepancy.[91]

Following the demonstrated failure of mathematicians' work with natural proofs to make any progress on the P = NP solution front, an entirely new proof method, known as *algebrizing proofs,* arose from the ashes of previous efforts. A major result where a subcategory of algebrizing proofs - *arithmetization* - was used was in proving that two classes of problems, *IP* and *PSPACE,* were equal. (IP is the class of problems that can be solved using an *interactive proof system;* i.e. a machine that models computation as the exchange of information between two parties, and PSPACE is the class of problems that can be solved by a Turing machine with an amount of memory resources, or *space,* that can be modeled as a polynomial function.)

At its most basic level, arithmetization treats the AND, OR, and NOT logic gates of a Boolean function as arithmetic operations - extending said function to a polynomial - and thereby more effectively segregates complexity classes. However, this technique, too, is inconclusive with regard to proving either P = NP or P $\neq$ NP. In a 2008 paper titled "Algebrization: A New Barrier in Complexity Theory," authors Scott Aaronson and Avi Wigderson showed that, due to the infeasibility of solving all NP-complete problems in polynomial time, it is also impossible to use arithmetization alone to resolve whether P = NP.

The current consensus of mathematicians and computer scientists seems to be that P $\neq$ NP; however, no one as yet has been able to provided an explanation grounded in sufficiently rigorous or generally accepted reasoning. According to a poll conducted on computer science researchers in 2002, 87% of the 70 respondents indicated that they thought P $\neq$ NP. William Gasarch, the University of Maryland professor who conducted the poll, himself predicted that developing and verifying a full proof would take between 200 and 400 years![92]

---

[91](Razborov & Rudich 1994)
[92]Gasarach, 2002.

Aside from popular opinion, there are very solid mathematical reasons to believe P $\neq$ NP (none of these establishing this for certain, of course). Thus far, in the decades since the problem was initially formally proposed in 1971, no one has been able to find an example of a polynomial-type algorithm for any one of the thousands of known NP-complete problems, which points more to its non-existence rather than the inadequacies of our mathematical and computer theoretic knowledge. Additionally, P = NP would imply many results that are currently believed to also be false, including P = PH and NP = co-NP.[93]

More intuitively speaking, if it were the case that P = NP, then how would we explain the discontinuity between actually solving a problem and merely verifying its solution? Dr. Scott Aaronson, ex-MIT professor who co-authored the 2008 algebrization paper with Dr. Avi Wigderson, has proclaimed that "the world would be a profoundly different place than we usually assume it to be," and that "there would be no special value in 'creative leaps.'"

This is not to rule out the possibility that P = NP, or to exile it away from consideration; indeed, though proof by exhaustion has not proven fruitful, we cannot make any definitive claims as to whether or not a theorem or conjecture is true just by testing cases, no matter how long we do it. Fermat's Last Theorem is a famous example of a theorem that was commonly believed to be true before it was finally proven, but which had had several "near-miss" counterexamples with large numbers (some of these being $1782^{12} + 1841^{12} = 1922^{12}$, $6107^6 + 8919^6 = 9066^6$, and $3987^{12} + 4365^{12} = 4472^{12}$, all of which by no measure discouraged mathematicians from believing the theorem was true[94]). In the words of Dr. Anil Nerode of Cornell University, "being attached to a speculation is not a good guide to research planning. One should always try both directions of every problem."

Finally, we will discuss the implications of coming to a conclusive resolution. First, let's think about what might happen if mathematicians' and computer scientists' suspicions are confirmed, and P $\neq$ NP. Our modern world is heavily structured around this very premise, and so it is not likely that much will significantly change. However, the proof of P $\neq$ NP would pave the way for great advances in the field of computational complexity theory in that it would be a reminder (perhaps a grim one) that not all common problems can be solved as simply or as elegantly as we think they can be, or how we want them to be. Although a P $\neq$ NP proof would probably lack the effect on current computational processes that a hypothetical proof of P = NP would have, it would nonetheless allow researchers to divert their current focus on the P = NP problem elsewhere.

Should it turn out that P = NP, on the other hand, we can be sure that there is an algorithm that runs in polynomial time. This could have catastrophic effects on the modern operations of private information that need to remain absolutely secure, lest there be a data leak or mishandling. In the public-key cryptography systems ubiquitous in today's world, the widely-held assumption that P $\neq$ NP is what secures financial transactions conducted through Internet-based banking services, ciphers, and cryptographic hashing. If P = NP, then all of the widely-implemented security measures for these purposes would need to be heavily revamped to accommodate polynomial time. Cybercriminals could much more easily hack ciphers and cryptography measures through brute force, unless the degree of coefficients of the polynomial for private data are exorbitantly large. For example, if data is protected in polynomial time by a function such as x^4, the repercussions of P = NP could be vastly devastating, but if we were dealing with a polynomial like $10^{2019}x^{2019}$, then the degree alone (or even the leading coefficient) would ensure that current safeguards against hacking, identity theft, and other criminal activity remain feasible, and that breaking the systems in place is impractical; all but impossible under the time constraints of the human lifespan.

---

[93]PH is the union of all complexity classes in the polynomial hierarchy, which applies the core ideas behind the formation of the P and NP classes to oracle Turing machines. Co-NP is the space of all problems whose complements are in NP. We say that the *complement* of a decision problem is the problem which reverses all yes/no answers. For a problem with a single yes/no answer, its complement is the problem whose answer is *no* when the answer to the original problem is *yes*, and vice versa. As an example, the complement of the decision problem "Is a real number r rational?" is "Is r irrational?"

[94]Modular arithmetic immediately disproves all of these - the first one is trivially false as the sum of an even and an odd number cannot be even!

A concrete demonstration of this lies with the Traveling Salesman problem, which is known to be in NP. We wish to traverse the most efficient path through all 50 states, but the polynomial-time algorithm can also be applied to smaller numbers of states. Say the algorithm used to solve the problem by P = NP allows for the problem to be solved for n states in n^{10} nanoseconds; for $n = 50$, the time required is $50^{10} \approx 9.766 \cdot 10^{16}$ nanoseconds, or approximately 3 years. And the problem only grows more unfathomable, even when we increase the degree by just 1 or 2; consider n^{12}, in which case the problem would take approximately 7742 years to solve!

Machine learning, in general, would be vastly simplified by Occam's Razor - the notion that the simplest solution is the correct or best one. All a machine would need to do to solve a problem is to determine the simplest solution out of all available, or find the smallest program compatible with a set of data. In this regard, P = NP would have dramatically beneficial effects on predictive capabilities in fields such as meteorology, machine translation, transportation, and OCR text recognition.

Finally, and foremost, P = NP has the potential to drastically change the way we approach mathematics itself. Problems that have stumped mathematicians for centuries - including the other Millennium Prize Problems - may well become trivial to Turing machines. Efficient solutions to NP-complete problems could be derived by computers without any human effort, as a machine could simply run through all possible proofs, given that the problem has a proof of a reasonable length. Could P = NP be the last mathematical problem mathematicians ever need to solve on their own? Only time will tell...

16.1.3 Riemann Hypothesis

Rivaling the P vs. NP problem as the most famous of the Millennium problems (and possibly the Millennium problem of the greatest current mathematical interest as well), the *Riemann hypothesis* (RH for short), much like that problem, revolves around one deceptively unyielding question. In order to study the Riemann Hypothesis to any degree of depth, first we must understand the *Riemann zeta function,* a function with intrinsically interesting properties that help to define the field of complex analysis, and which has shaped much of mathematical history (as well as its future). This function has a wide berth of applications in not only pure mathematics, but also physics, statistics, and probability theory.

> **Definition 16.11: Riemann Zeta Function**
>
> The Riemann zeta function $\xi : \mathbb{C}\setminus\{1\} \mapsto \mathbb{C}, \xi(s) = \sum_{n=1}^{\infty} \frac{1}{n^s}$ is the analytic continuation in a complex variable s ($s \neq 1$) of the *Dirichlet series:*
>
>> **Definition 16.11: Dirichlet Series**
>>
>> A *Dirichlet series* is any series of the form
>>
>> $$\sum_{n=1}^{\infty} \frac{a_n}{n^s}$$
>>
>> where $s \in \mathbb{C}$ and a_n is a complex-valued sequence.

When $\operatorname{Re}(s) > 1$, $\xi(s)$ converges. (If $\operatorname{Re}(s) = 1$, the series becomes the harmonic series, which diverges.)

Next we must define an *analytic continuation* of a function:

> **Definition 16.13: Analytic Continuation**
>
> Before defining an analytic continuation of a function f, we must ensure that f is *analytic.*

> **Definition 16.13: Analytic Functions**
>
> A function $f : \mathbb{C} \mapsto \mathbb{C}$ is *analytic* if it is smooth everywhere (i.e. $f^n(x)$ exists for all $n \in \mathbb{N}, x \in \mathbb{C}$), and its Taylor series about $x = x_0$ converges to f in some neighborhood about x_0 for all $x_0 \in \mathbb{C}$.
>
> If f is an analytic function defined on a non-empty subset U of $\mathbb{C}$, and $V \supset U, V \subset \mathbb{C}$, and F is an analytic function defined on V such that, for all $z \in U$, $F(z) = f(z)$, then F is said to be an *analytic continuation* of f.

Example 16.1.3.1. *Examples of analytic functions are all elementary functions (all polynomials, exponential functions, trigonometric and logarithmic functions, and e^x and the gamma function (defined by $\Gamma(n) = (n-1)!$ ($n \in \mathbb{N}$) and $\Gamma(z) = \int_0^\infty \frac{x^{z-1}}{e^x}\, dx$ otherwise). Examples of non-analytic functions are the absolute value function $f(x) = |x|$ and the complex conjugate function $f(a+bi) = a - bi$. (Showing that these are and are not analytic is left as a (quite challenging) exercise to the reader.)*

The following table lists some of the specific values of the zeta function:

s	$\xi(s)$
-1	$-\frac{1}{12}$
0	$-\frac{1}{2}$
1	∞ (note 1)
2	$\frac{\pi^2}{6}$ (note 2)
3	≈ 1.202 (note 3)
4	$\frac{\pi^4}{90}$
$2n, n \in \mathbb{N}$	$(-1)^{n+1} \cdot \frac{B_{2n} \cdot (2\pi)^{2n}}{2 \cdot (2n)!}, n \in \mathbb{N}$ (note 4)

Table 16.1: A table of common values of the zeta function.

Note 1: Unless we take into account the *Cauchy principle value*, given by $\lim_{\epsilon \to 0} \frac{\xi(1+\epsilon) - \xi(1-\epsilon)}{2} = \gamma \approx 0.577$, the Euler-Mascheroni constant, as the value of the zeta function at $s = 1$.

Note 2: Proving this is the Basel problem. The reciprocal of $\xi(2)$, $\frac{6}{\pi^2}$, turns out to be the probability that two randomly selected positive integers are relatively prime.

Note 3: The precise value of $\xi(3)$ is known as *Apéry's constant*, and was proven to be irrational in 1978 by its namesake, Roger Apéry.

Note 4: B_n denotes the nth Bernoulli number.

For odd positive integers s, there is no known explicit form or recurrence relation, though the values of $\xi(s)$ are known to be closely tied to the Bernoulli numbers (which themselves are defined recursively). It is, however, known that infinitely many of the $\xi(s)$, for s a positive odd integer, are irrational.

Of particular interest about this function is that Euler had proven the following identity in his 1737 thesis, *Variae observationes circa series infinitas*:

$$\xi(s) = \prod_{p \text{ prime}} \frac{p^s}{p^s - 1} \tag{16.1}$$

Even more interesting is that Euler's proof utilized only the basic ideas of geometric series and the fundamental theorem of arithmetic. His work on the zeta function is a paragon of simplicity from complexity -

an embodying idea of all of mathematics (and, really, all of life itself). In this manner, one could argue that Euler was far ahead of his time, and not just mathematically!

There is one more major point of interest regarding the zeta function before we even begin to discuss the Riemann Hypothesis itself: its value at $s = -1$. At $s = -1$, it would appear that $\xi(s)$ simplifies to the clearly divergent series

$$\sum_{n=1}^{\infty} n = 1 + 2 + 3 + \ldots$$

yet it analytically evaluates to...$-\frac{1}{12}$?! How is that possible? It seems completely absurd for this series to evaluate to anything other than infinity, yet there it is: a value that is not only finite, but pretty much as neat of a value as one could reasonably expect from a function like this. So what could be going on here? Indeed, one would be absolutely right to be wary of such a tempting equality - and of course, the sum $1 + 2 + 3 + \ldots$ is not *actually* equal to $-\frac{1}{12}$ - at least, not in the traditional context of infinite series. You haven't missed something big here: $1 + 2 + 3 + \ldots$ diverges (cue the sighs of relief). But $-\frac{1}{12}$ is still just as perfectly valid a value assignment to this nebulous quantity - remember, it is the result obtained from the analytic continuation of a complex function. Additionally, we can obtain this value through some heavy-handed manipulation.

Non-rigorous algebraic manipulation. Define three seemingly divergent sums:

$$S_1 = \sum_{n=1}^{\infty} (-1)^{n+1} = 1 - 1 + 1 - 1 + \ldots$$

$$S_2 = \sum_{n=1}^{\infty} (-1)^{n+1} \cdot n = 1 - 2 + 3 - 4 + \ldots$$

$$S_3 = \sum_{n=1}^{\infty} n = 1 + 2 + 3 + 4 + \ldots$$

The first sum, S_1, is *Grandi's series*, which has an assigned value of $\frac{1}{2}$. This can be seen by taking the negative of S_1, then adding it to $1 - S_1$, to obtain $1 - 2S_1 = 0$ (since all of the $+1$ and -1 terms cancel), or $S_1 = \frac{1}{2}$. (This is by no means a rigorous technique; however, it suffices to illustrate the point that algebraic manipulations such as these are no substitute for analytic continuation.)

Henceforth, consider $S_1 - S_2$. We have $S_1 - S_2 = (1-1) + (2-1) + (-3+1) + (4-1) + \ldots = 1 - 2 + 3 - 4 = S_2$, which implies $S_1 = 2S_2$, or $S_2 = \frac{1}{4}$. Finally, we have $S_2 - S_3 = -4 - 8 - 12 - 16 - \ldots = -4(1 + 2 + 3 + \ldots) = -4S_3$, so $S_2 = -3S_3 \implies S_3 = -\frac{1}{12}$. □

Remark. The equal signs here should not be thought of as equal signs in the traditional sense, but rather as "assignment" signs that *associate* a sum with a real value. All of these sums do, in fact, diverge in the classical sense.

Now that we have some background on the zeta function, we can begin to touch upon the subject of centuries of toiling, and the catalyst of many developments of modern mathematical fields: the Riemann Hypothesis, which is enshrouded by mystery, fame, and glory, all at the same time. Initially proposed in 1859, the problem has survived 160 years of frenzied mathematical collaboration without being resolved - more than 50 years longer than the Poincaré Conjecture! The mystery: what does it take to either prove or disprove the hypothesis, and why has it proven so insurmountable to even the world's top-tier mathematicians? First of all, what is the question asking? The statement itself is straightforward enough:

> **Theorem 16.5: Riemann Hypothesis**
>
> The Riemann zeta function $\xi(s)$ has its zeros at $s = -2k, k \in \mathbb{N}$, and $s = a + bi$, $a = \dfrac{1}{2}$. (The zeros of the form $s = 2k, k \in \mathbb{N}$ are known as the *trivial zeros,* and the complex zeros with real part $\dfrac{1}{2}$ are known as the *non-trivial zeros.*)

Should the hypothesis be correct, all non-trivial zeros lie on a *critical line* consisting of all complex numbers of the form $\dfrac{1}{2} + bi$, $b \in \mathbb{R}$.

The difficulty, needless to say, lies not in understanding the problem, but in determining and executing the necessary methods to solve it. Fortunately, we do have a few solid connections with applications that, at first glance, may seem totally unrelated; one notable example is the distribution of prime numbers.

Introduced by Riemann himself alongside his infamous hypothesis, the explicit formulae for *L-functions*[95] relate the sums of complex number zeros of an *L*-function to sums of powers of prime numbers. In his 1859 paper "On the Number of Primes Less Than a Given Magnitude," Riemann had sketched a formula for the prime-counting function $\pi(x)$ in terms of a sum over all ρ, where ρ is a non-trivial zero of $\xi(s)$. In 1976, using Riemann's definitions of the explicit formulae, Schoenfeld was able to show that $|\pi(x) - \mathrm{li}(x)| < \dfrac{1}{8\pi} \cdot \sqrt{x} \log(x)$ where $\mathrm{li}(x)$ denotes the *logarithmic integral* of x; i.e. $\int_0^x \frac{1}{\ln t} dt$. A later result by Dudek, in 2014, showed that, for all $x \geq 2$, there exists a prime p satisfying the inequality $x - \dfrac{4}{\pi} \cdot \sqrt{x} \log(x) < p \leq x$.

Most importantly, the Riemann Hypothesis relates to the prime gap conjecture, which posits that the limit of the mean of the gaps between consecutive primes is $\log p$ (where log denotes the exponential logarithm, i.e. $\ln = \log_e$). Some gaps are much larger than this; however, Cramér proved that, should the Riemann hypothesis hold, the upper bound for *all* prime gaps is $O(\sqrt{p} \cdot \log p)$. Cramér had also established *Cramér's conjecture,* which posits a much stronger upper bound of $O((\log p)^2)$.

But in the way of actual proofs, so far, while several mathematicians have broken new ground, none have actually managed to provide an exhaustive proof or disproof of the hypothesis. One path to a solution that is still teeming with activity and potential avenues lies in *operator (spectral) theory:* the study of (bounded) linear operators, or linear transformations between normed vector spaces. In the early 1910s, David Hilbert and George Pólya formally developed the *Hilbert-Pólya conjecture,* which approaches the Riemann Hypothesis by positing that, if the imaginary parts of the zeros s of the zeta function correspond to the eigenvalues of an unbounded self-adjoint operator, then the RH must be true.[96]

At the same time, this conjecture was somewhat unfounded, and was thus met with large amounts of skepticism; however, in the 1950s, Selbert's work with Riemann surfaces led credence to the proposed connections with Riemann's explicit formulae. Furthermore, in the 1970s and 1980s, Hugh Montgomery analyzed the zeros on the critical line from a quantum mechanical perspective, proposing a statistical distribution of zeros that, to this day, is commonly used in determining the energy levels of a nucleus. At the turn of the 21st

---

[95] An *L-function* is the analytic continuation of an *L-series,* or an infinite Dirichlet series in the complex plane.

[96] A *self-adjoint* operator T on a finite-dimensional complete vector space V satisfies the equation $\langle Tv, w \rangle = \langle v, Tw \rangle$ for all $v, w \in V$.

century, even more developments arose relating to the *Hamiltonian,* an operator that defines the total energy of a physical system in quantum mechanics.

In 2008, an upshot of Hilbert's and Pólya's original conjecture had been established, known as the *Berry conjecture* after one of its co-authors, Michael Berry. It relates to Hermitian operators corresponding to the classical Hamiltonian, as well as the quantization of the Hamiltonian. To date, it has not been used constructively to fully solve the RH, but non-trivial progress has been made regarding the construction of a satisfactory Hamiltonian operator (with the latest advancement in March 2017, in a paper co-authored by Bender, Brody, and Muller).

Not only does the Riemann hypothesis have a slew of applications in quantum mechanics, the assumption of its truth or falsehood also belies the foundation for several crucial results in pure mathematics. Consider the link between the RH and the distribution of prime numbers. Should the RH turn out to be true, we will have sharper estimates of the remainder in the prime number distribution formulas proposed by Cramér and Euler himself, a complete resolution of the *odd* Goldbach conjecture,[97] and an exponentially faster way to test for the primality of large numbers in polynomial time, to name just a few more prominent examples. Countless other conjectures would follow immediately, assuming the RH is true, and be disproven immediately, assuming the RH is false. (Modern understanding of these conjectures will likely remain unchanged, much like with the P vs. NP problem; the general consensus there is that P $\neq$ NP, upon which numerous algorithms and theorems depend.)

16.1.4 Hodge Conjecture

It's here where we begin to delve into truly uncharted and unfamiliar territory. While the P vs. NP problem and Riemann Hypothesis are both household names, to the point that they draw attention from amateurs and lead them to become professional mathematicians in their attempts to crack them, these other four Millennium Problems, though equally significant, are relatively unknown to those who have not dedicated their lives to solving them, or at least largely obscured by a wall of prerequisite mathematical (and perhaps interdisciplinary, as well) understanding.

The *Hodge Conjecture*, a cornerstone of the field of *algebraic geometry* (the study of the roots of multivariate polynomials, as applied to geometric problems), relies on the prior development of *algebraic varieties*, mathematical structures that yield the solutions to systems of multivariate polynomial equations. The precise statement of the Conjecture is extremely complicated, but in simplest terms, it asks whether, for all non-singular complex *Hodge cycles* within *projective* algebraic varieties, said Hodge cycle can be expressed as a sum of rational linear combinations of *algebraic cycles*.

First of all, what is our motivation? That is, why do we study the Hodge Conjecture in the first place, and where did it come from? The answer is surprisingly relatable to the math student experiencing the joy of discovery from the ground up. As geometers were just beginning to study geometric shapes in the standard Euclidean 2-dimensional plane, they began to encounter more distinct, unique, and interesting shapes, such as non-linear curves (especially circles). Any shape in 3 or more dimensions, however, was limited by the boundaries of geometry at the time - such shapes could only be visualized by the human imagination.

In particular, focus began to shift from pictorial representations on 2-dimensional paper to more abstract shapes in an unlimited number of dimensions. From there on out, shapes were no longer merely "shapes," but rather *algebraic cycles* (likely named for their "cyclic" nature, as they loop around much like circles in 2 dimensions and spheres in 3 dimensions). If that shape obeyed a set of properties that made it well-defined, neat, and smooth, it was also deemed a *manifold*.

From here, topology had really begun to take off. The discovery of manifolds was the factor that ignited the

---

[97] The variant of the Goldbach conjecture that states that all odd integers ≥ 7 are a sum of 3 (not necessarily distinct) primes. The even version of the conjecture states that all even integers ≥ 4 are the sum of two primes. Note that the even version implies the odd version.

rocket of topology and began its race through space, as topologists began to consider the consequences from the distortion of shapes, addition of other spaces onto manifolds, and the addition of holes in shapes. They consequently formed the idea of *homology classes* - sorts of "generalized" shapes from their 2-D Euclidean equivalents that would represent distinctions between the different types of holes found in various algebraic cycles.

Meanwhile, diverging from their fellow topologists, algebraists began to seek generalizations of their existing degree-2 and degree-3 equations that yielded already interesting curves in 2-D and 3-D space, extending them to algebraic cycles that manifested themselves within manifolds. It came to the point that topologists' and algebraists' work were seen as equivalent; that is, homology classses onto manifolds and algebraic cycles in manifolds were, for all intents and purposes, identical. It was at this point that the Hodge Conjecture, as we know it today, was formed. That is to say, can any random shape drawn onto a manifold be generalized to a shape that obeys the properties of a nice, smooth algebraic cycle?

In short, and in much more technical terms, in studying manifolds in algebraic topology, mathematicians often encounter the question of which cohomology classes come from complex subvarieties. Hence, they have defined several objects such as harmonic and differential forms, fundamental classes, and cap products to resolve the conjecture.

To even begin to understand any bit of the most complex conjecture we have covered thus far, we must establish our context and go through a laundry list of terminology:

Definition 16.14: Classifications of Manifolds

Let X be a manifold in an n-dimensional topological space. Then X is *complex* if, for all points $x \in X$, x has a neighborhood homeomorphic to $\mathbb{C}^n$ (rather than just $\mathbb{R}^n$). Furthermore, X is *compact* if it is both closed and bounded.

These two definitions in particular are prerequisites to defining the rest of the terminology, as they provide the necessary context to understand what specific types of manifolds we are working with. From here, we can define *orientability* and *smoothness* with respect to such a manifold X that is also differentiable at x for all $x \in X$:

Definition 16.15: Orientability and Smoothness

A compact, complex, (everywhere-)differentiable[a] manifold $X \subseteq \mathbb{C}^n$ is an *orientable manifold* iff, $\forall x \in X$, we can make a consistent choice of surface vector at x. Id est, for all loops in X, we can choose a clockwise orientation from x with respect to the origin $(\underbrace{0, 0, 0, \ldots, 0}_{n \text{ 0s}}) \in \mathbb{C}^n$.

In addition, we say X is *smooth* or a C^∞-*manifold* (as in the class of infinitely-differentiable, or *smooth*, functions) iff, $\forall x \in X$, all homeomorphisms to $\mathbb{C}^n$ at x are smooth.

---
[a]We say a manifold is a *differentiable* manifold if it is everywhere differentiable; "differentiable", in this context, is shorthand for "everywhere differentiable."

Theorem 16.6: Dimension Assuming Smoothness and Orientability

If X is a complex, compact, differentiable n-manifold, then it is an orientable, smooth $2n$-manifold.

> **Corollary 16.6: Cohomology Groups of X**
>
> As $\dim X = 2n$, X has cohomology groups in each of degrees 0 through $2n$, inclusive.

First things first: what is a cohomology group? For the proof of this theorem to make any sense to us, we must understand how such groups function and how they behave on smooth, orientable manifolds:

> **Definition 16.16: Cohomology Groups**
>
> First we must define the general notion of *cohomology*. Cohomology refers in general to a property of a sequence of abelian groups associated to a topological space. In particular, a *singular cohomology* is a topological invariant that associates to a topological space a *graded-commutative ring* (an abelian ring that is a direct sum of abelian groups).
>
> A singular cohomology is formed beginning from a *singular chain complex*, which relates all *free* abelian groups (i.e. abelian groups that have a basis) to each other in sequence. We define from this chain complex a set of homeomorphisms, and in turn, dual homeomorphisms d_i ($i \in \mathbb{N}$).
>
> A *cohomology group* of an n-manifold is i-indexed for $i \in \mathbb{N}$ (see above). More specifically, the ith cohomology group of the n-manifold X (with coefficients in the abelian group $A \subseteq X$), denoted $H^i(X, A)$, is given by $\dfrac{\ker(d_i)}{\operatorname{im}(d_{i-1})}$, where ker denotes the *kernel* (as in the kernel of a matrix) and im denotes the *image*. (For $i < 0$, $H^i(X, A) = 0$.)

Though these definitions may seem long-winded and inaccessible, it is thankfully not of vital importance to understand their origins to understand what the Hodge Conjecture is asking. (However, they are, needless to say, crucial in actually making progress in solving the conjecture.)

Finally, here are a few examples of cohomologies to solidify the concept:

Example 16.1.4.1. *The cohomology ring of a point in n-space is $\mathbb{Z}$.*

Example 16.1.4.2. *The cohomology ring of $\mathbb{R}^n$ itself is also $\mathbb{Z}$ (as is the case with any contractible space.*[98]

Example 16.1.4.3. *The cohomology ring of real projective space $\mathbb{RP}^n$ is $\dfrac{\mathbb{Z}/2[x]}{x^{n+1}}$ (with coefficients in $\mathbb{Z}/2$).*

Now we can prove Theorem 16.6:

Proof. Let X be a complex, compact, everywhere-differentiable n-manifold. This is equivalent to the (weak) *Whitney embedding theorem*.

> **Theorem 16.7: Weak Whitney Embedding Theorem**
>
> Any continuous function $f : X \mapsto Y$, where X is an n-manifold and Y is an m-manifold, can be approximated by a smooth embedding iff $m > 2n$.

> **Definition 16.17: Smooth Embedding**
>
> A *smooth embedding* is a mapping f between two differentiable manifolds M_1 and M_2 such that f is injective, f is everywhere-differentiable, and at any point $p \in M_1$, the mapping $df_p : T_p(M_1) \mapsto T_{(f(p))}(M_2)$ is injective.

---

[98] A contractible space is a space X for which the identity map on X is *homotopic* (or able to be continuously deformed) to some constant map.

The proof of the Whitney embedding theorem proceeds as follows:

Proof. Let $f : M_1 \mapsto M_2$ be a differentiable mapping between two differentiable manifolds M_1, M_2. We then have the induced mapping $df : T_p M_1 \mapsto T_{f(p)} M_2$.

> **Lemma 16.7: Mapping Lower to Higher Dimensions (16.7A)**
>
> Let $\dim M_1 < \dim M_2$. Then the image of M_1 in M_2 is a set with measure zero.[a]
>
> ---
> [a] The *measure* μ of a set is intuitively a quantification of "size"; namely, a function that serves as a generalization of length, area, and volume to a topological space that preserves the properties of non-negativity, having $\mu(\emptyset) = 0$, and countable additivity. When we say that "almost all" of a set has a certain property, we really mean that the complement of that set in the universal set has measure zero.

Proof. Because $\dim M_1 < \dim M_2$, the mapping df cannot be surjective (recall that surjectivity requires that each element in the codomain be mapped to at least one element in the domain). We now apply Sard's Theorem:

> **Lemma 16.7: Sard's Theorem (16.7B)**
>
> Let $f : M_1 \mapsto M_2$ be a differentiable map between differentiable manifolds M_1, M_2. Then the set of *critical values*[a] of f has measure zero in M_2.
>
> ---
> [a] Images of critical points, which are points $p \in M_1$ that do *not* make df a surjective mapping.

Proof. A proof of Sard's Theorem is presented in the paper "Sard's Theorem" (Wright 2008). □

This completes the proof of the lemma. □

Thus, we now need only ensure injectivity, and thus, the existence of a smooth embedding. The crux of the argument is to show that, if there is an embedding in $\mathbb{R}^N$, then an embedding also exists for $\mathbb{R}^{N-1}$ iff $N > 2n + 1$.

> **Lemma 16.7: Full-Measure Set of Hyperplanes (16.7C)**
>
> Let $N > 2n + 1$. Assume there exists an embedding E of an n-manifold $M \in \mathbb{R}^n$. Then there exists a full-measure set[a] of hyperplanes $\Pi \in \mathbb{R}^n$ such that the mapping $p_{E(\Pi)} : M \mapsto \Pi$ is injective.
>
> ---
> [a] A set whose complement has measure zero.

Proof. Observe that the mapping $p_{E(\Pi)}$ is injective iff every line orthogonal to Π intersects M at one and only one point. This is equivalent to requiring the mapping from pairs of distinct points in M to vectors (i.e. directions) in $\mathbb{R}^N$ have measure zero.

Let v be the direction in M^2, which has dimension $2n$. Note that we are working in the real projective plane of dimension $N - 1$ $\mathbb{RP}^{N-1}$, which has dimension greater than $(2N + 1) - 1 = 2N$. Thus, the image of the mapping from $M^2 - V$ to $\mathbb{RP}^{N-1}$ has measure zero, and its complement, along with Π, has full measure. □

To ensure a smooth embedding, we modify Lemma 16.7 slightly:

> **Lemma 16.7: Injectivity of df (16.7D)**
>
> Let $N > 2n + 1$. Assume there exists an embedding E of an n-manifold $M \in \mathbb{R}^n$. Then there exists a full-measure set[a] of hyperplanes $\Pi \in \mathbb{R}^n$ such that the mapping $p_{E(\Pi)} : M \mapsto \Pi$ has injective differential

df everywhere.

---
[a] A set whose complement has measure zero.

Proof. If df is injective, then the differential mapping does not map to a full-measure set of vectors. The total space thus has dimension $2n - 1$, while the dimension of $\mathbb{RP}^{N-1}$ is still greater than $2n$. Thus, the image again has measure zero, and so its complement will have full measure. □

To finish off the proof of the weak Whitney embedding theorem, and in turn, the proof of Theorem 16.6, we consider the intersection of the full-measure sets obtained from Lemmas 16.7C and 16.7D. Then both $p_{E(\Pi)}$ and the differential df are injective, from which it follows that the embedding in $\mathbb{R}^{N-1}$ is differentiable (smooth), beginning with an embedding in $\mathbb{R}^N$ (where $N > 2n + 1$). □

This completes the proof of the Whitney embedding theorem, and it follows that any complex n-manifold is also a real, smooth $2n$-manifold by way of the existence of a smooth embedding. □

We have now proven that if X is a complex, compact, and smooth n-manifold, it is automatically an orientable, real smooth $2n$-manifold. With this derivation, we can now make another assumption that leads to the Hodge Conjecture: that X is a *Kähler manifold*.

> **Definition 16.18: Kähler Manifolds**
>
> A *Kähler manifold* is a complex manifold with both a *Riemannian* and a *symplectic* structure.

> **Definition 16.19: Riemannian Structure**
>
> A *Riemannian structure* is the structure that ensures that a manifold is real, smooth, and is equipped with an inner product.

> **Definition 16.20: Symplectic Structure**
>
> A *symplectic structure* is the structure that ensures that a manifold is smooth and equipped with a closed, non-degenerate, differential *2-form*, known as the *symplectic form*.

> **Definition 16.21: Differential Forms**
>
> A *differential form* in n-dimensional space, or *n-form* for brevity, is any object that can be integrated over $\mathbb{R}^n$, and is homogeneous of degree n in the coordinate differentials (e.g. dx, dy, dz, etc.)

Thus, X has a decomposition into cohomology classes with complex coefficients on its cohomology:

$$H^k(X, \mathbb{C}) = \bigoplus_{p+q=k} H^{p,q}(X) \tag{16.2}$$

where $H^{p,q}(X)$ denotes those cohomology classes which can be represented by *harmonic forms*.[99]

Note that, for all $n \in \mathbb{N}$, $H^{2n}(X, \mathbb{C}) = H^{n,n}(X)$. This implies that the cohomology class $[Z]$ of a complex submanifold $Z \subseteq X$ lies in $H^{n-k,n-k}(X, \mathbb{C})$. This allows us to ask the question of which cohomology classes in $H^{n,n}(X)$ are subclasses of Z.

From here, after we establish one more relevant object, we can finally formulate the statement of the Hodge Conjecture:

---
[99] Differential forms that operate under a Riemannian metric M.

> **Definition 16.22: Hodge Classes**
>
> Define $\text{Hdg}^k(X) = H^{2k}(X, \mathbb{Q}) \cap H^{k,k}(X)$. $\text{Hdg}^k(X)$ is the group of *Hodge classes* on X with degree $2k$.

> **Theorem 16.8: Hodge Conjecture**
>
> Let X be a *non-singular*,[a] complex, projective manifold.[b] Then every Hodge class on X is a rational linear combination of the cohomology classes of complex subvarieties Z of X.
>
> [a]Such that it does not have any singular points; i.e. points for which the tangent space to that point is undefined.
> [b]A complex manifold that can be embedded in complex projective space.

Well. What a long, strange trip it's been...

...and that was just to formulate the *conjecture*. The proof itself remains yet to be unearthed, shrouded in a cloak of darkness. But mathematicians, undaunted and determined to unveil the truth or falsehood of the long-standing conjecture, have pressed on ever since the Great Depression,[100] and have achieved much successful, significant progress.

The first inkling of non-trivial progress, shockingly, came about before the conjecture itself was ever formulated, back in 1924. Perhaps more stunningly, the paper by Lefschetz actually motivated Hodge, in part, to make the conjecture! Lefschetz's work on the conjecture revolved around a low-dimension special case, namely the case of $(1,1)$-classes. The *Lefschetz theorem on $(1,1)$-classes* formalized the known relationship between compact Kähler manifolds to their cohomologies:

> **Theorem 16.9: Lefschetz Theorem on $(1,1)$-Classes**
>
> Let X be a compact Kähler manifold. The cohomology group $H^2(X, \mathbb{C})$ (which is a *de Rham cohomology group*) decomposes as $H^{0,2}(X) \oplus H^{1,1}(X) \oplus H^{2,0}(X)$. Then the map to $H^2(X, Z) \cap H^{1,1}(X)$ is surjective.

(Interestingly, Lefschetz's proof of this theorem itself relied on techniques introduced by Henri Poincaré - of Poincaré Conjecture fame! Perhaps the Millennium Problems are more related than we thought...)

---

[100]Hodge initially formulated the conjecture in the 1930s, though it did not receive widespread attention until 1950, when he presented it at the International Congress of Mathematicians in Cambridge, MA. Evidently, even the economic downturn couldn't stop the mathematical community!

Alongside the main conjecture, Hodge had also proposed the *integral Hodge conjecture:*

> **Theorem 16.10: Integral Hodge Conjecture**
>
> Let X be a projective complex manifold. Then every cohomology class in $H^{2k}(X, Z) \cap H^{k,k}(X)$ is the cohomology class of an algebraic cycle in X with integer coefficients.

This conjecture was shown to be false via a counterexample constructed in 1961, however. Essentially, this counterexample consisted of a *torsion cohomology class;* i.e. a cohomology class C such that, for some $n \in \mathbb{N}$, $nC = 0$. Such a class is automatically a Hodge class, but cannot be the class of an algebraic cycle.

In addition to the integral Hodge conjecture, Hodge proposed the *generalized Hodge conjecture,* which extended the idea of cohomology classes to c-codimensional subvarieties of a manifold X. This, too, was shown in 1995 to be false, but a corrected version - issued by the same mathematician (Grothendieck) who disproved the original form of the conjecture - is still open as of March 2019.

Consequences of resolving the Hodge Conjecture may not be as readily versatile in real life applications as, say, the P vs. NP problem or the Riemann Hypothesis, but they are nevertheless far-reaching in both their home field of algebraic topology, and in other mathematical fields such as number theory and pure algebra. Should the conjecture prove to be true, we would be able to construct direct correspondences between algebraic varieties (in simpler terms, the solutions to systems of multi-variable polynomial equations). This would open several doors in terms of number-theoretic applications; for instance, modular forms. Furthermore, the implications of the Hodge Conjecture would be instrumental in the development of Galois representations; if the conjecture turns out to be true, we could use algebraic varieties to bridge the gap between modular forms and Galois representations!

But what about more applicable, real-world consequences? The P vs. NP problem, if resolved, could have devastating consequences in cryptography and security methods, but on a more positive note, could be the golden ticket to resolving several future problems and deriving once extremely complicated proofs by way of Occam's Razor. Similarly, the Riemann Hypothesis could resolve the distribution of primes problem, thereby allowing for the expedient estimation of large quantities. In what capacity could the truth (or falsehood) of the Hodge Conjecture be used to improve the efficiency of modern mathematical methods? Perhaps the applications, if they do exist, are admittedly less prominent than those of the other Millennium Prize problems, but in other fields altogether, they are quite abundant; a prime example is string theory.

The offshoot of the Hodge conjecture - the entirety of *Hodge theory* - can be used to transform every computation down to the solution of an algebraic variety, which would describe the universe at both microscopic and macroscopic scales, and thereby contribute massively to the "theory of everything" - Einstein's theory of unification. Algebraic geometry and symplectic geometry may well be tied together by this grand unified theory!

de Rham Cohomology

Before we move on to the next of the Millennium Prize Problems, there is one loose end left to discuss that pertains to the Hodge Conjecture: the *de Rham cohomology,* an invaluable tool in formalizing smoooth manifolds in terms of their cohomology classes.

> **Definition 16.23: de Rham Complex**
>
> Consider some smooth manifold M. A *de Rham complex* is a chain of differential forms of M:
>
> $$0 \to \Omega^0(M) \to \Omega^1(M) \to \ldots \qquad (16.3)$$
>
> where $\Omega^n(M)$ is the space of n-forms on M (and, in particular, $\Omega^0(M)$ denotes the space of all

smooth/everywhere-differentiable functions on M).

If a differential form is the image of another form under the *exterior derivative* (the generalization of a differential to n-forms where $n \geq 3$), or is the zero function, it is said to be *exact*; if a differential form has exterior derivative zero, it is said to be *closed*.

> **Corollary 16.10: Relationship of Exact/Closed Differential Forms**
>
> All exact forms are closed, but not all closed forms are exact. (Consider the unit circle in $\mathbb{R}^2$, and the 1-form $d\theta$, where θ is the angle measure of a sector of the circle).

Our main motivation behind working with de Rham cohomologies is to establish equivalencies between closed forms of a manifold; we can do this by way of defining two closed forms in some $\Omega^n(M)$ as *cohomologous* if they differ by an exact form.

We then define the *de Rham cohomology group*:

> **Definition 16.24: de Rham Cohomology Group**
>
> The kth *de Rham cohomology group* $H_{dR}^k(M)$ is the set of equivalence classes in $\Omega^k(M)$, i.e. the set of closed forms modulo the exact forms.

We now present examples of de Rham cohomology groups for two common manifolds:

Example 16.1.4.4. *The n-sphere S^n (passing through the points $(1,0,0,\ldots,0)$, $(0,1,0,0,\ldots,0)$, $(0,0,1,\ldots,0)$, ..., $(0,0,0,\ldots,0,1)$). Let $n > 0$, $m \geq 0$, and let $I \in \mathbb{R}$ be an open interval. Then*

$$H_{dR}^k(S^n \times I^m) = \mathbb{R} \mid k = 0, n \tag{16.4}$$

and 0 otherwise.

Example 16.1.4.5. *Punctured Euclidean n-space; i.e. $\mathbb{R}^n - \underbrace{(0,0,0,\ldots,0)}_{n\ 0s}$. For all $n \in \mathbb{N}$,*

$$H_{dR}^k(\mathbb{R}^n - \{\vec{0}\}) = \mathbb{R} \mid k = 0, n-1 \tag{16.5}$$

and 0 otherwise, which is equivalent to $H_{dR}^k(S^{n-1})$. (Proving this is left as an exercise to the reader.)

16.1.5 Navier-Stokes Existence and Smoothness

The Navier-Stokes existence and smoothness problem that reflects the true extent to which math is ever-present in our world. Mathematical interest has burrowed so deep within not only the abstractions of math itself, but also in other fields altogether, such as statistics, pscyhology, linguistics, and the natural sciences. In particular, one of the greatest unsolved problems in physics revolves around the *Navier-Stokes equations* - equations that govern the mechanics of fluid substances. The application of Newton's second law to fluid motion, combined with the assumption that the stress of a fluid is equal to the sum of a pressure term and a diffusing viscosity term, give rise to these monumentally important equations. The Navier-Stokes equations have a veritably diverse set of applications, including weather forecasting, modeling air flow and wind turbulence, ocean currents, water flow in a pipe, blood flow, vehicular aerodynamics, air pollution, and heat transfer. For now, we will take a look at the more mathematical side of these landmark equations, and examine the core of the existence and smoothness Millennium Prize Problem.

Essentially, the Navier-Stokes equations describe how the velocity, pressure, density, and temperature of a fluid are all inter-related. The classical setup of the equations defines several variables:

> **Definition 16.25: Variables in the Navier-Stokes Equations**
>
> Consider a fluid in the 3-dimensional plane. There are four independent variables: the three coordinates $x, y, z \in \mathbb{R}^3$, and the time $t \geq 0$. In addition, there are six dependent variables: the pressure p, the density ρ,[a] the temperature T, and the three components u, v, w of the velocity vector.
>
> [a] Rho, occasionally replaced with r to avoid confusion with the pressure p.

The Navier-Stokes equations themselves consist of three sets of partial differential equations (five equations in total): one *continuity equation* for the conservation of mass, three conservation of momentum equations, and one conservation of energy equation, all of which depend on the time t.

> **Theorem 16.11: The Navier-Stokes Equations**
>
> Continuity equation:
> $$\frac{\partial \rho}{\partial t} + \frac{\partial \rho u}{\partial x} + \frac{\partial \rho v}{\partial y} + \frac{\partial \rho w}{\partial z} = 0 \tag{16.6}$$
>
> Conservation of mass equations:
>
> $$\frac{\partial(\rho u)}{\partial t} + \frac{\partial(\rho u)^2}{\partial x} + \frac{\partial(\rho uv)}{\partial y} + \frac{\partial(\rho uw)}{\partial z} = -\frac{\partial p}{\partial x} + \frac{1}{\text{Re}}\left(\frac{\partial \tau_{xx}}{\partial x} + \frac{\partial \tau_{xy}}{\partial y} + \frac{\partial \tau_{xz}}{\partial z}\right) \tag{16.7}$$
>
> $$\frac{\partial(\rho v)}{\partial t} + \frac{\partial(\rho uv)}{\partial x} + \frac{\partial(\rho v^2)}{\partial y} + \frac{\partial(\rho vw)}{\partial z} = -\frac{\partial p}{\partial y} + \frac{1}{\text{Re}}\left(\frac{\partial \tau_{xy}}{\partial x} + \frac{\partial \tau_{yy}}{\partial y} + \frac{\partial \tau_{yz}}{\partial z}\right) \tag{16.8}$$
>
> $$\frac{\partial(\rho w)}{\partial t} + \frac{\partial(\rho uw)}{\partial x} + \frac{\partial(\rho vw)}{\partial y} + \frac{\partial(\rho w^2)}{\partial z} = -\frac{\partial p}{\partial z} + \frac{1}{\text{Re}}\left(\frac{\partial \tau_{xz}}{\partial x} + \frac{\partial \tau_{yz}}{\partial y} + \frac{\partial \tau_{zz}}{\partial z}\right) \tag{16.9}$$
>
> for x-momentum, y-momentum, and z-momentum respectively.
>
> Conservation of energy equation:
>
> $$\frac{\partial E_T}{\partial t} + \frac{\partial(uE_T)}{\partial x} + \frac{\partial vE_T}{\partial y} + \frac{\partial wE_T}{\partial z} =$$
> $$-\frac{\partial(up)}{\partial x} - \frac{\partial vp}{\partial y} - \frac{\partial wp}{\partial z} - \frac{1}{\text{Re}_r \text{Pr}_r}\left(\frac{\partial q_x}{\partial x} + \frac{\partial q_y}{\partial y} + \frac{\partial q_z}{\partial z}\right)$$
> $$+ \frac{1}{\text{Re}_r} \cdot [\frac{\partial}{\partial x}(u\tau_{xx} + v\tau_{xy} + w\tau_{xz}) + \frac{\partial}{\partial y}(u\tau_{xy} + v\tau_{yy} + w\tau_{yz})$$
> $$+ \frac{\partial}{\partial z}(u\tau_{xz} + v\tau_{yz} + w\tau_{zz})] \tag{16.10}$$
>
> Here τ denotes stress, q denotes heat flux, Re denotes the *Reynolds number*, a constant equal to the ratio of the inertia of the flow to its viscous force, Pr denotes the *Prandtl number*, a constant equal to the ratio of the viscous stress to the heat stress, and E_T denotes the total energy.

Solving the equations is not the Millennium Prize Problem (moreover, this had already been done by the time the Problems were set, in 2000) - obtaining a complete theoretical understanding of them is. More specifically, we seek smooth solutions to the equations in three dimensions given certain initial conditions. If we are working in three spatial and chronological dimensions, and if we are given an initial velocity field, then does there exist a vector velocity and scalar pressure field which is smooth and defined everywhere, that is a solution to the Navier-Stokes equations?

There are actually two different contexts in which this problem may be construed. One is the more standard, common interpretation: the unbounded space $\mathbb{R}^3$. The other is the *periodic* version, which limits our field of interest to the 3-dimensional torus $\mathbb{T}^3 = \mathbb{R}^3/\mathbb{Z}^3$. We will henceforth examine the unbounded version of the problem.

We assume as part of our hypothesis for the unbounded conjecture that the initial condition $v_0(x)$ is itself a smooth function such that, for every n-tuple of non-negative integers $\alpha = (\alpha_1, \alpha_2, \alpha_3, \ldots, \alpha_n)$, and $K > 0$ a constant, there exists another constant $C > 0$ such that

$$|\partial^\alpha v_0(x)| \leq \frac{C}{(1+|x|)^K}, \forall x \in \mathbb{R}^3 \tag{16.11}$$

The external force function $f(x,t)$ satisfies a similar inequality, with the RHS being $\frac{C}{(1+|x|+t)^K}$ for all $(x,t) \in \mathbb{R}^3 \times [0,\infty)$ (where $t \in [0,\infty)$).

In order for $\{v(x,t), p(x,t)\}$ to be considered a satisfactory solution (where $v(x,t)$ is the time-dependent velocity vector and $p(x,t)$ is the time-dependent pressure scalar), we must have $f(x,t) \equiv 0$. Then for any $v_0(x)$ there exist smooth solutions to the Navier-Stokes equations.

The periodic variant is similar, but the definition of the velocity vector must be modified slightly. Set basis vectors $e_1 = (1,0,0)$, $e_2 = (0,1,0)$, $e_3 = (0,0,1)$; then $v(x,t)$ is periodic if for any $i = 1, 2, 3$,

$$v(x+e_i, t) = v(x,t), \forall (x,t) \in \mathbb{R}^3 \times [0,\infty) \tag{16.12}$$

(Note that this only works on the torus $\mathbb{T}^3$, where $\mathbb{T}^3 = \{(x,y,z) \mid 0 \leq x,y,z < 2\pi\}$.) Similarly to the previous variant, we have the necessary conditions that $f(x,t) \equiv 0$.

Remark. In both variants of the problem, the kinetic energy of the solution is globally bounded.

What progress has been made so far? Unsurprisingly, and as with the other Problems, quite a bit. In two dimensions, the Problem has already been entirely resolved, with a positive outcome: there always exist smooth and globally defined solutions to the Navier-Stokes equations. Furthermore, if we have a sufficiently small initial velocity $v(x,t)$, then the conjecture is also true. In 1934, Jean Leray showed that the conjecture that the Navier-Stokes equations always have smooth, globally-defined solutions holds true for *weak solutions* to the equations. Weak solutions satisfy the Navier-Stokes equations not pointwise, but rather in mean value.

In more recent years, mathematicians, engineers, and physicists have collectively made a further development regarding the Navier-Stokes equations: the notion of *blowup time*.

> **Definition 16.26: Blowup Time**
>
> A *blowup time* T ($0 < T < \infty$) is a finite time, dependent on the initial velocity $v_0(x)$, such that the Navier-Stokes equations on $\mathbb{R}^3 \times (0,T)$ have smooth, globally-bounded solutions in terms of a velocity vector and a pressure scalar.

It is still an open question as to whether solutions necessarily exist beyond the blowup time T for any given initial velocity $v_0(x)$. However, very recently, in 2014,[101] Terence Tao established a result regarding blowup time for the weak solutions to the Navier-Stokes equations.[102] His results have paved the way for blowup results to be formalized for the strong solutions to the equations.

---

[101] With the corresponding paper published in 2016.
[102] Source: https://terrytao.wordpress.com/2014/02/04/finite-time-blowup-for-an-averaged-three-dimensional-navier-stokes-equation/

16.1.6 Birch and Swinnerton-Dyer Conjecture

Mathematicians have long contemplated the solutions to *elliptic curves*, which are non-intersecting curves given by the general equation $y^2 = x^3 + ax + b$ where a and b are constants. These particular curves are of special interest due to their relationship with group theory, more specifically abelian groups and abelian varieties. Over finite fields, elliptic curves are often used for cryptography purposes, as well as for determining the prime factorizations of large numbers (a large motivating factor behind Wiles' proof of Fermat's Last Theorem).

With this Problem in particular, it would perhaps be more helpful to begin with the background behind the problem and then segue into more abstract mathematical terms. The starting point of the conjecture was in 1922, with the proof of *Mordell's Theorem*:

> **Theorem 16.12: Mordell's Theorem**
>
> The group of rational points on an elliptic curve has a finite basis (and is a finitely-generated abelian group).

As a corollary of Mordell's Theorem, any set of rational points can finitely generate another set of rational points on the curve. Proceeding, we must define a few more pieces of terminology relating to elliptic curves:

> **Definition 16.27: Elliptic Curve Definitions**
>
> The *rank* of a finite elliptic curve; i.e. a curve with a finite number of rational points, is always zero. For infinite elliptic curves (elliptic curves with infinite numbers of rational points), the rank of the curve is the number of independent basis points with infinite order.
>
> The *L-function* $L(E, s)$ of an elliptic curve E can be defined in terms of a parameter s of the Euler product (an expansion of a Dirichlet series into an infinite product with prime numbers as the indices). This function $L(E, s)$ is analogous to the Riemann zeta function $\xi(s)$ (and converges iff $\text{Re}(s) > \frac{3}{2}$). N_p is the number of points modulo p for large prime p on an elliptic curve E with known rank.

In 1965, Birch and Swinnerton-Dyer themselves conjectured that N_p obeys a formula in terms of the natural logarithm asymptotically as n approaches infinity; namely:

$$\lim_{x \to \infty} \prod_{p \leq x} \frac{N_p}{p} = c \cdot \ln^r(x)$$

where r is the rank of the curve and c is a real constant.

Improvements in numerical analysis methods over time have backed up these claims, lending them increasing credibility. This added credence led Birch and Swinnerton-Dyer to propose the current form of their conjecture: that at $s = 1$, for any elliptic curve E, the curve's L-function $L(E, s)$ would have a zero at order r, where r is the rank of E. The conjecture can also be alternatively stated as the conjecture that the rank of an abelian group $E(K)$ of points of an elliptic curve E is the order of the L-function at $s = 1$, as well as the first nonzero coefficient in the Taylor series expansion of the L-function about $s = 1$.

Thus far, the Birch and Swinnerton-Dyer Conjecture (henceforth BSDC for brevity) has only been proven for edge cases. All other cases remain unproven, even on a purely computational level. In 1977, twelve years after the conjecture was initially proposed, Andrew Wiles, along with John Coates, had proven that, if E is an elliptic curve over a field $\mathbb{F}$ equipped with complex multiplication, the imaginary quadratic field K has class number 1, and $\mathbb{F} = K$, $\mathbb{Q}$ (where $\mathbb{Q}$ assumes its standard definition), then $E(F)$ is a finite group.

In 1986, the rise of the *Gross-Zagier Theorem*, which conclusively showed that, if a *modular elliptic curve* (i.e. an elliptic curve which admits a parametrization via a *modular curve* (not the same as a modular form).

> **Theorem 16.13: Gross-Zagier Theorem**
>
> If a modular elliptic curve E has a first-order zero at $s = 1$, then it has a rational point of infinite order.

Two sequential developments, close together in 1989 and 1991, then showed that certain types of curves always had rank zero or one, depending on the zeroes of $L(E, 1)$, and also showed that the BSDC would predict the *p-part*, or equality of the main equation within, the *Tate-Shafarevich group*. Another development in 2001 involved Andrew Wiles, and in fact extended his landmark work on Fermat's Last Theorem, modular forms, and the modularity theorem (more commonly known as the Taniyama-Shimura conjecture before Wiles had completed his proof). Breuil proved that all elliptic curves defined over $\mathbb{Q}$ are modular, which builds upon the work done in the 1980s to show that all elliptic curves over the *rationals* also have those results. Additionally, Breuil's work had proven that all L-functions of elliptic curves over $\mathbb{Q}$ are defined for $s = 1$. Finally, the most recent result of all, in 2015, established an upper bound for the mean rank of elliptic curves over $\mathbb{Q}$, namely $\frac{7}{6}$.

In particular, regarding the real-world applications of this conjecture, we present two: one that is still mathematical in nature, but much less abstract and can be interpreted and understood in the syntax of everyday geometrical objects, and one that is outright a pure extension of the mathematical language of the conjecture.[103] The first relates to none other than the almighty Pythagorean Theorem, the attempted extension of which led to Fermat's Last Theorem - yes, that 300-odd year can of worms! (Isn't it amazing? Math really *is* a small world!) Namely, we can say the following assuming the truth of the BSDC:

> **Theorem 16.14: Rational Side Lengths of a Right Triangle**
>
> If n is an odd *square-free* integer (i.e. an integer that is not a multiple of the square of any prime number), then n is the area of a right triangle with rational side lengths if and only if the number of triplets of integers such that $2x^2 + y^2 + 8z^2 = n$ is equal to the number of triples of integers such that $2x^2 + y^2 + 32z^2 = n$.

There is a relationship of this statement with the zeros of the L-function at $s = 1$, and it is the statement of *Tunnell's theorem*, which had been proven in 1983.

The second of these applications relies on analytic methods, assuming the BSDC. Estimations of order zero can be used to determine the ranks of families of elliptic curves. Of particular interest is what happens when we also assume the generalized Riemann hypothesis: the average rank of *all* elliptic curves is smaller than 2!

---

[103] This conjecture is *extremely* abstract, esoteric, and generally difficult to understand (so don't feel bad if your eyes glazed over reading this section - or, for that matter, any of these last four), and as such, doesn't have many real-world applications, unlike the Riemann Hypothesis or the P = NP problem, which are relatively uncomplicated to grasp (at least on a surface level) as compared with the BSDC, the Navier-Stokes Equations, or the following Problem, the Yang-Mills Existence and Mass Gap (YMEMG). Note that the Millennium Problems were not chosen for their real-world applicability and/or versatility; they were chosen for their crucial importance within the field of mathematics, and for their highly specialized offshoots in the natural sciences.

16.1.7 Yang-Mills Existence and Mass Gap

At last, we come to the *Yang-Mills Existence and Mass Gap,* or YMEMG Problem. This Problem is especially deeply-rooted in application, as it revolves around the extension of mathematical physics to four-dimensional space.

At the core of this Problem is *Yang-Mills theory,* which aims to describe the behavior of elementary particles via non-abelian Lie groups. Yang-Mills theory is used to describe the strong, electromagnetic, and weak forces, and so forms an integral part of the foundation of modern particle physics. Yang-Mills theory has its roots in a six-dimensional generalization of Einstein's field equations, which themselves came about in 1953. The following year, Yang and Mills extended the existing work on field equations in $\mathbb{R}^6$ to non-abelian groups from abelian groups, so as to make progress toward the unification of the strong force with the other three electromagnetic forces. This idea was briefly set aside, until the 1960s, when breakthroughs were made regarding the mass of subatomic particles. From there, *quantum chromodynamics* (QCD), the theory of the interactions between quarks and gluons, began to take rise. The *confinement* of QCD is a subject that still has much progress to be made within it, and as such, the Clay Institute had decided to devote a Millennium Problem to it due to its topical relevance.

A cornerstone idea of Yang-Mills theory is *gauge theory,* which we can loosely sum up and relate back to the YMEMG Problem:

Definition 16.28: Gauge Theory (in one paragraph)

Though gauge theory is not rigorously defined from the quantum perspective in four dimensional space-time (which is, in fact, the entire crux of the YMEMG), we can nevertheless sum it up from a more general view. A *gauge theory* is any type of field theory (a construction of the dynamics of a field, such as a set of laws dictating the change of a field with respect to time or some other parameter) which has an invariant *Lagrangian* under certain Lie groups. Gauge transformations within this theory form *gauge groups,* which give rise to *gauge fields*. Gauge theories can either have *local* or *global* symmetry (or both), depending on where the Lagrangians are invariant. Gauge theories are critical in explaining field dynamics, and the dynamics of the elementary particles that comprise those fields (including photons, bosons, and gluons). They also help to explain gravitational phenomena in terms of Einstein's laws of general relativity, and, furthermore, can be re-framed in terms of classical electromagnetism.

(Note: The term *gauge* comes from the mathematical term, referring to a section of a *principal bundle,* or a mathematical object that essentially formalizes the properties of $X \times G$, where X is a space and G is a group).

Despite the very definition of the similar Yang-Mills theory relying on the definitions of gauge theories, the Clay Institute themselves have stated that "one does not yet have a mathematically complete example of a quantum gauge theory in four-dimensional space-time, nor even a precise definition of quantum gauge theory in four dimensions"[104]. This is where the first part of the Problem comes in! Indeed, the YMEMG Problem consists of two parts: the first is *existence*. In other words, does a cohesive Yang-Mills theory actually exist in the first place? Furthermore, does such a theory satisfy a set of rigorous conditions that characterizes mathematical physics, specifically constructive quantum field theory?[105] Secondly, the Problem attempts to tackle the *mass gap;* the notion that there exists a "smallest" particle (as predicted by the theory) whose mass is still positive, and bounded below by a positive value. That is, particles cannot be arbitrarily light.

The precise statement of the Problem is as follows:

---

[104] Clay Institute 2000.
[105] Constructive quantum field theory is the branch of quantum field theory that shows that quantum theory and special relatively are mathematically compatible.

> **Theorem 16.15: Yang-Mills Existence and Mass Gap**
>
> Prove that for any compact simple gauge group G, a non-trivial quantum Yang-Mills theory exists on $\mathbb{R}^4$ and has a mass gap $\Delta > 0$. Existence includes establishing axiomatic properties at least as strong as those cited in Streater & Wightman (1964), Osterwalder & Schrader (1973) and Osterwalder & Schrader (1975).

To be more specific, a proposed solution to the Problem would require it to abide by the *Wightman axioms* (or similarly rigorous ones). In particular, we will focus on the Wightman axioms:

> **Definition 16.29: Wightman Axioms**
>
> There are four axioms, W0 through W3:
>
> - W0: These are based on the assumptions belying relativistic quantum mechanics. According to von Neumann, quantum mechanics as a whole can be described as consisting of pure states that are given by the 1-dimensional subspaces of a separable, complex Hilbert space X. If v and w are vectors in X, then let the scalar product of v and w be denoted by $\langle v, w \rangle$, and let the norm of a vector be denoted by $||\cdot||$. We can define a *transitional probability* between two states $[v]$ and $[w]$ to be
>
> $$P([v],[w]) = \frac{|\langle v, w \rangle|^2}{||v||^2 \, ||w||^2}$$
>
> This transitional probability allows for the formulation of Wigner's theory of symmetry; namely, that all observers under a single transformation of relativity observe the same transition probability between two states. Quantum theory becomes invariant under a Poincaré group G as a result, which yields the equality
>
> $$\langle v(a,L), w(a,L) \rangle = \langle v, w \rangle$$
>
> for any element (a,L) in the group G. Under these conditions, the spin of a particle is always a multiple of π, which implies further conditions relating to the observability of certain particles depending on their spin. Furthermore, the group of translations in four-dimensional space-time commutes, meaning that we have four self-adjoint operators: P_0, P_1, P_2, P_3. Each of these operators transform as an energy-momentum vector in four dimensions. Additionally, $P_0 \geq 0$, and $P_0^2 - P_j^2 \geq 0$ for $j = 1, 2, 3$. Finally, there exists a unique state (called a *vacuum*) in the Hilbert space such that, acted upon by the Poincaré group, it remains invariant.
>
> - W1: This axiom sets assumptions regarding the domain and continuity of the field. Define a *test function* f to be a smooth function with a compact support; i.e. $f \in C_0^\infty$. Then for each test function f, there exists a set of operators $\{A_i(f) \mid i = 1, 2, 3, \ldots, n\}$, which, together with their adjoints, contain the vacuum of a Hilbert space. The fields A are *tempered distributions*, distributions[a] which have a Fourier transform. Then the set of polynomials acting on the vacuum is the span of the Hilbert state space.
>
> - W2: The fields A are covariant under the action of the Poincaré group. Additionally, they transform under a representation of the Lorentz group.
>
> - W3: If the supports of two fields are space-like separated,[b] then those fields either commute or anti-commute.
>
> ---
> [a]Not statistical distributions, as in probability distribution functions, but rather generalizations of functions that are not everywhere-differentiable to smooth analogues of those functions.
> [b]In layman's terms, we say two events in space-time are *space-like separated* if there exists some frame of reference through which both events occur at the same time, but in different places.

Before we proceed with our discussion of the YMEMG Problem itself, we'll take a quick look at some of the important consequences of the Wightman axioms.

One theorem that follows directly from the Wightman axioms is the *CPT theorem,* where CPT stands for *charge, parity, time reversal.* The theorem states that CPT symmetry is an innate property of all physical phenomena. None of charge, parity, or time reversal can exist by themselves, but together with the other two properties, they form the basis for symmetry - i.e. the invariance of an object and its physical properties under a mirror-image transformation. This even applies to the grand scale of the universe; if we were to flip the universe so that we were to obtain its "mirror image," *per se,* then this would correspond to an inversion of the parity of each point - each object - in our universe. All instances of time would also be reversed, and all matter would be replaced by its corresponding form of anti-matter (which would correspond to an inversion of charge).

Another direct consequence serves to confirm a major tenet of Einstein's theory of relativity - that faster-than-light travel is impossible under the usual formulation of special relativity. If two observers are space-like separated, then the actions of one observer would not affect the actions or observations of the other observer, since they could not possibly travel quickly enough to reach the other observer in time. In this manner, the Wightman axioms, and the YMEMG Problem by extension, could perhaps be the most important constructs in mathematical physics - right up there with the Navier-Stokes equations!

Given the Wightman axioms, we can henceforth consider the *mass gap* in a quantum field theory context. We define the mass gap as the difference in the energy levels of the vacuum and the next-lowest energy state. Since we can assume that the energy of the vacuum is zero *a priori,* the mass gap, by definition, is the mass of the lightest particle.

So how exactly would we go about determining whether there exists a mass gap, assuming the existence of a cohesive Yang-Mills theory to begin with? Such a question is extremely difficult - even impossible - to answer with our current knowledge. Unlike the other six Millennium Problems, little progress has been made on this Problem ever since its inception in the 1970s. However, the situation is far from hopeless. As with the Riemann Hypothesis and the P = NP problem, many derivations in physics, and other branches of math, rely on these conjectures being true. Furthermore, it has already been shown that a quantum Yang-Mills theory would have the confinement property (the property of color-charged particles not being isolated). It is *expected* that there is a unifying Yang-Mills theory, and it is *expected* that the mass gap exists, similarly to how it is *expected* that P = NP. This Problem, however, appears to be one of the most frustrating among not only the Millennium Prize Problems, but all open problems in general, since we have the intuition; the *gut feeling* that said intuition is correct; but we have hitherto been unable to actually formalize it with sufficient accuracy and rigor.[106]

16.2 Hilbert's 23 Problems

In 1900, famed German mathematician David Hilbert, prolific contributor to several areas of mathematics that are still of modern interest (including mathematical physics, algebraic number theory, functional analysis, and the spectral theory of operators), posed a set of 23 problems that set the course for not only the research of future mathematicians, but *his very own research* as well! It is likely that Hilbert proposed the problems out of sheer passion for his areas of study, and his desire to impart this passion onto future mathematical generations.

For reasons of space,[107] we will eschew the level of detail that we have gone into for the Millennium Prize Problems, but we will list the Problems, and provide a cursory overview of each. The following descriptions

---

[106]Don't worry, professional mathematicians and physicists - we're right there with you. And so is every student. Don't you hate the feeling of knowing what a question's asking, and knowing the answer, but not being able to prove it? It's likely a very familiar feeling...

[107]And printing costs. And the authors' energy levels. And their looming publication deadlines. And the reader's time. Seriously, we don't want this chapter to be its own book!

are paraphrased versions of Hilbert's originals; indeed, Hilbert's descriptions were deemed by mathematicians to be vague, in some cases to the point where the vagueness obscured the problem's solvability.

1. Prove the *continuum hypothesis:* i.e. prove that there does not exist a set whose cardinality is strictly between those of the real numbers and the integers. (Outcome: proven to be impossible to either prove or disprove under Zermelo-Fraenkel set theory, with or without the Axiom of Choice. It is not known whether this is a solution to the problem as Hilbert originally intended it.)

2. Prove that the axioms of arithmetic are internally consistent. (Outcome: no consensus as to whether current progress counts as a solution.)

3. Given two polyhedra with equal volume, is it always possible to disassemble one of them into finitely many polyhedral pieces and then reassemble those pieces into the other polyhedron? (Outcome: no; proven using *Dehn invariants.*)

4. Construct all metrics where lines are geodesics. (Outcome: too vague to solve as stated.)

5. Are continuous groups automatically differential groups? (Outcome: depending on the interpretation of the statement, this problem is either solved or still open, the latter if this is understood to be a variant on the Hilbert-Smith Conjecture.)

6. Mathematically axiomatize the laws of physics. (Outcome: partially resolved, up to interpretation.)

7. For any algebraic number $a \neq 0, 1$ and any irrational algebraic number b, is a^b transcendental? (Outcome: Yes; this in in fact the Gelfond-Schneider Theorem, which was proven in 1934.)

8. Prove the Riemann hypothesis. (Outcome: this is a (now famously) open problem.)

9. Generalize the reciprocity theorem to any algebraic number field. (Outcome: partially resolved.)

10. Find an algorithm to compute the integer solutions of a polynomial Diophantine equation with integer coefficients. (Outcome: proven to be impossible by Matiyasevich's theorem.)

11. Solve all quadratic forms with algebraic coefficients. (Outcome: partially resolved.)

12. Extend the Kronecker-Weber theorem on abelian extensions of $\mathbb{Q}$ to any base number field. (Outcome: unresolved.)

13. Solve all polynomial equations of degree seven using algebraic (in some variants, continuous) two-variable functions. (Outcome: partially solved.)

14. Is the ring of invariants of an algebraic group acting on a polynomial ring always finitely generated? (Outcome: no, shown via a counterexample.)

15. Provide a rigorous foundation for the Schubert enumerative calculus. (Outcome: partially resolved.)

16. Describe the relative positions of ovals originating from real algebraic curves and as limit cycles of a polynomial vector field on the plane. (Outcome: unresolved.)

17. Can we express any non-negative rational function as the quotient of sums of squares? (Outcome: resolved; yes.)

18. (a) Does there exist a three-dimensional polyhedron that only admits an anisohedral tiling?

    (b) Find the densest possible sphere packing.

    (Outcome: both parts resolved; the answer to (a) is yes, and the answer to part (b) is approximately 74% of the total volume of the containing space.)

19. Are all solutions of regular problems in the calculus of variations analytic? (Outcome: yes; proven independently in 1957 by both Ennio de Gorgi and John Forbes Nash.)

20. Do all variational problems with certain boundary conditions have solutions? (Outcome: yes.)

21. Prove the existence of linear differential equations with a prescribed monodromic group. (Outcome: partially resolved, as originally stated by Hilbert. Other variants are either fully resolved or fully unresolved.)

22. Uniformize analytic relations via automorphic functions. (Outcome: unresolved.)

23. Further develop the calculus of variations. (Outcome: too vague to solve as stated.)

Let's now take a brief look at what each of the problems is actually asking. (Don't worry, this won't be as long as last time with the Millennium Prize Problems.)

Problem 1, the proof of the continuum hypothesis, has been officially deemed impossible to prove *or* disprove (at least using current methods). However, it has been shown that the answer would be independent of *ZFC set theory,* where ZFC is an abbreviation for **Zermelo-Fraenkel** "Axiom of **Choice**."

Problem 2 revolves around showing that all axioms in arithmetic are internally consistent with each other; that is, showing that they do not produce any contradictions amongst themselves. The proof of Gödel's incompleteness theorem in 1931 showed that no proof of consistency could be derived from arithmetic itself, but to date, no one is sure as to whether this would have constituted a satisfactory solution according to Hilbert.

The third problem is notable in that it was actually the first Hilbert problem to be solved, in 1900, the same year it was proposed; and by one of Hilbert's own students, no less - Max Dehn! It doesn't end there; Dehn had actually *invented his own mathematical construct,* the *Dehn invariant,* for the sake of solving Hilbert's third problem! The Dehn invariant can be described most briefly as a function of a polyhedron's lengths and dihedral angles that determines whether it can be dissected into another polyhedron. Noteworthily, Dehn's approach in 1900 differs slightly from the modern approach to solving the problem, which uses the tensor product.

In Problem 4, we wish to construct all metric (spaces), or topological spaces equipped with a definition of distance, such that all straight lines are the shortest distances between any two points in the metric. A *geodesic* is a generalization of a straight line in $\mathbb{R}^n$ to curved spaces, such as the surface of a sphere. A common example is the spherical geometry of the Earth's surface (hence the name *geo*desic). However, in its original form as stated by Hilbert, the consensus is that the problem is too vague to be completely resolved.

The fifth problem revolves around the assumptions and conjecturing of world-renowned Norwegian mathematician Sophus Lie (who is the namesake of the *Lie algebra* and *Lie group*, two mathematical objects that are of profound interest even to this day). Lie had proposed a set of geometric axioms, and had himself proven that they sufficed for all geometrical purposes using group theory. However, he had accomplished this primarily through the aid of *continuous groups,* and was unsure as to whether or not he could avoid making the assumption that the functions defining his group are differentiable. Hence, Hilbert asks whether continuity implies differentiability. This Problem, much like the third Problem, is interesting in that its outcome differs based on how it is interpreted. If the question is taken on its own terms, then it is considered to be resolved; if it is understood as a paraphrasing of the Hilbert-Smith conjecture, on the contrary, it remains an open problem.[108]

Upon initial inspection, Problem 6 may seem to be the vaguest problem yet, but mathematicians have made a considerable amount of progress. Essentially, the problem boils down to two much more specific goals: firstly, to treat probability axiomatically in the context of limit theorems so as to provide a rigorous foundation for the field of statistical physics; and secondly, to formalize the theory of limiting processes "which [would] lead from the atomistic view to the laws of motion of continua." In both regards, the establishment of Kolmogorov's axioms in 1933 is now accepted as standard, though it is still unclear whether these constitute a comprehensive lead from the "atomistic view" to the "laws of motion of continua" as originally desired.

---

[108]The Hilbert-Smith conjecture states that, if G is a locally compact topological group acting faithfully upon a topological manifold M, then G must be a Lie group. So far, the latest progress made on this conjecture was in 2013 by John Pardon, who proved the three-dimensional case.

For problem 7, we must first define a piece of terminology: an *algebraic number*.

> **Definition 16.30: Algebraic Numbers**
>
> An *algebraic number* is any complex number that is the root of a non-zero polynomial whose coefficients are all rational numbers.

Note that *algebraic* is not the same as *rational* - any n^{th} root of a positive integer is most likely irrational, but it is most definitely algebraic, since we can just raise this quantity to the nth power and then subtract the integer under the root to obtain zero. These are, in fact, the simplest examples of algebraic numbers, aside from the integers and rational numbers themselves. Another famous example of an algebraic, but irrational, number is the golden ratio, $\phi = \dfrac{1 + \sqrt{5}}{2}$ (as ϕ is a root of the polynomial $x^2 - x - 1$; similarly, $\dfrac{1}{\phi} = \dfrac{\sqrt{5} - 1}{2}$ is a root of $x^2 + x - 1$).

A complex number that is not algebraic is called *transcendental*. Examples of transcendental numbers include π, e, e^π, $\gamma \approx 0.5772$ (the Euler-Mascheroni[109] constant), any value of $\sin x$, $\cos x$, $\tan x$, or their multiplicative inverses for x a non-zero algebraic number, and $\ln x$ whenever x is algebraic and $x \neq 0, 1$.

Problem 8 needs no introduction, and Problem 9 is quite brief in its statement, and, most likely, its solution, compared to those of the other Problems. The Problem amounts to proving for any algebraic field of numbers that the law of reciprocity holds for any p^{th} power, where p is a power of 2, an odd prime, or a power thereof. Mathematicians are currently studying the pth roots of unity and attempting to make a connection with Hilbert's Problem, so the problem is considered partially resolved.

The tenth Problem asks whether there exists an algorithm that can determine for any Diophantine polynomial with integer coefficients whether that polynomial has an integer solution. The work of mathematician and computer scientist Yuri Matiyasevich, along with the earlier work of Julia Robinson, Martin Davis, and Hilary Putnam, led to the proof of the *MRDP theorem*, or Matiyasevich's theorem, which states that every *computably enumerable set* is Diophantine, along with the converse. First we must define what it means for a set to be computably enumerable:

> **Definition 16.31: Computable Enumerability**
>
> A set S is *computably enumerable* iff there exists an algorithm such that, for all $n \in \mathbb{Z}$, if $n \in S$, the algorithm is finite (i.e. at some point, it will halt), and if $n \notin S$, the algorithm will continue to run indefinitely, and will list all members of S.

Furthermore, we define what it means for a set to be Diophantine:

> **Definition 16.32: Diophantine Sets**
>
> A set S is *Diophantine* iff there exists a polynomial $P(x)$ with integer coefficients $x_1, x_2, x_3, \ldots, x_k$ such that an integer $n \in S$ iff there exists a sequence $\{x_i\}_{i=1}^k \subset \mathbb{Z}^k$ such that $P(n) = 0$.

We can also show that every Diophantine set is computably enumerable (and, in fact, this is much easier to prove than the converse, or the main MRDP theorem):

Proof. Consider a Diophantine set S and a polynomial $P(x)$ with integer coefficients $x_1, x_2, x_3, \ldots, x_k$ such that, for all $n \in S$, $P(n) = 0$. Our ideal algorithm will try the values $x_1, x_2, x_3, \ldots, x_k$ for n, and print n whenever $P(n) = 0$. This algorithm will print all n for which $P(n) = 0$ has a solution in $x_1, x_2, x_3, \ldots, x_n$, and so will run forever. □

---
[109] Insert tired joke about macaroni here.

As for the MRDP theorem itself, its proof, along with the work of Robinson, Davis, and Putnam, led directly to the solution of Hilbert's tenth Problem.

Problem 11, which requests a solution method for all quadratic forms with algebraic coefficients, is partially solved.

> **Definition 16.33: Quadratic Forms**
>
> A *quadratic form* is a homogeneous polynomial with degree two in any number of variables. (A *homogeneous* polynomial is a polynomial whose nonzero terms all have the same degree (in this context, the sum of exponents, not the highest exponent). For instance, $x + y + z$ and $2x^3y^2z + y^6 + z^4x^2$ are homogeneous, but not $x^2y^2 + y^2z + z^2x^2$.)

Example 16.2.0.1. *For example, $x^2 + 2xy + 3y^2$ is a quadratic form, as are $10x^2$ and $xy + xz + yz$, but not $x + y^2 + z^2$ (since one of its terms only has degree 1).*

In the cases of having one, two, and three variables, they are called *unary*, *binary*, and *ternary* quadratic forms, respectively. For four or more variables, they are simply called n-variable quadratic forms.

Some progress has been made from the front of p-adic number systems, but this avenue has thus far not yielded an exhaustive mode of classification for quadratic forms that states which quadratic forms are equal to each other (i.e. have the same solutions).

The twelfth Problem deals with the *Kronecker-Weber theorem*, which concerns abelian extensions of the field $\mathbb{Q}$:

> **Theorem 16.16: The Kronecker-Weber Theorem**
>
> Every finite abelian extension of $\mathbb{Q}$ is contained in some cyclotomic field.[a] That is, every algebraic integer (a complex number that is a root of a *monic* polynomial with integer coefficients, not just any polynomial with integer coefficients) whose Galois group is abelian can be written as the sum of complex roots of unity with rational coefficients.
>
> ---
> [a] A *cyclotomic field* is a number field obtained by adjoining to $\mathbb{Q}$ a primitive root of unity in $\mathbb{C}$.

Example 16.2.0.2. *We can represent the algebraic integer $\sqrt{-7}$ as $2(e^{\frac{2\pi i}{7}} + e^{\frac{4\pi i}{7}} + e^{\frac{8\pi i}{7}}) + 1$.*

Proof. For brevity, let $\xi = e^{\frac{2\pi i}{7}}$. Let $S = \xi + \xi^2 + \xi^4$, so that $S^2 = \xi^2 + \xi^4 + \xi^8 + 2(\xi^3 + \xi^5 + \xi^6) = -2 - S$ (since $\xi^8 = \xi$). Let $A = 2S + 1$; then $A^2 = 4S^2 + 4S + 1 = 4(-2 - S) + 4S + 1 = -7$ and $A = \pm\sqrt{-7}$. □

Hilbert's Problem asks for an extension of the Kronecker-Weber theorem to any base number field, which is as yet unresolved. Nevertheless, mathematicians have made inspiring progress on this problem in recent years. Complex multiplication of abelian varieties became a much more accessible area of study due to the work of Goro Shimura and Yutaka Taniyama (of Shimura-Taniyama Conjecture[110] fame).

Problem 13 entails whether or not a solution exists for certain degree 7 polynomials, namely polynomials of the form $x^7 + ax^3 + bx^2 + cx + 1$ (where a, b, and c are nonzero constants), that are functions of a, b, and c, and whether they could be written as the composition of a finite number of two-variable functions. This Problem's resolution status, like that of many other Problems, depends on its precise wording. If the Problem is understood to reference *continuous* functions, then it was resolved in 1957, by Vladimir Arnold. Arnold, a student of Hilbert, initially showed that any satisfactory function could be constructed from three-variable functions, then proceeded to extend his proof to two-variable functions, thereby disproving Hilbert's own conjecture that it was not always possible to find a solution! (That's the second time your students

---
[110] The conjecture that proved most instrumental in finally proving Fermat's Last Theorem.

have solved one of your problems, Hilbert!)

In solving Problem 14, we wish to determine whether the ring of *invariants*, or those mathematical objects which are unchanging under transformations of a linear group, of an algebraic group that acts upon a polynomial ring is necessarily finitely generated. In (rough) terms, an algebraic group, which is a group that is also an algebraic variety (where an *algebraic variety* is a set of solutions to a system of polynomial equations over $\mathbb{R}$ or $\mathbb{C}$), has a ring of functions that remain constant, in a sense, over some group G. More formally:

> **Definition 16.34: Rings of Invariants**
>
> f is a *G-invariant* if, for all $g \in G$, $f^g = f$ (where f^g denotes $f(f^{g-1})$). $R^G = \{f \mid f^g = f, \forall g \in G\}$ is the *ring of G-invariants*.

The question: is this ring always finitely generated? First, we must understand what it means for a ring to be *generated* over a field $\mathbb{F}$:

> **Definition 16.35: Generation of Algebras**
>
> An associative algebra A over a field $\mathbb{F}$ is *finitely generated* if there exists a finite set $\{a_1, a_2, a_3, \ldots, a_n\}$ where $a_i \in A$ for all $i \in [1, n]$ for which every $a \in A$ can be expressed as a polynomial in terms of the a_i, with coefficients in $\mathbb{F}$. If A is not finitely generated, it is *infinitely generated*.

Example 16.2.0.3. *The polynomial algebra, $\mathbb{F}[x] = a_n x^n + a_{n-1} x^{n-1} + \ldots + a_1 x + a_0$, where $a_i \in \mathbb{F}$ for $i \in [0, n]$, is finitely generated. (Proving this is left as an exercise.)*

In order to solve Hilbert's 14th Problem, Hilbert himself managed to prove that invariant rings are finitely generated in the field of complex numbers for certain types of Lie groups and actions on polynomial rings. Other mathematicians have attempted to come up with counterexamples; Nagata Masayoshi succeeded in doing so in 1959.

His counterexample consisted of a field $\mathbb{F}$ (or k, as he denoted it) containing 48 elements of the form a_{1i}, a_{2i}, a_{3i}, ..., a_{16i} where $i \in \{1, 2, 3\}$ (a dummy index, not to be confused with the imaginary unit $\sqrt{-1}$). These elements were all algebraically independent over the prime field, meaning that the set consisting of these elements did not satisfy any non-trivial (non-zero) polynomial equation with coefficients that were prime numbers. The ring R is the polynomial ring in 32 variables, and the vector space V is a 13-dimensional vector space cover k containing all vectors of the form $\langle b_1, b_2, b_3, \ldots, b_{16} \rangle$ that are orthogonal to all vectors $\langle a_{1i}, a_{2i}, a_{3i}, \ldots, a_{16i} \rangle$ for $i \in \{1, 2, 3\}$. As it turns out, R is invariant under the action of V, which acts as a group. However, R is in this case an infinitely-generated k-algebra, solving the 14th problem!

Problem 15 is more of an "investigation" of a relatively new (in 1900) area of math than it is a problem proper: it requests a rigorous foundation for Schubert's enumerative calculus (or just *Schubert calculus*). Schubert calculus is an offshoot of algebraic geometry that studies counting problems in *projective geometry* (which is itself a branch of *enumerative geometry*). Schubert, in constructing the analysis of linear subspaces of projective spaces, defined an object known as the *Schubert cell*.

> **Definition 16.36: Schubert Cells**
>
> A *Schubert cell* is a locally closed set (i.e. the intersection of an open and closed subset of a topological space) of a *Grassmannian*, a space which parametrizes k-dimensional linear subspaces of the vector space V^n. The generalized Grassmannian is denoted $\text{Gr}(k, V^n)$.

Example 16.2.0.4. *The Grassmannian $Gr(1, V^2)$ is an analogue to the space of lines passing through the origin in the Cartesian plane. Similarly, $Gr(2, V^3)$ can be thought of as the set of two-dimensional*

> *planes passing through the origin in three-dimensional space.*

Otherwise, the problem itself gives little guidance in that it does not define a "rigorous foundation." To partially combat this problem, Hilbert had elaborated on the problem statement considerably in a fuller version of his list; the following is the official English translation of his elaboration by Dr. Maby Winton Newson, as found in *Bulletin of the American Mathematical Society*, Volume 8:

> The problem consists in this: To establish rigorously and with an exact determination of the limits of their validity those geometrical numbers which Schubert especially has determined on the basis of the so-called principle of special position, or conservation of number, by means of the enumerative calculus developed by him.
>
> Although the algebra of today guarantees, in principle, the possibility of carrying out the processes of elimination, yet for the proof of the theorems of enumerative geometry decidedly more is requisite, namely, the actual carrying out of the process of elimination in the case of equations of special form in such a way that the degree of the final equations and the multiplicity of their solutions may be foreseen.

Still, the problem remains partially unsolved.

Problem 16 is one of the more divisive problems among Hilbert's 23, since it has two wildly differing interpretations: one shared by those who specialize in the field of real algebraic geometry, and the other among those who specialize in dynamics. Moreover, it actually consists of two distinct parts, which is where the dichotomy comes from. Algebraic geometers tend to place more focus on the first part of the problem, while dynamicists tend to study the second part of the problem more than the first.

The first part of the Problem concerns the relative positions of ovals, or branches, of degree-n real algebraic curves. The second part revolves around determining a relative position of, and upper bound for, the *limit cycles* in two-dimensional polynomial vector fields of degree n. (The study of limit cycles, set forth by none other than Henri Poincaré, involves analyzing a closed trajectory in a phase space to which other trajectories also converge as time approaches infinity or negative infinity.) By "relative positions," Hilbert's intended meaning appears to be that the first portion of the Problem entails placing an upper bound on the number of inter-connected arcs of an algebraic curve. Back in 1876, Carl Gustav Axel Harnack showed that a degree-n curve had an upper bound of $\frac{n^2-3n+4}{2}$ connected arcs in the real projective plane. It was then that Hilbert proposed a generalization to *M-curves*; those curves with exactly that number of connected arcs, or *components*. This generalized form (what we know today as the first part of the 16th Problem) is, as yet, unsolved for $n = 8$.

The second part of the Problem has also perplexed the mathematical community ever since its inception. It entails finding, and proving, an upper bound on the number of limit cycles in fields of degree-n polynomials. Though it has been known since the early 1990s that every polynomial vector field has only finitely many limit cycles, it is still unknown whether an upper bound exists for every *planar* polynomial vector field.

Problem 17, regarding the representation of polynomials that always assume non-negative values (*positive-definite* polynomials), asks whether we can always write said polynomials as the (quotients of) sums of squares of other rational functions (functions that are the quotients of two other polynomials). The problem statements, with or without the "quotient of," are both equivalent (since we can merely substitute the reciprocal of a function and square it to obtain the square of its reciprocal).

Interestingly, Hilbert himself had proven this for degree-2 polynomials (in 1893; even before he proposed the problems), meaning that he proposed a problem that he had already made non-trivial progress on. Still, the problem remained unsolved until 1927, when Emil Artin proved that the assertion was true. Later solutions expanded on his work by providing algorithmic solution methods and an upper bound for the number of square terms required.

Problem 18, which pertains to sphere packing in Euclidean space, also asks whether an *anisohedral tiling*, or a tiling which tiles $\mathbb{R}^3$ but does not admit an *isohedral* (tile-transitive, meaning that all pairs of tiles are equivalent under all symmetries of the tiling) tiling, exists for some 3-dimensional polyhedron. Intriguingly, the first example for three dimensions came before the first example for two dimensions; Karl Reinhardt and Heinrich Heesch had discovered them in 1928 and 1935, respectively. A related problem, the aptly-named *Einstein problem,* concerns the existence of a shape that can tile/tessellate Euclidean space, but not in a periodic manner. That is, the shape's tiling should not repeat after any interval.

The second part of the problem is the perfect testament to the enduring willpower and passion of the mathematical community: the proof did not fully arise until 1998, and it was a computer-assisted proof. For almost a full century, the finest mathematicians in the world were unable to lay the problem to rest, despite their continued collaboration. It was only until the advent of computer technology that the problem become remotely solvable and the bleakness of leaving an unsolved problem in the annals of mathematical history began to dissipate. Sphere packing, despite its intuitive appearance on the surface, had proven to be one of the most confounding among Hilbert's solved Problems.

As it turns out, the densest sphere packing is in a pyramidal shape; this configuration occupies about 74% of the containing space (whereas a random packing will, on average, occupy only about 64% of the available space), with $\frac{\pi}{3\sqrt{2}}$ being the exact ratio for an optimal *lattice packing* (i.e. a packing with spheres corresponding to regular lattices), as proven by Gauss. However, the case for non-lattice packings was not verified until 1998 via computer-assisted proof. Only in 2014 did the proof actually reach verified status, via automated proof checking!

The nineteenth Problem deals with the solutions in the calculus of variations to *regular problems;* that is, variational problems whose Euler-Lagrange equation is a partial differential equation with analytic coefficients. As stated by Hilbert in 1900 as part of his speech where he proposed the problems, one of the most interesting parts of analytic functions[111] is that there exists partial differential equations for which such functions are the only solutions.

Significant progress on this Problem began in 1904, when Sergei Bernstein presented his thesis on the topic: his work showed that solutions of 2-variable non-linear elliptic analytic equations are themselves analytic. However, methods in the study of the calculus of variations seemed to imply a contradictory result; hence, a disconnect began to form between the two schools of thought. This gap was eventually resolved in 1957 by John Forbes Nash and Ennio de Georgi, who showed that the solutions to the partial differential equations in question had first derivatives that were Hölder continuous (i.e. having constants $C, \alpha \geq 0$ such that $|f(x) - f(y)| \leq C \, ||x - y||^{\alpha}$ for all x, y in the domain of the function (where $||\cdot||$ denotes the d-metric distance function)), thus resolving Hilbert's 19th Problem in the affirmative.

The twentieth Problem again concerns differential equations in the general context of variations, asking whether or not all boundary value problems (problems involving differential equations equipped with additional conditions) are solvable. Hilbert was able to prove the existence of solutions for boundary value problems with function values given at the boundary, but challenged the mathematical community to determine the existence (or non-existence) of solutions for more complicated conditions at the boundary. This was a topic teeming with activity in the 1900s, and it was solved with the exception of non-linear cases. (For this reason, the problem, as stated, is considered solved.)

The 21st Problem, more commonly known as the *Riemann-Hilbert problem,* concerns the existence of a class of linear differential equations[112] with given singularities (points where a function is undefined) and *monodromic group* (groups defined in terms of singularities). To date, this problem is only partially resolved by general consensus; interestingly, though, a solution proposed in 1908 was accepted as a proper solution until 1990, what with doubt rising over its validity from the results of previous work in the middle of the century. Indeed, Andrey A. Bolibrukh eventually found a counterexample, both to the solution proposed

---

[111]Infinitely differentiable functions that are given locally by convergent power series.

[112]Three problems in a row about differential equations? Hilbert must've been *really* obsessed with those things!

in 1908 and to Hilbert's problem statement. It is still not agreed upon as to whether this constitutes a resolution to the 21st Problem.

The penultimate of Hilbert's Problems, the 22nd Problem, is decidedly brief: it requests the complete uniformization of analytic relations via automorphic functions. Poincaré proved before 1900 that any algebraic relation of two variables could be reduced to uniformity, or the condition that all surfaces in Riemann's topology are equivalent to one of three basic types of surfaces: the open unit disk (the intersection of a circle centered at the origin and its boundary), the complex plane, or the Riemann sphere. In general, these spaces all give way to the three common properties of *completeness, uniform continuity,* and *uniform convergence.*

The final problem in Hilbert's original list is even shorter, yet has mystified mathematicians for 119 years and counting: it simply asks for further development of the calculus of variations (see the exposition on the 15th Problem for a better idea of what the calculus of variations is). In short, the problem is much too vague as stated to be "solved," *per se,* but it is at least vague *a priori;* mathematicians, thankfully, did not need to invest heavily into the Problem to understand that it was nebulous and somewhat intangible. Rather, Hilbert was decidedly upfront in proposing this conclusion to his list as a coax into exploring the newfound field of calculus of variations, rather than as a problem to be solved.

It is quite similar, in this respect, to the 15th Problem, except that Problem does give some more direction as to what is considered a fully adequate solution. Concerted efforts have nevertheless been made on this Problem despite the common knowledge of its open-ended abstractness; Emmy Noether,[113] Henri Lebesgue,[114] and Hilbert himself had all majorly contributed to the field. Today, the field has several branching applications in optimal control theory, dynamic programming, and *Morse theory,* named not after Samuel Morse, the namesake of Morse code, but after Marston Morse, who introduced many of the fundamental ideas behind differential topology, and became the namesake of the *Thue-Morse sequence.*[115]

16.2.1 The 24th Problem: Meta-Mathematics at its Finest

Though the fittingly open-ended 23rd Problem may be the last of Hilbert's celebrated list, it is not actually the last problem Hilbert proposed. Indeed, his true swan song, only unveiled in 2000 from Hilbert's original draft of the list, was a problem that he scrapped from the final list; unofficially, his (arguably even more open-ended and nebulous) 24th Problem. This Problem reflects a beautiful self-awareness on the part of mathematicians, and their desire to simplify rampant, unnecessary complication in the world.

The meat of this Problem deals with specific, rigorous criteria for determining whether a mathematical proof is "simple," along with a development of a theory to prove that other proofs are as simple as possible. Namely, it seeks the one "simplest" proof to each and every mathematical problem, regardless of the complexity of the manner in which the problem itself may be constructed. If a theorem or problem has multiple proofs, one must determine which precise methods were used in both proofs, and simplify them by determining equivalences in the methods and subsequently eliminating any redundancies between them.

This Problem is the perfect embodiment of Occam's Razor, and a prime demonstration of how mathematicians think of their own work. Even when an elegant, seemingly simple proof is found, mathematicians always attempt to make the proof simpler, clearer, more concise, and better in every way imaginable. This tireless spirit of hard-working mathematicians is what ultimately drives so much of the progress they have made to this day, and what will continue to push us forward.

---

[113] Described by Albert Einstein, among several others, as the woman who had hitherto made the biggest impact on mathematics as a whole.

[114] Who introduced the concept of the *Lebesgue measure* and corresponding *Lebesgue integral.*

[115] The Thue-Morse sequence t_n has $t_n = 1$ if the number of ones in the binary expansion of n is odd, and $t_n = 0$ otherwise. Positive integers n for which $t_n = 1$ are called *odious* (for *odd*) and positive integers n for which $t_n = 0$ are called *evil* (for *even*). They weren't kidding about that one!

16.3 The Twin Prime Conjecture

While not a Millennium Prize Problem, the *Twin Prime Conjecture* is nonetheless one of the most well-known, enigmatic, and captivating open problems in all of mathematical history. Like several of the greatest problems, its statement is exceedingly straightforward, but its execution is beyond confounding. The only requisite definition is that of a *twin prime*:

> **Definition 16.37: Twin Primes**
>
> A *twin prime* p is a prime number such that $p+2$ (or $p-2$) is also a prime number.

The Twin Prime Conjecture simply asks whether there are an infinite number of pairs of twin primes.

The proof that there are an infinite number of prime numbers has been known since antiquity; however, twin primes are another story entirely. To get started, we can analyze the problem from a modular arithmetic perspective. First note that all prime numbers, with the exception of 2, are odd, so all twin prime pairs must consist of odd numbers. ((2, 3) is not generally considered to be a twin prime pair.) In addition, if we consider the positive integers (mod 3), then it becomes apparent that any prime number larger than 3 is equivalent to either 1 or 5 (mod 6) (and it immediately follows that all twin prime pairs, save for (3, 5), are of the form $(6k-1, 6k+1)$ for some positive integer k).

Another interesting intrinsic property of twin primes is that the sum of their reciprocals converges, much like the sum of squares of positive integers to $\frac{\pi^2}{6}$. Here, however, the limit is not a nice, clean expression in terms of a familiar constant like π, but an invented constant known as *Brun's constant* - the crux of Brun's Theorem.

> **Theorem 16.17: Brun's Theorem**
>
> Let $S = \{p \in \mathbb{N} \mid p \text{ is a twin prime}\}$. Then
>
> $$\sum_{p \in S} \frac{1}{p} = B_2 \approx 1.902$$

Brun's constant B_2 includes the term $\frac{1}{2} + \frac{1}{3}$, even though $(2, 3)$ is not a twin prime pair by definition. This is yet another mathematical quirk, probably best explained by factors of convenience.[116]

The subscript 2 indicates that we are summing the expression $\frac{1}{p} + \frac{1}{p+2}$ over all pairs of twin primes $(p, p+2)$. Similar constants $B_3, B_4, \ldots$ exist which also converge. However,

$$B_1 = \sum_{p \text{ prime}} \frac{1}{p}$$

does not converge.[117] The proof of Brun's Theorem relies on the fact that the density of twin primes can be approximated as less than the density of prime numbers by a logarithmic factor.

"Weaker" forms of the twin prime conjecture have seen considerable progress. In 1940, Paul Erdős showed that there exist infinitely many intervals between increasingly large twin primes that grow slowly in size.

---

[116] This is similar to how we define $0! = 1$ for combinatorial purposes, and in certain contexts, "assign" the value of 1 to the expression 0^0, or $-\frac{1}{12}$ to the divergent sum $1 + 2 + 3 + \ldots$

[117] Note that the infinitude of prime numbers immediately follows from the divergence of the sum of reciprocals of the primes. Talk about solving a simple problem in a complicated manner!

Erdős defined "slow" growth as the existence of a constant $c < 1$ such that $(p' - p) < c \ln p$, where p' is the next prime after p. This result was later improved upon, with $\frac{1}{4}$ as the next-smallest upper bound for c. After that, the bound was reduced to ≈ 0.08, and eventually arbitrarily close to zero (where "arbitrarily" means that it cannot actually be zero, but the limit of dividing the difference between twin primes by the natural logarithm of the first prime is zero).

We can actually generalize this already-inscrutable problem to *Polignac's Conjecture*, which posits that, for any even positive integer k (not just $k = 2$, as with the Twin Prime Conjecture), there are infinitely many ordered pairs of positive integers $(p, p+k)$ such that both p and $p+k$ are prime. So far, this conjecture has not been proven or disproven for *any* value of k, making this perhaps one of the "most open" problems out of all open problems! (If that makes any sense...)

Progress on this conjecture *really* began to take off, however, with the advent of 21st century computing technology. As the Internet became a mainstream part of everyday life and computing power grew exponentially by the year, online twin prime-finding projects have begun to take rise. In 2007, the Twin Prime Search and PrimeGrid projects were launched, officially kicking off the world's search for record-breakingly large twin primes. As of 2018, the largest twin prime pair (which was discovered two years prior) is $(2996863034895 \cdot 2^{1,290,000} - 1, 2996863034895 \cdot 2^{1,290,000} + 1)$, which has 388,342 digits - enough to fill about 15 manuscript pages with just digits![118]

Though the Twin Prime Conjecture and its generalizations may be unproven, prime number enthusiasts have established firm upper bounds on the gaps that exist between consecutive prime numbers. The difference between two consecutive prime numbers is a *prime gap*. The nth prime gap, $g_n = p_{n+1} - p_n$, where $g_1 = 3 - 2 = 1$ and $g_2 = 5 - 3 = 2$, has been shown to be less than p_n (the nth prime number), as a corollary of Bertrand's postulate (which states that, for any positive integer k, there is always at least one prime number between k and $2k$). Furthermore, the prime number theorem posits that the average value of g_n tends towards $\ln(p_n)$, so it follows that

$$\forall \epsilon > 0, \exists N \text{ s.t. } \forall n > N, g_n < \epsilon \cdot p_n \qquad (16.13)$$

and also that

$$\lim_{n \to \infty} \frac{g_n}{p_n} = 0 \qquad (16.14)$$

Further results focused on a constant $\theta < 1$ such that

$$\pi(x + x^\theta) - \pi(x) \sim \frac{x^\theta}{\ln x} \qquad (16.15)$$

as $x \to \infty$. As a consequence of these results, in 2013, it could be shown that $g_n < 7 \cdot 10^7$ for all $n \in \mathbb{N}$. From here, Internet-based collaborative efforts had taken over, to reduce the bound to 4,680 in July of that year, then to 600 in November of that year, and finally to 246 in the following year. If we assume the Elliott-Halberstam Conjecture, then N is further reduced to 12, and if we assume the generalized form, then N becomes a mere 6.[119]

As with several of the Millennium Prize Problems, proving the conjecture one way or another would not have a dramatic effect on the current operations of practical algorithms, since they most likely assume the conjecture to be true as a base assumption. But we would open up valuable connections to other areas and open problems in math, such as the Riemann Hypothesis (though the proof of this may actually come first, in which case its affirmative result could be used to aid with the proof of the Twin Prime Conjecture). At the very least, it would open doors to new knowledge of primes, and allow us to view them in a way that we have never before imagined.

---

[118] This still pales in comparison to the largest (non-twin) prime number (discovered in December 2018 via the Great Internet Mersenne Prime Search (GIMPS), namely $2^{82,589,933} - 1$, which clocks in at just shy of a whopping 25 million digits!

[119] "Bounded gaps between primes," PolyMath 2013.

16.4 The Collatz Conjecture

This problem is another prime example of the contrast between simplicity and beautiful elegance, and complete and total imperviousness to all but the most advanced methods. The *Collatz Conjecture* was chosen for this chapter on account of its very accessible and straightforward problem statement which has great potential to *inspire* the amateur to solve it, and at the same time for the devious manner in which it leads them astray almost instantly.

> **Theorem 16.18: Collatz Conjecture**
>
> Consider an infinite sequence (a_n) of positive integers, with $a_1 = n$ for some positive integer n. We define a_{n+1} such that, if a_n is even, $a_{n+1} = \frac{1}{2}a_n$, and if a_n is odd, $a_{n+1} = 3a_n + 1$. The Collatz Conjecture posits that, regardless of the initial value n, (a_n) will always reach 1.

Introduced by Lothar Collatz in 1937, the problem remains unsolved as of April 2019. Paul Erdős has said of the problem: "Mathematics may not be ready for such problems," but offered \$500 for its solution.[120] In 2010, Jeffrey Lagarias described the problem as "extraordinarily difficult" and "completely out of reach of present day mathematics," despite its unassuming appearance. Much akin to the likes of Fermat's Last Theorem, the Collatz Conjecture is a paragon of deception in simplicity, and shows we have a ways to go in *completely* and *thoroughly* understanding how numbers work.

In order to construct a counterexample to the conjecture, one would need to find a sequence (a_n) such that the terms either increase without bound or be periodic after a certain point, excluding 1 in that cycle. To date, no such sequence has been found, let alone had the possibility of its existence excluded. Note that a cycle in a counterexample sequence would need to be a cycle other than the trivial $(4, 2, 1)$ cycle; thus, the conjecture is equivalent to stating that the sequence eventually reaches 4 (and, by, extension, any power of 2).

We can define a *shortcut map* of the Collatz sequence (a_n):

$$f(n) = \frac{1}{2}(3n+1), n \text{ odd} \tag{16.16}$$

$$f(n) = \frac{1}{2}n, n \text{ even} \tag{16.17}$$

Using this shortcut definition of (a_n), we can define a generalized form of the cycle, known as the k-cycle:

> **Definition 16.38: k-Cycles**
>
> A k-cycle of the sequence (a_n) is a cycle that can be partitioned into $2k$ subsequences that are all contiguous. k of these subsequences should be increasing and consisting of odd elements, and the other k should be decreasing and consisting of even elements, in alternating order.

We have that $(1, 2)$ is the only 1-cycle. In 2004, Simons used Steiner's work from 1977, which showed that $(1, 2)$ was the only 1-cycle, to show that no 2-cycles exist. The following year, Simons and de Weger made astounding progress, proving that no k-cycles existed for all $k \leq 68$. For $k \geq 69$, there are proven upper bounds for the elements in any k-cycle that may exist.

This is yet another instance in which there is no conclusive evidence to prove or disprove the conjecture, yet mathematicians have come to a consensus that the conjecture is, more likely than not, true. Experimental evidence (as of 2017) seems to indicate that the conjecture is likely true, as all initial values up to $87 \cdot 2^{60}$

---

[120]This shows better than anything else that mathematics isn't all about glory, fame, and material reward - in fact, it's hardly about those things at all. Nowadays, \$9000 (the 2019 equivalent of \$500 in 1937, adjusting for inflation) could barely cover the cost of the cheapest automobiles, if at all!

eventually lead to the trivial cycle $(4, 2, 1)$. We also have a probability heuristic at our disposal, albeit a non-rigorous one. If we consider the odd numbers in the Collatz sequence (a_n), and form a subsequence (b_n) from the sequence of odd numbers in (a_n), then we can observe that the limit

$$\lim_{n \to \infty} \frac{b_{n+1}}{b_n} = \frac{3}{4} < 1$$

and so the sequence should converge to 1 in the long run.

Proof. (The following argument is originally from Lagarias, 1985.)

Pick a random odd integer k from the subsequence (b_n), and continue applying the definition of the sequence (a_n) until another odd integer is obtained. $\frac{1}{2}$ of the time, this odd integer will be equal to $\frac{3k+1}{2}$; $\frac{1}{4}$ o the time, this integer will be equal to $\frac{3k+1}{4}$; $\frac{1}{8}$ of the time, it will be equal to $\frac{3k+1}{8}$; and so forth. If the odd integers behave as though they were drawn uniformly at random, then the expected ratio of two consecutive odd integers in Collatz's sequence (a_n) is

$$\left(\frac{3}{2}\right)^{\frac{1}{2}} \cdot \left(\frac{3}{4}\right)^{\frac{1}{4}} \cdot \left(\frac{3}{8}\right)^{\frac{1}{8}} \cdots = \frac{3}{4} < 1$$

Hence, the sequence (b_n) converges, and thus (a_n) should converge as well. $\square$

Remark. This reasoning is not air-tight, and even it it were, it would not guarantee that the sequence (a_n) would converge to 1 for *all* initial values n.

How, then, could we possibly approach Collatz's conjecture? Perhaps we could consider it in reverse. Instead of proving that all positive integers lead to 1, we can show that 1 leads to all positive integers. Since 1 is an odd number, it is equivalent to 1 (mod 2), and if $n \equiv 1$ (mod 2), then $3n + 1 \equiv 4$ (mod 6) (and the converse holds true as well). Additionally, it follows that $n \equiv 4$ (mod 6) $\iff \frac{n-1}{3} \equiv 1$ (mod 2). Similarly, $n \equiv 1$ (mod 2) $\iff \frac{3n+1}{2} \equiv 2$ (mod 3).

The idea of the Collatz parity function also shows promise. Here, we set

$$\begin{cases} \frac{n}{2} & n \equiv 0 \pmod{2} \\ \frac{3n+1}{2} & n \equiv 1 \pmod{2} \end{cases}$$

Remark. This setup works because, whenever n is odd, $3n + 1$ is always even.

Let $P(n)$ be the parity function, such that $P(2n) = 0$ and $P(2n + 1) = 1$ for some $n \in \mathbb{Z}$. Then define the *Collatz parity sequence* (p_n) as $p_0 = n$, and $p_k = P(f(p_{k-1}))$. Using this sequence, it can be shown that the parity sequences for two initial values x and y will have the same i first terms iff $x \equiv y \pmod{2^i}$.

What is the outcome of applying the Collatz function f some number of times k to an initial value n? Assume n is of the form $a \cdot 2^k + b$, where a and b are constants. We claim $f^k(n) = a \cdot 3^c + d$, where c is the number of times the sequence increases and $d = f^k(b)$. (In this manner, f is a recursively defined function.) Demonstrating and proving this is left as an exercise to the reader. (Hint: Certain forms of numbers will always lead to smaller numbers; consider a linear function for one example.)

Finally, what if we extended the Collatz sequence to *all* integers, not just positive integers? If we were to consider a starting term in $\mathbb{Z}$ instead of just $\mathbb{N}$, we now have four known cycles, rather than just one: $(1, 4, 2, 1, \ldots)$, $(-1, -2, -1, \ldots)$, $(-5, -14, -7, -20, -10, -5, \ldots)$, and a cycle of length 18 and initial value -17.

16.5 Fermat's Last Theorem: Not an Open Problem (Anymore), But...

Think of the most frustrating thing that's ever happened to you. Making one stupid mistake on a test that brought you down a whole letter grade in a very important class, getting stuck in traffic on the way to work the day you have to attend an important meeting, having your computer crash when you were just about to save that big term paper, seeing your football team lose the game of the season on a last-minute touchdown... Got it? Now picture the mathematical community back in 1637, when Pierre de Fermat wrote the following sentence down in the margins of his *Arithmetica*: "I have discovered a truly marvelous proof of this, which this margin is too small to contain."

...Words cannot do justice to the unadulterated frustration of mathematicians everywhere upon Fermat taking his proof with him to the grave. This single comment was the shot heard round the world: the bullet that started the war to prove *Fermat's Last Theorem* - three words which are all but guaranteed to send shivers down the spine of any professional mathematician who has devoted a major chunk of their life to cracking this solitary nut, and for perfectly good reason. Then, and for what seemed like an eternity until 1995, it was not a theorem at all, but rather one of the most infamous open problems in all of mathematics. Ironically, Fermat's Last Theorem is not an open problem as of the 21st century, but is a problem that infuriatingly and tantalizingly remained open for the better part of four centuries!

On the surface, this figurative "nut" appears quite easy to crack, even trivially so; the problem merely asks for a proof that the Diophantine equation $a^n + b^n = c^n$ has no solutions in positive integers for $n \geq 3$. (The cases $n = 1$ and $n = 2$ are indeed trivial; $n = 1$ is obvious, and $n = 2$ is the Pythagorean Theorem.) But like all of the other problems we have seen thus far, Fermat's Last Theorem is far, *far* from trivial.

The problem takes the deceptive mystique from the Millennium Problems and cranks it to the nth degree; not only is the proof among the most complicated and involved in all of mathematical history (requiring the invention of several new fields and branches of mathematics to complete), its problem statement, at a superficial level, can be "understood" with little more than middle school mathematics.[121] Thus, it may trick some eager students with enough innocent naiveté into attempting a proof, but truth be told, despite its non-assuming appearance, Fermat's Last Theorem is perhaps the single hardest solved problem in all of mathematics to date. It even held a Guinness world record for it!

Fermat's Last Theorem, like many of the other notoriously difficult unsolved problems in mathematics, had humble beginnings. The problem statement arose from none other than the familiar Pythagorean Theorem - the clean, sleek, and beautifully straightforward statement that, in any right triangle with leg lengths a and b, and hypotenuse c, the equation $a^2 + b^2 = c^2$ always holds. Fermat then wondered if there existed any solutions to the general equation $a^n + b^n = c^n$ for $n \geq 3$, and conjectured (correctly) that there were none.

Despite his infamous comments in the margins, Fermat never actually found a proof; in fact, the problem statement itself was only discovered 30 years after Fermat's death! However, Fermat did manage to prove the case $n = 4$. All that remained was to prove his conjecture for prime numbers n (since if n is a composite number greater than 4, then we can write $n = pq$, where $p > 2$, and obtain $a^n = a^{pq} = (a^p)^q$, which requires that there exist a solution for an exponent of p as well).

Over the two centuries following the discovery of Fermat's conjecture, mathematicians, including Sophie Germain, collaborated to provide a proof for $n = 3, 5$, and 7. Shortly thereafter, Ernst Kummer proved Fermat's Last Theorem for all *regular* primes, leaving only the *irregular primes*. Using Kummer's work as a foundation, other mathematicians had managed to prove the theorem for irregular primes into the millions, but had not yet found a way to prove it irrefutably for all primes.

---

[121] Note the quotation marks around the word *understood;* though we say "understood" in typical contexts to mean that one has grasped the general intuition behind a problem and what it is asking, "understanding" Fermat's Last Theorem is completely different from actually knowing how to prove it at a deep, thorough, and rigorous level.

In the mid-20th century, however, Japanese mathematicians Yutaka Taniyama and Goro Shimura began to suspect a connection between two fields of math that were previously thought to be wholly unrelated: *elliptic curves* and *modular forms*.[122] Shimura, building on the previous work of Taniyama, posed along with Taniyama the *Taniyama-Shimura conjecture*, now known as the *modularity theorem*. For decades, it was admittedly viewed as an important conjecture in and of itself, but its connection with Fermat's Last Theorem went unnoticed.

This all changed in 1984, however, when Gerhard Frey observed a potential link between the two seemingly-unrelated fields, and provided a sketch of a proof that they were, in fact, interrelated. The following year, Jean-Pierre Serre partially proved that all Frey curves were non-modular, and the year after that, Ken Ribet completed the proof, in the process solving the *epsilon conjecture* (now referred to as *Ribet's theorem* in his honor). Collectively, the work of Frey, Serre, and Ribet showed that the modularity theorem holding for *semi-stable* elliptic curves would imply Fermat's Last Theorem. (Of course, the contrapositive is also true: a contradiction to Fermat's Last Theorem would also contradict the modularity theorem.)

And this is where Andrew Wiles' work comes into the picture. Wiles, upon hearing of Ribet's proof of the epsilon conjecture, decided to embark on his years-long attempt to finally hammer in the nail in the coffin of Fermat's Last Theorem. Prior to beginning work on the proof, Wiles had had a considerable background in elliptic curves, so he (thankfully) had somewhat of a headstart.

The proof itself occupied seven years of Wiles' research time, and ran over 100 pages in its final form. In his landmark paper proving the theorem, Wiles utilized several techniques that were discovered and/or invented in the 20th century, and thus could not possibly have been available to Fermat. (This is a point of interest to this day; many contend that Fermat *did* prove the theorem, albeit with unknown tools.)

Even then, in 1993, when he had initially announced his proof finished, peer reviewers had discovered a crucial flaw in the proof, leading him to spend another year revising it. During this time he collaborated with Richard Taylor, a former student of Wiles'. On September 19, 1994, one of the most momentous days in mathematical history, Wiles had the epiphanic stroke of insight necessary to finish off the proof, which he himself described as "the most important moment of [his] working life." In 1995, Wiles announced his corrected proof complete, and it was universally accepted, to the rejoicing of the mathematical community! Wiles himself was rewarded with knighthood, as well as the 2016 Abel Prize.

Almost a quarter of a century after the resolution of Fermat's Last Theorem, its notoriety still precedes it. Arguably more than any other problem in mathematical history, it serves to illustrate that looks can be deceiving. Just because a problem *looks* simple does not indicate that it actually is. We can, of course, also look at the situation more optimistically. Just because a problem looks difficult, that does not necessarily mean that it is - it may be a problem that falls prey to one creative trick![123]

---

[122]An *elliptic curve* is a curve given by the general equation $y^2 = x^3 + ax + b$ which is not self-intersecting. A *modular form* is an analytic function in the region of the complex plane given by $H = \{a + bi \mid b \geq 0, a, b \in \mathbb{R}\}$ (i.e. the *upper half-plane*) which satisfies a functional equation with respect to the group action of a *modular group*.

[123]Putnam problems, in particular, are great examples of these. But one could argue those *are* difficult anyway.

Chapter 17

Just for Fun...

As with any intense field of study, math can often be stress-inducing - even overwhelmingly so. But fear not; in this chapter we'll be taking a bit of a break from the frenzy of pure, abstract math and take a slightly less serious look at some bogus proofs, some entertaining factoids, and even some interesting paradoxes. So with that, let the fun begin!

17.1 The Sea of Pythagorean Theorem Proofs

The Pythagorean Theorem (or PT for short) is perhaps the most widely proven theorem in mathematical history. In fact, *entire books* have been written containing the *hundreds upon hundreds* of proofs of this deceptively simple theorem! Obviously, we can't and won't list every single one here, but we will analyze several of the most elegant ones[124] in this section.[125] (And we will be numbering them as we go along, so that you can attain some sense of how many there actually are - we're only scraping the surface here!)

17.1.1 Proof 1: Squares Along the Edges

Arguably the most well-known among the vast multitude of PT proofs, the "squares along side lengths" technique is commonplace in math classrooms to demonstrate the theorem, but the actual proof involves a great amount of geometric detail. Actually the first of Euclid's own two proofs of the PT, this proof begins by constructing three squares sharing edges with a right triangle, as in the familiar diagram below:

---

[124] At least in the authors' opinion, anyway.
[125] Which is so named because Hippasus was drowned in the *sea* after proving $\sqrt{2}$ was irrational. *Sea* what we did there? :P (Apologies for the bad punnery.)

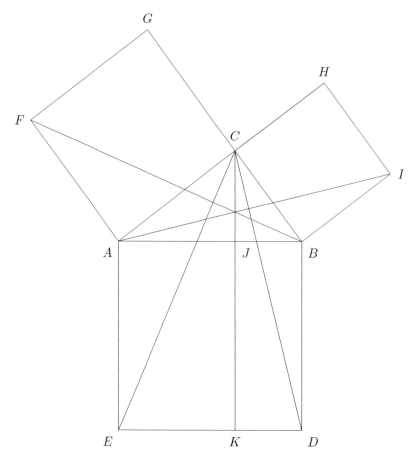

Figure 17.1: A typical depiction of the Pythagorean Theorem with labels and auxiliary lines drawn. We'll be going to town on this, don't worry!

In the figure above, we have labeled each vertex of each square, as well as two auxiliary points J and K and auxiliary line segments $\overline{AI}, \overline{BF}, \overline{CD}, \overline{CE}, \overline{CK}$.

Observe that $\triangle ABF$ and $\triangle AEC$ are SAS-congruent (since $AB = AE$, $AC = AF$, and $BAF = BAC + CAF = CAB + BAE = CAE$). Note that $\triangle ABF$ has base $\overline{AF}$ and altitude equal in length to $\overline{AC}$, so its area is $\dfrac{AF \cdot AC}{2} = \dfrac{AC^2}{2}$. Now consider $\triangle AEC$, which has base $\overline{AE}$ and altitude equal in length to $\overline{AJ}$. Hence, the area of $\triangle AEC$ is $\dfrac{AE \cdot AJ}{2}$, which is half of the area of rectangle $AEKJ$. Thus, the areas of $AFGC$ and $AEKJ$ are equal.

Similarly, the area of square $BCHI$ is equal to that of rectangle $JBDK$. Then the sum of the areas of $AEKJ$ and $JBDK$ is the area of square $ABED$, which is equal to AB^2. We can then conclude that $AC^2 + BC^2 = AB^2$, as desired.

17.1.2 Proof 2: Taking Advantage of the Binomial Expansion

Proof 2A: Tilted Square

Consider four "copies" of our right triangle, and arrange them as follows, with a square hole in the center:

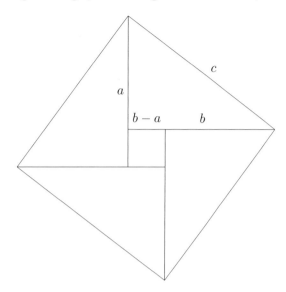

Figure 17.2: An illustration of the proof with four copies of an a-b-c right triangle.

Each triangle has area $\dfrac{ab}{2}$ (so that the total area of the four triangles is $2ab$), and the square "hole" has area $(b-a)^2 = b^2 - 2ab + a^2$; summing the areas yields $a^2 + b^2$. Note that the sum of the areas is also the area of the largest square with side length c, or c^2. This implies $a^2 + b^2 = c^2$, as desired.

17.1.3 Proof 2B: Tilted Hole

Begin with the same four copies of the right triangle. This time, the triangles should form a square with a side length of $a + b$, rather than c. (Figure on the next page.)

The area of the largest square is $(a+b)^2 = a^2 + 2ab + b^2$, which is also equal to $4 \cdot \dfrac{ab}{2} + c^2 = 2ab + c^2$. Setting these expressions equal yields $c^2 = a^2 + b^2$.

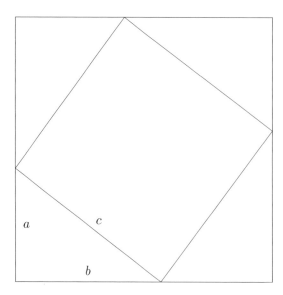

Figure 17.3: An illustration of Proof 2B with the same copies arranged to fit in a square of side length $a+b$, rather than c.

17.1.4 Proof 3: Constructing the Altitude

Begin with right triangle $\triangle ABC$, with $B = 90°$, and construct altitude $\overline{BD}$ (with D on hypotenuse $\overline{AC}$). Then $\triangle ABC$, $\triangle DAB$, and $\triangle DBC$ are all similar triangles. We then obtain the ratios $\dfrac{AB}{AC} = \dfrac{AD}{AB}$, $\dfrac{BC}{AC} = \dfrac{DC}{BC}$, so $AB^2 = AC \cdot AD$ and $BC^2 = AC \cdot DC$. Summing both equations yields $AB^2 + BC^2 = AC \cdot (AD + DC) = AC \cdot AC = AC^2$ as desired.

17.1.5 Proof 4: Drawing the Circumcircle

Circumscribe a circle with radius c about the right triangle (with the hypotenuse as the radius). Draw a diagram such as the following page. Henceforth, observe that $\triangle DAE$ is a right triangle, with $\overline{AC}$ as the altitude of the triangle. Then diameter $\overline{DE}$ of the circle is the hypotenuse of $\triangle DAE$, and is split into $\overline{DC}$ and $\overline{CE}$. It follows that $AC^2 = DC \cdot CE$, or $AC^2 = b^2 = (c+a)(c-a) = c^2 - a^2 \implies a^2 + b^2 = c^2$.

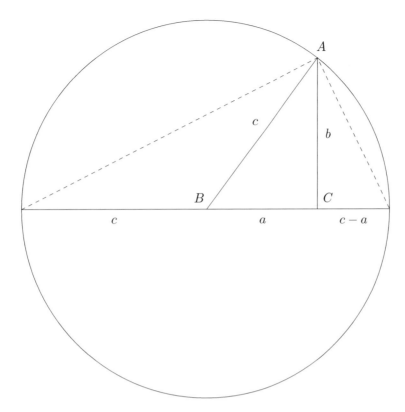

Figure 17.4: Drawing the circumcircle of the triangle for Proof 4.

17.1.6 Proof 5: The da Vinci Co- I Mean, Proof

Attributed to the great innovator Leonardo da Vinci, this particular proof invokes a similar strategy as our first proof in this showcase - namely, adjoining three squares onto the sides of our right triangle. From here, draw the line segments $\overline{DG}$ and $\overline{BH}$, which pass through B and O (the center of square $ACJI$) respectively. Observe that $ABH = 45°$; thus, $ABHI$, $JHBC$, $ADGC$, and $EDGF$ are all equal. Then $[ABHI] + [JHBC] = [ADGC] + [EDGF]$, from which the PT directly follows. (Diagram on the next page.)

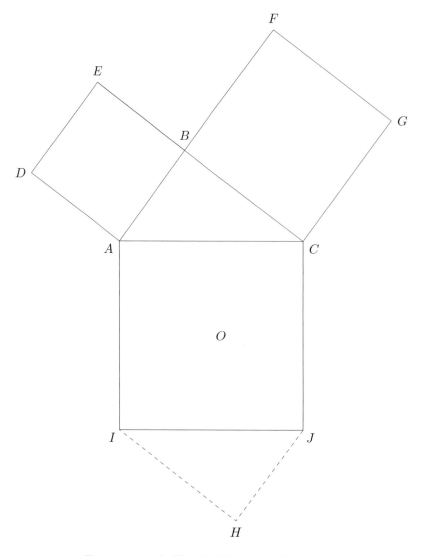

Figure 17.5: da Vinci's PT proof, illustrated.

17.1.7 Proof 6: Trivial by Ptolemy's Theorem

Exactly what it says on the tin. This proof is in a similar vein to Proof 5, where we construct the circumcircle of the triangle, except now we copy the right triangle onto its own hypotenuse (effectively reflecting it about the line $y = -x$) to form a rectangle with side lengths a and b, and diagonal length c. Ptolemy's Theorem states that the sum of the products of the opposite side lengths of the rectangle is equal to the product of the diagonal lengths, from which the desired result immediately follows.

Here's a diagram of our construction:

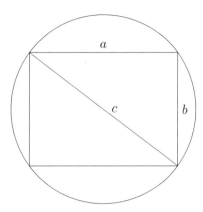

Figure 17.6: Proof of PT using Ptolemy's Theorem. By Ptolemy, $a \cdot a + b \cdot b = c \cdot c \implies a^2 + b^2 = c^2$.

17.1.8 Proof 7: Constructing the Incircle

Consider the following diagram of a right triangle with its incircle drawn:

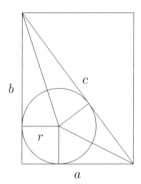

Figure 17.7: A right triangle with its incircle drawn, along with the inradii.

Let $s = \dfrac{a+b+c}{2}$ be the semi-perimeter of the triangle. Note that $c = (b-r) + (a-r) = b+a-2r$, or $r = \dfrac{a+b-c}{2} = s - c$. (Proving that c can be split into $b - r$ and $a - r$ is left as an exercise.) The area of the triangle is $\dfrac{ab}{2}$, but it is also equal to $rs = s(s-c)$. Thus, $(a+b+c)(a+b-c) = 2ab$, or equivalently $(a+b)^2 - c^2 = 2ab$ by the difference of squares identity. This finally simplifies to $a^2 + b^2 = c^2$, the Pythagorean Theorem.

17.1.9 Proof 8: Scaling the Triangle

Consider the right triangle with side lengths a, b, c. Let T_a be the triangle with side lengths a^2, ab, and ac; define triangles T_b and T_c analogously. (Note that all three triangles are similar.) Then let side length bc of T_c be the hypotenuse length of T_b, and let side length ac of T_c be the hypotenuse length of T_a (with c^2 being the hypotenuse length of T_c). The three triangles form a rectangle with an upper base length of $a^2 + b^2$, which is the same as its lower base length of c^2.

17.1.10 Proof 9: Heron's Formula

Proof 9A: Reflecting the Triangle

Reflect the triangle about its leg with length a, to obtain an isosceles triangle with side lengths c, c, and $2b$ (where $2b$ is the base length). Then the area of the isosceles triangle is ab. Using Heron's Formula, we can confirm this (using $s = b + c$):

$$\sqrt{(b+c)(b)(b)(-b+c)} = b\sqrt{c^2 - b^2} = ab$$

$$a = \sqrt{c^2 - b^2} \implies a^2 = c^2 - b^2 \implies a^2 + b^2 = c^2$$

Proof 9B: Derivative of Heron's Formula

In general, for a triangle with side lengths a, b, and c, Heron's formula says that its area A is given by

$$A = \frac{1}{4}\sqrt{(a+b+c)(a+b-c)(a-b+c)(-a+b+c)}$$

Instead of differentiating this expression directly with respect to c,[126] we can first drop the $\frac{1}{4}$ at the front, then rewrite A as $\sqrt{((a+b)^2 - c^2)(c^2 - ((a+b)^2)}$. Taking the derivative of the radicand (since it and its square root have the same zeros) in c yields $A' = 4c(a^2 + b^2 - c^2)$ via the Product Rule, and setting $A' = 0$ yields the Pythagorean Theorem.

---

[126] Here, we differentiate WRT c, as the angle C is an argument in the Law of Sines, which determines area.

17.1.11 Proof 10: Inscribed Semicircle

If we inscribe a semicircle in the triangle, such that we obtain a figure such as the following:

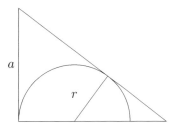

Figure 17.8: The inscribed semicircle of a right triangle.

then we obtain $\dfrac{r}{c-a} = \dfrac{a}{b}$ by triangle similarity, so $r = \dfrac{a}{b}(c-a)$. Now we apply Power of a Point to obtain $(c-a)^2 = b(b-2r)$, or $r = \dfrac{b^2 - (a-c)^2}{2b}$ after some simplification. Then $2a(c-a) = b^2 - (a-c)^2$, which yields $a^2 + b^2 = c^2$.

17.1.12 And One More for Good Measure!

As the swan song for this section, we present a sort of generalization of the Pythagorean Theorem to any triangle with angles α, β, γ. If we define a *signum function* $\text{sgn}(x)$ such that $\text{sgn}(x) = -1$ when $x < 0$, $\text{sgn}(x) = 0$ when $x = 0$, and $\text{sgn}(x) = 1$ when $x > 0$, then $\text{sgn}(\alpha + \beta - \gamma) = \text{sgn}(a^2 + b^2 - c^2)$ (where a, b, and c are the side lengths). Proving this we leave to you, the reader!

17.2 Paradoxes and Bogus "Proofs": The Art of Mathematical Deception

Bogus proofs are often perpetuated to trick unsuspecting victims into believing a non-existent paradox, or that the "performer" of the mathematical "trick" has somehow broken all of mathematics with a shocking discovery! Though they are primarily for entertainment purposes, they can be used to hone one's fallacy-finding skills, and to develop in a person a healthy amount of skepticism, which is a vital trait for any mathematician to have.

Similarly, paradoxes, though not outright trickery in the sense that false proofs are, often require approaching the problem at a different angle to resolve the apparent contradiction or inconsistency. Here is a compendium of several mathematical oddities ranging from the fairly common to the positively quirky and exquisite.

17.2.1 Unequal Numbers are Equal

The "unequal numbers are equal" bogus proof is one of the more prevalent absurd proofs. The variant that is most often presented is a fallacious demonstration that $2 = 1$:

Traditional false proof. Let $a = b$. Then multiply both sides by a to get $a^2 = ab$, and then subtract b^2 from both sides to obtain $a^2 - b^2 = ab - b^2$. It follows from $a^2 - b^2 = (a+b)(a-b)$ and $ab - b^2 = b(a-b)$ that $(a+b)(a-b) = b(a-b)$, and so $a + b = b$, implying that $2b = b$, or $2 = 1$ after dividing through by b. □

Here is another version of the $2 = 1$ "proof":

A variation on the "proof". Let $a = b$. Then $a^2 = ab$, so $a^2 + a^2 = a^2 + ab$, and $2a^2 = a^2 + ab$. We can then subtract $2ab$ from both sides to obtain $2a^2 - 2ab = a^2 - ab$; dividing both sides by $a^2 - ab$ yields the desired result, $2 = 1$. □

What Happens When You Aren't Careful with Your Operations...

In a similar vein is the "proof" that $5 = 4$, which has a mistake that is hidden a little more deeply. Can you spot the error?

False proof.
$$-20 = -20$$
$$25 - 45 = 16 - 36$$
$$5^2 - 45 = 4^2 - 36$$
$$5^2 - 45 + \frac{81}{4} = 4^2 - 36 + \frac{81}{4}$$
$$\left(5 - \frac{9}{2}\right)^2 = \left(4 - \frac{9}{2}\right)^2$$
$$5 - \frac{9}{2} = 4 - \frac{9}{2}$$
$$5 = 4$$

Thus, it seems that any number can be made equal to any other, simply by subtracting one side from itself to get an equality with 0, and then by multiplying by any real number! Have we broken mathematics?! □

No - we haven't. As always, there is a glaring hole in the "proof" that restores the balance of mathematics once and for all. The error here lies in the sixth line; it does not follow from $a^2 = b^2$ that $a = b$, as we could also have $a = -b$! Indeed, the equation $5 - \frac{9}{2} = \frac{9}{2} - 4$ would be correct.

Remark. A similar line of thought can be used to "show" that $2 + 2 = 5$, beginning from $2 + 2 = 4$; the fallacy here lies in treating the square root as a single-valued function instead of a dual-valued function (much like what we have done above).

Just Because it's Calculus, Doesn't Make it Right

For those less easily convinced, there is yet another variation on the $2 = 1$ "proof" that involves calculus on a surface level, but which contorts it in a way that is removed from all mathematical truth:

Derivative "proof". Consider the true equation $\underbrace{x + x + x + \ldots + x}_{x \text{ times}} = x^2$. Taking the derivative of the LHS and RHS (ordinarily a perfectly valid operation in solving a differential equation) yields $\underbrace{1 + 1 + 1 + \ldots + 1}_{x \text{ times}} = 2x$, but this becomes $x = 2x$, which implies $1 = 2$. $\square$

Here, x is not even a variable; it is a constant, but is nevertheless treated as a variable! The derivative of a constant is always zero, so the derivation step does not hold. Furthermore, in the first equation, the number of x terms depends on x, so we cannot take the derivative term-by-term.

Just Because it's Calculus, Doesn't Make it Right: Part 2

A more obscure paradox rooted in calculus revolves around the integration by parts formula. Recall that we have
$$\int uv = u \, dv - \int v \, du$$

If we substitute $u = \dfrac{1}{x}$, $dv = dx$, $du = -\dfrac{dx}{x^2}$, and $v = x$, then we obtain

$$\int \frac{1}{x} \, dx = \frac{1}{x} \cdot x - \int x \left(-\frac{1}{x^2}\right) dx$$

$$\int \frac{1}{x} \, dx = 1 + \int \frac{1}{x} \, dx$$

$$0 = 1$$

Whoops...we've made a terrible mistake! Or have we...? Have we actually broken math again?

No, we haven't broken anything (cue the sighs of relief)! As it turns out, the mistake is actually an extremely common first-year calculus student mistake: forgetting the constant of integration. If we make sure to include the constant after both integrals, then we obtain the correct equation

$$F(x) + c = 1 + F(x) + C \implies c = 1 + C$$

which allows for the equation to hold when the two sides differ by a constant.

Another resolution to this "paradox" is to take definite integrals, rather than indefinite integrals:

$$\int_a^b \frac{1}{x} \, dx = 1 \Big|_a^b + \int_a^b \frac{1}{x} \, dx$$

$$\int_a^b \frac{1}{x} \, dx = 0 + \int_a^b \frac{1}{x} \, dx$$

Here, the 1's on both sides vanish since we are taking the integral from a to b (and the integral of any constant over an interval is zero).

17.2.2 All Horses Are the Same Color

This is a classic example of a "paradox" that arises from a false premise: a logical leap that lies buried somewhat cleverly in the induction step. This, along with the afore-discussed "unequal numbers" paradoxes, is known as a *falsidical paradox,* which establishes a false result, as opposed to a *veridical paradox,* which establishes a true result that merely *appears* false, an *antinomy* which reaches an apparent self-contradiction through correct reasoning, and a *dialetheia,* which is a statement that is both true and false at the same time.

For this paradox, we want to try to show that all horses in a group of any number of horses will always have the same color.

"Proof" by induction.

Base case ($n = 1$): This is trivial.

Inductive step: Assume that some n horses all have the same color. We wish to show that $n + 1$ horses will also have the same color. Consider the set of every horse but the first; by the inductive hypothesis, these horses will all have the same color. Similarly, consider the set of every horse but the second; these horses will also be monochrome. Continue until the set of every horse but the final horse, which again has every horse the same color. Since the first horse, the last horse, and all horses in between are of the same color, by the Transitive Property, it follows that all horses in the set of $n + 1$ horses have the same color, which completes the inductive step. This "proves" that all horses in a set of n horses will have the same color for any number of horses n. □

The flaw here lies in the false assumption that there *will* exist at least one horse in between the first and last horse. If there are only two horses, then excluding the first and second horses will not show that they have the same color (since each horse always has the same color as itself; this says nothing about the color of the other horse). Hence, the reasoning of the "proof" is fundamentally flawed.[127]

17.2.3 Howlers: or Why Mathematicians Show Their Work

Keeping with the animal theme, mathematicians have named their more amusing errors *howlers;* probably after the howling laughter of an observer upon seeing the error, but perhaps also due to the "howling" nature of the glaring error. These errors, unlike paradoxes such as the "unequal numbers" and horse conundrums, yield a mathematically sound result, albeit through unsound means.

Anomalous Cancellations

An *anomalous* or *accidental* cancellation is an ill-advised trick where one strikes out identical digits in the numerator and denominator of an un-simplified fraction to obtain a correctly simplified form of the fraction. (We do not consider cancelling trailing zeros to be anomalous, since this is a perfectly valid operation.) This may seem quite appealing to the fraction newbie,[128] and may even be encouraging if one happens to use the right values for the numerator and denominator! Some famous examples of anomalous cancellations include $\frac{16}{64} = \frac{१\!\!\!6}{6\!\!\!4} = \frac{1}{4}$, $\frac{98}{49} = \frac{9\!\!\!8}{4\!\!\!9} = \frac{8}{4} = 2$, and $\frac{165}{462} = \frac{1\!\!\!6 5}{4 6\!\!\!2} = \frac{15}{42}$.

Let's now analyze what makes certain anomalous cancellations work miraculously, and others fall flat on their face. First we'll consider anomalous cancellations with 2-digit numbers. Consider the un-simplified fraction $\frac{\overline{ab}}{\overline{cd}} = \frac{10a + b}{10c + a} = \frac{b}{c}$, where a, b, c, d are base-10 digits. Whenever $a = c$, we require $a = b$; i.e. the numerator and denominator must be equal. Cases where $a \neq c$ can be tested via trial-and-error. Thus,

---

[127]You may be asking - and rightfully so - "But why *horses* specifically, of all animals?" The seemingly arbitrary choice is said to have come about from a 1930s figure of speech, "That's a horse of a different color!" Fun fact of the day!

[128]I like to call it the *elementary schooler's dream;* like the freshman's dream, only for fifth graders! Much like the freshman's dream (that's the so-called "identity" $(a + b)^n = a^n + b^n$), this can be true, but only in certain special cases. Of note is the related *sophomore's dream,* which unlike anomalous cancellation or the freshman's dream, is always true. For that matter, if we're working in a commutative ring of characteristic n, then the freshman's dream is true after all! But I digress...

we arrive at just four anomalous cancellations with value less than 1, and two digits in the numerator and denominator: $\frac{16}{64} = \frac{1}{4}$, $\frac{19}{95} = \frac{1}{5}$, $\frac{26}{65} = \frac{2}{5}$, and $\frac{49}{98} = \frac{4}{8}$. (Clearly, the reciprocals also work.)

We now generalize to all cancellations of the form $\frac{ap+b}{cp+a} = \frac{b}{c}$, where we assume WLOG that the digit a is the one being canceled, and a, b, and c are digits in base p (with $0 \leq a, b, c < p$). We can observe through some experimentation that, whenever the base p is prime, no correct anomalous cancellations exist. We can actually prove this, as follows:

Proof. The given form of a solution simplifies to $c(ap+b) = b(cp+a) \implies pc(a-b) = b(a-c)$ It follows that $p \mid b(a-c)$, yet $p > a, b, a-c$, so we must have $a = c$. However, we have shown above that this is impossible, since $a = b$ - contradiction! □

If p is not prime, what can we say about the number of solutions that do exist? As it turns out, the number of correct cancellations in a base p is odd iff p is an even perfect square:

Proof. Suppose that a solution exists in the form $\frac{\overline{ab}}{\overline{ca}} = \frac{b}{c}$ where $0 \leq a, b, c < p$. We again obtain $pc(a-b) = b(a-c)$, but here we suppose instead that $a > b, c$. Then observe that, if (a, b, c) is a solution, then so is $(a, a-c, a-b)$. If $b = a-c$, $c = a-b$, then substituting into our equation yields $p(a-b)^2 = b^2$, which forces p to be a perfect square; let $p = k^2$ for some integer k. This yields $ak = b(k+1)$. Furthermore, since k and $k+1$ are relatively prime, let $a = x(k+1)$, $b = kx$. $a, b < k^2$, so $x \in [1, k-1]$, which yields an odd number of solutions iff k is even, and so $p = k^2$ is an even perfect square. We can prove the converse by verifying that all of these solutions do indeed work. □

Furthermore, for anomalous cancellations with three or more digits in the numerator and denominator, under what conditions will there or will there not exist cancellations. If there do exist any, how many? What happens if we try anomalous cancellations with differing numbers of digits in the numerator and denominator - will this ever yield a satisfactory result? We leave you to explore this topic further and seek out the answers to all of these atypical, intriguing questions!

"Proving" the Cayley-Hamilton Theorem in Linear Algebra

The Cayley-Hamilton Theorem is covered in section §6.8, "Matrix Algebra, Part 2." We have re-stated the theorem here for convenience:

> **Theorem 17.1: Cayley-Hamilton Theorem**
>
> Let A be an $n \times n$ matrix, and let I_n be the $n \times n$ identity matrix. The *characteristic polynomial* of A is given by $p(\lambda) = \det(\lambda I_n - A)$. Then $p(A) = 0$.

As a refresher, det denotes the determinant of the matrix $\lambda I_n - A$, where λ is a scalar constant. This determinant will always be a degree-n polynomial in λ.

Bogus proof. Simply substitute A in place of λ to get $p(A) = \det(AI_n - A) = \det(A - A) = 0$. □

The fault here lies in the fact that A is an $n \times n$ matrix, but λ must be a scalar! If we set λ equal to a matrix, we would obtain an un-parseable expression that we could not even interpret at all, let alone compute the determinant of. This is because $p(\lambda)$ is in fact a *polynomial* that takes real values of λ and outputs the determinant of the matrix $\lambda I_n - A$; this polynomial cannot take a matrix as input. Hence, it makes no sense to say anything about the determinant of $AI_n - A$.

As a concrete example, let $A = \begin{bmatrix} 0 & 1 \\ 2 & 3 \end{bmatrix}$:

$$p(A) = \det(AI_n - A) = \det \begin{bmatrix} \begin{bmatrix} 0 & 1 \\ 2 & 3 \end{bmatrix} - 0 & -1 \\ -2 & \begin{bmatrix} 0 & 1 \\ 2 & 3 \end{bmatrix} - 3 \end{bmatrix} \quad (17.1)$$

but this is a nonsensical expression, since we cannot subtract a scalar from a matrix.

Furthermore, assuming the proof were true, a pair of even more absurd results would follow.

> **Lemma 17.1: Faulty Singularity Lemma**
>
> Let A and B be $n \times n$ matrices. Let p_A be the characteristic polynomial of A. If $B - A$ is singular (non-invertible, i.e. $\det(B - A) = 0$), then B is a zero of p_A.

Bogus proof. $p_A(B) = \det(BI_n - A) = \det(B - A) = 0$. □

Again, this suffers from the fatal flaw of trying to take the determinant of a nonsensical expression, namely $BI_n - A$. We cannot multiply the identity matrix by another matrix B, then subtract a scalar and expect a meaningful result!

In particular, this "lemma" has an equally-nonsensical corollary:

> **Corollary 17.1: Faulty Nilpotence Corollary**
>
> Every singular matrix is nilpotent.[a]
>
> ---
> [a] Recall that a *nilpotent* matrix A is a matrix such that, for some positive integer n, A^n is the zero matrix.

Bogus proof. Let A be the zero matrix, and let B be a singular matrix. We have $p_A(\lambda) = \lambda^n$, as well as $p_A(B) = 0$ (since $B - A$ is singular). Thus, $B^n = 0$ and B is nilpotent. □

It is certainly not true to say that for every matrix whose determinant is zero, it will be nilpotent! *Any singular matrix with positive elements is a valid counterexample.*

17.2.4 The Missing Square Paradox

Imagine that you have a square, 8 × 8 chocolate bar. Pretend that you absolutely love chocolate, and want to find a way to generate as much as you could ever want from a single $1 convenience store bar. Can you find a way to do it? Apparently so! You decide to break up the bar in the following configuration:

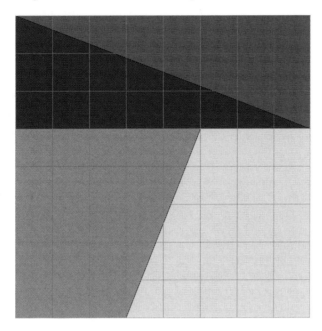

Figure 17.9: Dividing the original chocolate bar into four pieces to be rearranged.

While doing this, though, reservations start to creep into your head. You can just feel the rational part of your brain screaming at you, "No, that's not going to work! How is it possible to just make a square of chocolate appear out of thin air just by rearranging pieces?!"

And your brain would be right. Surely this is no surprise! But at first glance, this is a decently convincing demonstration of an apparent way to attain infinite resources and destroy the global economy - until one takes a closer, more analytical look.

Supposedly, we can rearrange the pieces to produce the following 5 × 13 chocolate bar. In reality, however, it will end up looking like this:

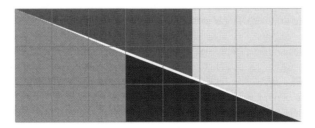

Figure 17.10: The rearranged, 5 × 13 chocolate bar (drawn to scale).

By this same faulty logic, it is also possible to create an irregular shape with an area of only 63, seemingly destroying a grid square's worth of chocolate in the process:

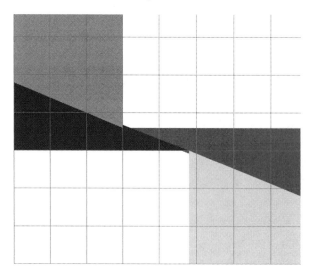

Figure 17.11: Dividing the original chocolate bar into four pieces to be rearranged.

But at the end of the day, remember the Law of Conservation of Energy: energy can neither be created nor destroyed, only transformed or transferred! The chocolate bar is no different: we cannot magically create chocolate out of nothing, nor can we destroy it without eating it. Notice that, in Figure 17.10, there is a sliver of chocolate in the middle that is absent. As it turns out, this sliver has area exactly 1, which accounts for the discrepancy. This is because the line from $(0,5)$ to $(5,3)$, and the line from $(0,5)$ to $(8,2)$, do not have the same slope (and similarly, the lines from $(13,0)$ to $(8,2)$ and to $(5,3)$).

As an extra challenge, for what side lengths of the rectangle is the missing area exactly 1, or some other integer? Can you prove that the missing area is an integer for only these values?

A 3-Dimensional Generalization: The Banach-Tarski Paradox

A classic example of a veridical paradox - an apparent paradox that is actually true and logically sound - is the *Banach-Tarski Paradox,* which is actually known as the Banach-Tarski *Theorem* in the mathematical literature! This theorem is a generalization of the missing square chocolate bar problem in two dimensions to any solid sphere in 3-dimensional Euclidean space.

> **Theorem 17.2: Banach-Tarski Paradox**
>
> Given a solid ball (the union of a sphere and its boundary) B in $\mathbb{R}^3$, there exists a decomposition of B into disjoint subsets $B_1, B_2, B_3, \ldots, B_i \subset B$, which can then be re-assembled to form two identical copies of the original ball B.

A stronger version of the theorem states that this can also be achieved with any two "reasonably close" solid objects (for instance, a sphere the size of a pea and a sphere the size of the Sun), which is why the paradox is also known (informally) as the *pea and the Sun paradox.*

Though the Banach-Tarski Theorem is called a "paradox" due to its extremely counter-intuitive nature, and its first impressions of being absurd, it can in fact be proven rigorously using methods such as the axiom of choice. Under ordinary contexts, translations ought to preserve volume; however, if we consider the pieces of the original ball as not rigid, solid shapes, but collections of points with no well-defined volume, then the axiom of choice allows for a construction of a non-measurable set.

For now, we will examine the weak version of the theorem (that a ball in $\mathbb{R}^3$ can be duplicated to two balls of the same size). We can use group theory as the instrumental tool in defining the theorem statement. Let G be a group acting on $\mathbb{R}^3$ such that G consists of all isometries of $\mathbb{R}^3$ (i.e. transformations of $\mathbb{R}^3$ onto itself that preserve distance). If two pieces can be transformed into each other, we say they are *congruent*. Call two subsets $A, B \subset X$ *G-equidecomposable* iff A and B can both be partitioned into the same number of G-congruent pieces. (This defines an equivalence relation among the subsets of X; proving this is left as an exercise to the reader.) We can finally restate the theorem as follows:

> **Theorem 17.3: Restatement of Banach-Tarski Theorem**
>
> A ball $B \in \mathbb{R}^3$ is equidecomposable with two copies of itself. Furthermore, B must consist of at least five disjoint pieces.

17.2.5 Zeno's Paradoxes

You might be familiar with the ancient Greek philosopher Zeno of Elea - famed for his mind-boggling and seemingly impossible conundrums. Though none of Zeno's works remain in written form, other philosophers, such as Aristotle, have thankfully rephrased and reconstructed Zeno's writings. It's a good thing, too - these are some of the most sense-defying paradoxes yet (and in a good way)!

Achilles and the Tortoise

This is probably the most well-known of Zeno's paradoxes by far; it involves Achilles (ironic considering his weak spot was his heel!) and a tortoise in a 100-meter footrace. Achilles allows the tortoise a headstart - of, say, 10 meters - and begins to run at ten times the tortoise's speed. In some finite amount of time, Achilles will reach the 10 meter mark, but in that same time, the turtle will move 1 meter further, to 11 meters. But by the time Achilles catches up to the turtle at 11 meters, the tortoise will have gone 0.1 meters further. Once Achilles reaches 11.1 meters, the tortoise will be ever so slightly ahead of him, at 11.11 meters, and so on, continuing indefinitely.

Of course, the reader who is well-versed in geometric series will suspect instantly that something is awry. The paradox is no paradox at all - all that is happening is that time is being slowed down by a factor of 10 each time! If we let time run at its normal speed, Achilles and the tortoise will reach a common point at $\frac{100}{9}$ meters; beyond that, Achilles will be far ahead of the tortoise, and will easily win the race. If we examine the amount of time that has elapsed between our "stillshots" of the race, then we notice that - assuming we set Achilles' speed to 1 meter per second and the tortoise's to 0.1 meters per second - the first part of the race takes 10 seconds. The second part takes 1 second, the third part takes 0.1 seconds, and so forth. Eventually, these values do sum to a finite amount of time, namely $\frac{100}{9}$ seconds into the race. So ultimately, there is no paradox here; just a slowing down of time until an infinitesimally small amount of time passes, thereby not allowing Achilles and the tortoise to actually go beyond that instant in time where the two meet.

The Dichotomy Paradox

The *dichotomy paradox* is very similar to the Achilles and the tortoise paradox, in that it, too, entertains the idea of motion being limited and an illusion. Suppose Homer (Simpson, if you like) is headed to the donut shop. But before he can get there, he must first get halfway there. Before he can get *there*, he first needs to get one-quarter of the way there, and so on. This requires Homer to perform an infinite number of motions, which Zeno maintains is impossible; hence, Homer can never reach his beloved donut shop. ("D'oh!") By extension, he can never get anywhere at all!

Like with the Achilles/tortoise paradox, all that is really happening is that time is being slowed down, and the "paradox" lies in the false assumption that an infinite number of terms in sequence cannot sum to a finite value. In fact, the very idea of limits may well have been established to combat such paradoxes!

Is Motion an Illusion? The Arrow Paradox

The *arrow paradox,* or *fletcher's paradox,* also seems to imply that motion as a whole is impossible, and is really just an illusion. Consider an arrow that is flying in the air. At any given moment in time, the arrow is neither moving to where it already is, nor moving to where it is not (since no time elapses in that instant). Since time consists of an infinite number of instants, it is not moving at all, and so all motion is but an illusion. Here, the paradox stems from dividing time into an infinite number of points - but this is much like saying that

$$\lim_{n \to \infty} \frac{1}{n} \cdot n = 0$$

since $\frac{1}{n}$ tends to zero as n grows arbitrarily large!

17.2.6 The Interesting Number Paradox

We close out this chapter with a rather arbitrary and self-referential paradox: the *interesting number paradox*. The paradox first very loosely defines *interesting* and *dull* natural numbers according to our intuitive definitions of those words, without any objective mathematical backing. It then states that all natural numbers are interesting; the proof is by contradiction.

Proof. Suppose otherwise - that there exists at least one "dull" natural number. Then the smallest dull number would be interesting in and of itself, which is a contradiction. □

But what do we really mean by "interesting" or "dull"? One workaround to the lack of rigor in this paradox is to define an *interesting* natural number as one that makes an appearance in the Online Encyclopedia of Integer Sequences (OEIS); under this definition, the smallest dull number is 17,843. Perhaps we could also define an interesting number based on its presence on other popular websites, such as Wikipedia's repository of articles on the integers and "What's special about this number?" Under the former definition, 262 is the smallest dull number (in that it doesn't have its own, dedicated article; however, it is nested within the article for the number 260), and under the latter, 391 is the smallest dull number (since it does not have a special property listed on the site).

A more interesting (pun intended) definition has an origin story: G. H. Hardy riding in a taxi cab to see the ill Ramanujan and noticing its number, 1729. Hardy supposedly remarked to Ramanujan that the number 1729 was completely uninteresting - so much so that he feared it was a bad omen for Ramanujan's health - but Ramanujan replied that 1729 is actually very interesting, as it is the smallest natural number that can be written as the sum of two positive cubes in two different ways (this also explains why such numbers are referred to as *taxicab numbers*).

Yet, at the same time, one could also make the counter-argument that, if every number is interesting, then no number is interesting at all, since by definition, an interesting number must not have any property that some set of other numbers do. If we define an interesting number more leniently by saying that an interesting number only need not have any property that is common to *all* other natural numbers, then our paradox remains intact, but if we say that each and every interesting number must be truly unique, then the idea of interesting and dull numbers becomes doubly paradoxical!

Part VII

End Matter

Closing Remarks

Well, congratulations on making it to the end! We hope you've had an absolutely wonderful, magical, thrilling, and most of all, meaningful, experience! Whether you came in a first-timer to competition math (or math in general), or a seasoned veteran, it's been a pleasure taking you along with us on the glorious journey that is mathematics!

Keep in mind, too - *this is only the beginning*. If this book is a river of knowledge, then mathematics as a whole is the Marianas Trench. We hope that we've given you a welcoming, inspiring, and motivating introduction to the proof-based side of math, and competitive math in general.

But besides that, we hope that our mathematical journey has also given you a deeper understanding of, and appreciation for, the world around you as a whole. Not only is life on Earth teeming with bottomlessly fascinating interactions, so is the entire universe! Nothing could be more natural than mathematics - the driving force behind the sciences, the humanities, the arts, and all of nature. Even on a purely humanistic level, the understanding of natural phenomena may very well finally get you more motivated to begin actively seeking out an answer to the question many of you have probably had since the very beginning of your school days: *When will I ever need this in my life?*

Indeed, it is through the *applications* of math to nature - not so much the formulas themselves, but their motivations and what makes them work together so seamlessly, as well as the thought and logical organization that goes into understanding what exactly makes it all work, that is truly the most useful in life. Though it is certainly true, to an extent, that simply memorizing formulas will not prove immediately or obviously useful, it is what led you to truly *understand* them, along with the drive, resilience, tenacity, courage, willpower, and thirst for knowledge you channeled to fine-tune that understanding, that will get you farthest in the end.

Closing Exam

At last, it's time for the Closing Exam! This is an opportunity for you to determine how much you've learned along our mathematical journey.

Directions: Answer all of the following 15 questions, which you will have 4 hours to complete. Unless otherwise specified, all answers must be complete, simplified, and fully justified with supporting work and mathematically accurate reasoning. (The difficulty level is intended to aggregate everything we have learned; as such, you can consider this as roughly Intermediate-level in general. Note that problems are not necessarily in increasing order of difficulty, nor are the parts within a numbered problem.)

The test begins on the following page. It is recommended to either print this test out and write your answers on separate scratch paper, or type your responses out and score them using the scoring rubric after you are finished.

Note: Throughout this test, "compute", "find", or "evaluate" all mean that you should provide a numerical answer *along with justification*. A correct answer without supporting work will not receive credit.

Finally, you may be wondering why some letters are randomly bolded throughout the test. No, this isn't a printing error - it's intended to be that way. Best of luck!

1. (12 pts.) Vincent and his twin brother Paul are both very interested in twin primes! They all notice the following properties about twin primes, and your job is to either prove or disprove their observations.

    (a) Vincent and Paul claim that there does *not* exist a pair of prime numbers that differ by 7. Are they correct? If so, provide a proof of the statement. If not, provide a counterexample.

    (b) The twins then claim that $(3, 5, 7)$ is the only triple of consecutive odd prime numbers. Are they correct?

    (c) Vincent and Paul have found the largest positive integer that divides the sum of any two primes greater than 3 in a twin prime pair. With proof, find this integer.

2. (10 pts.) If x and y are complex numbers and z is a **real** number such that $x+y+z = 3$, $x^2+y^2+z^2 = 7$, $xy = 5$, and $z > 0$, compute the positive imaginary part of $x - y$. Full justification is required.

3. (13 pts.) **Yo**landa has an ample supply of 2-inch-tall blocks, 3-inch-tall blocks, and 5-inch-tall blocks. She wishes to use exactly ten of these blocks to create a stack of blocks. Each block is randomly oriented, with an equal probability of having a height of 2 inches, 3 inches, or 5 inches. (She also happens to know exactly how to balance the tower so as to prevent any blocks from falling over.) If she selects each block uniformly at random, what is the probability that her stack is exactly 30 inches tall? Justify your answer.

4. (10 pts.)

    (a) State and prove the Angle Bisector Theorem.

    (b) $\triangle ABC$ has $AB = 13$, $BC = 14$, and $CA = 15$. Points D, E, and F lie on $\overline{AB}$, $\overline{BC}$, and $\overline{CA}$ respectively such that $\overline{CD}$, $\overline{AE}$, and $\overline{BF}$ bisect angles C, A, and B, respectively. Compute the area of $\triangle DEF$ as a common fraction.

5. (15 pts.)

    (a) Give a careful and complete statement of the Power of a Point theorem. Be sure to include all cases, not just one.

    (b) Using any method, prove the Intersecting Chords Theorem.

    (c) Using Power of a Point, prove Euler's relation: In a triangle with circumcenter O, circumradius R, incenter I, and inradius r, $OI^2 = R(R - 2r)$.

6. (10 pts.) Evelyn and Lou are playing a game in which Evelyn has the first turn. She picks a number between 1 and 100, inclusive, and Lou, on his first turn, divides that number by one of its prime factors. Evelyn then divides the result by another one of its prime factors (not necessarily distinct from the first choice). The winner is the first person to reach either 1 or a perfect square. For how many initial choices does Evelyn have a winning strategy? (It may be helpful to know that there are 25 prime numbers between 1 and 100, inclusive.)

7. (15 pts.)

    (a) Determine all solutions x to the equation $\left(x^3 + \frac{15}{16}x\right)^3 + \frac{15}{16}\left(x^3 + \frac{15}{16}x\right) - x = 0$ over the real numbers.

    (b) Show that these solutions include all possible real values of x.

8. (10 pts.) Kendra draws $\triangle ABC$ with $AB = a$, $BC = b$, $CA = c$. At vertices A, B, and C, Kendra draws a circle centered at the vertex with radii $|b - c|$, $|a - c|$, $|a - b|$. She then wraps a piece of string as tightly around the circles as possible. If the length of Kendra's piece of string is 2019, with proof, compute the number of possible integer ordered triples (a, b, c).

9. (10 pts.) Call a graph G on v vertices *tropical* if, for any vertex $V_i \in G$, the distance between V_i and another vertex V_j is bounded below by 1 and above by 2.

    (a) Let G_B be a bipartite graph consisting of vertices in G. When is G_B tropical?
    (b) Show that the complete graph on n vertices K_n is tropical for all natural numbers n.

10. (15 pts.) **So**crates (no, not the philosopher) is selling video games in bulk. He is tracking his daily sales figures, when he notices some peculiarities about them...

    (a) Let $f(n)$ be the number of copies that Socrates sells on day n. He notices that $f(1) = 1$, $f(2) = 2$, and for all positive integers $n \geq 3$, $f(n) = f(n-1)^2 + f(n-2)^2$. Is there an ordered pair of positive integers (m, k) for which $f(n)$ is k-periodic (mod 10) for $n \geq m$; i.e. $f(n) \equiv f(n+k)$ (mod 10) for all $n \geq m$? If so, find all such pairs; if not, prove that no such pairs exist.
    (b) Socrates is entering his sales figures into a spreadsheet program, but alas, he is using outdated technology that only stores the last digit of each day's sales figure. *According to the spreadsheet,* how many games does Socrates sell in the first 2019 days?
    (c) For what positive integer values of $n \leq 100$ is Socrates' cumulative total, up to and including the nth day, a multiple of 15?

11. (18 pts.) Let M be a $k \times k$ **matrix** with real entries (i.e. $M \in \mathcal{M}_{k \times k}(\mathbb{R})$).

    (a) For all $k \in \mathbb{N}$:
        i. Show that $\det(c \cdot M) = c^n \cdot \det M$ for all $n \in \mathbb{N}$, $c \in \mathbb{R}$.
        ii. Show that iff M is singular, then M^T is singular.
        iii. Let $\text{adj}(M)$ denote the adjoint matrix of M. Prove that $M\text{adj}(M) = \det M \cdot I_k$.
    (b) i. Briefly describe the dot product of the column vectors of a 2×2 matrix M in terms of the column vectors.
        ii. Briefly explain why the dot product is only a valid operation in two arguments.
    (c) What is the volume of the parallelepiped with edges given by the vectors $\mathbf{u} = \langle 0, 2, 0 \rangle$, $\mathbf{v} = \langle 1, 9, 0 \rangle$, $\mathbf{w} = \langle 4, 0, 5 \rangle$?

12. (12 pts.) Let $z_1 = a + bi$ and $z_2 = c + di$ be complex numbers with $a, b, c, d \in \mathbb{R}$ such that $a, c > 0$ (where $i = \sqrt{-1}$).

    (a) Set up (but *do not solve*) an equation to represent the values of a, b, c, d that satisfy the equation $\arg z_1 + \arg z_2 = \arg(z_1 + z_2)$.
    (b) As much as possible, simplify the equation that relates resulting from $\tan(\arg z_1) + \tan(\arg z_2) = \tan(\arg(z_1 + z_2))$ in terms of a, b, c, d.

13. (15 pts.) Circles C_1 and C_2 are drawn in the plane. Respectively, they have radii r_1 and r_2, and centers at O_1 and O_2. Construct the two external tangents to C_1 and C_2, and let P be their intersection point.

    (a) Show that P, O_1, and O_2 are collinear unless $r_1 = r_2$ (in which case P is undefined).
    (b) Prove Monge's Theorem:

    > **Theorem 17.4: Monge's Theorem**
    >
    > If we have three circles in the plane, and draw the external tangents to each pair of circles, then the three points of intersection of each pair of external tangents will be collinear.

14. (15 pts.)

    (a) i. Describe the generating function associated to the outcome of a fair six-sided die (with labels 1-6, inclusive).

ii. Describe the generating function associated to the outcome of an *unfair* six-sided die (with labels 1-6, and probability of coming up with the side labeled n proportional to the value of n).

iii. Describe the generating function associated to the outcome of three fair six-sided dice (with labels 1-6, inclusive) being rolled simultaneously.

(b) Suppose you flip a fair coin 100 times. Let N be the number of heads flipped. The probability that N is divisible by 3 is equal to $\frac{1}{3} - \epsilon$ for some positive real number ϵ. Find ϵ.

(c) Prove that, if $f(x)$ and $g(x)$ are the generating functions of two independent random variables X and Y, respectively, then $f(x)g(x)$ is the generating function corresponding to $X + Y$.

15. (20 pts.) Let $\triangle ABC$ have points D, E, F on $\overline{AB}$, $\overline{BC}$, $\overline{CA}$ respectively.

(a) Show that the circumscribing circles of $\triangle ADF$, $\triangle BDE$, and $\triangle CEF$ will pass through a common point P. Prove any theorems cited (or you may lose marks!)

(b) Let G, H, and I be the circumcenters of $\triangle ADF$, $\triangle BDE$, and $\triangle CEF$, respectively. Prove that $\triangle ABC$ and $\triangle GHI$ are similar, i.e. prove that $\triangle ABC$ and $\triangle GHI$ are spiral similar about P.

End of exam.

Solutions to Exams

Beginner Exam

1. (a) A permutation of a set is an arrangement of its elements with respect to order. (More precisely, it is a bijection from itself to itself.)

    (b) S has $\frac{4!}{2!} = \boxed{12}$ permutations, since there are $n!$ ways to rearrange n elements, but we must divide by $2! = 2$ to account for the two C's being indistinguishable.

    (c) T has $\frac{8!}{2!4} = \frac{8!}{16} = \frac{7!}{2} = \boxed{2520}$ permutations.

2. (a)
    - $\forall$: the "for all" quantifier.
    - $\exists$: the "exists" quantifier.
    - $\in$: denotes membership in a set.
    - $\supset$: superset of.
    - $\subset$: subset of.
    - $\emptyset$: the empty set (set with no elements).
    - $\mathbb{R}$: set of real numbers.
    - $\mathbb{Z}$: set of integers.
    - $\mathbb{Q}$: set of rational numbers.
    - $\mathbb{C}$: set of complex numbers.

    (b) Yes. By definition, if a is rational, then it can be written as the quotient of two integers. Indeed, *for all* rational numbers a, *there exist* b, c with $a = \frac{b}{c}$.

    (c) No. This does not hold for negative a, since if $b, c > 0$ (as implied by $b, c \in \mathbb{N}$), a would be positive as well. Hence it does not hold for *all* a, which contradicts the statement.

3. (a) Let $S(n)$ be the statement for some $n \in \mathbb{N}$. $S(1)$ clearly holds (base case). Assume $S(n)$, and show $S(n+1)$. We have $\sum_{k=1}^{n+1} k = \sum_{k=1}^{n} k + (n+1) = \frac{n^2 + n}{2} + (n+1) = \frac{n^2 + 3n + 2}{2} = \frac{(n+1)(n+2)}{2}$ so $S(n+1)$ is true and the inductive step is complete.

    (b) Same approach as above: $S(1)$, the base case, is trivial, and show $S(n+1)$ assuming $S(n)$. Adding $(n+1)^2 = n^2 + 2n + 1 = \frac{6n^2 + 12n + 6}{6}$ to $\frac{n(n+1)(2n+1)}{6}$ yields the desired result.

    (c) Yet again, use induction (it may help to first expand $\frac{n^2(n+1)^2}{4}$). This is largely left as an exercise to the reader.

    (d) The general formula for the sum is
    $$\sum_{k=1}^{n} k(k^2 - k + 1) = \frac{n(n+1)(3n^2 - n + 4)}{12}$$

Here, we need not use induction at all; simply expand the expression to obtain

$$\sum_{k=1}^{n} k^3 - k^2 + k$$

and apply linearity of the sum to obtain

$$\sum_{k=1}^{n} k^3 - k^2 + k = \frac{n^2(n+1)^2}{4} - \frac{n(n+1)(2n+1)}{6} + \frac{n(n+1)}{2}$$

$$= \frac{3n^2(n+1)(n+1) - 2n(n+1)(2n+1) + 6n(n+1)}{12}$$

$$= \frac{(n+1)\left(3n^2(n+1) - 2n(2n+1) + 6n\right)}{12}$$

$$= \frac{n(n+1)\left(3n(n+1) - 2(2n+1) + 6\right)}{12}$$

$$= \frac{n(n+1)(3n^2 - n + 4)}{12}$$

as desired.

4. (a) By the Pythagorean Theorem, the diagonal length of $ABCD$ is $\sqrt{5}$, which is also the diameter of O (since $ABCD$ is cyclic). Then the radius of O is $\frac{\sqrt{5}}{2}$, so its area is $\boxed{\frac{5\pi}{4}}$.

(b) If $ACP = 15°$, then $PCB = 30°$, since $ACD = 45°$ and BCD is right (in turn, because $\overline{AC}$ is a diagonal of $ABCD$). It follows that $BPC = 60°$ and that $\triangle PCB$ is a 30-60-90 right triangle, with $BC = 1$ opposite $P = 60°$. Thus, $PB = \frac{1}{\sqrt{3}}$ and the area of $\triangle PBC$ is $\frac{1}{2\sqrt{3}} = \frac{\sqrt{3}}{6}$. Subtracting this from 2, the area of $ABCD$, yields the area of trapezoid $APCD$, $\boxed{\frac{12 - \sqrt{3}}{6}}$.

(c) We have $d_0 = \sqrt{5}$, $d_1 = \sqrt{2^2 + \sqrt{5}^2} = \sqrt{9} = 3$, $d_2 = \sqrt{\sqrt{5}^2 + 3^2} = \sqrt{14}$, $d_3 = \sqrt{3^2 + \sqrt{14}^2} = \sqrt{23}$, and so forth. In general, $d_i^2 = d_{i-1}^2 + d_{i-2}^2$ for all $i \geq 2$. Then the u_i follow a pattern that repeats with period 20 (for completeness purposes: 0, 4, 4, 3, 2, 0, 2, 2, 4, 1, 0, 1, 1, 2, 3, 0, 3, 3, 1, 4, 0, 4, 4, …). (Interestingly, for any starting point (u_0, u_1), we will eventually observe the beginnings of the Fibonacci sequence. Can you figure out why?)

(d) Since the sum repeats in "blocks" of 20, and each block contributes 40 to the total, we can add 101 blocks, then subtract 4, to obtain $40 \cdot 101 - 4 = \boxed{4036}$.

5. (a) The expansion of $(x+y)^n = \binom{n}{0}x^n + \binom{n}{1}x^{n-1}y + \binom{n}{2}x^{n-2}y^2 + \ldots + \binom{n}{n-1}xy^{n-1} + \binom{n}{n}y^n$

$$= \sum_{k=1}^{n} \binom{n}{k} x^k y^{n-k}$$

(b) The coefficient is $\binom{5}{2} = \binom{5}{3} = \boxed{10}$.

(c) Since we have

$$(ax + by)^n = a^n x^n + \binom{n}{1} a^{n-1} b x^{n-1} y + \ldots + \binom{n}{n-1}$$

the coefficient of $x^{20} y^{18}$ is given by $\boxed{20^{20} 18^{18} \binom{38}{18}}$.

6. (a) The number of possible codes is $10^4 = \boxed{10,000}$, by the Multiplicative Principle.

   (b) The number of codes becomes $10 \cdot 9 \cdot 8 \cdot 7 = \boxed{5,040}$.

   (c) Observe that this is equivalent to finding how many sequences contain all 0's before all 1's, which come before all 2's, and so forth, with all 9's coming last. Let the digit d appear n_d times in a length 4 password. This is essentially a stars-and-bars problem, where the number of sequences (n_1, n_2, n_3, n_4) of digits from 0-9 such that $n_1 + n_2 + n_3 + \ldots + n_9 = 4$ is $\binom{13}{4} = \boxed{715}$.

7. Note that $x^4 + y^4 = (x^4 + 2x^2y^2 + y^4) - 2x^2y^2 = (x^2 + y^2)^2 - 2(x^2y^2)$, so we want to maximize $x^2 + y^2 = (x+y)^2 - 2xy$ while minimizing $x^2y^2 = (xy)^2$. (To maximize $x^2 + y^2$, in turn, we maximize $x + y$ and minimize xy.) Given that $(x^2y^2 + 1)(x^2y^2 - 1) = 48$, we have $(x^2y^2)^2 = 49 \implies x^2y^2 = 7$ (as x, y are real numbers, $x^2y^2 = (xy)^2$ cannot equal -7). By AM-GM, $x + y$ is maximized when $x^2 = y^2 = \sqrt{7}$, or $x^2 + y^2 = 2\sqrt{7}$ and $x^2y^2 = 7$. Finally, $x^4 + y^4 \leq (2\sqrt{7})^2 - 2(7) = \boxed{14}$.

Intermediate Exam

1. (a) By Vieta's formulas, the product of the roots of f is 1, so the other root is $\boxed{\dfrac{1}{17}}$.

   (b) In general, by the same logic as above, the other root will be $\boxed{\dfrac{1}{r}}$.

2. (a) A quadratic residue modulo p is an integer in the interval $[0, p-1]$ that is congruent to a perfect square modulo p. A quadratic nonresidue is an integer in the interval $[0, p-1]$ that is not a quadratic residue.

   (b) The set of QRs mod 19 is $\{0, 1, 4, 5, 6, 7, 9, 11, 16, 17\}$, and the set of QNRs is its complement in the universal set $\{0, 1, 2, \ldots, 18\}$, i.e. $\{2, 3, 8, 10, 12, 13, 14, 15, 18\}$.

   (c) The Legendre symbol (with top argument q and bottom argument p for prime p) returns 1 if q is a QR mod p, -1 if q is a QNR mod p, and 0 if $q = 0$. Note that the Legendre symbol is defined only for prime p.

   The Jacobi symbol, on the other hand, is the extension of the Legendre symbol to all odd integers p, not just primes.

   (d) The Law of Quadratic Reciprocity states that
   $$\left(\frac{q}{p}\right)\left(\frac{p}{q}\right) = (-1)^{\frac{(p-1)(q-1)}{4}}$$
   where $\left(\dfrac{q}{p}\right)$ in this context denotes the Legendre symbol.

3. (a) The answer is $\boxed{\text{never}}$. Since p is prime, $\tau(p^p) = p + 1$ (directly from the prime factorization). But $\tau(p + 1^{p+1})$ has a prime factorization with at least two distinct prime factors, as $p + 1$ is composite if p is prime. The product of these prime factors is guaranteed to be greater than $p+1$, by AM-GM. Hence no such primes p exist that satisfy the inequality.

   (b) The answer is $p = \boxed{\text{evens from 4 to 24 excluding 14, 15}}$. If $p + 1$ is prime, then $\tau((p+1)^{p+1}) = p + 2$, which is less than $\tau(p^p)$ for p composite by the same logic as in part (a). In addition, if $p + 1 = q^2$ is the square of a prime, then $\tau((p+1)^{p+1}) = \tau(q^{2(p+1)}) = 2p + 3$, which will be less than $\tau(p^p)$ if p has at least two prime factors (as $2\tau(p) > 2p + 3$), unless p is a perfect power. For sufficiently large p, $(p+1)^{p+1}$ begins to overwhelm p^p (try verifying this empirically). (In particular, if p is even, then it has 2 as a prime factor.)

4. WLOG assign coordinates of $A = (7, 24), B = (0, 0), C = (39, 0)$. Then using the formula $[\triangle ABC] = rs$, where $[\triangle ABC]$ denotes the area of $\triangle ABC$, r is the inradius, and s is the semi-perimeter, by Heron's formula, we obtain $468 = 52r \implies r = 9$. Hence, $I = (9, 9)$ and $O_1 = \left(\dfrac{16}{3}, 11\right), O_2 = \left(\dfrac{55}{3}, 11\right), O_3 = (16, 3)$. Then setting up equations of lines, we get $D = \left(\dfrac{7}{2}, 12\right), E = (23, 12), F = \left(\dfrac{39}{2}, 0\right)$ from which we can use Shoelace Theorem to obtain $[\triangle DEF] = \boxed{117}$.

5. (a) $\det M = (4x)(3-x) - (2)(x^2) = \boxed{12x - 6x^2}$.

   (b) M singular $\iff \det M = 0 \implies 12x = 6x^2 \implies x = 0, 2$. Since
   $$M^2 = \begin{bmatrix} (4x)(4x) + (x^2)(2) & (4x)(x^2) + x^2(3-x) \\ 2(4x) + (3-x)(2) & (2)(x^2) + (3-x)(3-x) \end{bmatrix} = \begin{bmatrix} 18x^2 & 3x^3 + 3x^2 \\ 6x + 6 & x^2 + 9 \end{bmatrix}$$
   we require $18x^2(x^2 + 9) = (6x+6)(3x^3 + 3x^2)$, or $18x^4 + 162x^2 = 18x^4 + 36x^3 + 18x^2 \implies 144x^2 = 36x^3 \implies x = 0, 4$.

(c) Since $\det M = 12x - 6x^2$, we have

$$M^{-1} = \frac{1}{12x - 6x^2} \begin{bmatrix} 3-x & -x^2 \\ -2 & 4x \end{bmatrix} = \begin{bmatrix} \dfrac{3-x}{12x-6x^2} & \dfrac{x}{6x-12} \\ \dfrac{1}{3x^2-6x} & \dfrac{2}{6-3x} \end{bmatrix}$$

(d) The eigenvalues of

$$\begin{bmatrix} 4 & 1 \\ 2 & 2 \end{bmatrix}$$

are the roots of the characteristic polynomial $(4-\lambda)(2-\lambda) - 2 \cdot 1 = 6 - 6\lambda + \lambda^2$, which are $\lambda = 3 \pm \sqrt{3}$. Then the eigenvectors are $\left(\dfrac{\sqrt{3}+1}{2}, 1\right)$ and $\left(\dfrac{-\sqrt{3}+1}{2}, 1\right)$.

6. (a) Let's examine the set of all possible movesets for the ant. Consider the up/right directions as "positive" directions and the down/left directions as "negative" directions. Then we need two positive and two negative directions - but we also cannot have combinations such as U/U/L/L. Indeed, if we also define up/down as "height" directions and right/left as "length" directions, we also require two height and two length directions. Essentially, the ant can move in a permutation of ULDR (24), of LLRR (6), or of UUDD (6). The desired probability is therefore $\dfrac{36}{4^4} = \boxed{\dfrac{9}{64}}$.

(b) In the gambler's ruin problem, we effectively prove that the casino indeed has a house edge by considering a hypothetical scenario in which two gamblers, Alpha and Beta, begin with pre-determined amounts of money and equal probability of one taking money from the other per round of a gambling game. Namely, if Alpha starts with d dollars and Beta starts with $N-d$ dollars, then the probability Alpha wins is $\dfrac{d}{N}$, and the probability Beta wins is $\dfrac{N-d}{N}$. This is because, when we consider the probability p_d that Alpha wins with a starting balance of d dollars, we end up with a recursive equation that yields the characteristic polynomial $px^2 - x + (1-p) = 0$ with roots $r_1 = 1$ and $r_2 = \dfrac{1-p}{p}$.

Advanced Exam

1. (a) Let X be a convex set over a real vector space, and let $f : X \mapsto \mathbb{R}$ be a function. f is convex if, $\forall x_1, x_2 \in X, \forall t \in [0,1]$, we have $f(tx_1 + (1-t)x_2) \leq tf(x_1) + (1-t)f(x_2)$.

   (b) We can phrase this in terms of convex sets. Let (S_i) be a convex set for $i \in [1, n]$. For $x, y \in \bigcap_{i=1}^n S_i$, and $t \in [0,1]$, we have that, for $i \in [1, n]$, $x \in S_i$ and $y \in S_i$ implies that $tx + (1-t)y \in S_i$ by the convexity of S_i. Hence, $tx + (1-t)y \in \bigcup_{i=1}^n S_i$, and $\bigcup_{i=1}^n S_i$ is convex, as desired.

   (c) i. Jensen's Inequality: Let the secant line of a function $f : X \mapsto \mathbb{R}$ (for X a real vector space) be defined by $tf(x_1) + (1-t)f(x_2)$, and let the graph of the function be given by $f(tx_1 + (1-t)x_2)$. Then $f(tx_1 + (1-t)x_2) \leq tf(x_1) + (1-t)f(x_2)$ for all $t \in [0,1]$.

   ii. At $x = \mu$, let $g(x) = a + bx$ be the equation of the tangent line to $f(x)$ at $x = \mu$ (where $\mathbb{E}(X) = \mu$). Then for all x, $f(x) \geq g(x) = a + bx$. Thus it follows that

   $$\mathbb{E}[f(x)] \geq \mathbb{E}[g(x)] = \mathbb{E}[a + bx] = a + b\mathbb{E}(X) = g(\mu) = f(\mu) = f(\mathbb{E}(X))$$

   Note that, if $f(x)$ is linear, then we have equality. Suppose $f(x)$ is nonlinear, but convex, so that $\mathbb{E}[f(x)] = f(\mathbb{E}(X)) = f(\mu)$. Then we have

   $$f(x) - g(x) \geq 0 \implies \int (f(x) - g(x))dx > 0$$

   Hence, we have $\mathbb{E}[f(x)] > \mathbb{E}[g(x)]$. Since $g(x)$ is linear, we have $\mathbb{E}[g(x)] = a + b\mathbb{E}(X) = f(\mu)$, which yields $\mathbb{E}[f(x)] > f\mathbb{E}(x)$. But this is a contradiction, since the two must be equal. Jensen's inequality follows.

   (d) Taking the logarithm of the left and right sides, we obtain the inequality

   $$\frac{x_1 + x_2 + x_3 + \ldots + x_n}{n} \log\left(\frac{x_1 + x_2 + x_3 + \ldots + x_n}{n}\right)$$
   $$\leq \frac{x_1 \log x_1 + x_2 \log x_2 + x_3 \log x_3 + \ldots + x_n \log x_n}{n}$$

   This holds from applying Jensen's to the convex function $f(x) = x \log x$ with $\alpha_i = \frac{1}{i}$ for all i.

   (e) We can write

   $$x_1^2 + x_2^2 + \ldots + x_5^2 = \left(x_1^2 + \frac{1}{3}x_2^2\right) + \left(\frac{2}{3}x_2^2 + \frac{1}{2}x_3^2\right) + \left(\frac{1}{2}x_3^2 + \frac{2}{3}x_4^2\right) + \left(\frac{1}{3}x_4^2 + x_5^2\right)$$

   in a form that is conducive to the Trivial Inequality. Specifically, since $a^2 + b^2 \geq 2ab$ for all $a, b \in \mathbb{R}$, we have

   $$x_1^2 + x_2^2 + x_3^2 + x_4^2 + x_5^2 \geq \frac{2}{\sqrt{3}}(x_1 x_2 + x_2 x_3 + x_3 x_4 + x_4 x_5)$$

   Thus, $c \geq \frac{2}{\sqrt{3}}$. To prove equality, observe that $(x_1, x_2, x_3, x_4, x_5) = (1, \sqrt{3}, 2, \sqrt{3}, 1)$ satisfies the inequality condition, so $c = \boxed{\frac{2\sqrt{3}}{3}}$.

   (f) We can rewrite the inequality as

   $$a(a-b)(2a-b) + b(b-c)(2b-c) + c(c-a)(2c-a) \geq 0$$

   WLOG assume a is the largest of a, b, c. Then we have two disjoint cases: $a \geq b \geq c$ and $a \geq c \geq b$.

$a \geq b \geq c$: The inequality is trivial for $a \geq 2c$, so assume $a < 2c$. Then $2a-b \geq 2c-a, 2b-c \geq 2c-a$ together imply
$$a(a-b)(2a-b) + b(b-c)(2b-c) + c(c-a)(2c-a)$$
$$\geq a(a-b)(2c-a) + b(b-c)(2c-a) + c(c-a)(2c-a)$$
$$= (2c-a)(a^2 + b^2 + c^2 - ab - bc - ca) \geq 0$$

$a \geq c \geq b$: We have
$$a(a-c)(2a-b) + c(c-a)(2c-a) + a(c-b)(2a-b) + b(b-c)(2b-c) \geq 0$$
so it suffices to show that $a(a-c)(2a-b) + c(c-a)(2c-a) \geq 0$. Again, if $a \geq 2c$, this is trivial. Otherwise, this follows from the assumption that $2a - b \geq 2c - a$. Moreover, the second term of the inequality is trivial if $c \geq 2b$, so if $c < 2b$, we have from $2a - b \geq 2b - c$ that
$$a(c-b)(2a-b) + b(b-c)(2b-c)$$
$$\geq a(c-b)(2b-c) + b(b-c)(2b-c) = (c-b)(2b-c)(a-b) \geq 0$$

(g) i. For any real numbers $a_1, \ldots, a_n$ and $b_1, \ldots, b_n$,
$$\left(\sum_{i=1}^n a_i b_i\right)^2 \leq \left(\sum_{i=1}^n a_i^2\right)\left(\sum_{i=1}^n b_i^2\right),$$
with equality when $a_i = cb_i$ for all $1 \leq i \leq n$, for $c \in \mathbb{R}$.

ii. Consider two functions f, g defined on $[0,1]$. Define $F = \dfrac{f}{||f|| + ||g||}, G = \dfrac{g}{||f|| + ||g||}$. We have $||F|| + ||G|| = 1$, and therefore, we can consider them as parameters in a convex linear combination. By Jensen's Inequality applied to the convex function $f(x) = x^2$,
$$(f+g)^2 = \left(||F|| \cdot \frac{f}{||F||} + ||G|| \cdot \frac{g}{||G||}\right)^2$$
$$\leq ||F|| \cdot \frac{f^2}{||F||^2} + ||G|| \cdot \frac{g^2}{||G||^2} = \frac{f^2}{||F||} + \frac{g^2}{||G||}$$

If we integrate both sides, we obtain
$$||f+g||^2 \leq \frac{||f||^2}{||F||} + ||g||^2 ||G|| = (||f|| + ||g||)^2$$

This yields
$$||f||^2 + 2\langle f, g\rangle + ||g||^2 \leq ||f||^2 + 2\,||f||\,||g|| + ||g||^2$$
which is equivalent to the Cauchy-Schwarz Inequality.

2. (a) Multinomial Theorem: For $m \in \mathbb{N}$, $n \in \mathbb{N} \cup \{0\}$, we have
$$(x_1 + x_2 + x_3 + \ldots + x_m)^n = \sum_{k_1 + k_2 + k_3 + \ldots + k_m = n} \binom{n}{k_1, k_2, k_3, \ldots, k_m} \prod_{t=1}^m x_t^{k_t}$$

(b) $\binom{-4}{2} = \dfrac{-4 \cdot -5}{2} = \boxed{10}$, $\binom{4.5}{3} = \dfrac{4.5 \cdot 3.5 \cdot 2.5}{3!} = \boxed{\dfrac{105}{16}}$, $\binom{2\sqrt{2}}{2} = \dfrac{2\sqrt{2} \cdot (2\sqrt{2} - 1)}{2} = \boxed{4 - \sqrt{2}}$.

(c) We claim that this product is inextricably tied with the *convolution operator,* i.e. the product of terms taken in pairs whose exponents sum to an incrementally-higher value.

That is, let
$$A(x) = \sum_{k=0}^{\infty} a_k x^k, B(x) = \sum_{k=0}^{\infty} b_k x^k$$

Then we have
$$A(x)B(x) = \sum_{k=0}^{\infty} \left(\sum_{i=0}^{k} a_i b_{k-i} \right) x^k$$

If $A(x)$ counts objects of group A with "size" k, and $B(x)$ counts objects of group B with "size" k, then $A(x)B(x)$ counts those pairs (A, B) with total size k. This is a direct corollary of Vandermonde's Identity.

(d) $f(x) = \dfrac{2\arctan x}{x^2+1} = 2\dfrac{\arctan x}{1-(-x^2)} = 2 \cdot \dfrac{1}{1-(-x^2)} \cdot \arctan x$. The power series for $\dfrac{1}{1-x}$ is $1+x+x^2+x^3+\ldots$ by the infinite geometric series formula, so that of $\dfrac{2x}{1-(-x^2)}$ is $2x(1-x^2+x^4-x^6+\ldots) = 2x - 2x^3 + 2x^5 - 2x^7 + \ldots$. In addition, note that $(\arctan x)' = \dfrac{1}{1+x^2} = 1 - x^2 + x^4 - x^6 + \ldots$, so $\arctan x = x - \dfrac{x^3}{3} + \dfrac{x^5}{5} - \dfrac{x^7}{7} + \ldots = x\left(1 - \dfrac{x^2}{3} + \dfrac{x^4}{5} - \dfrac{x^6}{7} + \ldots\right)$. Then we have
$$f(x) = \left(\sum_{n=0}^{\infty} (-1)^n x^{2n}\right)\left(\sum_{n=0}^{\infty} \frac{(-1)^n x^{2n+1}}{2n+1}\right)$$

(e) This follows from Taylor's Theorem; let $\deg f = n$. Then for all derivatives $d \geq n+1$, $f^{(d)}(x) = 0$. Hence, for any $x \in \mathbb{R}$, $\lim_{n\to\infty} R_n(x) = 0$, as required. (Here $R_n(x)$ denotes the remainder of the nth Taylor polynomial at x.) Indeed, such functions are known as *analytic functions*.

(f) Consider $f(x) = \dfrac{1}{1+x}$, defined on $\mathbb{R}\setminus\{-1\}$. This is an example of a function that is infinitely differentiable but non-analytic, since we end up with a geometric series that clearly diverges outside of $x \in (-1, 1)$.

3. (a) Chebyshev Inequality: Let X be an (integrable) random variable with expected value $\mu \in \mathbb{R}$ and non-zero variance $\sigma^2 \in \mathbb{R}$. Then for any real number $k \geq 0$,
$$P(|X - \mu| \geq k\sigma) \leq \frac{1}{k^2}$$
where $P(A)$ denotes the probability of A.

Proof. From the probabilistic definition of expected value,
$$\mathbb{E}(X) = \int_{-\infty}^{\infty} xf(x)dx = \int_{0}^{\infty} xf(x)dx$$
as X is non-negative. It follows that
$$\mathbb{E}(X) = \int_{0}^{a} xf(x)dx + \int_{a}^{\infty} xf(x)dx$$
$$\geq \int_{a}^{\infty} xf(x)dx \geq \int_{a}^{\infty} af(x)dx$$
$$a\int_{a}^{\infty} f(x)dx = a \cdot P(X \geq a)$$

Hence, $P(X \geq a) \leq \dfrac{\mathbb{E}(x)}{a}$, as desired. □

This is in fact a proof of the Markov inequality, from which the Chebyshev inequality follows as a direct inequality.

(b) The probability of X lying *outside* the interval is $1 - \frac{1}{n} \leq \frac{1}{a^2}$, so $a^2 \leq \frac{n}{n-1}$ and $a \leq \sqrt{\frac{n}{n-1}}$.

(c) Let G be a finite group that acts on a set X. $\forall g \in G$, denote by X^g the set of elements in X fixed by g. Then
$$|X/G| = \frac{1}{|G|} = \sum_{g \in G} |X^g|$$

For Burnside's Lemma to apply, G must satisfy the axioms of a group, namely closure, associativity, the existence of an identity element, and the existence of an inverse element.

- Closure: For all $a, b \in G$, $ab \in G$.
- Associativity: For all $a, b, c \in G$, $(ab)c = a(bc)$.
- Identity: $\exists e \in G$ such that, $\forall a \in G$, $ae = a = ea$.
- Inverse: $\exists a^{-1} \in G$ such that, $\forall a \in G$, $aa^{-1} = e = a^{-1}a$.

(d) Let X be the set of 3^6 colorings *without* respect to rotation, and let G be the rotation group acting upon X. There is an identity element (leaving the colorings unchanged), as well as six $90°$ rotations which leave 3^3 colorings unchanged, three $180°$ rotations which leave 3^4 colorings unchanged, six $180°$ rotations which leave 3^3 colorings unchanged, and finally, eight $120°$ rotations leaving 3^2 colorings unchanged. By Burnside's, we have $\frac{3^6 + 3 \cdot 3^4 + 12 \cdot 3^3 + 8 \cdot 3^2}{24} = \boxed{57}$.

Closing Exam

1. (a) Yes; as 7 is odd, we would need the primes to be of different parities. But $(2,9)$ is the only possibility, and 9 is not prime. If we start with an odd prime and add 7, we end up with an even number, which clearly cannot be prime (if it is not 2).

   (b) Again, yes. Same logic as above; use the fact that $(n, n+2, n+4)$ has one of the numbers a multiple of 3.

   (c) This integer is $\boxed{12}$. Let the primes be $p, p+2$; then $12 \mid (2p+2)$, since $4 \mid (2p+2)$ (as $2 \mid (p+1)$, which holds since p is odd; $(2,3)$ is not considered a twin prime pair), and furthermore, $3 \mid (2p+2)$, since primes are $\pm 1 \pmod 3$, and adding 2 to $-1 \pmod 3$ yields $1 \pmod 3$, whereas adding 2 to $+1 \pmod 3$ yields $0 \pmod 3$, which contradicts the definition of a prime. Anything larger is not guaranteed to divide $2p+2$ (take $2p+2 \pmod k$ for $k \geq 5$).

2. We can subtract the z terms from both sides of the equations to get $x + y = 3 - z$, $x^2 + y^2 = 7 - z^2$. Squaring the first equation yields $x^2 + 2xy + y^2 = 9 - 6z + z^2 \implies x^2 + 10 + y^2 = 9 - 6z + z^2 \implies 17 - z^2 = 9 - 6z + z^2 \implies 2z^2 - 6z - 8 = 0 \implies z^2 - 3z - 4 = 0 \implies z = 4, -1$. We then have $z = 4$, so $x + y = -1$ and $x^2 + y^2 = -9$. Solving yields $(x, y) = \left(\dfrac{-1 + i\sqrt{19}}{2}, \dfrac{-1 - i\sqrt{19}}{2} \right)$, so $\text{Im}(x - y) = \boxed{\sqrt{19}}$.

3. We perform casework on the number of 5-inch-tall blocks. If there are none, then Yolandas stack must consist solely of 3-inch blocks. There is clearly only one way to do this. If there is one 5-inch-tall block, then henceforth, let the number of 2-inch blocks be x, and the number of 3-inch blocks be y. We have $2x + 3y = 25, x + y = 9 \implies (x, y) = (2, 7)$. Then the number of ways is $\dfrac{10!}{7!2!1!} = 360$. If there are two 5-inch blocks, then $2x + 3y = 20, x + y = 8 \implies (x, y) = (4, 4)$. The number of ways is $\dfrac{10!}{4!^2 \cdot 2!} = 3150$. If there are three 5-inch blocks, we have $2x + 3y = 15, x + y = 7 \implies (x, y) = (6, 1)$. Thus, the number of ways is $\dfrac{10!}{6!3!1!} = 840$. The stack cannot have four or more 5-inch-tall blocks, as even having the rest of the blocks be 2 inches tall would lead to a stack of at least 32 inches. Hence, we have found all possibilities, and the desired probability is $\dfrac{1 + 360 + 3150 + 840}{3^{10}} = \boxed{\dfrac{4351}{3^{10}}}$.

4. (a) Angle Bisector Theorem: In $\triangle ABC$ with side lengths $AB = c, BC = a, CA = b$, the internal bisectors of angles ABC, BCA, CAB split their corresponding line segments into the ratios $c : b$, $a : c$, and $a : b$, with each ratio corresponding to its adjacent side length.

   Proof. Applying Law of Sines on triangles $\triangle BCD, \triangle ACD$ yields $\dfrac{a}{\sin BDC} = \dfrac{x}{\sin BCD}$ and $\dfrac{b}{\sin ADC} = \dfrac{y}{\sin ACD}$, where x, y denote the split line segments of c. Using the fact that angles ADC and BDC are supplementary yields $BCD = ACD$, since $\overline{CD}$ is the angle bisector by construction, we get $\dfrac{a}{x} = \dfrac{\sin BDC}{\sin BCD} = \dfrac{\sin ADC}{\sin ACD} = \dfrac{b}{y}$, as desired. $\square$

   (b) WLOG set $B = (0, 0)$, so that $A = (5, 12), C = (14, 0)$. Then $D = \left(\dfrac{65}{29}, \dfrac{156}{29} \right), E = \left(\dfrac{13}{2}, 0 \right), F = \left(\dfrac{28}{3}, \dfrac{56}{9} \right)$. Applying Shoelace Theorem yields $\boxed{\dfrac{5447}{261}}$.

5. (a) The Power of a Point Theorem has three cases: the intersecting chords theorem (in which case chords $\overline{AB}, \overline{CD}$ intersect at E), in which case $AE \cdot BE = CE \cdot DE$; the tangent-secant theorem (in which case $\overline{AB}, \overline{BD}$ meet at B outside circle O, and C lies on O, with $AB^2 = BC \cdot BD$), and the secant-secant theorem (in which case two secants $\overline{AC}, \overline{CE}$ of circle O intersect at C, with B, D lying on O; we have $AC \cdot BC = CD \cdot CE$).

(b) *Proof.* Let chords $\overline{AB}$ and $\overline{CE}$ of circle Ω with center O intersect at E. If $E = O$, we are done. Otherwise let F be the foot of the perpendicular from O to $\overline{AE}$, and G the foot of the perpendicular from O to $\overline{DE}$. Then F is the midpoint of $\overline{AC}$, and similarly G is the midpoint of $\overline{BD}$. Hence, $AE \cdot EC + EF^2 = FC^2$. Add FO^2 to both sides, and apply the Pythagorean Theorem to obtain $AE \cdot EC + EO^2 = CO^2$. Similarly, $DE \cdot EB + EO^2 = BO^2$. As $BO = CO$, with both radii of Ω, it follows that $AE \cdot EC = DE \cdot EB$, as desired. $\square$

(c) Let the triangle be $\triangle ABC$. Then extend $\overline{AI}$ to hit the circumcircle at M, the midpoint of arc BC. Let $\overline{MO}$ intersect the circumcircle at L. Draw the perpendicular from I to $\overline{AB}$ to the foot D, with $ID = r$. Then $\triangle ADI \sim \triangle LBM$, i.e. $\dfrac{ID}{BM} = \dfrac{AI}{LM} \implies ID \cdot LM = BM \cdot AI$. Hence, $2R \cdot r = AI \cdot BM$, and drawing $\overline{BI}$ yields the congruence of angles BIM, IBM. Thus, $BM = IM$ and $AI \cdot IM = 2R \cdot r$. Extend $\overline{OI}$ to hit the circumcircle at P, Q. Then $PI \cdot QI = AI \cdot IM = 2R \cdot r \implies (R+d)(R-d) = 2R \cdot r \implies d^2 = R(R-2r)$, where $d = IO$.

6. Observe that Evelyn can win when she picks a number that is 1, has an even number of *distinct* prime factors, is a perfect square, or is equal to a square times a number with an even number of distinct prime factors. In the case that her number is not a perfect square, nor a multiple of one, numbers with an even number of distinct prime factors include $2 \cdot 3, 2 \cdot 5, 2 \cdot 7, \ldots, 2 \cdot 47$, which number 14. (Note that $2 \cdot 3 \cdot 5 \cdot 7 = 210 > 100$.) The perfect squares number 10 (including 1), and as for the last case, we have $2^2 \cdot 3, 2^2 \cdot 5, 2^2 \cdot 7, \ldots, 2^2 \cdot 23$ (which number 8), $3^2 \cdot 5, 3^2 \cdot 7, 3^2 \cdot 11$, so her grand total is $14 + 10 + 8 + 3 = \boxed{35}$ winning numbers.

7. (a) Observe that the function $f : \mathbb{R} \mapsto \mathbb{R}$, $f(x) = x^3 + \dfrac{15}{16}$ is monotonically increasing for $x \geq 0$, and furthermore, that $f(x) = x$ is a solution to the equation. Solving the equation $x^3 + \dfrac{15}{16}x = x$ yields $x = 0, \pm\dfrac{1}{4}$.

(b) By the Fundamental Theorem of Algebra, we note that the equation has exactly 9 complex roots, at least 3 of which are real. Using synthetic division, we can write $\dfrac{\left(x^3 + \dfrac{15}{16}x\right)^3 + \dfrac{15}{16}\left(x^3 + \dfrac{15}{16}x\right) - x}{x\left(x + \dfrac{1}{4}\right)\left(x - \dfrac{1}{4}\right)}$

as $x^6 + \dfrac{23}{8}x^4 + \dfrac{721}{256}x^2 + \dfrac{31}{16}$. Re-writing this as $x^2\left(x^2\left(x^2 + \dfrac{23}{8}\right) + \dfrac{721}{256}\right) + \dfrac{31}{16}$ and setting this resultant expression equal to zero, we observe that this forces $x^2 < 0$, which yields six complex solutions (again by the Fundamental Theorem of Algebra). Thus, we have obtained all real solutions to the equation: namely, $x = 0, \dfrac{1}{4}, -\dfrac{1}{4}$, as desired.

8. The length of Kendra's string is given by $(a + b + c) + \pi(2a - 2c)$, since it consists of the sum of the triangles edge lengths and half the circumferences of each of the circles. Hence, we require $a + b + c = 2019, 2a - 2c = 0 \implies a = c$. At the same time, the triple (a, b, c) must satisfy the Triangle Inequality, so $a + c > b \implies 1009 \geq a, c \geq 505$. Therefore, there are $1009 - 505 + 1 = \boxed{505}$ possibilities for the ordered triple (a, b, c).

9. (a) Note that $G = (V, E)$ is tropical iff it both inscribes a circle with radius 1, and can be inscribed within a circle of radius 2 (concentric with the circle of radius 1). Hence, the bipartite graph G_B is tropical precisely when half the vertices lie in the larger circle but outside the smaller one, and half lie inside the smaller circle (with all distances at least 1 as well).

(b) It suffices to show that all complete graphs K_n, $n \in \mathbb{N}$, can be inscribed in a circle with radius 2 and contain a circle with radius 1. This is a consequence of the vertices of the representation of K_n as the vertices of a regular n-gon (WLOG with side length 1). Clearly, this regular n-gon inscribes a circle with radius 1. It can furthermore be circumscribed by a circle with radius 2, as twice the side length never exceeds the circumradius.

10. (a) Yes; we claim $\boxed{m = 13, k = 12}$. This can be seen by listing out the first few terms of the sequence $f(n)$ by hand; namely, $1, 2, 5, 9, 6, 7, 5, 4, 1, 7, 0, 9, 1, 2, 5, \ldots$

(b) From (a), we have blocks of size 12 that sum to 56 each. Hence, after $\lfloor \frac{2019}{12} \rfloor = 168$ blocks, the sum is $56 \cdot 168 = 9408$. After three more entries, the sum becomes $9408 + 1 + 2 + 5 = \boxed{9416}$.

(c) It suffices to consider the total sales mod 3 and mod 5, by CRT. Modulo 3, we have $f(n) \equiv 1, 2, 2, 2, \ldots$, and modulo 5, we have $f(n) \equiv 1, 2, 0, 4, 1, 2, 0, 4, \ldots$ Hence, the sum of the $f(n)$ mod 3 is $1, 0, 2, 1, 0, 2, 1, 0, \ldots$, i.e. $\sum_{i=1}^{n} f(i) \equiv 0 \pmod{3} \iff n \equiv 2 \pmod{3}$. The sum of the $f(n)$ mod 5 is $1, 3, 3, 2, 3, 0, 0, 4, \ldots, 4, 1, 1, 0, 1, 3, 3, 2, \ldots$ with a period of 20. The zeroes occur at $n \equiv 0, 6, 7, 9 \pmod{20}$. Via another application of CRT, we get that $n \equiv 20, 26, 29, 47 \pmod{60}$ are the desired n. Thus, $n = \boxed{20, 26, 29, 47, 80, 86, 89}$.

11. (a) i. Note that $\det M$ is defined by the RREF form of the matrix; when we multiply the matrix M by a scalar c, we are effectively multiplying each row by c, hence multiplying the determinant by c^n (via n elementary row operations).

ii. This is a consequence of M and M^T having the same determinant. M is singular iff $\det M = 0$, and $\det M = 0 \iff \det M^T = 0 \implies M^T$ is singular (the reverse is also true). We have proven this in the lesson material.

iii. We make the equivalent claim that $\det M = M_{1,j} \cdot \text{adj}(M)_{1,j} + M_{2,j} \cdot \text{adj}(M)_{2,j} + \ldots$ Note that $\text{adj}(M)_{i,j}$ is linear in all rows of M except the ith row, and in all of the columns as well (except for column j). Hence, $\text{adj}(M)_{i,j} = 0$ if two of these rows and columns are identical, so it must be proportional to the determinant of the matrix that is obtained by removing the ith row and j^{th} column from M.

(b) i. Let $M = \begin{bmatrix} a & b \\ c & d \end{bmatrix}$. Then its column vectors are $\begin{bmatrix} a \\ c \end{bmatrix}$ and $\begin{bmatrix} b \\ d \end{bmatrix}$ whose dot product is, by definition, $\boxed{ab + cd}$.

ii. The dot product of two arguments produces a scalar as output, but the dot product of a scalar with a vector is undefined. Hence, the dot product is not a valid operation for three or more arguments.

(c) This is given by the absolute value of the determinant of the matrix $\begin{bmatrix} 0 & 1 & 4 \\ 2 & 9 & 0 \\ 0 & 0 & 5 \end{bmatrix}$, which is

$$\left| 0 \cdot \det \begin{bmatrix} 9 & 0 \\ 0 & 5 \end{bmatrix} - 1 \cdot \det \begin{bmatrix} 2 & 0 \\ 0 & 5 \end{bmatrix} + 4 \cdot \det \begin{bmatrix} 2 & 9 \\ 0 & 0 \end{bmatrix} \right| = |0 - 10 + 4 \cdot 0| = \boxed{10}.$$

12. (a) Since $\arg(z)$, by definition, is the angle spanned by z in the complex plane, its tangent is the ratio of $\text{Im}(z)$ to $\text{Re}(z)$. As such, $\arg(z_1) = \arctan\left(\frac{b}{a}\right)$, $\arg(z_2) = \arctan\left(\frac{d}{c}\right)$. As $z_1 + z_2 = (a + c) + (b + d)i$, we have

$$\arg(z_1 + z_2) = \arctan\left(\frac{b+d}{a+c}\right)$$

Hence,

$$\arctan\left(\frac{b}{a}\right) + \arctan\left(\frac{d}{c}\right) = \arctan\left(\frac{b+d}{a+c}\right)$$

(b) We have $\tan(\arg(z_1)) = \frac{b}{a}$ and $\tan(\arg(z_2)) = \frac{d}{c}$, as well as $\tan(\arg(z_1 + z_2)) = \frac{b+d}{a+c}$, so $\frac{b}{a} + \frac{d}{c} = \frac{b+d}{a+c}$. This implies that $\frac{(a+c)(bc+ad)}{ac(a+c)} = \frac{ac(b+d)}{ac(a+c)}$, i.e. $(a+c)(bc+ad) = ac(b+d) = abc + acd \implies abc + bc^2 + a^2d + acd = abc + acd \implies a^2d + bc^2 = 0$.

13. (a) Let Q, R be points on circles C_1, C_2 respectively such that $\overline{O_1Q}$ and $\overline{O_2R}$ are radii of C_1, C_2, and Q, R, P are collinear. Then the desired conclusion is a consequence of the similarity of triangles

$\triangle PRO_2$ and $\triangle PQO_1$, which implies that $\dfrac{O_1T}{O_2T} = \dfrac{O_1Q}{O_2R} = \dfrac{r_1}{r_2}$ (hence showing that P is undefined if $r_1 = r_2 \implies \dfrac{r_1}{r_2} = 1$).

(b) We can apply Menelaus' Theorem here to show that $\dfrac{AC}{CB} \cdot \dfrac{BA}{AC} \cdot \dfrac{CB}{BA} = 1$, where A, B, C are the centers of the circles, and A, B, C are the specified points of intersection. From part (a), $\dfrac{AC}{CB} = \dfrac{r_A}{r_B}$ (where r_A, r_B denote the radii of the circles with centers A, B respectively), and similarly, $\dfrac{BA}{AC} = \dfrac{r_B}{r_C}$, $\dfrac{CB}{BA} = \dfrac{r_C}{r_A}$. Thus it follows that their product is equal to $\dfrac{r_A}{r_B} \cdot \dfrac{r_B}{r_C} \cdot \dfrac{r_C}{r_A} = 1$, from which Monge's Theorem holds.

14. (a) i. The OGF of a fair six-sided die is $\dfrac{x + x^2 + \ldots + x^6}{6}$, as each of 1 through 6 has an equal probability of occurring, and we can apply the definition of the OGF from there.

   ii. The probability of landing on side n is $\dfrac{n}{21}$, so $\dfrac{x + 3x^2 + 6x^3 + 10x^4 + 15x^5 + 21x^6}{21}$ is the OGF of this die.

   iii. The possible sums are 3 through 18, with coefficients of the x^n terms corresponding to their probabilities (sum n). The OGF is then

$$\sum_{i=3}^{18} P_n x^i$$

where P_n denotes the probability of rolling a sum of exactly n.

(b) We can actually solve this problem by pattern-finding. Suppose we were dealing with only 4 coin flips. Then the probability of getting either 0 or 3 heads is $\dfrac{5}{16} = \dfrac{1}{3} - \dfrac{1}{3 \cdot 2^4}$. Now suppose we were dealing with 7 coin flips. The probability of satisfying the condition is $\dfrac{1}{3} + \dfrac{1}{3 \cdot 2^7}$. If we do the same with 10 flips, we get $\epsilon = \dfrac{1}{3 \cdot 2^{10}}$. Note that ϵ is being added for odd numbers of flips and subtracted for even numbers of flips (exploit the symmetry of the binomial coefficients to prove this). We thus subtract $\boxed{\dfrac{1}{3 \cdot 2^{100}}}$ from $\dfrac{1}{3}$ to get the desired probability.

(c) In general, let $G_S(x) = \prod_{i=1}^n G_{x_i}(x)$, where x_i is a random variable from a set of n random variables $\{x_1, x_2, \ldots, x_n\}$, and $G_S(x)$ denotes the generating function of $S = x_1 + x_2 + \ldots + x_n$. We prove this by observing that

$$G_S(x) = \mathbb{E}(x^S) = \mathbb{E}(x^{x_1} x^{x_2} \cdots x^{x_n})$$

$$= \mathbb{E}(x^{x_1}) \mathbb{E}(x^{x_2}) \cdots \mathbb{E}(x^{x_n})$$

by virtue of the x_i's being independent random variables, as required.

15. (a) This is the precise statement of Miquel's Theorem. Let $AFO = \theta$, so that quadrilaterals $AFOD$ and $BEOD$ are cyclic. Then $ADO = 180° - \theta$, $BDO = \theta$, $BEO = 180° - \theta$, so $CFO = 180° - \theta$ and $CEO = \theta$. Hence quadrilateral $CQPD$ is cyclic as well.

(b) Consider two cases: one where ERM and MRF are adjacent angles, and one where they are not. In the first case, $\triangle MRE$ is isosceles, and that $\overline{QR}$ bisects angle MRE. Similarly, isosceles $\triangle MRF$ is bisected by $\overline{PR}$.

Let $MRE = 2a°$, $MRF = 2b°$. Then angle ECF has measure $(a + b)°$, from which it follows that $ECF = QRP$ are congruent angles, and similarly, angles DAF, QPR are congruent. Therefore, $\triangle ABC \sim \triangle PQR$.

In the second case, where angles EIP, MIF are *not* adjacent, we have a superposition of $\triangle PIE, \triangle PIF$ such that angle $EIF = 2(b-a)°$. Note that $HIF = (b-a)°$, $ECF = (b-a)°$, as twice the measure of angle ECF is EIF by the subtended angle theorem. Thus, angles ECF and HFI are congruent. As such, $DAF \cong HGI$, and so we have a spiral similarity of $\triangle ABC$ with $\triangle GHI$.

By the way, in the closing exam, you may have noticed something interesting regarding the haphazardly bolded letters. If you don't want to spoil it, feel free to carry on. Otherwise, it's in the footnote at the bottom of this page (flip upside down)![129]

---

[129] We admire your dedication to learning mathematics. Thank you so much for reading this book!

Scoring Rubrics

Beginner Exam

1. 15 points:

    (a) 5 points: 5 points for any accurate definition.

    (b) 5 points: 2 points for the setup, 3 for the answer.

    (c) 5 points: See above.

2. 15 points:

    (a) 5 points: 0.5 for each correctly defined symbol.

    (b) 5 points: 1 point for the answer, 2 points for using $\forall, \exists$ correctly (1 each), 2 points for an example.

    (c) 5 points: See above.

3. 20 points:

    (a) 4 points: 1 point for using induction, 3 for using it *correctly*.

    (b) 4 points: See above.

    (c) 5 points: See above (4 points for using it correctly to finish the proof).

    (d) 7 points: 3 points for splitting up the sum, 2 points for applying sum identities, 2 points for the complete proof.

4. 15 points:

    (a) 3 points: 1 point for using the fact that $ABCD$ is cyclic to conclude that the diagonal is a diameter, 1 point for the Pythagorean Theorem, 1 point for the conclusion/answer.

    (b) 3 points: 1 point for the angle insight, 1 point for the special right triangle, 1 point for the answer.

    (c) 5 points: 3 points for proving periodicity by listing out terms, 2 points for finding the period.

    (d) 4 points: 2 points for using the "block" tactic, 1 point for computation, 1 point for the answer.

5. 12 points:

    (a) 4 points: 2 points for the correct pattern in the coefficients, 2 points for the correct powers of x, y.

    (b) 3 points: 1 point for using binomial coefficients properly, 1 point for the computation of the choose function, 1 point for final answer.

    (c) 5 points: 4 points for generalizing successfully, 1 point for final answer.

6. 13 points:

    (a) 3 points: 1 point for multiplicative principle, 2 points for answer.

    (b) 4 points: 1 point for re-invoking multiplicative principle, 3 points for answer.

    (c) 6 points: 2 points for drawing the equivalent problem, 1 point for defining d, n_d, 1 point for the stars-and-bars connections, 2 points for completion.

7. 10 points: 2 points for writing $x^4 + y^4$ in the form given, 1 point for maximizing the desired quantity, 1 point for relating back to x^2+y^2, 1 point for writing $x^2y^2 = 7$, 3 points for AM-GM to get $x^2+y^2 = 2\sqrt{7}$, 2 points for finishing the solution (1 point for answer among these 2).

Score	Recommendation
70-100	Intermediate
50-69	Beginner (extension exercises)
30-49	Beginner
0-29	An introductory competition math course.

Intermediate Exam

1. 12 points:

    (a) 6 points: 1 point for mentioning Vieta's formulas, 2 points for proving Vieta's for quadratics, 1 point for applying Vieta's to the problem, 2 points for a complete solution.

    (b) 6 points: 2 points for re-applying Vieta's, 4 points for the answer and solution.

2. 20 points:

    (a) 4 points: 2 points for each correct definition.

    (b) 6 points: 3 points for the QRs, and 3 points for the QNRs. If no more than 1 value is incorrect, award 2 points. If no more than 2 values are incorrect, award 1 point. Otherwise, award 0 points.

    (c) 5 points: 3 points for defining the Legendre symbol correctly, 2 points for defining the Jacobi symbol correctly.

    (d) 5 points: 1 point for the initial condition of p, q odd primes, $p \neq q$, 4 points for the statement (all-or-nothing).

3. 15 points:

    (a) 7 points: 1 point for observing $\tau(p) = p + 1$ for prime p, 2 points for the prime factorization, 1 point for the product of prime factors, 2 points for the application of AM-GM, 1 point for the answer (never).

    (b) 8 points: 1 point for considering when $p + 1$ is prime, 1 points for extending this to p composite, 2 points for the conclusions, 1 point for the square of a prime, 1 point for "two prime factors," 1 point for dominance, 1 point for the final answer.

4. 15 points: 1 point for a reasonable diagram, 2 points for the WLOG setup, 2 points for demonstrated understanding of the incenter, 3 points for demonstrating understand of the centroid, 4 points for setting up equations of lines and their intersections, 3 points for the final answer.

5. 20 points:

    (a) 3 points: 2 points for the proper setup of the determinant, 1 point for the answer.

    (b) 5 points: 2 point for the definition of singular, 1 point for the algebraic setup, 2 points for the answer.

    (c) 4 points: 1 point for the determinant setup, 2 points for the algebra computation, 1 point for the final answer.

    (d) 8 points: 2 points for identifying eigenvalues as the roots of the characteristic polynomial, 2 points for computing the characteristic polynomial, 2 points for the eigenvalues/eigenvectors (1 point for each).

6. 18 points:

    (a) 6 points: 2 points for +/- directions, 2 points for a viable casework approach, 2 points for the final answer and correct solution.

    (b) 12 points: 2 points for an adequate description of the problem, 1 point for the observation "predetermined" amounts of money, 1 point for "equal probability," 3 points for setting probabilities, 5 points for justification of those probabilities (2 points for the recursion, 3 points for obtaining the resulting closed form).

Score	Recommendation
65-100	Advanced
40-64	Intermediate (extension exercises)
25-39	Intermediate
0-24	Beginner

Advanced Exam

1. (a) 3 points: 1 point for considering x_1, x_2 and the parameter t, 2 points for the full, accurate definition. Award 1 point only for a non-rigorous, but correct, intuition (e.g. "a function is convex if, for all points on the curve, the straight line between them is always above the curve"). Other rubric items are not applicable in this case.

    (b) 5 points: 1 point for considering the problem in terms of convex sets, 1 point for defining x, y, t appropriately, 2 points for observing the convexity of S_i, 1 point for the desired conclusion.

    (c) i. 3 points: 1 point for considering the secant line, 2 points for the actual statement.

    ii. 7 points: 1 point for considering the tangent line, 1 point for writing $f(x) \geq g(x) = a + bx$, 2 points for writing $\mathbb{E}[f(x)] \geq f(\mathbb{E}(X))$ with proper justification, 1 point for observing that we have equality iff f is linear, 1 point for observing the relationship with the integral, 1 point for obtaining the final contradiction.

    (d) 7 points: 2 points for taking the log of both sides, 2 points for writing the resulting inequality, 1 point for applying Jensen's inequality, 1 point for $f(x) = x \log x$ and $\alpha_i = \dfrac{1}{i}$, 1 point for showing f is indeed convex.

    (e) 7 points: 2 points for writing the expression in the form conducive to Trivial Inequality, 1 point for applying Trivial Inequality, 2 points for writing the resulting inequality, 1 point for the lower bound of $\dfrac{2}{\sqrt{3}}$, 1 point for proving the equality case with an example.

    (f) 7 points: 1 point for rewriting the inequality, 1 point for observing the WLOG assumption and splitting into two cases, 2 points for the first case, 1 point for rewriting the second case, 2 points for finishing the proof.

    (g) i. 3 points: 2 points for the inequality statement, 1 point for the equality case.

    ii. 8 points: 1 point for considering functions f, g, 1 point for defining F, G, 2 points for applying Jensen's on $f(x) = x^2$, 2 points for writing $(f+g)^2$ in the desired form, 1 point for integrating both sides, 1 point for the final conclusion of equivalence with Cauchy-Schwarz.

2. (a) 4 points: 1 point for the initial conditions, 3 points for the rest of the statement (all-or-nothing here).

    (b) 3 points: 1 point for each correct value.

    (c) 7 points: 2 points for considering the convolution, 2 points for proposing a actual, correct, concrete formula, 3 points for finishing the solution.

    (d) 6 points: 1 point for splitting up the function into manageable parts, 1 point for computing the power series of $\dfrac{1}{1-x}$, 1 point for relating this to $\dfrac{1}{1+x^2}$, 1 point for the derivative of $\arctan x$, 2 points for a full solution.

    (e) 6 points: 1 point for mentioning Taylor's Theorem, 2 points for making the derivative connection, 3 points for making a conclusion about the remainder from Taylor's Theorem.

    (f) 4 points: 1 point for the example itself, 1 point for showing it is infinitely differentiable, 2 points for showing it is non-analytic by considering the geo series (1 point only if the geo series is mentioned, but the solution is not completely correct).

3. (a) 5 points: 1 point for variable definitions, 1 point for the correct statement, 3 points for the proof (1 point for correctly considering the probabilistic definition of expected value, 1 point for taking the integral, 1 point for the correct conclusion).

    (b) 3 points: 1 point for the initial observation of "flipping" the problem around, 1 point for the answer, 1 point for full justification.

    (c) 4 points: 2 points for the statement of Burnside's. 2 points (0.5 per correct axiom) for the axioms of a group.

(d) 8 points: 1 point for defining X and G, 1 point for recognizing the identity element, 2 points for recognizing the other cases, 4 points for a complete and correct solution (1/4 points only for the answer with no supporting work).

Score	Recommendation
60-100	An Olympiad-level prep course.
50-59	Advanced (extension exercises)
30-49	Advanced
15-29	Intermediate (extension exercises)
0-14	Intermediate

Closing Exam

Note: Points for the answers do not apply if justification is absent or completely specious. In such cases your answer should be graded as a zero.

1. (NT) 12 points:

    (a) 3 points: 1 point for answering yes, 1 point for considering parity, 1 point for a complete solution.

    (b) 4 points: 1 point for yes, 2 points for considering modulo 3, 1 point for a complete solution.

    (c) 5 points: 1 point for the answer (12), 2 points for using the CRT/GCD properly, 1 point for each factor.

2. (Algebra) 10 points: 1 point for isolating z, 2 points for noticing $\left(x + \frac{1}{x}\right)^2 - 2 = x^2 + \frac{1}{x^2}$, 1 point for squaring the first equation, 1-3 points for reaching $z = -1, 4$ depending on level of detail and general progress, 1 point for obtaining $z = 4$, 1 point for solving for x and y, 1 point for the correct answer ($\sqrt{19}$).

3. (Discrete) 13 points: 1 point for any potentially viable casework approach, 1 point for selecting a block height to do the casework on, up to 8 points for doing the casework correctly (6/8 for minor errors, such as calculation mistakes, 3/8 for a major misunderstanding but partial progress, 1/8 for almost no progress), 1 point for observing the upper limit on the number of blocks, 1 point for conclusion, 1 point for the final answer.

4. (Geometry) 10 points:

    (a) 7 points: 1 point for specifying all necessary givens, 1 points for any correct theorem statement. For the proof, 1 point for using Law of Sines, 1 point for the correct choice of triangles, 1 point for setting up the proper equations, 1 point for supplementary angles, 1 point for observing that $\overline{CD}$ is the angle bisector, 1 point for a complete solution and answer.

    (b) 3 points: 1 point for partial progress, 1 point for good progress, 1 point for the final answer.

5. (Geometry) 15 points:

    (a) 3 points: 1 point for each case correctly stated.

    (b) 5 points: 1 point for considering the equality case, 2 points for designating the feet and obtaining $AE \cdot EC + EF^2 = FC^2$, 1 point for reducing this to $AE \cdot EC + EO^2 = CO^2$, $DE \cdot EB + EO^2 = BO^2$, 1 point for observing $BO = CO$ and subsequently finishing the proof.

    (c) 7 points: 1 point for the line extension and defining M, 1 point for the point L, 2 points for the triangle similarity, 1 point for angle congruence and $BM = IM, AI \cdot IM = 2R \cdot r$, 1 point for extending $\overline{OI}$ to hit the circumcircle, 1 point for completion.

6. (Discrete/NT) 10 points: 3 points for the three non-trivial cases, 2 points for the correct amount in the first case, 2 points for the correct amount in the second case, 2 points for the correct amount in the third case, 1 point for the final answer (35).

7. (Algebra) 15 points:

    (a) 6 points: 2 points for observing that f is monotonically increasing for $x \geq 0$, 1 point for $f(x) = x$, 2 points for solving the resulting equation, 1 point for the answer.

    (b) 9 points: 2 points for using the Fundamental Theorem of Algebra to determine that there are at least 3 real roots, 3 points for obtaining a sixth-degree polynomial, 2 points for realizing that this forces $x^2 < 0$, 1 point for writing that this forces 6 complex roots, 1 point for the conclusion.

8. (Geometry) 10 points: 2 points for the expression of the length of the string, 2 points for justification of the expression, 2 points for the resulting equations, 1 point for the Triangle Inequality constraint, 2 points for the setup of the final answer, 1 point for the actual answer (505).

9. (Discrete) 10 points:

    (a) 4 points: 2 points for the concentric circle setup, 1 point for a demonstrated understanding of a bipartite graph, 1 point for a correct construction.

    (b) 6 points: 2 points for equating K_n to the regular n-gon, 2 points for inscribing/circumscribing it, 2 points for writing that twice the side length does not exceed the circumradius.

10. (NT) 15 points:

    (a) 4 points: 2 points for m and k (1 each), 1 point for the listing terms tactic, 1 point for showing 10-periodicity.

    (b) 5 points: 2 points for the block sizes and sums (1 each), 1 point for the number of blocks, 1 point for the sum up to 2016 entries, 1 point for the last three entries and final answer (9416).

    (c) 6 points: 1 point for considering mod 3, mod 5 by CRT, 1 point for the mod 3 sequence, 1 point for the mod 5 sequence, 1 point for $n \equiv 2 \pmod 3$, 1 point for $n \equiv 0, 6, 7, 9 \pmod{20}$, 1 point for the final set of n (all must be correct to earn the point).

11. (Algebra) 18 points:

    (a) 9 points:

        i. 2 points: 1 point for recognizing scalar multiplication as an elementary row operation, 1 point for the correct relation.

        ii. 3 points: 2 points for writing $\det M = \det M^T$, 1 point for observing that $\det M = 0$.

        iii. 4 points: 1 point for any equivalent claim, 1 point for the linearity of the adjoint matrix, 1 point for observing that $\text{adj}(M)_{i,j} = 0 \iff$ two rows/columns are identical, 1 point for the correct conclusion.

    (b) 5 points:

        i. 2 points: 1 point for setting up the column vectors, 1 point for the correct definition of the dot product.

        ii. 3 points: 1 point for writing that the output is a scalar, 1 point for writing that the other element is a vector, 1 point for concluding that we cannot take the dot product of a scalar and a vector.

    (c) 4 points: 2 points for making the determinant connection, 2 points for actually evaluating the determinant (1/2 for correct procedure, but wrong final answer).

12. (Algebra) 12 points:

    (a) 8 points: 2 points for defining $\arg(z)$, 1 point for writing $z_1 + z_2 = (a+c) + (b+d)i$, 3 points for the arguments of $z_1, z_2, z_1 + z_2$, 2 points for the final equation.

    (b) 4 points: 2 points for the expressions for the tangents (all or nothing), 2 points for algebraic simplification to a reasonable degree.

13. (Geometry) 15 points:

    (a) 7 points: 1 point for defining Q, R, 1 point for observing Q, R, P are collinear, 3 points for triangle similarity, 1 point for the resulting equation, 1 point for mentioning the case where P is undefined at all.

    (b) 8 points: 2 points for Menelaus, 1 point for defining A, B, C and A, B, C, 3 points for invoking the result of (a) to get $\dfrac{AC}{CB} = \dfrac{r_A}{r_B}$ and the other two equations, 2 points for finishing.

14. (Algebra) 15 points:

    (a) 6 points:

        i. 2 points: 1 point for the expression, 1 point for reasonable justification/understanding.

ii. 2 points: see above.

iii. 2 points: 1 point for considering proability of sum n, 1 point for the sum in any reasonable format (simplified or not).

(b) 4 points: 1 point for an attempt at an inductive approach, 1 point for considering the changing sign, 1 point for manipulating the binomial coefficients, 1 point for the final answer.

(c) 5 points: 1 point for defining $G_s(x)$ (along with the random variables and S, 1 point for writing $G_S(x) = \mathbb{E}(x^S)$, 1 point for writing x^S in expanded product form, 1 point for the last equation, 1 point for the conclusion (must use independence somehow to earn this point).

15. (Geometry) 20 points:

    (a) 8 points: 1 point for writing that this is Miquels Theorem. If this point is not earned, 1 extra point to observing that $AFOD, BEOD$ are cyclic (usually 3, but 4 in this case); 3 points for 5/5 angle measures (2 points for 4/5, 1 point for 2/5 or 3/5), 1 point for concluding that $CQPD$ is cyclic.

    (b) 12 points: 2 points for the two cases, 2 points for writing which segments are angle bisectors of which angles, 1 point for the measure of ECF, 1 point for congruence of ECF and QRP, 1 point for subsequent congruence of DAF, QPR, 1 point for similarity, 1 point for $EIF = 2(b-a)°$ in the second case, 1 point for the subtended arc/angle theorem, 1 point for congruence of ECF, HFI, 1 point for $DAF \cong HGI$ and the conclusion of spiral similarity.

Score	Interpretation
160-200	Well prepared for the Olympiad and proof-intensive classes
120-159	Prepared for proof-based contests (e.g. USAMTS) and classes
100-119	Somewhat qualified for proof-based contests and classes
70-99	May benefit from additional review of fundamental concepts
50-69	May require additional review of fundamental concepts
0-49	Not yet fully prepared for rigorous, proof-based study

Algebra Subscore	Interpretation
55-70	Well prepared for the Olympiad and proof-intensive classes
45-54	Prepared for proof-based contests (e.g. USAMTS) and classes
35-44	Somewhat qualified for proof-based contests and classes
25-34	May benefit from additional review of fundamental concepts
15-24	May require additional review of fundamental concepts
0-14	Not yet fully prepared for rigorous, proof-based study

Geometry Subscore	Interpretation
55-70	Well prepared for the Olympiad and proof-intensive classes
45-54	Prepared for proof-based contests (e.g. USAMTS) and classes
35-44	Somewhat qualified for proof-based contests and classes
25-34	May benefit from additional review of fundamental concepts
15-24	May require additional review of fundamental concepts
1-14	Not yet fully prepared for rigorous, proof-based study

Discrete/NT Subscore	Interpretation
50-60	Well prepared for the Olympiad and proof-intensive classes
40-49	Prepared for proof-based contests (e.g. USAMTS) and classes
30-39	Somewhat qualified for proof-based contests and classes
20-29	May benefit from additional review of fundamental concepts
15-19	May require additional review of fundamental concepts
0-14	Not yet fully prepared for rigorous, proof-based study

Appendix A: Counting by Infinities

When we toss around the idea of "infinity," usually, we like to think of it as an intangible abstraction representing the end-all-be-all of large quantities: a quantity so large that we can never actually reach it, even if we count for all eternity. There will always be *some* number larger than even the largest number one can think of. (And yes, that includes Graham's number!) But what exactly do we mean by "infinity?" The answer is surprisingly deep and interesting in both historical and current mathematical study.

As it turns out, there exists a whole tier of different infinities, given by the cardinalities of sets upon which we can establish bijections with them. For instance, consider the set of natural numbers, $\mathbb{N} = \{1, 2, 3, 4, \ldots\}$. It should be intuitively clear that this set does not have a finite number of elements, as it never ends; there is no such thing as a "largest" natural number. So the cardinality of $\mathbb{N}$ must be infinite. And indeed it is; more specifically, we denote its cardinality by $\aleph_0$, pronounced "aleph-null."

One might think that "aleph-null" is where things must logically end. But our story hasn't even started yet! There is an entire class of "aleph" numbers, denoted by $\aleph_n$ for all the positive integers n. $\aleph_1$ represents the next-largest cardinality after $\aleph_0$, $\aleph_2$ represents the cardinality after $\aleph_1$, and so forth. More specifically, n should be an *ordinal number* (a generalization of the natural numbers that incorporates the idea of collections of objects in order, with one object following another. (Here it should be noted that $\aleph$ does not quite convey the same idea as does ∞. Whereas $\aleph$ numbers measure set cardinalities (sizes), ∞ is more of a conceptual abstraction; the *limit* of the real line.)

Going from our base - that $\aleph_0$ counts the cardinality of $\mathbb{N}$ - logically it should follow that $\aleph_1$ enumerates the cardinality of ordinal numbers. To define further $\aleph$ numbers, we must define a *successor cardinal operation*, which assigns to each cardinal number n the next largest cardinal number n^+. Then $\aleph_{n+1} = \aleph_n^+$, with

$$\lim_{t \to \infty} \aleph_t = \bigcup_{k \in \mathbb{N}} \aleph_k$$

But what about sets other than the natural numbers? Is there such a thing as a *truly* infinite set, or does there exist an infinity that is simply too large to accommodate even other infinities? If we take a look at the set of rationals from a unique and perspective, we can actually show that the cardinality of $\mathbb{Q}$ is larger than that of $\mathbb{R}$ - infinitely so, in fact - as Cantor did in his *diagonal argument!* Essentially, Cantor's diagonal argument considered a set of infinite sequences of digits (binary in his formulation, but any base works equally well) $\{s_1, s_2, s_3, \ldots, s_n\}$. If we define a sequence s_{n+1} that differs from each of the other sequences in at least one position, then the new sequence of digits (comprising a decimal number) will clearly not be equal to any of the numbers $s_1, s_2, s_3, \ldots, s_n$. Hence, the set of all infinite sequences of binary digits is uncountable, whereas the set of reals is countable; therefore, the cardinality of the set of infinite binary sequences must havea a greater cardinality than $\aleph_0$!

Appendix B: Studying the Roots of nth-Degree Polynomials for $n \geq 3$

For polynomials of degree 2 or less, it is very straightforward to find their roots (using Quadratic Formula in the case of quadratics, and trivial for degree 1 (linear) or especially degree 0 (constant) polynomials). However, how do we go about finding the exact roots of polynomials with degree 3 or higher?

In the particular case of cubics, we can use *Cardano's formula,* a not-very-easy to memorize or use formula that nevertheless yields the useful technique of *cubic depression.* In order to depress a cubic of the form $ax^3 + bx^2 + cx + d = 0$ into a more managable form, we make the subtitution $x = y - \dfrac{b}{3a}$, which is motivated in part by Vieta's formulas. It follows through expansion that $ay^3 + \left(c - \dfrac{b^2}{3a}\right)y + \left(d + \dfrac{2b^3}{27a^2} - \dfrac{bc}{3a}\right) = 0$ hence eliminating the y^2 term altogether. To solve the resulting depressed equation (of the general form $y^3 + Ay = B$), make the substitutions $A = 3st$, $B = s^3 - t^3$, with $y = s - t$ (due to the difference of cubes identity).

For quartic polynomials of the form $ax^4 + bx^3 + cx^2 + dx + e = 0$, we can exploit the inherent symmetry of even-degree polynomials by dividing through by x^2, then using the fact that $\left(x + \dfrac{1}{x}\right)^2 - 2 = x^2 + \dfrac{1}{x^2}$ to simplify the expression, make a substitution, solve for the resulting quadratic, and then back-substitute.

The degree 5 polynomial, however, is a major turning point in that we cannot solve for the exact roots in terms of radicals. This is a landmark consequence of *Galois theory,* more specifically the *Abel-Ruffini Theorem.* Note that the theorem does *not* posit that degree-5 or higher equations have no solutions at all; quite the opposite. However, they do have all transcendental solutions (real) or complex solutions. The basic outline of the proof relies on the fact that Galois theory holds for any field of characteristic 0, as well as the fact that a polynomial in said field is solvable in algebraic numbers iff its splitting field has a solvable Galois group (which we take as fact here). We henceforth compute the fifth-degree Galois group, and show it is not solvable by way of expanding the ensuing symmetric sums. Indeed, this approach works to disprove the existence of algebraic solutions to an nth degree polynomial for *any* $n \geq 5$.

Appendix C: Proving the Fundamental Theorem of Algebra

In tandem with our prior observation that no polynomial with degree 5 or greater has algebraic roots, we may be motivated to prove the fundamental theorem of algebra: that each and every degree-n polynomial does, in fact, have exactly n roots, real or complex (accounting for multiplicity).

Proof. We make extensive use of two primary facts: firstly, that every polynomial with odd degree and real coefficients has at least one real root, and secondly, that every non-negative real number has a square root that is also a real number. This second fact, together with the quadratic formula, proves the theorem for quadratics with real coefficients. Thus, generalizing to any real-closed field R, $C = R(\sqrt{-1})$ is closed.

It then suffices to prove that every polynomial $p(x)$, $\deg p \geq 1$, with real coefficients has a root in $\mathbb{C}$. To show this, we induct upon positive integers k such that $2^k \mid \deg p$. Let p have a term ax^n, and let C have splitting field $\mathbb{F}$; then $\exists x_1, x_2, x_3, x_n \in \mathbb{F}$ such that $p(x) = a\prod_{i=1}^{n}(x - x_i)$. If $k = 0$, then n is odd and $p(x)$ has a real root. Otherwise, let $n = m \cdot 2^k$, m odd. For $t \in \mathbb{R}$, let

$$q_t(x) = \prod_{1 \leq i \leq j \leq n} (x - x_i - x_j - tx_ix_j)$$

Then this polynomial has the symmetric polynomials in terms of x_i for its coefficients, or as polynomials with real coefficients in the elementary symmetric polynomials. Therefore, $q_t(x)$ has real coefficients. Additionally, $\deg q_t = \frac{1}{2}n(n-1) = m(n-1) \cdot 2^{k-1}$, with $m(n-1)$ odd. By the inductive step, q_t has a complex root, and so $x_i + x_j + tx_ix_j \in \mathbb{C}$ for $i, j \in \{1, 2, 3, \ldots, n\}$, $i \neq j$. There are fewer ordered pairs (i, j) than there are real numbers, so there exist t_1, t_2 with $x_i + x_j + t_1x_ix_j, x_i + x_j + t_2x_ix_j \in \mathbb{C}$ given the same ordered pair (i, j). Hence, $x_i + x_j, x_ix_j \in \mathbb{C}$, and so every quadratic polynomial with complex coefficients has a complex root (by Quadratic Formula). Then x_i, x_j themselves are complex, as desired. $\square$

Appendix D: The Many, Many Centers of a Triangle

In the Geometry sections, we covered some of the most commonly-used centers of a triangle (such as the incenter, circumcenter, orthocenter, nine-point center, and centroid). Indeed, there is a veritable abundance of centers of a triangle - thousands of triangle centers that all technically qualify as the "center"! The actual amount is staggeringly high, and all centers known to date are logged in the Encyclopedia of Triangle Centers. As of December 2018, 30,451 triangle centers have been found. Each center is denoted X_n, where n is a positive integer. Some of the more traditional centers have also made the cut, but the encyclopedia primarily focuses on more unique and esoteric types of centers that we usually do not work with in a competition setting.

In particular, X_1 is the incenter, X_2 is the centroid, X_3 is the circumcenter, X_4 is the orthocenter, X_5 is the nine-point center, and X_6 is the symmedian point - all points that we have hitherto discussed. But from X_7 onward, the points gradually become less well-known, until eventually we reach a point where the study of these points is in itself an interesting area of focus. There are points such as X_7 and X_8 (the Gergonne and Nagel points respectively) that are still in the knowledge base and reserved for specialized contexts, but triangle centers such as X_{22} (called the *Exeter point*) are mostly inaccessible to those who have not explicitly memorized them (for instance, they do not have derivations that are commonly learned - or at all!) Indeed, the entire list (aside from the first 6 and select others) is more of a curiosity than anything else, mainly to demonstrate that triangles and their study are far less limited than we may think.

Appendix E: Exploring Projective Geometry

In Euclidean geometry, all lines intersect except for parallel lines. Hence, parallel lines are the exception case to certain key geometry theorems, such as *any two non-parallel lines intersect at a unique point*. To combat this exception, we created a new type of geometry called **projective geometry**, in which parallel lines do intersect at *points at infinity*.

Here's a little bit of intuition for visualizing points of infinity: Imagine having two intersecting lines, and changing the slope of one of them little by little, moving the intersection point, until the lines are parallel, as shown in the diagram below.

Now let's go back and consider the theorem we brought up in the beginning and rephrase it in "projective geometry language." This means that we remove the exception of parallel lines since they, too, intersect now. The theorem states that *any two lines intersect at a unique point in the projective plane*.

Proof. Firstly, I should note that this proof is more of a basic construction of the projective geometry universe than a proof. However, this theorem is proven in the process, so I still retain the header of "proof."

There are two types of lines: parallel lines and intersecting lines (note: intersecting lines refer to those that intersect in the Euclidean plane). We will consider each case individually.

As shown in the diagram on the previous page, two parallel lines intersect at a unique point at infinity. Since they don't intersect in the Euclidean plane, parallel lines have a unique intersection in the projective plane. (The picture on the next page depicts a line m and its point at infinity. All lines parallel to m intersect at that point at infinity.)

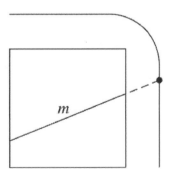

Intersecting lines intersect uniquely in the Euclidean plane, but they intersect at different points at infinity, as shown in the diagram below:

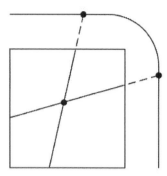

Therefore, any two lines have a unique intersection in the projective plane. □

Now we will consider a theorem that is also fundamental to Euclidean geometry: Two points construct a unique line. Taking the fact that we are working in projective geometry into account (in other words, points at infinity exist unlike in Euclidean geometry), we state that *any two points lie in a unique line in the projective plane*.

Proof. We will proceed via casework on the identities of the two points; the three cases are two points at infinity, two Euclidean points, and one of each.

Every point at infinity is on the unique line at infinity, so two points at infinity lie on a unique line in the projective plane, as shown below:

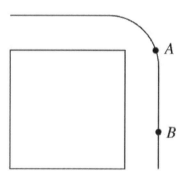

As we know from Euclidean geometry, two Euclidean points lie on a unique line (projective geometry is very similar in that regard).

Now we must consider one of each (one point at infinity and one point in the Euclidean plane). Note that this may be confusing because a whole family of parallel lines intersects at the given point at infinity, as shown below:

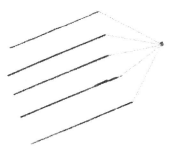

However, only one member of the family of parallel lines goes through the given Euclidean point (diagram shown below). Therefore, since this case is proven, any two points create a unique line in the projective plane.

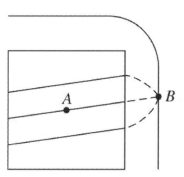

□

Homogeneous Coordinates

Recall from the introduction that we were often referring to something called the **projective plane** as, in informal terms, the Euclidean plane plus points at infinity. Now let us formally define it as the set of points determined by the parametric equations $x = ta, y = tb, z = tc$ such that $a, b, c, t \in \mathbb{R}$ and $t \neq 0$. The ordered triples (ta, tb, tc) present in the projective plane are called **homogeneous coordinates** (the term "homogeneous" refers to the fact that the constants a, b, and c are multiplied by the same amount of multiples of variable t, and, therefore, are considered the same point).

For example, the points $(3, 4, 5); (6, 8, 10);$ and $(0.6, 0.8, 1)$ all represent the same point. The point $(0.6, 0.8, 1)$ lies on the $z = 1$ plane, which is considered to be the Euclidean plane; in other words, points on that plane are considered two-dimensional. A diagram of the general case is shown below:

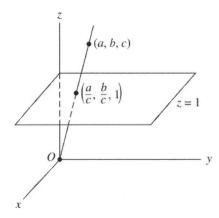

However, we cannot convert any point with a z-coordinate of 0 into a Euclidean point (make the z-coordinate equal to 1). Therefore, these remaining points are classified as the *points at infinity* and the origin is associated (equal to, in fact) with all points.

Just like in Cartesian coordinates, we can also write equations in homogeneous coordinates. However, just because equations are graphed in homogeneous coordinates does not mean that they are **homogeneous equations**. In order for an equation to be homogeneous, the sum of the exponents of all the variables in the equation must be the same. For example, $y = x^3$ is not a homogeneous equation, but $yz^2 = x^2$ is.

How do homogeneous equations connect to homogeneous coordinates? Let's consider the non-homogeneous equation $y = x^3$ again and plug in our parameterization for homogeneous coordinates: $tb = (ta)^3 = t^3 a^3$. The problem here is that the degrees of the variable t are not equal on both sides. The homogeneous equation, $yz^2 = x^2$, combats this problem as shown: $tb * (tc)^2 = (ta)^3$ or $t^3 bc^2 = t^3 a^3$ as desired.

The example that I just gave actually feeds into the next part of this section: how to convert non-homogeneous equations into homogeneous equations. Notice that we multiplied the left side of the non-homogeneous equation above by powers of z so that the degrees of both sides matched up. In short, this is what we do in order to convert non-homogeneous equations (2-dimensional; if there were a different amount of dimensions, we would just multiply by the next variable not in use instead) into homogeneous equations, which are one dimension higher.

Let's analyze the example equations geometrically. We can consider the normal (non-homogeneous) equation $y = x^3$ to reside in the Euclidean plane; that is, it resides in the plane $z = 1$.

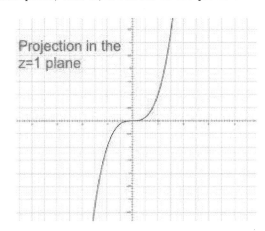

Now let us consider the homogeneous version: $yz^2 - x^3$. Below are three cross-sections of the 3-dimensional graph.

The xy-plane cross section. Note that it is a cubic like its projection onto the $z = 1$ plane. (To write this in equation format, $1^2 y - x^3 = 0$ or $y = x^3$.)

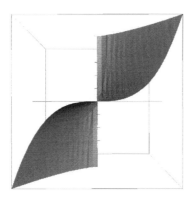

The xz-plane cross section. $z^2 1 - x^3 = 0$ or $z^2 = x^3$

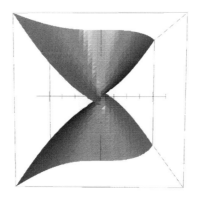

The yz-plane cross section. $z^2 y - 1^3 = 0$ or $yz^2 = 1$ or $y = \dfrac{1}{z^2}$

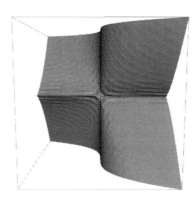

In fact, each cross section shown looks like the homogeneous graph's projection onto one of the Euclidean planes (i.e. we set one of the variables equal to 1). Each projection equation is shown in the heading preceding the graph of its cross-section.

If we consider lines in the $z = 0$ plane, then we see a cool representation of points at infinity (remember that $z = 0$ is the projective plane). In fact, we can see them in the $x = 1$ and $y = 1$ planes, as shown above. These planes, unlike $z = 1$, are not parallel to $z = 0$, and thus intersect to show points at infinity. For example, in the top diagram on this page, at $z = 0$ the curve has a cusp.

Conclusion

There are a lot of different directions that projective geometry can be carried to. It includes a full theory about conic sections, it can be applied to show perspective in artwork, and much more. This paper provides a mere introduction to such a vast area of mathematics, not just limited to geometry.

Part VIII

Solutions to Exercises

Part 4: Applied Discrete Math

Chapter 10: Beginning Discrete Math

Section 10.1: Introduction to Set Theory

Exercise 1. Let $S_1 = \{1\}$, $S_2 = \{1, \mathcal{P}(S_1)\}$, $S_3 = \{1, \mathcal{P}(S_2)\}$, and so forth. What is S_n in terms of n?

Solution 1. We have $\mathcal{P}(S_1) = \{\emptyset, \{1\}\}$, so $S_2 = \{1, \{\emptyset, \{1\}\}\}$. Similarly, $S_3 = \{1, \{\emptyset, \{1\}, \{\emptyset, \{1\}\}, \{1, \{\emptyset, \{1\}\}\}\}\}, \ldots$ and in general, S_n is the set with a 1, and a second set consisting of the empty set, the element 1, the empty set and the set with only 1, the set consisting of 1 and all the aforementioned subsets, and so forth. In particular, S_n will always have 2 elements, but the second element will have 2^{n-1} elements, as $|\mathcal{P}(S)| = 2^{|S|}$.

Exercise 2. How many distinguishable Venn diagrams (in the sense that one can tell them apart visually based on which parts of the diagrams are shaded) are there that consist of 2 sets? What about 3 sets?

Solution 2. A 2-set Venn diagram has 3 disjoint regions, so there are $2^3 = 8$ possible diagrams. A 3-set diagram has $2^3 - 1 = 7$ disjoint regions, leading to $2^7 = 128$ possible diagrams.

Exercise 3. In the following definition:

A metric space is an ordered pair (M, d) where M is a set and d is a metric, often called a distance function, on M. d is defined to be a function from $M \times M$ to $\mathbb{R}$ such that, for any $x, y, z \in M$, we have

(a) $d(x, y) \geq 0$, and $d(x, y) = 0$ iff $x = y$.

(b) $d(x, y) = d(y, x)$.

(c) $d(x, y) + d(y, z) \geq d(x, z)$ (Triangle Inequality).

Show that the first part of condition (a) follows from the second part of condition (a), and conditions (b) and (c).

Solution 3. If $x \neq y$, then $d(x, y) > 0$, since distance cannot be negative. The reverse direction of the second part of (a) is trivial. This also follows from (b) and (c), which serve to ensure that, if $d(x, y) \geq 0$ on some subset of $M \times M$, $d(x, y) \geq 0$ on the entire set $M \times M$.

Exercise 4. Verify that all of the following are examples of metric spaces:

1. The real line; $d(x, y) = |x - y|$.

2. $\mathbb{R}^2$ equipped with its usual "distance function": $d((x_1, y_1), (x_2, y_2)) = \sqrt{(x_2 - x_1)^2 + (y_2 - y_1)^2}$, often called the 2-metric d_2.

3. The taxicab metric $d((x_1, y_1), (x_2, y_2)) = |x_2 - x_1| + |y_2 - y_1|$, often called the 1-metric d_1.

4. The p-metric d_p for any $p \in \mathbb{N}$ (essentially the Distance Formula in p dimensions).

5. The supremum metric or maximum metric d_∞, given by $d((x_1, y_1), (x_2, y_2)) = \max(|x_2 - x_1|, |y_2 - y_1|)$.

6. The discrete metric: $d(x,y) = 0$ if $x = y$, and $d(x,y) = 1$ otherwise.

Solution 4. *(1) is a metric space, as $|x-y| \geq 0$ for all $x, y \in \mathbb{R}$, $|x-y| = |y-x|$, and $|x-y| + |y-z| \geq |x-z|, \forall x, y, z \in \mathbb{R}$ (which itself can be proven by casework on the signs of the arguments).*

(2) is also a metric space; (a) is clear, and (b) follows immediately from symmetry. (c) holds in the standard two-dimensional Euclidean geometry.

(3) is a metric space as well. This is similar to (1).

(4) is a metric space; this is just a generalization of (2).

Furthermore, (5) is a metric space, as $d((x_1, y_1), (x_2, y_2)) \geq 0$ for all $(x_1, y_1), (x_2, y_2)$. In addition, it has symmetry in $(x_1, y_1), (x_2, y_2)$, and obeys the triangle inequality (not difficult to check, but tedious enough to not be instructive here).

Finally, (6) is indeed a metric space, as $d(x,y) \geq 0$ for all x, y, $d(x,y) = 0 \iff x = y$ a priori, $d(x,y) = d(y,x)$, and $d(x,y) + d(y,z) \geq d(x,z)$ (since $d(x,z) = 0$ if both $d(x,y) = 0$ or $d(y,z) = 0$, by the Transitive Property).

Note, however, that $d(x,y) = 1$ if $x = y$ and $d(x,y) = 0$ is not a metric space, as it fails the last condition! A counterexample is $(x, y, z) = (1, 0, 1)$; here, $d(x,y) = 0, d(y,z) = 0$, but $d(x,z) = 1$, thereby failing to satisfy the Triangle Inequality.

Exercise 5. Each of the following sets is either described in English, or has its (first few) elements listed; write each in set-builder form.

(a) $A = \{\text{the set of letters in the word "mathematics"}\}$

(b) $B = \{1, 1, 2, 4, 7, 13, 24, \ldots\}$

(c) $C = \{1, 2, 4, 8, 16, 32, 64, \ldots\}$

(d) $D = \{\text{the set of all perfect squares that have exactly 3 positive integral divisors}\}$

(e) $E = \{\text{the set of positive integers that can be written as } 1 + 2 + 3 + \ldots + k \text{ for some positive integer } k\}$

(f) $F = \{2, 3, 5, 7, 11, 13, 17\}$

Solution 5.

(a) $A = \{l \mid l \in \{m, a, t, h, e, i, c, s\}\}$

(b) $B = \{n \mid n = S_k, k \in \mathbb{N}\}$ (Here S_k denotes the recursive sequence with $S_1 = S_2 = S_3 = 1$, and $S_k = S_{k-1} + S_{k-2} + S_{k-3}$ for all $k \geq 4$.)

(c) $C = \{2^n \mid n \in \mathbb{N} \cup \{0\}\}$

(d) $D = \{p^2 \mid p \text{ prime}\}$

(e) $E = \{\frac{k(k+1)}{2} \mid k \in \mathbb{N}\}$

(f) $F = \{p \mid p \text{ prime}, p \leq 17\}$

Exercise 6. *Using a Venn Diagram, demonstrate both of De Morgan's Laws.*

Solution 6.
First law: $(A \cup B)^c = A^c \cap B^c$

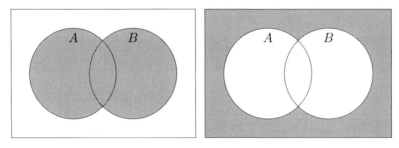

Second law: $(A \cap B)^c = A^c \cup B^c$

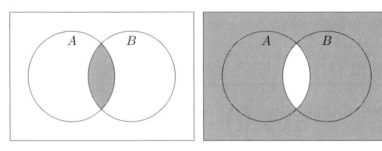

Exercise 7. *Show that $S = T$ iff $S \subseteq T$ and $S \supseteq T$.*

Solution 7. If $S \subseteq T$, then T contains all elements in S. At the same time, S contains all elements in T. For contradiction, assume that $S \neq T$; i.e. at least one element in S is not in T, without loss of generality. But then S is not a subset of T - contradiction.

Exercise 8. *Prove that if $S \cap T = S$, then $S \subseteq T$.*

Solution 8. By definition, all elements in S are contained in both S and T. Hence, T contains all elements of S, which implies $S \subseteq T$.

Exercise 9. *Prove that $(A^c)^c = A$ for all sets A in some universal set U.*

Solution 9. Let $A \subseteq U$. We show that $A \subseteq (A^c)^c$ and $(A^c)^c \subseteq A$, using the fact that if $s \in S, s \notin S^c$.

Let $a \in A$. Then $a \notin A^c$, and subsequently, $a \in (A^c)^c$. Thus, $A \subseteq (A^c)^c$. In the reverse direction, take $a \in (A^c)^c$. Then $a \notin A^c \implies a \in A$, so $(A^c)^c \supseteq A$. Hence $A = (A^c)^c$.

Exercise 10.

(a) *Prove that the interval $(1, 2)$ is open.*

(b) *Prove that the interval $[0, 1]$ is closed.*

(c) *Prove that the interval $[0, 2)$ is neither open nor closed.*

Solution 10.

(a) Let $c \in \mathbb{R}, c \in (1, 2)$. Let $0 < \epsilon < \min(2 - c, c - 1)\epsilon \in \mathbb{R}^+$. Denote by $B_\epsilon(c)$ the open ϵ-ball about c. Then $c + \epsilon < 2$ and $1 < c - \epsilon$. Hence, $B_\epsilon(c) \subseteq (1, 2)$ and $(1, 2)$ is a neighborhood of c, which is the definition of an open set.

(b) *By definition, a set in a universal set U is closed iff its complement in U is open; that is, in this case, $(-\infty, 0) \cup (1, \infty)$ is open. This amounts to showing that the union of two open sets is open (see part (a) above). Denote $S_1 = (-\infty, 0), S_2 = (1, \infty)$. Take $x \in S_1 \cup S_2$. Then $x \in S_1$ or $x \in S_2$ by the definition of the union. If $x \in S_1$, since S_1 is open, $\exists \epsilon > 0$ such that $B_\epsilon(x) \subseteq S_1 \subseteq S_1 \cup S_2$, and $B_\epsilon(x)$ is an open set. Hence $S_1 \cup S_2$ is also open.*

(c) *Consider $B_\epsilon(2)$. For any $\epsilon > 0$, $B_\epsilon(2)$ contains two points, one of which is in $[0, 2)$ and one of which is not in $[0, 2)$. Hence $[0, 2)$ is neither open nor closed.*

Exercise 11.

(a) *Prove that the union of a finite number of closed sets is also closed.*

(b) *Prove that the intersection of a finite number of open sets is also open.*

Solution 11.

(a) Let S_1, S_2 be two closed sets, and let x be a limit point of $S_1 \cup S_2$. Given this, $x \in S_1 \cup S_2$ by definition. Suppose that x is not a limit point of both S_1 and S_2. Then there exist a, b such that $B_a(x), B_b(x)$ do not intersect with S_1, S_2 (respectively) except for x. Let $m = \min(a, b)$; then $B_m(x)$ will not intersect with $S_1 \cup S_2$ except for x; this contradicts x being a limit point, implying the result. We can extend this to finitely many closed sets as needed.

(b) Let $X = S_1 \cap S_2$. Then $x \in X \implies x \in S_1 \land x \in S_2$, which implies the existence of some open ball $B_r(x)$ with radius r centered at x.

For contradiction, assume X is not open. Then there exists $x \in X$ such that x is on the boundary of X, i.e. it lies on the boundary of either S_1 or S_2. But this is a contradiction; x must be an interior point of both S_1, S_2. Hence, all $x \in X$ are interior points, and $X = S_1 \cap S_2$ is open by definition. This can be extended similarly to any finite number of open sets. (Note that the "finite" is important here: the intersection of all sets of the form $S_n = \left(-\frac{1}{n}, \frac{1}{n}\right)$ is $\{0\} = [0, 0]$, which is a closed set!)

Exercise 12. *For any two sets A and B, show that $\mathcal{P}(A \cap B) = \mathcal{P}(A) \cap \mathcal{P}(B)$. Is it necessarily true that $\mathcal{P}(A \cup B) = \mathcal{P}(A) \cup \mathcal{P}(B)$?*

Solution 12. Let $X \in \mathcal{P}(A \cap B)$. Then $X \subseteq A \cap B \implies X \subseteq A \land X \subseteq B$, which further implies $X \in \mathcal{P}(A) \land X \in \mathcal{P}(B)$ by the definition of the power set. Finally, $X \in \mathcal{P}(A) \cap \mathcal{P}(B)$ by the definition of the set intersection. QED.

It is not necessarily the case, however, that $\mathcal{P}(A \cup B) = \mathcal{P}(A) \cup \mathcal{P}(B)$. It is true, though, that $\mathcal{P}(A \cup B) \supseteq \mathcal{P}(A) \cup \mathcal{P}(B)$. This follows by setting $X \in \mathcal{P}(A) \cup \mathcal{P}(B) \implies X \subseteq A \lor X \subseteq B \implies X \subseteq A \cup B \implies X \in \mathcal{P}(A \cup B)$.

Exercise 13. *Is it possible for a set to be both open and closed? If not, provide a disproof. If so, provide an example of such a set.*

Solution 13. Yes; the empty set and $\mathbb{R}$ both qualify. The empty set is simultaneously open and closed, as its complement, $\mathbb{R}$, is open; also, $\emptyset$ itself is open because it does not contain any of its boundary points (by virtue of being empty!) $\mathbb{R}^c = \emptyset$ is open, as we have just shown, so $\mathbb{R}$ is closed. $\mathbb{R}$ is also open, as for any $\epsilon > 0, \exists x \in \mathbb{R}$ such that $(x - \epsilon, x + \epsilon) \in \mathbb{R}$, which is clearly true.

Exercise 14.

(a) *Prove or disprove the following: $A \times (B \cup C) = (A \times B) \cup (A \times C)$ for all sets A, B, C.*

(b) *Prove or disprove the following: $A \times (B \cap C) = (A \times B) \cap (A \times C)$ for all sets A, B, C.*

327

Solution 14.

(a) Note that $(S_1 \cup S_2) \times (T_1 \cup T_2) = (S_1 \times T_1) \cup (S_1 \times T_2) \cup (S_2 \times T_1) \cup (S_2 \times T_2)$ Let $S_1 = S_2 = A, T_1 = B, T_2 = C$. Then $A \times (B \cup C) = (A \cup A) \times (B \cup C) = (A \times B) \cup (A \times C) \cup (A \times C) \cup (A \times B) = (A \times B) \cup (A \times C)$. The result follows.

(b) We prove that $(S_1 \cap S_2) \times (T_1 \cap T_2) = (S_1 \times T_1) \cap (S_2 \times T_2)$ (with $A = A \cap A$, the result follows). To prove this, take $(x, y) \in (S_1 \cap S_2)$. Then $(x \in S_1 \wedge x \in S_2) \wedge (y \in T_1 \wedge y \in T_2) \implies (x \in S_1 \wedge y \in T_1) \wedge (x \in S_2 \wedge y \in T_2)$ implies $(x, y) \in S_1 \times T_1 \wedge (x, y) \in S_2 \times T_2 \implies (x, y) \in (S_1 \times T_1) \cap (S_2 \times T_2)$.

Exercise 15. *For each of the sets X and binary relations R on X, list whether or not R is a reflexive, symmetric, transitive, or equivalence relation on X, and justify your answers.*

(a) (i) $R =$ *"is a spouse of," $X = \{Michael, Steven, Nadia, Francis, Julia\}$. Nadia is married to Francis, and Julia is married to Michael.*

(ii) $R =$ *"is a husband of," $X = \{Michael, Steven, Nadia, Francis, Julia\}$. Nadia is married to Francis, and Julia is married to Michael.*

(b) $R =$ *"is relatively prime with," $X = \mathbb{N}$*

(c) (i) $R =$ *"is at least 1 (Cartesian coordinate) unit away from," $X = \mathbb{N}^2$*

(ii) $R =$ *"is at most 1 (Cartesian coordinate) unit away from," $X = \mathbb{R}^2$*

Solution 15.

(a) (i) R is not reflexive, since a person is not his or her own spouse. R is symmetric, since no person is polygamous (each person married to another person is also the spouse of that other person). R is not transitive, since $aRb \wedge bRc$ does not imply aRc (once again, no one is married to two people, and rightfully so!) Hence, R is not an equivalence relation.

(ii) R is not reflexive; one cannot be one's own husband. R is also not symmetric, since it will always be false for a female; e.g. for Francis it holds true, but for Nadia it is false. R is again non-transitive; finally, it is not an equivalence relation.

(b) R is not reflexive (and thus not an equivalence relation); not every element of $\mathbb{N}$ is relatively prime with itself (indeed, only 1 is). R is symmetric, however, since if $\gcd(a, b) = 1$, then $\gcd(b, a) = 1$. R is not transitive; consider $(a, b, c) = (2, 3, 4)$. aRb is true, as is bRc, but aRc is false.

(c) (i) R is not reflexive (and is not an equivalnece relation); all points are 0 units away from themselves. It is symmetric, since distance remains the same regardless of order, but not transitive (we can take a different point, then return back to the same point).

(ii) Similarly, R is not reflexive (and hence not an equivalence relation). It is symmetric, but isn't transitive (same logic as above, not to mention not all points in $\mathbb{R}^2$ are at most 1 unit away from each other).

Extension Exercise 1. *Prove the cardinality relationships in Definition 10.4.*

Extension Solution 1. *If there is a bijection between sets A and B, clearly, $|A| = |B|$ by the definition of a bijection. Furthermore, a function is injective iff it is one-to-one; i.e. every element in the domain is mapped to exactly one element in the codomain. Since no two elements in the codomain may be mapped to the same element in the domain, the desired conclusion directly follows.*

Extension Exercise 2. *(2015 Berkeley Math Tournament) Find two disjoint sets N_1 and N_2 with $N_1 \cup N_2 = \mathbb{N}$, so that neither set contains an infinite arithmetic progression.*

Extension Solution 2. *One configuration that works is $N_1 = \{1, 4, 5, 6, 11, 12, \ldots, \}$, $N_2 = \{2, 3, 7, 8, 9, 10, \ldots\}$ with $N_1 \cup N_2 = \mathbb{N}$. Any arithmetic progression would need to be of the form $(a, a + d, a + 2d, \ldots)$. But regardless of the chosen value of d, if we split at each triangular number, there is no way we can continue the arithmetic progression infinitely, since we can form arbitrarily large intervals with gaps in between $a + dk$ and $a + d(k + 1)$ for some k.*

Extension Exercise 3. *Prove that the intersection of two equivalence relations on a nonempty set X is also an equivalence relation.*

Extension Solution 3. *Let the equivalence relations be R and S. Then they are relations on the same set, and so are reflexive, symmetric, and transitive on that set. We have that $R \cap S$ is the set of all pairs of elements in the set with $(x, y) \in R \wedge (x, y) \in S$. Elements that are necessarily included in $R \cap S$ are those that are reflexive, symmetric, and transitive themselves. Then check to see that they are in both R and S.*

Extension Exercise 4. *Let $f : \mathbb{R} \mapsto \mathbb{R}$. Define a binary relation R on f by xRy iff $x = y \iff f(x) = f(y)$. Show that R is an equivalence relation. What are the equivalence classes for $f(x) = x$, $f(x) = x^2$, $f(x) = x^2 + 20x + 18$, and $f(x) = \sin x$? (Here, x is in radians, not degrees.)*

Extension Solution 4. *R is reflexive, as xRx is true for all x (i.e. $x = x \iff f(x) = f(x)$ is trivially true). R is symmetric, since $y = x \iff f(y) = f(x)$ (equality itself is symmetric, hence this must be). Finally, R is transitive; consider x, y, z such that $y = z \iff f(y) = f(z)$. Then we indeed have $x = z \iff f(x) = f(z)$ by Transitive Property.*

The equivalence class for $f(x) = x$ is $\mathbb{R}$, since $x = x$ trivially holds. For $f(x) = x^2$, it is the set $\{0, 1\}$. For $x^2 + 20x + 18$, after writing $x^2 + 20x + 18 = 0 \implies x^2 + 19x + 18 = 0$ and factoring as $(x + 1)(x + 18) = 0$, we get $\{-1, -18\}$. Finally, for $\sin x$, $x = 0$ is the only solution.

Extension Exercise 5. *Let $n \in \mathbb{N}$, and let $x, y \in \mathbb{Z}$. The equivalence relation $x \sim y$ of modulo n has an associated quotient set, $\mathbb{Z}_n = \mathbb{Z}/\sim\, = \{[0], [1], [2], \ldots, [n-1]\}$. For an integer $a \in \mathbb{Z}$, define $f_a : \mathbb{Z}_n \mapsto \mathbb{Z}_n$, $f_a([x]) = [ax]$. Prove that f_a is a bijection iff a and n are relatively prime.*

Extension Solution 5. *As it turns out, this is actually wholly equivalent to the Chinese Remainder Theorem! Let $\gcd(a, b) = 1$; consider the map $\phi : \mathbb{Z}/ab\mathbb{Z} \mapsto (\mathbb{Z}/a\mathbb{Z}) \times (\mathbb{Z}/b\mathbb{Z})$ defined by $\phi(x) = (x \pmod{a}, x \pmod{b})$. Then $\phi(x)$ is uniquely determined, as well as surjective (since the system of modular congruences formed by the CRT algorithm always has a solution). Finally, ϕ is a bijection, as its domain and codomain have equal cardinalities.*

Section 10.2: Introduction to Statistics

Exercise 16. *A researcher selects 50 freshmen from a pool of 500 high school students, who attend the same high school, in order to evaluate their athletic capabilities. What types of sampling is this an example of?*

Solution 16. *This is an instance of simple random sampling, since each student has an equal likelihood of being chosen. It is also a prime example of stratified random sampling, since we are selecting purely from the pool of freshmen (as opposed to all high school students). Finally, it is an example of quota sampling, as the researchers stop after reaching their quota of 50 freshmen.*

Exercise 17. *For each of the following situations, determine, with explanation, whether there has been a Type I error or a Type II error.*

(a) *In a village during the dark of night, a boy cries out, "Wolf, wolf!" but amidst all the ensuing panic and commotion, there is no wolf to be found (and the people lose trust in the boy as a result).*

(b) *A few minutes later, a wolf barges through the gates out of nowhere and begins viciously attacking the villagers! However, this time, the boy has gone home. He is unaware of the wolf's presence, and so does not report anything.*

Solution 17.

(a) *This is a Type I error (false positive).*

(b) *This is a Type II error (false negative).*

Exercise 18.

(a) *What is the five-number summary of $S = \{2, 2.5, 3.5, 6, 8, 10, 12\}$?*

(b) *On the box plot drawn in part (a), also plot any outliers (elements below the 9th and above the 91st percentiles, using the "compromise" definition). Are these any different from outliers obtained from the > 1.5 IQR definition?*

Solution 18.

(a) $\{2, 2.5, 6, 10, 12\}$ *(minimum, first quartile, median, third quartile, maximum).*

(b) *S only has 7 elements, so it is not hard to verify that 2 and 12 are the sole outliers (at the $\frac{50}{7}^{th}$ and $\frac{650}{7}^{th}$ percentiles, respectively). As the IQR of S is $10 - 2.5 = 7.5$, 1.5 times the IQR is 11.25, which produces a drastically different result (median is 6; 1.5 IQR below the median is actually negative, and 1.5 IQR above is not in the range of the set at all!)*

Exercise 19. *A standardized test is graded with a real number ranging from 0 to 6. The mean score is 4.8, and the standard deviation is 0.9. To the nearest whole number, in what percentile does a score of 5.4 fall?*

Solution 19. *The score of 5.4 is $\frac{5.4 - 4.8}{0.9} = \frac{2}{3}$ standard deviations above the mean. According to a z-score table, this is approximately in the $\boxed{75^{th}}$ percentile.*

Exercise 20. *In the fictitious country of Mathlandia, there are two candidates for President: Sophie Noether and Emmy Germain. 51% of the voting population in one of Mathlandia's provinces, Pascalia, have a Noetherian party affiliation, while 49% of the Pascalian voting population have a Germainian party affiliation. In another Mathlandian province, Pythagorea, 52% of the voting population side with Noether, while the other 48% of the population side with Germain. A survey is conducted on a simple random sample of 100 people from each of Pascalia and Pythagorea. What is the probability that the survey will indicate that Pascalia has a smaller percentage of Noetherian voters than does Pythagorea? (Hint: Use the formula $\sigma_d = \sqrt{\frac{P_1(1-P_1)}{n_1} + \frac{P_2(1-P_2)}{n_2}}$ for the standard deviation of the difference (where P_1 and P_2 denote the proportions of Noetherian voters in Pascalia and Pythagorea respectively, and n_1 and n_2 denote the sample populations of Pascalia and Pythagorea respectively))*

Solution 20. We have $\sigma_d = \sqrt{\dfrac{0.51(0.49) + 0.52(0.48)}{100}} \approx 0.0707$. The difference in the proportions of Noetherians is 1%, so our z-score is $\dfrac{0 - 0.01}{0.0707} \approx -0.141$, which corresponds to an approximate probability of $\boxed{0.4443}$.

Exercise 21. *Mike is a participant in a study (of thousands of participants) where the researchers test his reaction time to a stimulus via a computer program. He is given 5 different stimuli to react to (and does not know when the stimulus will be triggered). His reported result will be the mean of his 5 reaction times.*

(a) *Given that the mean reaction time is 215 ms, the standard deviation of the sample population is 50 ms, and Mike's reported mean reaction time is 205 ms, approximately in what percentile is Mike's result? (Lower reaction times have lower percentiles; lower percentiles are considered "better" results.)*

(b) *Suppose that the researchers' hypothesis is that Mike's reaction time is the result of chance (and that he is actually perfectly average). What is the p-value associated with the hypothesis? Interpret this in terms of statistical significance.*

(c) *What is the interval of reported mean reaction times that would be considered statistically insignificant?*

(d) *What problems, if any, does this hypothetical study have that may affect the validity of the results? What parts of the study are properly conducted?*

Solution 21.

(a) Mike's result is at $\dfrac{205 - 215}{50} = -0.2$ standard deviations relative to the mean, which is his z-score. So his percentile is $\boxed{42\%\text{ile}}$, using a z-table.

(b) The p-value is the percentile, which is greater than/equal to 0.05, so the result is not statistically significant.

(c) Any result with a z-score corresponding to a percentile between the 5th and 95th percentiles is statistically insignificant; using a table, these are $z \in [-1.65, 1.65]$, so the reaction time should be in the interval $[250 - 1.65 \cdot 50, 250 + 1.65 \cdot 50] = [167.5, 332.5]$ ms.

(d) There is only one trial, which could fail to reflect the participants' actual reactive capabilities. Furthermore, a participant may have excellent reaction time with respect to one stimulus but slower reaction time with respect to another; the stimuli are all different, and so cannot possibly account for this given that each stimulus is only used once. On the other hand, it is proper procedure that the study is conducted with 5 stimuli as opposed to one, as this reduces variability in the results (i.e. the likelihood is lessened of the result being merely up to chance).

Extension Exercise 6.

(a) *Prove that, if X and Y are variables in a linear relationship, then the correlation coefficient of X and Y is ± 1.*

(b) *Prove the converse of the statement above.*

Extension Solution 6.

(a) This is by definition. The slope of a line is constant, so the ratio between x_i and $\overline{x}$ never changes (same with y_i and $\overline{y}$).

(b)
$$\left(\sum_{i=1}^n (x_i - \overline{x})(y_i - \overline{y})\right)^2 = \sum_{i=1}^n ((x_i - \overline{x})(y_i - \overline{y}))$$

for any linear relationship, as $x_i - \overline{x}$ and $y_i - \overline{y}$ are in a constant ratio (since the slope never changes in a line).

Extension Exercise 7. *Let X and Y be independent random variables, and let a and b be constants. Compute $Var(aX + bY)$.*

Extension Solution 7. *Recall the linearity of variance/expectation, which states that $Var(aX + bY) = Var(aX) + Var(bY) = a \cdot Var(X) + b \cdot Var(Y)$.*

Section 10.3: Introduction to Probability

Exercise 22. *What is the probability that, if you select two elements with replacement from the set $\{1, 2, 3, 4, 5\}$, at least one will be prime?*

Solution 22. *We use complementary counting. This is equal to 1 minus the probability that neither is prime, i.e.* $1 - \frac{2}{5} \cdot \frac{1}{4} = \boxed{\frac{9}{10}}$.

Exercise 23. *Richard and Harvey both pick two random numbers between 18 and 20, inclusive. What is the probability that Richard's number is at least twice Harvey's number, minus 21?*

Solution 23. *This is a geometric probability problem, since we can plot Richard and Harveys choices as points on the grid in the Cartesian plane given by $[18, 20] \times [18, 20]$. We want $R \geq 2H - 21$, which occurs when $R \geq 15$ at $H = 18$ - trivially guaranteed. At $H = 20$, $R \geq 19$, which occurs with probability $\frac{1}{2}$. For all $H \leq 19.5$, we only require $R \geq 18$, which is guaranteed. Hence the probability is the portion of the square that is also part of the area above the curve $y = 2x - 21$, which is $\frac{3}{4} + \frac{1}{4} \cdot \frac{1}{2} = \boxed{\frac{7}{8}}$.*

Exercise 24. *Three unfair coins, A, B, and C, are flipped in that order. Coin A lands on heads $\frac{2}{3}$ of the time, coin B lands on tails $\frac{3}{4}$ of the time, and coin C lands on heads $\frac{9}{10}$ of the time if coins A and B both landed on heads, and $\frac{3}{10}$ of the time otherwise. What is the probability that at least two coins land on tails?*

Solution 24. *If A lands on heads, then B lands heads with probability $\frac{1}{4}$. Assuming that B lands heads-up, the outcome of C is irrelevant, since there is no way to get two flips of heads at this point. However, B comes up tails with probability $\frac{3}{4}$, in which case C has a $\frac{3}{10}$ chance of landing heads, i.e. a $\frac{7}{10}$ chance of landing tails. Thus the probability of getting 2 or 3 tails here is $\frac{2}{3} \cdot \frac{3}{4} \cdot \frac{7}{10} = \frac{42}{120}$.*

In the event that A lands tails, however, if B lands tails, then we are already done (probability of $\frac{1}{3} \cdot \frac{3}{4} = \frac{1}{4}$). If B lands heads, we need C to be tails, which it is with probability $\frac{7}{10}$. Thus the probability of THT is $\frac{1}{3} \cdot \frac{1}{4} \cdot \frac{7}{10} = \frac{7}{120}$. Summing these probabilities yield the final probability of $\boxed{\frac{79}{120}}$.

Exercise 25. *(2017 AMC 12B) Real numbers x and y are chosen independently and uniformly at random from the interval $(0, 1)$. What is the probability that $\lfloor \log_2 x \rfloor = \lfloor \log_2 y \rfloor$?*

Solution 25. *First assume both are equal to -1. Then $\frac{1}{2} < x, y \leq 1$, which occurs with probability $\frac{1}{2} \cdot \frac{1}{2} = \frac{1}{4}$. Next, assume both are -2; then $\frac{1}{4} < x, y \leq \frac{1}{2}$, which happens with probability $\frac{1}{4} \cdot \frac{1}{4} = \frac{1}{16}$. Evidently, this is an infinite geometric series with initial term and common ratio $\frac{1}{4}$, and hence sum $\boxed{\frac{1}{3}}$.*

Exercise 26. *(2013 Stanford Math Tournament) Nick has a terrible sleep schedule. He randomly picks a time between 4 AM and 6 AM to fall asleep, and wakes up at a random time between 11 AM and 1 PM of the same day. What is the probability that Nick gets between 6 and 7 hours of sleep?*

Solution 26. *Another geo probability problem. If Nick falls asleep at 4 AM, he cannot get between 6-7 hours of sleep. But if he drifts off between 4 AM and 5 AM, he has a linearly increasing chance, from 0 to $\frac{1}{2}$, of being able to get 6-7 hours of sleep. Between 5 AM and 6 AM, he can still have a $\frac{1}{2}$ chance. Hence, his overall probability of getting 6-7 hours of sleep is $\frac{1}{2} \cdot \frac{1}{4} + \frac{1}{2} \cdot \frac{1}{2} = \boxed{\frac{3}{8}}$.*

Exercise 27. *(2006 AMC 10B) Bob and Alice each have a bag that contains one ball of each of the colors blue, green, orange, red, and violet. Alice randomly selects one ball from her bag and puts it into Bob's bag. Bob then randomly selects one ball from his bag and puts it into Alice's bag. What is the probability that after this process the contents of the two bags are the same?*

Solution 27. *The color of the ball that Alice puts in Bobs bag is irrelevant, so assume WLOG that Alice puts the red ball into Bobs bag. For both bags to end up with the same balls, Bob must also pick a red ball from his bag; he will do so with probability* $\boxed{\dfrac{1}{3}}$.

Exercise 28. *Compute the probability that a randomly chosen positive integral factor of 10^{2019} is not also a factor of 10^{2017}.*

Solution 28. *Observe that all factors of 10^{2017} are all factors of 10^{2019}, since the two have the same prime factors (2 and 5). Out of $(2019+1)^2 = 2020^2$ factors of 10^{2019}, $(2017+1)^2 = 2018^2$ of them are factors of 10^{2017}. Hence the desired probability is* $\dfrac{2020^2 - 2018^2}{2020^2} = \dfrac{4038 \cdot 2}{2020^2} = \boxed{\dfrac{2019}{1010^2}}$.

Exercise 29. *(2016 AIME II) There is a 40% chance of rain on Saturday and a 30% chance of rain on Sunday. However, it is twice as likely to rain on Sunday if it rains on Saturday than if it does not rain on Saturday. The probability that it rains at least one day this weekend is $\dfrac{a}{b}$, where a and b are relatively prime positive integers. Find $a+b$.*

Solution 29. *If it rains on Saturday, then let the probability of rain on Sunday be 2p; if it doesn't rain on Saturday, then let the probability of rain on Sunday be p. Then* $\dfrac{2}{5} \cdot 2p + \dfrac{3}{5} \cdot p = \dfrac{3}{10} \implies p = \dfrac{3}{14}$. *We want to compute* $1 - \dfrac{3}{5} \cdot \dfrac{11}{14} = 1 - \dfrac{33}{70} = \dfrac{37}{70} \implies \boxed{107}$.

Exercise 30. *An unfair coin has probability p of landing heads and $1-p$ of landing tails. Given that the probability of it landing on heads twice and tails three times out of five flips is equal to $20p^2$, what is the value of p?*

Solution 30. *There are* $\binom{5}{2} = 10$ *ways for the coin to land on heads twice and tails thrice. Hence,* $10p^2(1-p)^3 = p^2 \implies 1-p = \sqrt[3]{\dfrac{1}{10}} \implies p = \boxed{1 - \sqrt[3]{\dfrac{1}{10}}}$.

Exercise 31. *(2006 AIME II) When rolling a certain unfair six-sided die with faces numbered 1, 2, 3, 4, 5, and 6, the probability of obtaining face F is greater than $\dfrac{1}{6}$, the probability of obtaining the face opposite is less than $\dfrac{1}{6}$, the probability of obtaining any one of the other four faces is $\dfrac{1}{6}$, and the sum of the numbers on opposite faces is 7. When two such dice are rolled, the probability of obtaining a sum of 7 is $\dfrac{47}{288}$. Given that the probability of obtaining face F is $\dfrac{m}{n}$, where m and n are relatively prime positive integers, find $m+n$.*

Solution 31. *The probabilities of the cube landing with each face up are* $\dfrac{1}{6}, \dfrac{1}{6}, \dfrac{1}{6}, \dfrac{1}{6}, \dfrac{1}{6} + x, \dfrac{1}{6} - x$ *for some* $x < \dfrac{1}{6}$. *Multiplying opposite pairs out and doing some algebra yields* $x = \dfrac{5}{24} \implies \boxed{29}$.

Extension Exercise 8. *A giant parking lot has 2018 consecutive parking spaces, each with the same width. 2000 cars park in a random space. Then, a 2-space-wide RV rolls into the parking lot. What is the probability that it can park?*

334

Extension Solution 8. *Essentially, we want to compute the probability that, of the 18 non-occupied spaces, some 2 of them will be adjacent. This is the complement of the probability that none are adjacent, which is given by* $\dfrac{\binom{2001}{18}}{\binom{2018}{18}}$ *by stars-and-bars; hence, our answer is* $\boxed{1 - \dfrac{\binom{2001}{18}}{\binom{2018}{18}}}$.

Extension Exercise 9. *(2016 ARML Individual Round #9) An n-sided die has the integers between 1 and n (inclusive) on its faces. All values on the faces of the die are equally likely to be rolled. An 8-sided die, a 12-sided die, and a 20-sided die are rolled. Compute the probability that one of the values rolled is equal to the sum of the other two values rolled.*

Extension Solution 9. *Let a, b, c be the outcomes of the 8-, 12-, and 20-sided die rolls respectively. Fix a, b; then $c = a + b, c = a - b,$ or $c = b - a$. In the first case, the probability is $\dfrac{1}{20}$ for any such pair (a, b). In the second case, we must have $a > b$, and in the third, $b > a$. Note that $P(a > b \vee b > a) = 1 - P(a = b) = \dfrac{11}{12}$, which implies $\dfrac{11}{240}$ for cases 2 and 3. Adding this to $\dfrac{1}{20}$ yields the answer,* $\boxed{\dfrac{23}{240}}$.

Extension Exercise 10. *Andrew and Beth both flip a certain amount of fair coins. Andrew flips n coins in succession, while Beth flips $n + 1$ coins in succession. Show that Andrew always has exactly a $\dfrac{1}{2}$ chance to flip at least as many heads as Beth, independently of the value of n.*

Extension Solution 10. *This probability is given by*

$$\sum_{i=0}^{n} \dfrac{\binom{n}{i}}{2^{n+1}} = \dfrac{2^n}{2^{n+1}} = \dfrac{1}{2}$$

for all n.

Extension Exercise 11. *(2016 ARML Team Round #7) Real numbers x, y, and z are chosen at random from the unit interval $[0, 1]$. Compute the probability that $\max(x, y, z) - \min(x, y, z) \leq \dfrac{2}{3}$.*

Extension Solution 11. *Now we make the leap to 3D geo probability! Here, we consider points in the first octant of a unit cube with a corner at the origin. For some z, the x- and y-values must satisfy $0 \leq x, y \leq 1, -\dfrac{2}{3} \leq y - x \leq \dfrac{2}{3}, z - \dfrac{2}{3} \leq x, y \leq z + \dfrac{2}{3}$. The resulting area is either that of a square (area $\dfrac{4}{9}$, when $z = 0, 1$), a hexagon with area $\dfrac{4 + 12z}{9}$ (when $0 < z < \dfrac{1}{3}$), a hexagon with area $\dfrac{8}{9}$ (when $\dfrac{1}{3} \leq z \leq \dfrac{2}{3}$), or a hexagon with area $\dfrac{16 - 12z}{9}$ (when $\dfrac{2}{3} < z < 1$). The probability is thence $\dfrac{2}{3} \cdot \dfrac{2}{3} + \dfrac{1}{3} \cdot \dfrac{8}{9} =$* $\boxed{\dfrac{20}{27}}$.

Extension Exercise 12. *(1992 AHSME) An "unfair" coin has a $\dfrac{2}{3}$ probability of turning up heads. If this coin is tossed 50 times, what is the probability that the total number of heads is even?*

Extension Solution 12. *We have*

$$P = \left(\dfrac{1}{3}\right)^{50} + \binom{50}{2}\left(\dfrac{1}{3}\right)^{48}\left(\dfrac{2}{3}\right)^2 + \ldots + \left(\dfrac{2}{3}\right)^{50}$$

which is just the binomial expansion of $1 = \left(\dfrac{2}{3} + \dfrac{1}{3}\right)^{50}$ with all odd-exponent terms removed. This essentially means adding $\left(\dfrac{2}{3} - \dfrac{1}{3}\right)^{50}$ and then dividing by 2, to get $\boxed{\dfrac{1 + 3^{50}}{2 \cdot 3^{50}}}$.

Extension Exercise 13.

(a) Compute the probability that a random 10-digit string of 0's and 1's will not contain more than 2 consecutive 0's or 1's.

(b) In general, compute the probability that a random d-digit string of 0's and 1's will not contain more than 2 consecutive 0's or 1's.

Extension Solution 13.

(a) This is a recursive sequence s_n with $s_{n+2} = s_{n+1} + s_n$, $s_1 = 2$, $s_2 = 4$. Thus, $s_{10} = 178$ and the probability is $\boxed{\dfrac{89}{512}}$.

(b) The general probability is $\dfrac{s_n}{2^n}$, which requires a closed form for s_n. Hint: for this use the characteristic polynomial method outlined in the advanced section.

Extension Exercise 14. Compute the probability that a solution $(x, y, z) \in \mathbb{R}^3$, $0 < x, y \leq 3$ to the equation $\log_4 x + \log_4 y = \log_{16} z$ is also a solution to the equation $x^2 + y^2 + z \leq 3$. $\left(\dfrac{8\ln(2) - 3}{9}\right)$

Extension Solution 14. We require $\log_2 \sqrt{xy} = \log_2 \sqrt[4]{z} \implies z = x^2 y^2$. Thus, $x^2 + y^2 + x^2 y^2 \leq 3 \implies (x^2 + 1)(y^2 + 1) \leq 4$ by Simon's Favorite Factoring Trick. We also require $0 \leq x, y \leq 3$, and can integrate over these intervals using a double integral to get the answer of $\boxed{\dfrac{8\ln 2 - 3}{9}}$.

Section 10.4: Combinatorics and Probability: Part 1

Exercise 32. *Roll two fair six-sided dice. What is the probability that you roll at least one prime number?*

Solution 32. *This is yet more complementary counting; 1 minus the probability of not rolling a prime on either roll, i.e.* $1 - \frac{1}{2} \cdot \frac{1}{2} = \boxed{\frac{3}{4}}$.

Exercise 33. *At a dance party, there are 7 boys and 5 girls. Each boy is paired with one or zero girls. How many different arrangements are possible?*

Solution 33. *We must pick 5 of the boys to be paired off with a girl; the specific girls for the boys can be chosen in 5! ways. Thus the answer is* $\binom{7}{5} \cdot 5! = \boxed{2,520}$.

Exercise 34. *A modified, single-person variation of the card game Blackjack has the dealer deal to the bettor two cards that are either the Ace, or 2-10 (not including Jacks, Queens, or Kings), and where Aces are always worth 1. A blackjack in this version is a hand with two cards summing to 18. If the sum is less than 18, the bettor wins a dollar amount equal to his sum. If the sum is exactly 18, the bettor wins $100. Otherwise, if the sum is over 18, the bettor busts and wins nothing. Suppose that the dealer is generous and wants to make the game truly fair (and not slightly advantageous to the casino, like in real life). To the nearest dollar, how much should he charge to play this version of Blackjack?*

Solution 34. *For $2 \leq i \leq 11$, the probability of being dealt a sum of i is $\frac{i-1}{100}$; for $12 \leq i \leq 17$, the probability is $\frac{21-i}{100}$, for $i = 18$, the probability is $\frac{3}{100}$, and the probability of busting is $\frac{3}{100}$ as well. Then the expected winnings are $\sum_{i=2}^{11} \frac{i(i-1)}{100} + \sum_{i=12}^{17} \frac{i(21-i)}{100} + 3$, which evaluates to $\boxed{\frac{322}{25}} = \boxed{12.88}$ using the sum formulas.*

Exercise 35. *How many ways are there to arrange the letters in the word TARGET in a circle such that the T's are not adjacent?*

Solution 35. *This is equivalent to finding the permutations of TARGET that do not have the two Ts adjacent. There are $\binom{6}{2} = 15$ choices for the positions of the two Ts, and 6 choices for adjacent spots. Hence the answer is $6! \cdot \frac{15-6}{15} = \boxed{432}$ ways.*

Exercise 36.

(a) *Show that, if there are at least 367 people in a room, at least two of them are guaranteed to have the same birthday (assuming they were all born in a leap year).*

(b) *Show that, if there are at least $366n + 1$ people in a room, at least $n+1$ of them are guaranteed to have the same birthday.*

(c) *Let there be p people in a room, $1 \leq p \leq 366$. Find, with proof, the smallest value of p such that the probability of two or more people sharing the same birthday is at least $\frac{9}{10}$.*

(d) *Let there be p people in a room, $1 \leq p \leq 366$. Find, with proof, the smallest value of p such that the probability of two or more people sharing the same birthday is at least $\frac{k-1}{k}$, in terms of k.*

Solution 36.

(a) *This is an immediate application of Pigeonhole.*

(b) *Pigeonhole with $366n + 1$ pigeons and 366 holes. If we only had n people having each possible birthday, this would still leave 1 person unaccounted for.*

(c) With p people, the probability that none of them have the same birthday is $\frac{366}{366} \cdot \frac{365}{366} \cdot \frac{364}{366} \cdots \frac{366-p}{366}$. The smallest p-value turns out to be $\boxed{41}$.

(d) This is equivalent to finding the probability that no one shares a birthday is at most $\frac{1}{k}$, i.e. finding p such that $\prod_{i=1}^{p} \frac{367-p}{366} \leq \frac{1}{k}$.

Exercise 37. *(2016 AMC 12A) Each of the 100 students in a certain summer camp can either sing, dance, or act. Some students have more than one talent, but no student has all three talents. There are 42 students who cannot sing, 65 students who cannot dance, and 29 students who cannot act. How many students have two of these talents?*

Solution 37. *This is a direct application of PIE, where the size of the set of two-talented people is added to 100 minus the sizes of the sets of single-talented people. We have $100 - 42 - 65 - 29 + s = 0 \implies s = 36$, so $100 - 36 = \boxed{64}$ students have two talents.*

Exercise 38. *Find the value of*

$$\binom{3}{0} + \binom{3}{1} + \binom{4}{2} + \binom{5}{3} + \ldots + \binom{98}{96} + \binom{99}{97}.$$

Solution 38. *Repeatedly applying Pascals Identity $\binom{n}{k} + \binom{n}{k+1} = \binom{n+1}{k+1}$ yields a collapse of each initial term into the next, eventually leading to the single term $\binom{100}{97}$ that evaluates to $\boxed{161,700}$.*

Exercise 39. *Prove the Hockey Stick Identity (§10.4.5).*

Solution 39. *Using induction on n, the base case is $n = r$, which leads to $\sum_{i=r}^{n} \binom{i}{r} = \binom{n+1}{r+1}$. Next suppose for some $k \in \mathbb{N}, k > r$, that $\sum_{i=r}^{k} \binom{i}{r} = \binom{k+1}{r+1}$. Then*

$$\sum_{i=r}^{k+1} \binom{i}{r} = \binom{k+1}{r+1} + \binom{k+1}{r} = \binom{k+2}{r+1}$$

which is true by Pascal's Identity.

Extension Exercise 15. *We can also prove Pascal's Identity purely combinatorially. How so? (Hint: consider an arbitrary element x in a set with n elements.)*

Extension Solution 15. *As per the hint, construct a set of k elements containing x by choosing x and some other $k-1$ elements from the $n-1$ that remain; this yields $\binom{n-1}{k-1}$ choices. To construct a set of k elements that does not contain x, choose k elements from the $n-1$ elements other than x; this yields $\binom{n-1}{k}$ choices. Summing these yields Pascals Identity, as these cases are mutually exclusive, and the total is $\binom{n}{k}$ choices.*

Extension Exercise 16.

(a) Let $\{a_i\}_{i=1}^{n}$ be an integer sequence. Show that there exists an ordered pair of integers (k, m) with $k < m$ and $k + m \leq n$ such that $a_k + a_{k+1} + a_{k+2} + \ldots + a_{k+m}$ is divisible by n.

(b) Using the Generalized Pigeonhole Principle, prove the Chinese Remainder Theorem.

Extension Solution 16.

(a) *Consider the terms* (mod n). *If two of the terms were congruent* (mod n), *there is nothing to prove, since we could just take those two terms as the ordered pair. Otherwise, note that, in $m+1$ terms, the sum from 0 to m of modulo n will be greater than or equal to n, by the triangular number formula.*

(b) *Note that the sequence in (a) is actually a complete residue class modulo n. Hence, $b + kn \equiv a$ for some k, but by construction, this is also equivalent to b (mod m). Then $b + km$ satisfies this property.*

Extension Exercise 17. *(2014 Caltech-Harvey Mudd Math Competition) There is a long-standing conjecture that there is no number with $2n + 1$ instances in Pascal's triangle for $n \geq 2$. Assuming this is true, for how many $n \leq 100,000$ are there exactly 3 instances of n in Pascal's triangle?*

Extension Solution 17. *The only way for an element to occur exactly thrice in Pascals Triangle is for it to be in a diagonal twice and in the very center of a row once (provided, of course, that this center exists). Hence, it must be of the form $\binom{2k}{k}$ for some integer k. After $\binom{2}{1} = 2$ (which only occurs once), we have $6, 20, 70, 252, \ldots, \binom{18}{9} \implies \boxed{8}$.*

Extension Exercise 18. *Generalize Pascal's Identity to the p-variable case, then prove that generalized form of the identity.*

Extension Solution 18. *For any integer $p \geq 2$, with $k_i \in \mathbb{N}$ ($1 \leq i \leq p$), and $n = \sum_i k_i$, we have*

$$\binom{n-1}{k_1-1, k_2, k_3, \ldots, k_p} + \binom{n-1}{k_1, k_2-1, k_3, \ldots, k_p} + \ldots + \binom{n-1}{k_1, k_2, k_3, \ldots, k_p-1} = \binom{n}{k_1, k_2, k_3, \ldots, k_p}$$

Algebraically, we can write each term as a rational function of factorials, so that the sum becomes

$$\frac{(n-1)!}{(k_1-1)!k_2!k_3!\cdots k_p!} + \frac{(n-1)!}{k_1!(k_2-1)!k_3\cdots k_p} + \ldots + \frac{(n-1)!}{k_1!k_2!k_3!\cdots(k_p-1)!}$$

$$= \frac{k_1(n-1)! + k_2(n-1)! + k_3(n-1)! + \ldots + k_p(n-1)!}{k_1!k_2!k_3!\cdots k_p!}$$

$$= \frac{n(n-1)!}{k_1!k_2!k_3!\cdots k_p!} = \frac{n!}{k_1!k_2!k_3!\cdots k_p!} = \binom{n}{k_1, k_2, k_3, \ldots, k_p}$$

as desired.

Chapter 10 Review Exercises

Exercise 40. *Roll three fair six-sided dice. What is the probability of getting a triple-digit product?*

Solution 40. *The viable cases are 3-6-6, 4-5-5, 4-5-6, 4-6-5, 4-6-6, 5-4-5, 5-4-6, 5-5-4 through 5-5-6, 5-6-4 through 5-6-6, 6-3-6, 6-4-5, 6-4-6, 6-5-4 through 6-5-6, and finally 6-6-3 through 6-6-6, for a total of 23 cases out of $6^3 = 216$ total* $\implies \boxed{\dfrac{23}{216}}$.

Exercise 41. *(1970 AHSME) If the number is selected at random from the set of all five-digit numbers in which the sum of the digits is equal to 43, what is the probability that the number will be divisible by 11?*

Solution 41. *A positive integer is a multiple of 11 iff the alternating sums of digits in odd-numbered and even-numbered positions add to a multiple of 11. The odd-numbered positions can sum to 27 (or 5), and the even-numbered positions to 16, or 38. Casework in both cases yields* $\boxed{\dfrac{1}{5}}$.

Exercise 42. *In a length-k string of characters, define the polar of position p to be position $k - p + 1$ (from left to right). Call a string of A's and B's yin-yang if there is an A in the nth spot whenever the letter in the polar of the nth spot is a B, and vice versa. The probability that a randomly-chosen 10-letter string with 5 A's and 5 B's is yin-yang can be written in the form $\dfrac{m}{n}$, where m and n are relatively prime positive integers. Find $m + n$.*

Solution 42. *This is just $\binom{10}{5}$ divided by 2^{10}, i.e.* $\dfrac{252}{1024} = \boxed{\dfrac{63}{256}}$.

Exercise 43. *Let $S = \{-3, -3, -3, -1, 1, 1, 9, 16, 26, 41\}$.*

(a) *What are the mean, median, mode(s), and sample/population standard deviations of S?*

(b) *What is the five-number summary of S?*

(c) *Using each of the following definitions of an outlier, compute the outlier(s) of S.*

   (i) *A data point that lies more than 1.5 IQR away from the mean*

   (ii) *A data point that lies below the 9th percentile or above the 91st percentile (using the "compromise" definition)*

(d) *Suppose that the elements of S are each written down on a slip of paper, and the slips are shuffled and put into a hat. Josie picks three of the slips of paper (with replacement) uniformly at random. She knows that there are 10 slips of paper in the hat, but does not know what the numbers are beforehand. In her three picks, she comes up with -3 every time. Using a two-tailed test, she then concludes that S must contain an unusually large proportion of -3's. Is she correct in her conclusion, or is she making a statistical error? If so, what kind of error is she making, and why? Would her conclusion change if she instead used a single-tailed test? (Use the standard threshold for statistical significance, $\alpha = 0.05$.)*

Solution 43.

(a) The mean is $\dfrac{84}{10} = 8.4$, the median is 1, the mode is -3, the SSD is $\dfrac{58\sqrt{15}}{15}$, and the PSD is $\dfrac{29\sqrt{6}}{5}$.

(b) The five-number summary is $\{-3, -3, 1, 16, 41\}$.

(c) Using the IQR definition (with IQR = 19, 1.5 IQR = 28.5), we have anything outside the range $[-20.1, 36.9]$ as an outlier, so 41 is the only outlier. Using the percentile definition, -3 and 41 are the outliers.

(d) The probability of drawing three -3's in a row is $\left(\dfrac{3}{10}\right)^3 = \dfrac{27}{1000} = 0.027$, and using a two-tailed test, we double this to $0.054 \geq 0.05$, so Josie's conclusion is unfounded (false positive; Type I error). However, if she were to use a single-tailed test, her result would be statistically significant.

Exercise 44. *(2017 MathCounts Chapter Sprint) Yu has 12 coins, consisting of 5 pennies, 4 nickels and 3 dimes. He tosses them all in the air. What is the probability that the total value of the coins that land heads-up is exactly 30 cents? Express your answer as a common fraction.*

Solution 44. *Casework: we can have 3 dimes, 2 dimes and 2 nickels; 2 dimes, 1 nickel, and all 5 pennies; 1 dime and 4 nickels; 1 dime, 3 nickels, and 5 pennies. Use binomial coefficients and divide through by 2^{12} to get* $\boxed{\dfrac{23}{2048}}$.

Exercise 45. *(2003 AMC 12B) Let S be the set of permutations of the sequence $1, 2, 3, 4, 5$ for which the first term is not 1. A permutation is chosen randomly from S. The probability that the second term is 2, in lowest terms, is $\dfrac{a}{b}$. What is $a + b$?*

Solution 45. *There are $4! \cdot 4 = 96$ permutations for which the first term is not 1. Of those, of the permutations where the second term is 2, there are $3! = 6$ permutations with the first term 1, hence $4! - 3! = 18$ permutations with a first term other than 1. The desired probability is $\dfrac{3}{16} \implies \boxed{19}$.*

Exercise 46. *(2015 AIME II) Two unit squares are selected at random without replacement from an $n \times n$ grid of unit squares. Find the least positive integer n such that the probability that the two selected unit squares are horizontally or vertically adjacent is less than $\dfrac{1}{2015}$.*

Solution 46. *In an $n \times n$ grid, there are 4 corner squares, $4(n-2)$ edge squares, and $(n-2)^2$ inner squares. Each corner square is adjacent to 2 squares, each edge square to 3 squares, and each inner square to 4 squares. Then setting up the algebra and simplifying yields $\dfrac{4}{n(n+1)} \leq \dfrac{1}{2015} \implies n \geq \boxed{90}$.*

Exercise 47. *How many ways can the digits of 112233 be arranged such that no two identical digits are adjacent?*

Solution 47. *Consider just the 1's, without loss of generality. There are 10 ways to place them in non-adjacent spots, and half of them allow for $\binom{4}{2} - 2 = 4$ configurations, while the other half allow for just 2 configurations. Hence the total is $\boxed{30}$ arrangements.*

Exercise 48. *(2015 AIME I) The nine delegates to the Economic Cooperation Conference include 2 officials from Mexico, 3 officials from Canada, and 4 officials from the United States. During the opening session, three of the delegates fall asleep. Assuming that the three sleepers were determined randomly, the probability that exactly two of the sleepers are from the same country is $\dfrac{m}{n}$, where m and n are relatively prime positive integers. Find $m + n$.*

Solution 48. *There are $\binom{9}{3} = 84$ ways to pick 3 of the delegates. Consider the number of ways for exactly two sleepers to be from the same country through casework. If 2 are from Mexico, we have 7 choices; if 2 are from Canada, there are $3 \cdot 6 = 18$ ways, and $6 \cdot 5 = 30$ ways for 2 from the US. Hence, our answer is $\dfrac{55}{84} \implies \boxed{139}$.*

Exercise 49. *Jackson is taking a Probability class which has three equally weighted tests - two midterms and a final. In this class, an A equates to any percentage score 90 or above, a B to a score at least 80 but less than 90, and so forth, until an F which equates to any score under 60. On the first test, he scores a B, and on the second test, he scores an A. Jackson forgot his exact percentage scores on the first two tests, but wants to secure an A in the class, so he studies long and hard for the final. What is the probability that he will be able to do so, given that he scores an A on the final? (The test scores need not be integers.)*

Solution 49. *This is 3-dimensional geometric probability, with a right-angled tetrahedon with side length 10 and a vertex positioned at the origin of a cube with side length 10. The tetrahedron's volume is $\dfrac{1}{6}$ of the cube's, so the probability of Jackson being able to get an A is its complement, or $\boxed{\dfrac{5}{6}}$.*

341

Exercise 50. *A 5-question exam awards 5 points for a correct answer, 1 point for a blank, and 0 points for an incorrect answer. For each attainable score s, let $N(s)$ be the number of distinct answer patterns corresponding to that score. For example, $N(6) = 20$, since the only way to achieve a score of 6 is to answer 1 correctly, leave 1 blank, and answer 3 incorrectly; this can be done in $\frac{5!}{(3! \cdot 1! \cdot 1!)} = 20$ ways. Over all attainable scores s, the average value of $N(s)$ can be written in the form $\frac{p}{q}$, where p and q are relatively prime positive integers. Find $p + q$.*

Solution 50. *Note that there are $3^5 = 243$ possible answer patterns, and a total of 20 attainable scores (all except 14, 18, 19, 22-24), so the average is $\frac{243}{20} \implies \boxed{263}$.*

Exercise 51. *If we pick a positive integer between 1 and 1000, inclusive, the probability that it is expressible as $x\lfloor x \lfloor x \rfloor \rfloor$ (where $x > 0$ is a real number and $\lfloor x \rfloor$ represents the floor function/greatest integer function) can be written as $\frac{m}{n}$, where m and n are relatively prime positive integers. Find $m + n$.*

Solution 51. *Observe that, at the integers, the function "jumps" abruptly to a much higher number. For instance, at $x = 3$, the value of the function is 27, but just under, it is less than 15. Let $x = n - \epsilon$ for some integer n and infinitesimally small ϵ. For $x = 1$ to $x = 2 - \epsilon$, the function is always 1; for $x = 2$ to $x = 3 - \epsilon$, the range is from 8 to 14; for $\lfloor x \rfloor = 3$, the range is 27 to 43; for $\lfloor x \rfloor = 4$, the range is 64 to 94; and so forth. We end up getting $m + n = \boxed{781}$ at the end.*

Exercise 52. *(2018 AIME I) Kathy has 5 red cards and 5 green cards. She shuffles the 10 cards and lays out 5 of the cards in a row in a random order. She will be happy if and only if all the red cards laid out are adjacent and all the green cards laid out are adjacent. For example, card orders RRGGG, GGGGR, or RRRRR will make Kathy happy, but RRRGR will not. The probability that Kathy will be happy is $\frac{m}{n}$, where m and n are relatively prime positive integers. Find $m + n$.*

Solution 52. *We have possible card orders RRRRR, RRRRG, GRRRR, RRRGG, GGRRR, and twice those by symmetry. For RRRRR, we have $\frac{1}{\binom{10}{5}} = \frac{1}{252}$. For the permutations of RRRRG, we have $2 \cdot \frac{5}{252}$; and for the permutations of RRRGG, we have $2 \cdot \frac{10}{252}$. These sum to $\frac{31}{252}$, and multiplying by 2 to account for symmetry yields $\frac{31}{126} \implies \boxed{157}$.*

Exercise 53. *A prototype of an email spam filter is $p_1\%$ effective at filtering spam (which comprises 10% of all emails received): that is, it will correctly identify a spam email $p_1\%$ of the time. However, it will also incorrectly identify legitimate emails as spam $p_2\%$ of the time. In terms of p_1 and p_2, what is the probability that an email identified as spam by the filter actually is spam?*

Solution 53. *WLOG let there be 1000 emails: 100 spam, 900 legitimate. Then there are p_1 correctly identified spam emails, and $100 - p_1$ false negatives. There are also $900 - 9p_2$ correctly identified legitimate emails, as well as $9p_2$ false positives. Hence, the number of emails identified as spam by the filter is $p_1 + 9p_2$, but only p_1 of those are actually spam. Hence the desired probability is $\boxed{\dfrac{p_1}{p_1 + 9p_2}}$.*

Extension Exercise 19. *If a, b, c are real numbers between 0 and 10 inclusive, compute the probability that $a + 2b + 3c \leq 15$.*

Extension Solution 19. *Reprise of geometric probability! Hint to get you started: consider $a = b = 0$, then observe $c \leq 5$. Then let $b = 7.5$ and scale c down linearly. Continue until $a = 10$. This may be aided by calculus to obtain the final area under the curve; the answer is $\frac{1}{10} \int_0^{10} \frac{1}{48}(3-x)^2 \, dx = \boxed{\dfrac{13}{144}}$.*

Extension Exercise 20. *(2012 HMMT February Combinatorics Round) A frog is at the point $(0,0)$. Every second, he can jump one unit either up or right. He can only move to points (x,y) where x and y are not both odd. How many ways can he get to the point $(8,14)$?*

Extension Solution 20. *When the frog is at (x,y) for x,y even, then if he moves right, his next move is also forced to be right; the same holds for up. Reducing each set of two steps to a single step, the problem becomes finding the number of paths of length 11 from the origin to $(4,7)$, i.e. $\binom{11}{4} = \boxed{330}$.*

Extension Exercise 21. *(2018 AIME I) For every subset T of $U = \{1, 2, 3, \ldots, 18\}$, let $s(T)$ be the sum of the elements of T, with $s(\emptyset)$ defined to be 0. If T is chosen at random among all subsets of U, the probability that $s(T)$ is divisible by 3 is $\frac{m}{n}$, where m and n are relatively prime positive integers. Find m.*

Extension Solution 21. *Consider the subset of 1 modulo 3. We have just one way to choose zero numbers, 6 ways to choose one number, and so forth. There are 22 subsets congruent to 0 mod 3, and 64 possible subsets. By symmetry, there are 21 subsets for each possible sum of 1, 2 mod 3. The same holds for the subsets of 2 mod 3. For the subset consisting of multiples of 3, this is trivial; all 64 subsets sum to 0 mod 3. Hence the probability is $\frac{\boxed{683}}{2^{11}}$.*

Extension Exercise 22. *(1972 USAMO P3) A random number selector can only select one of the nine integers $1, 2, \ldots, 9$, and it makes these selections with equal probability. Determine the probability that after n selections ($n > 1$), the product of the n numbers selected will be divisible by 10.*

Extension Solution 22. *In order for the product to be a multiple of 10, it must have prime factors of 2 and 5. The probability of neither prime factor being in the factorization is $\left(\frac{4}{9}\right)^n$. The probability of there being no 5 is $\left(\frac{8}{9}\right)^n$, so the probability of there being a 2, but no 5, is $\left(\frac{8}{9}\right)^n - \left(\frac{4}{9}\right)^n$. Similarly, the probability that there is a 5, but no 2, is $\left(\frac{5}{9}\right)^n - \left(\frac{4}{9}\right)^n$. The desired probability is then $1 - \boxed{\left(\frac{8}{9}\right)^n - \left(\frac{5}{9}\right)^n + \left(\frac{4}{9}\right)^n}$.*

Extension Exercise 23. *(2018 USAJMO P1) For each positive integer n, find the number of n-digit positive integers that satisfy both of the following conditions:*

- *no two consecutive digits are equal, and*
- *the last digit is a prime.*

Extension Solution 23. *Let a_n denote the number of satisfactory n-digit numbers. We have $a_1 = 4, a_2 = 32$. For $n \geq 3$, we have the recurrence relation $a_n = 8a_{n-1} + 9a_{n-2}$.*

To prove this, take a casework approach in which we consider whether or not the second digit is 0. If the second digit is non-zero, then our number is formed from a_{n-1} with one fewer digit, multiplied with 8 to account for the leftmost digit. If the second digit is zero, then we have a_{n-2} with 2 fewer digits, and 9 possible choices (including 0) for the left digit.

Hence, the solution to the recurrence in closed form, namely $c_1 r_1^n + c_2 r_2^n$ for constants c_1, c_2, can be derived from the system of linear equations $9c_1 - c_2 = 4, 81c_1 + c_2 = 32 \implies (c_1, c_2) = \left(\frac{2}{5}, -\frac{2}{5}\right)$, completing the proof and yielding the answer of $\boxed{\frac{2}{5}(9^n - (-1)^n)}$.

Chapter 11: Intermediate Discrete Math

Section 11.1: Introduction to Game Theory and Rational Decision-Making

Exercise 54. *The essay portion of a standardized test is graded as follows. Two graders grade the essay simultaneously, and cannot collude with each other at any point during the grading process. Each grader assigns the essay an integer score from 1 to 6, inclusive. Then those scores are added together to obtain the student's final score.*

(a) Represent this scenario as a normal form game.

(b) Determine all Nash equilibria for this game.

The payoff function for a sum score of n ($2 \leq n \leq 12$, $n \in \mathbb{Z}$) is given by $P(n) = |(n-7)^2 - 49|$, so that the payoff is maximized for a sum score of 7 (at 49 utils, or \$49 in the graders' pockets) and minimized for a sum score of 2 or 12 (at a measly 24 utils, or \$24).[130]

Solution 54.

(a)

	1	2	3	4	5	6
1	24	33	40	45	48	49
2	33	40	45	48	49	48
3	40	45	48	49	48	45
4	45	48	49	48	45	40
5	48	49	48	45	40	33
6	49	48	45	40	33	24

(b) *The NEs are all sum scores of 7, since no grader is motivated to unilaterally change his/her score for a better payoff. That is, the 49s dominate all other payoffs in their respective rows and columns.*

Exercise 55. *The city of Forbes has an infamously dangerous intersection, so the city planners are considering adding some traffic lights to the intersection. Each year during their 10-year-long tenure, they may elect to have zero, one, or two traffic lights installed. Once a traffic light has been installed, it cannot be removed, and the intersection can only have up to two lights (any more would be prohibitively expensive!) The payoff of having no lights is 0 utils, the payoff of having one light is +50 utils, and the payoff of having two lights is +80 utils (where 1 util represents 10 annual accidents prevented). When should the Forbes city planners add the first and second traffic lights? Assume the city planners have a 50% chance of choosing to add one or two lights each year.*

Solution 55. *Hint: use backward induction. After year 9, the expected utility of having no lights is +65 utils, but installing 2 lights is still the superior option. Going back to the end of year 8, the waiting proposition now has +130 utils worth of merit, while installing one light immediately yields +100 utils and installing two lights yields +160 utils. Thus, he should install the one-car garage after year 7; before then, waiting it out is optimal (provided, of course, that he cannot build the 2-car garage immediately for whatever reason).*

Exercise 56. *On a math test with 30 multiple-choice questions (each with 5 answer choices), participants earn 4 points for each correct answer, 0 points for blanks, and lose 1 point for each wrong answer. Suppose that the passing score is 60 points out of a possible 120. What is the optimal strategy to pass this test (assuming each answer choice is equally likely to be correct)? Describe the best course of action for < 15, 15, 16, 17, and ≥ 18 confident answers.*

---

[130] Yes, in this imaginary universe, the graders receive bonuses according to the quality of the essays that they have scored! Seems preposterous, but fear not - each essay is also scanned by the higher-ups, and if they disagree excessively with the graders, the graders forfeit their bonuses and risk termination (so the graders can't just hand out 6's (or 1's) like candy). This (quite Orwellian, and thankfully hypothetical) testing bureaucracy wants most students to have average scores, even if their essays are all top-notch. Why? Because otherwise, they would be accused of making the test too easy or too hard (or just plain unfair)!

Solution 56. *For < 15 correct answers, it is impossible to pass, so it is ideal to guess 6 answers (at least 2 of them correctly). Any more (or fewer) is too risky a proposition. For exactly 15, do not guess whatsoever. For 16, feel free to guess on 4, but no more. For 17, guess on 8, and for 18, guess on all the rest to maximize one's score (passing is guaranteed at this point; even 18 correct and 12 wrong still equates to a score of 60).*

Exercise 57. *Describe an example of a symmetric two-player, two-decision game that is also a zero sum game and a generalized prisoner's dilemma, or show that no such game exists.*

Solution 57. *The matching pennies game is an example of such a game: it is zero-sum, as each payoff is balanced by the payoff of the other player. Furthermore, it is symmetric in its construction, and consists of two decisions in the decision set of each player. It also relates back to the prisoner's dilemma in that human randomization is imperfect; humans may attempt to randomize by switching their actions, albeit without realizing that their switching is taking place too often.*

Exercise 58. *(2016 Caltech-Harvey Mudd Math Competition) A gambler offers you a $2 ticket to play the following game. First, you pick a real number $0 \leq p \leq 1$. Then, you are given a weighted coin with probability p of coming up heads and probability 1p of coming up tails, and flip this coin twice. The first time the coin comes up heads, you receive $1, and the first time it comes up tails, you receive $2. Given an optimal choice of p, what is your expected net winning?*

Solution 58. *Fix p; the probability of HH is p^2, the probability of TT is $(1-p)^2$, and the probability of flipping either HT or TH is $2p(1-p)$. The expected winnings are hence $p^2 + 6p(1-p) + 2(1-p)^2 = -3p^2 + 2p + 2$, which is maximized at $p = \dfrac{1}{3}$ (maximal value $\dfrac{7}{3}$). The ticket costs $2, so the expected net winnings from this game are* $\boxed{\$\dfrac{1}{3}}$.

Exercise 59. *Show that the mixed-strategy equilibrium for Rock Paper Scissors against an unknown, unpredictable adversary is $\left(\dfrac{1}{3}, \dfrac{1}{3}, \dfrac{1}{3}\right)$, where (x, y, z) denotes the proportions x, y, z played of Rock, Paper, Scissors respectively.*

Solution 59. *Let $P(r)$ be the probability that player 1 picks rock, and define $P(s), P(p)$ analogously. Define $Q(r), Q(s), Q(p)$ similarly for player 2. (N.B.: none of these probabilities are 1, since a player will not repeat the same decision/move every single time.) Player 2s expected value, then, is $\mathbb{E}(Q(R)) = 0 \cdot P(r) + (-1) \cdot P(p) + 1 \cdot P(s)$, with similar equations set up for the winning outcomes with a play of scissors and of paper. Solving the resulting simultaneous system of equations yields the desired result, $\left(\dfrac{1}{3}, \dfrac{1}{3}, \dfrac{1}{3}\right)$ for both players.*

The "Bernardo and Silvia" series of problems from the AMC competitions:

Exercise 60. *(2010 AMC 10A) Bernardo randomly picks 3 distinct numbers from the set $\{1, 2, 3, 4, 5, 6, 7, 8, 9\}$ and arranges them in descending order to form a 3-digit number. Silvia randomly picks 3 distinct numbers from the set $\{1, 2, 3, 4, 5, 6, 7, 8\}$ and also arranges them in descending order to form a 3-digit number. What is the probability that Bernardo's number is larger than Silvia's number?*

Solution 60. *Case 1: If Bernardo's largest number is 9, he wins instantly (he does so with probability $\dfrac{1}{3}$).*

Case 2: If Bernardo's largest number is not 9, the probability of Bernardos largest number being larger than Silvia's is equal to the probability that Silvia's number is larger. The probability that the two will pick the same largest number is $\dfrac{3!}{8 \cdot 7 \cdot 6} = \dfrac{1}{56}$. Hence, the probability of Bernardo's number being strictly greater than Silvia's is $\dfrac{55}{112}$, and multiplying this by $\dfrac{2}{3}$ to account for the probability of the case occurring in the first place yields $\dfrac{55}{168}$. Therefore, our final answer is $\dfrac{1}{3} + \dfrac{55}{168} = \boxed{\dfrac{37}{56}}$.

Exercise 61. *(2012 AMC 12B) Bernardo and Silvia play the following game. An integer between 0 and 999 inclusive is selected and given to Bernardo. Whenever Bernardo receives a number, he doubles it and passes the result to Silvia. Whenever Silvia receives a number, she adds 50 to it and passes the result to Bernardo. The winner is the last person who produces a number less than 1000. Let N be the smallest initial number that results in a win for Bernardo. What is the sum of the digits of N?*

Solution 61. *Bernardo's last number must lie between 950 and 999, exclusive. Henceforth, note that the chain $x \to 2x \to 2x + 50 \to 4x + 100 \to 4x + 150 \to 8x + 300 \to 8x + 350 \to 16x + 700$ implies $950 < 16x + 700 < 999 \implies x \geq 16$. If $x = 16$, then $16x + 700 = 956$. Working backwards, we get a starting point of 16, whose sum of digits is $\boxed{7}$.*

Exercise 62. *(2016 AMC 12A) Let k be a positive integer. Bernardo and Silvia take turns writing and erasing numbers on a blackboard as follows: Bernardo starts by writing the smallest perfect square with $k+1$ digits. Every time Bernardo writes a number, Silvia erases the last k digits of it. Bernardo then writes the next perfect square, Silvia erases the last k digits of it, and this process continues until the last two numbers that remain on the board differ by at least 2. Let $f(k)$ be the smallest positive integer not written on the board. For example, if $k = 1$, then the numbers that Bernardo writes are $16, 25, 36, 49, 64$, and the numbers showing on the board after Silvia erases are $1, 2, 3, 4,$ and 6, and thus $f(1) = 5$. What is the sum of the digits of $f(2) + f(4) + f(6) + \cdots + f(2016)$?*

Solution 62. *First, consider $f(2)$. The numbers that remain will show the hundreds place at the end. For the hundreds place to differ by 2, the difference between two squares must be ≥ 101. This occurs for the first time at $(50^2, 51^2)$ (but, of course, not with this specific pair). As it turns out (the next few numbers can be written by hand), the actual pair is $(59^2, 60^2)$, so no four-digit perfect square begins with the digits 35.*

Next, consider $f(4)$. The difference of squares is now $10^4 = 10,000$, which first occurs at $(5000^2, 5001^2)$. Similarly, we actually encounter the pair at $(5099^2, 5100^2)$. Then the skipped value is 2600 in this case.

More generally, for all k even, $f(k)$ results in adding two digits to the 5 and one digit to the 1. Hence, for $k = 6$, Bernardo's final number is $(500,000 + 1,000)^2$; for $k = 8$, it is $(50,000,000 + 10,000)^2$, and so forth. This continues until $k = 2016$. Removing the final k digits removes $\frac{1}{2}k$ trailing zeros on the number to be squared. Hence, the last number on the board for $k = 6$ is 5001^2, for $k = 8$ is $50,001^2$, and so on. Then the first number that is missing is $5001^2 - 1$, the second is $50,001^2 - 1$, etc. Squaring forms a 25 with two more digits than the previous numbers, 10 with one extra digit, and a 1 at the end. Then we subtract the 1 from $f(k)$ to get $f(k) = 25 \cdot 10^{k-2} + 10^{\frac{k}{2}}$.

It follows that $f(2) = 25 + 10$, $f(4) = 2500 + 100$, $f(6) = 250,000 + 1,000$, etc. As such, $\sum_{k=1}^{1008} f(2k)$ is a number with 2016 copies of both 25 and 1. Thus, each $f(k)$ instance contributes 8 to the sum of digits. This means that the sum of all the digits is $8 \cdot 1,008 = \boxed{8,064}$.

Extension Exercise 24. *(2014 Stanford Math Tournament) A one-player card game is played by placing 13 cards (Ace through King in that order) in a circle. Initially all the cards are face-up and the objective of the game is to flip them face-down. However, a card can only be flipped face-down if another card that is '3 cards away' is face-up. For example, one can only flip the Queen face-down if either the 9 or the 2 (or both) are face-up. A player wins the game if they can flip all but one of the cards face-down. Given that the cards are distinguishable, compute the number of ways it is possible to win the game.*

Extension Solution 24. *Consider instead a simplified version of the game, in which the cards are arranged in a circle as A, 4, 7, 10, K, 3, ... (effectively 1 mod 3). We can flip a card face-down if an adjacent card is face-up. In this version of the game, we win if we have exactly one face-up card such that its two adjacent cards are face-down (we call this a loner).*

Clearly, we cannot have a loner on the very first turn. However, if we create one on the second turn, then we also need to create one more, and the game cannot be won in this manner. We then must create the loner card on the last (twelfth) turn.

As the cards are distinguishable, we have 13 choices for the very first card, and subsequently, 2 choices for each card thereafter, including the winning loner card. Hence our answer is $\boxed{13 \cdot 2^{11}} = \boxed{26,624}$.

Extension Exercise 25. *(2013 Stanford Math Tournament) Robin is playing notes on an 88-key piano. He starts by playing middle C, which is actually the 40th lowest note on the piano (i.e. there are 39 notes lower than middle C). After playing a note, Robin plays with probability $\frac{1}{2}$ the lowest note that is higher than the note he just played, and with probability $\frac{1}{2}$ the highest note that is lower than the note he just played. What is the probability that he plays the highest note on the piano before playing the lowest note?*

Extension Solution 25. *Denote by p_i the probability of Robin playing the highest note before the lowest note, given that he starts at note i. (Clearly, we have $a_1 = 0, a_{88} = 1$.) For $i \in [2, 87]$, $a_i = \frac{a_{i-1} + a_{i+1}}{2}$, so $a_i = \frac{i-1}{87}$ and $a_{40} = \boxed{\frac{13}{29}}$.*

Extension Exercise 26. *(2004 USAMO P4) Alice and Bob play a game on a 6 by 6 grid. On his or her turn, a player chooses a rational number not yet appearing on the grid and writes it in an empty square of the grid. Alice goes first and then the players alternate. When all squares have numbers written in them, in each row, the square with the greatest number is colored black. Alice wins if she can then draw a line from the top of the grid to the bottom of the grid that stays in black squares, and Bob wins if she can't. (If two squares share a vertex, Alice can draw a line from on to the other that stays in those two squares. Find, with proof, a winning strategy for one of the players.*

Extension Solution 26. *Bob can move as follows. If Alice writes a number in rows 3 through 6, then Bob also writes in one of those rows. Otherwise, Bob writes in row $3 - a$ and column $c + 3 \pmod 6$, where a is Alices row and c is her column. If Alices number dominates its row, then so will Bobs; the same goes for the columns. Thus, there will necessarily be two columns between the black squares in the first two rows. They then cannot touch, and so Bob will win.*

Extension Exercise 27. *(1999 USAMO P5) The Y2K Game is played on a 1×2000 grid as follows. Two players in turn write either an S or an O in an empty square. The first player who produces three consecutive boxes that spell SOS wins. If all boxes are filled without producing SOS then the game is a draw. Prove that the second player has a winning strategy.*

Extension Solution 27. *Label the squares with the positive integers from 1 to 2000, inclusive. Then the first player writes a letter in some square with the number k. Player 2 proceeds to select some empty square s with $\min(|s - 1|, |s - 2000|, |s - k|) > 3$, and writes in it an S. After this, player 1 may play in some other square m. If the second player can win, he will do so (as both players are rational); else, if $s > m$, he writes S in $l + 3$, and otherwise, he writes S in $l - 3$.*

We then have two Ss in the grid that are separated by empty space. If a player plays a space in between them, this is basically a surrender to the other player! At this point, player 2 will win if possible, but if not, he will place an O where there is an empty space surrounded by two filled or two empty squares. Then player 1 cannot yet win on the next turn.

Assuming player 2 has not yet won, the algorithm will terminate eventually. The empty squares will subsequently come in pairs (there being an even number of them). It follows that it is currently player 1s turn; if he plays on a pair of empty squares without Ss on both sides, player 2 will counter in the other square of that pair. Then the first player will need to play between the sides that remain, forcing player 2 to clutch the win!

Extension Exercise 28. *(2014 USAMO P4) Let k be a positive integer. Two players A and B play a game on an infinite grid of regular hexagons. Initially all the grid cells are empty. Then the players alternately take turns with A moving first. In his move, A may choose two adjacent hexagons in the grid which are* empty and place a counter in both of them. In his move, B may choose any counter on the board and remove it. If at any time there are k consecutive grid cells in a line all of which contain a counter, A wins. Find the minimum value of k for which A cannot win in a finite number of moves, or prove that no such minimum value exists.

Extension Solution 28. We claim that $k \geq 6$. A can clearly create 5 in a row, and B can take off any two counters that A places down. A must then form a triangle by placing two adjacent counters next to an existing one. Then regardless of Bs strategy, he will still be forced to leave two adjacent counters.

Then A should make a 4-in-a-row by positioning 2 counters at the ends of the line. B must then remove a counter in the middle, but then A could just as easily place a counter there, as well as adjacent to the hexagon. This time, B should not remove the middle counter again. Eventually, A is able to surround the middle counter by hexagons. From here, we claim he can actually win as follows.

Let S be the set of 2 adjacent counters, and let T be the set of the other two counters satisfying the condition. A ought to place two counters on the first two hexagons that are not touching any counter in S, so that B must then remove the outermost counter placed by A. Then A should position two counters adjacent to the remaining counter parallel to the line of counters in T. B then is forced to remove the outermost one again. A places 2 more counters in that same direction, while B removes the middle counter and A counters by placing counters on the two spaces adjacent to existing counters. B removes the second of those counters, and A can then win by replacing the removed counter, and placing another one adjacent to that counter.

To show that $k = 6$ doesnt work, however, we must tile the hexagon with A, B, C tiles. An A tile is a tile marked with the letter A; define B and C tiles similarly. Player B must remove counters on A tiles placed by player A. (Since A tiles are non-adjacent, player A may play at most one counter on an A tile at once.) A 6-in-a-row, requiring A-B-C-A-B-C, is then impossible; QED.

Extension Exercise 29. (2016 USAMO P6) Integers n and k are given, with $n \geq k \geq 2$. You play the following game against an evil wizard.

The wizard has 2n cards; for each $i = 1, ..., n$, there are two cards labeled i. Initially, the wizard places all cards face down in a row, in unknown order.

You may repeatedly make moves of the following form: you point to any k of the cards. The wizard then turns those cards face up. If any two of the cards match, the game is over and you win. Otherwise, you must look away, while the wizard arbitrarily permutes the k chosen cards and turns them back face-down. Then, it is your turn again.

We say this game is winnable if there exist some positive integer m and some strategy that is guaranteed to win in at most m moves, no matter how the wizard responds.

For which values of n and k is the game winnable?

Extension Solution 29. We claim the game is winnable iff $n > k$. After the initial step, shuffle cards 1 through n. We then have n cards whose position we do not know. Pick p cards which we do know, and observe that all are distinct. There is still a possibility here that the other cards complement the known cards, making the game unwinnable.

Suppose $n > k$. Then pick cards 1 through k, followed by 2 through $k+1$, adding the value of the last card to the set of values of cards. Repeat this until we pick cards $2n-k$ through $2n-1$; then we know the value of card 2n. Go back to cards 1 through k and repeat the process until we reveal the $(2n-1)^{th}$ card (or win in doing so). The process stops when we have $\leq k$ unknown cards, or if we win beforehand. We then know the values of $2n-k$ cards; by Pigeonhole, since $n > k$, at least two have the same value. Pick those two cards (and any $k-2$ others), and we force a win.

Section 11.2: Combinatorics and Probability: Part 2

Exercise 63. *(1997 AHSME) Two six-sided dice are fair in the sense that each face is equally likely to turn up. However, one of the dice has the 4 replaced by 3 and the other die has the 3 replaced by 4. When these dice are rolled, what is the probability that the sum is an odd number?*

Solution 63. *On the first die, the probability of rolling an odd number is $\frac{2}{3}$, and an even number is $\frac{1}{3}$. The reverse is true of the second die; multiplying and summing probabilities yields $\boxed{\frac{5}{9}}$.*

Exercise 64. *(2015 Caltech-Harvey Mudd Math Competition) There are twelve indistinguishable blackboards that are distributed to eight different schools. There must be at least one board for each school. How many ways are there of distributing the boards?*

Solution 64. *This is stars-and-bars with $n = 8, k = 12 - 8 = 4$ (since each school already gets at least one board); we have $\binom{n+k-1}{k} = \binom{11}{4} = \boxed{330}$.*

Exercise 65. *(2015 AMC 12B) An unfair coin lands on heads with a probability of $\frac{1}{4}$. When tossed n times, the probability of exactly two heads is the same as the probability of exactly three heads. What is the value of n ?*

Solution 65. *The probability of getting two heads is $\binom{n}{2}\left(\frac{1}{4}\right)^2\left(\frac{3}{4}\right)^{n-2}$, and the probability of getting three heads is $\binom{n}{3}\left(\frac{1}{4}\right)^3\left(\frac{3}{4}\right)^{n-3}$. Hence, $\binom{n}{3} = 3\binom{n}{2} \implies \frac{n(n-1)(n-2)}{6} = 3 \cdot \frac{n(n-1)}{2} \implies \frac{n-2}{3} = 3 \implies n = \boxed{11}$.*

Exercise 66. *Compute the number of non-decreasing sequences of positive integers with length 7, whose last term does not exceed 9.*

Solution 66. *This is a classic example of stars-and-bars, in this particular case with $n = 9, k = 7 \implies \binom{9+7-1}{7} = \binom{15}{7} = \boxed{6,435}$.*

Exercise 67. *(2018 ARML Team Round #4) A fair 12-sided die has faces numbered 1 through 12. The die is rolled twice, and the results of the two rolls are x and y, respectively. Given that $\tan(2\theta) = \frac{x}{y}$ for some θ with $0 < \theta < \frac{\pi}{2}$, compute the probability that $\tan\theta$ is rational.*

Solution 67. *Let $\tan 2\theta = \frac{x}{y} \implies \frac{2\tan\theta}{1 - \tan^2\theta} = \frac{x}{y}$, i.e. $x\tan^2\theta + 2y\tan\theta - x = 0$, or $\tan\theta$ rational $\iff (2y)^2 - 4x(-x) = 4(x^2 + y^2) = k^2$ for some integer k. But the only Pythagorean triples with leg lengths ≤ 12 are $(3,4,5), (5,12,13), (6,8,10), (9,12,15)$. The ordered pairs that make $\tan\theta$ rational are $(3,4), (4,3), (6,8), (8,6), (9,12), (12,9), (5,12), (12,5) \implies \frac{8}{12^2} = \boxed{\frac{1}{18}}$.*

Exercise 68. *(1983 AHSME) Three balls marked 1, 2, and 3, are placed in an urn. One ball is drawn, its number is recorded, then the ball is returned to the urn. This process is repeated and then repeated once more, and each ball is equally likely to be drawn on each occasion. If the sum of the numbers recorded is 6, what is the probability that the ball numbered 2 was drawn all three times?*

Solution 68. *Since we draw three balls, and the sum is 6, the only possibilities are 2, 2, 2 and the permutations of 1, 2, 3 - 7 possibilities. The probability that 2 was drawn all three times is then $\boxed{\frac{1}{7}}$.*

Exercise 69. *(2016 AMC 12A) Jerry starts at 0 on the real number line. He tosses a fair coin 8 times. When he gets heads, he moves 1 unit in the positive direction; when he gets tails, he moves 1 unit in the negative direction. The probability that he reaches 4 at some time during this process is $\frac{a}{b}$, where a and b are relatively prime positive integers. What is $a + b$? (For example, he succeeds if his sequence of tosses is HTHHHHHH.)*

Solution 69. From 6 to 8 heads, this is guaranteed. For 4 heads, we must have all of them immediately at the beginning (i.e. HHHHTTTT, probability $\frac{1}{256}$). For 5 heads, either we begin with 4 heads (4 cases), or 5 heads and 1 tail, which gives 4 more ways - 8 in total. Then we have 46 ways in total, yielding a final probability of $\boxed{\frac{23}{128}}$.

Exercise 70. *(1999 AHSME) A tetrahedron with four equilateral triangular faces has a sphere inscribed within it and a sphere circumscribed about it. For each of the four faces, there is a sphere tangent externally to the face at its center and to the circumscribed sphere. A point P is selected at random inside the circumscribed sphere. What is the probability that P lies inside one of the five small spheres?*

Solution 70. Let the largest sphere have radius R, and let the insphere have radius r; then let O be the center of tetrahedron $ABCD$. Furthermore, $[OABC] + [OABD] + [OACD] + [OBCD] = [ABCD]$; hence, $4[OABC] = [ABCD]$. It follows that $4r = h_{ABCD}$, the height of $ABCD$, i.e. $r + R$. Thus, $R = 3r$, so the radius of an external sphere is $\frac{R}{3}$. Hence, the ratio of volumes is $5 \cdot \left(\frac{1}{3}\right)^3 = \boxed{\frac{5}{27}}$.

Exercise 71. *How many ordered size-18 sets with elements from $\{1, 2, 3\}$ have elements summing to 50 or greater?*

Solution 71. Clearly, we will need at least 14 3s. In that case, only 14 3s and 4 2s will do. If there are 15 3s, we need at least two 2s. If there are 15 3s, 2 2s, and 1 1, this works; with 15 3s and 3 2s, this works as well. Similarly, 16 3s and above is a guarantee; continuing the casework in this manner yields $\boxed{685}$.

Exercise 72. *(2017 HMMT February Combinatorics Round) How many ways are there to insert +s between the digits of 111111111111111 (fifteen 1s) so that the result will be a multiple of 30?*

Solution 72. For a number to be a multiple of 30, it must be a multiple of both 3 and 10. 3 is already guaranteed, as the sum of digits is 15. For the number to end in 0, we must add in exactly nine plus signs in 14 spaces, so the answer is $\binom{14}{9} = \boxed{2002}$.

Exercise 73. *(2013 AIME I) Melinda has three empty boxes and 12 textbooks, three of which are mathematics textbooks. One box will hold any three of her textbooks, one will hold any four of her textbooks, and one will hold any five of her textbooks. If Melinda packs her textbooks into these boxes in random order, the probability that all three mathematics textbooks end up in the same box can be written as $\frac{m}{n}$, where m and n are relatively prime positive integers. Find $m + n$.*

Solution 73. Note that we can have math books in the size-3 box, which has a probability of $\frac{3!}{12 \cdot 11 \cdot 10} = \frac{1}{220}$, with 1320 possible orders. Furthermore, in the size-4 box, applying similar logic yields $\frac{1}{55}$, and for the 5-size, we have $\frac{1}{22}$, for a sum of $\frac{3}{44} \implies \boxed{47}$.

Exercise 74. *(1998 AIME) Let n be the number of ordered quadruples (x_1, x_2, x_3, x_4) of positive odd integers that satisfy $\sum_{i=1}^{4} x_i = 98$. Find n.*

Solution 74. Let $x_i = 2y_i - 1$; then $\sum_{i=1}^{4} y_i = 51$. This can be done via stars-and-bars; the answer is $\binom{50}{3} = \boxed{19,600}$.

Exercise 75. Let $a, b, c \in [-k, k] = \mathcal{I}$, $a, b, c, k \in \mathbb{R}$. In terms of k, determine the probability that the quadratic equation $ax^2 + bx + c = 0$ has a real solution, assuming a, b, c are selected at random from the interval $\mathcal{I}$.

Solution 75. We require $b^2 \geq 4ac$, which reduces to a geo probability problem. If $b = 0$, the probability is clearly 0; if $b = \pm k$, it becomes 1. (By symmetry, we need only consider $b \in [0, 1]$.) Take the double integral from 0 to 1 of xy with respect to x, y to obtain $\boxed{\dfrac{1}{4}}$.

Exercise 76. *(2018 AIME II)* A real number a is chosen randomly and uniformly from the interval $[-20, 18]$. The probability that the roots of the polynomial
$$x^4 + 2ax^3 + (2a - 2)x^2 + (-4a + 3)x - 2$$
are all real can be written in the form $\dfrac{m}{n}$, where m and n are relatively prime positive integers. Find $m + n$.

Solution 76. By Rational Root Theorem, we can factor the given polynomial as $(x - 1)(x + 2)(x^2 + x(2a - 1) + 1)$. Hence $(2a - 1)^2 - 4 \geq 0 \implies a \notin \left(-\dfrac{3}{2}, \dfrac{1}{2}\right)$. Thus, the interval where all roots are real is $\dfrac{18}{19}$ of the whole interval $[-20, 18]$, which yields the answer $\boxed{37}$.

Extension Exercise 30. *(2015 AIME I)* Consider all 1000-element subsets of the set 1, 2, 3, ..., 2015. From each such subset choose the least element. The arithmetic mean of all of these least elements is $\dfrac{p}{q}$, where p and q are relatively prime positive integers. Find $p + q$.

Extension Solution 30. Making extensive use of Hockey-Stick, we obtain $\binom{2015}{1000} \cdot \mu = \binom{2016}{1001}$, where μ is the desired mean. Hence $\mu = \dfrac{2016}{1001} = \dfrac{288}{143} \implies \boxed{431}$.

Extension Exercise 31. *(2015 Berkeley Math Tournament)* Evaluate
$$\sum_{k=0}^{37}(-1)^k \cdot \binom{75}{2k}.$$

Extension Solution 31. The Binomial Theorem says that the coefficients of $(1 + x)^{75}$ are given by $\sum_{k=0}^{75} \binom{75}{k}$. Substituting $x = i$ yields the desired sum as the real part of $(1 + i)^{75}$. Converting to polar form yields $(1 + i)^{75} = \left(\sqrt{2}, \dfrac{\pi}{4}\right)^{75} = \left(\sqrt{2^{75}}, \dfrac{3\pi}{4}\right)$, so the sum evaluates to $\sqrt{2^{75}} \cdot \cos\dfrac{3\pi}{4} = \boxed{-2^{37}}$. *(An alternate method is roots of unity filter.)*

Extension Exercise 32. *(2014 Stanford Math Tournament)* A composition of a natural number n is a way of writing it as a sum of natural numbers, such as $3 = 1 + 2$. Let $P(n)$ denote the sum over all compositions of n of the number of terms in the composition. For example, the compositions of 3 are 3, $1 + 2$, $2 + 1$, and $1 + 1 + 1$; the first has one term, the second and third have two each, and the last has 3 terms, so $P(3) = 1 + 2 + 2 + 3 = 8$. Compute $P(9)$.

Extension Solution 32. For $1 \leq k \leq n$, the number of compositions with k components is $\binom{n-1}{k-1}$, so that $P(n) = \sum_{k=1}^{n} k \binom{n-1}{k-1}$. By splitting up k into 1 and $k - 1$, we obtain a more convenient expression that simplifies to $(n + 1)2^{n-2}$. Therefore, $P(9) = \boxed{1280}$.

Extension Exercise 33. *(2018 HMMT February Combinatorics Test)* How many ways are there for Nick to travel from $(0, 0)$ to $(16, 16)$ in the coordinate plane by moving one unit in the positive x or y direction at a time, such that Nick changes direction an odd number of times?

Extension Solution 33. *Nicks first and final step must differ; perform casework based on each of the two scenarios. In both cases, we reduce the problem to a traditional grid-walking problem with no additional constraints, from $(1,0)$ to $(16,15)$. Hence, the answer is* $\boxed{2 \cdot \binom{30}{15}}$.

Extension Exercise 34. *(2010 USAMO P2) There are n students standing in a circle, one behind the other. The students have heights $h_1 < h_2 < \ldots < h_n$. If a student with height h_k is standing directly behind a student with height h_{k-2} or less, the two students are permitted to switch places. Prove that it is not possible to make more than $\binom{n}{3}$ such switches before reaching a position in which no further switches are possible.*

Extension Solution 34. *In a switch, we say that the taller student moves "forward (i.e. in the direction that all the students are facing). For $k \geq 3$, student h_k cannot swap with h_{k-1}, and h_k can make up to $k-2$ more moves forward than h_{k-1}. Hence, if student $k-1$ moves forward s_{k-1} times, then student k can move forward $s_{k-1} + k - 2$ times (where $s_1 = s_2 = 0$, $s_3 = 1$). It follows that $s_k = \binom{k-1}{2}$. By Pascals Identity, the sum is $\binom{n}{3}$.*

Extension Exercise 35. *(2018 Berkeley Math Tournament) Compute*
$$\sum_{i=0}^{\infty} \sum_{j=0}^{\infty} \binom{i+j}{i} \cdot 3^{-(i+j)}.$$

Extension Solution 35. *The sum is equivalent to*
$$\sum_{i=0}^{\infty} \frac{1}{3^i} \sum_{j=0}^{i} \binom{i}{j} = \sum_{i=0}^{\infty} \frac{2^i}{3^i} = \boxed{3}$$

Extension Exercise 36. *The odds of San Holo successfully navigating his way through an asteroid field are $3720:1$. As $n \to \infty$, compute the limit of the probability that Han navigates his way through at least n out of $3721n$ asteroid fields.*

Extension Solution 36. *We claim the answer is $\boxed{\dfrac{1}{2}}$. This is a remarkable result of the Central Limit Theorem, as well as the Law of Large Numbers. Hint: try smaller cases first.*

Section 11.3: Graph Theory: The Massive Web of Mathematics

Exercise 77. What is the chromatic number of the complete graph K_n on n vertices in terms of n? What about the chromatic polynomial of K_n?

Solution 77. Since each vertex in K_n is adjacent to all other vertices, the chromatic number is just n. The chromatic polynomial is given by the number of choices of colors out of k available, or $k(k-1)(k-2)\cdots(k-n+1) = \frac{k!}{(k-n)!} = n!\binom{k}{n}$.

Exercise 78. Determine the chromatic polynomial for a cycle of order 4.

Solution 78. Call the vertices A, B, C, D, with A adjacent to D. There are k choices for the color of A. Next, there are two cases.

Case 1: B and D are different colors. We have $(k-1)(k-2)$ ways to color B, D. Subsequently, there are $k-2$ ways to color vertex C.

Case 2: B and D are the same color. There are $k-1$ ways to color B, D simultaneously. There are then $k-1$ ways to color C.

It follows that the chromatic polynomial is given by $\boxed{k(k-1)(k-2)^2 + k(k-1)^2}$.

Exercise 79. How many cycles does K_4 have?

Solution 79. We can count the numbers of 3- and 4-cycles. The number of 3-cycles is just $\binom{4}{3} = 4$ while the number of 4-cycles is 3. Thus, K_4 has $\boxed{7}$ cycles in total.

Exercise 80. *(2012 AMC 10A/12A)* Adam, Benin, Chiang, Deshawn, Esther, and Fiona have internet accounts. Some, but not all, of them are internet friends with each other, and none of them has an internet friend outside this group. Each of them has the same number of internet friends. In how many different ways can this happen?

Solution 80. The group of friends is akin to a hexagonal graph. WLOG the first person may be anywhere on the hexagon, while also having $\binom{5}{2} = 10$ choices for his/her two friends. Then each of the three remaining people may be across from the other three, which yields $10 \cdot 3! = 60$ hexagonal configurations (70 in total for $n = 2$). $n = 3$ has 70 configurations, and $n = 4$ has 15, so the total is $2(70 + 15) = \boxed{170}$. (For $n = 1, 4$, we simply consider pairs of friends and non-friends.)

Exercise 81. *(2016 Berkeley Math Tournament)* How many graphs are there with 6 vertices that have degrees 1, 1, 2, 3, 4, and 5?

Solution 81. We cannot have a graph with two vertices of degree 1, as well as another vertex of odd degree, so the answer is $\boxed{0}$.

Exercise 82. Show that every tree has at least two leaves.

Solution 82. Consider the longest path $P = v_0 v_1 v_2 \cdots v_k$ in the tree (denote $x = v_0, y = v_k$). We claim that x, y must be leaves. Assume the contrary; then $\deg x \geq 2$ and x is adjacent to another vertex $z \neq v_1$. Then $z \notin P$, so we add the edge zx to P, but this is a contradiction. Hence, x is a leaf (and so is y).

Exercise 83. Show that a graph G is a tree if and only if G has no cycles but the addition of any edge (between two existing vertices) would create a cycle.

Solution 83. $\Rightarrow$ Suppose G is a tree. One characteristic of a tree is that there is exactly one pathway between any two vertices in the tree. If we were to add any edge to G, then that would create a second pathway between the two vertices that the edge connects, which would create a cycle.

⇐ *For a graph to have no cycles, there are two cases: (1) the graph consists of at least two connected components, each of which are trees or (2) the graph is a tree. If G belonged to Case (1), then an added edge could connect two of the connected components, which would not create a cycle. Therefore G belongs to Case (2); in other words, G is a tree. QED.*

Exercise 84. *There are n people at a meeting. Among any group of 4 participants, there is one who knows the other three members of the group. Prove that there is a person who knows everyone else.*

Solution 84. *Consider vertex v with maximum degree. We claim that this vertex is adjacent to every other vertex. For contradiction, assume the existence of some vertex w not adjacent to v. Then every pair of vertices is adjacent in the neighborhood $N(v)$ of v. Moreover, at least one of the vertices u in $N(v)$ is adjacent to w. Hence, $\deg u > \deg v$ - contradiction.*

Exercise 85. *Prove that in a group of 18 people, there is either a group of 4 mutual friends or a group of 4 mutual strangers.*

Solution 85. *In the complete graph K_{18}, color the edges with two colors. For a vertex v, there are 9 edges of the same color. In K_9, we either have a purely red C_3, which together with v, yields a wholly-red or wholly-blue subgraph K_4.*

Exercise 86. *There are n people at a meeting, some of whom are mutual friends. Demonstrate how we can split them into two groups such that each person has at least half of his/her friends in a different group.*

Solution 86. *Select subgraphs A and B with $A \cup B = G$, and such that we maximize the number of edges from a vertex in A to a vertex in B.*

Exercise 87. *Show that, if $v \in G$ has $\deg(v_1)$ odd, then there exists a path $P = v_1 v_2$ such that $\deg(v_2)$ is also odd.*

Solution 87. *We construct a path with no edge traversed more than once. Take vertex v as the starting point, and begin to diverge from it. If the path ever reaches a vertex of odd degree, we are done. However, when we reach a vertex of even degree, we can erase two edges that converge at that vertex. Since the vertex originally had even degree, its degree will remain even after the erasure. In this fashion, we can always proceed from an even-degree vertex. Since there are only a finite number of edges, the path algorithm eventually halts at an odd-degree vertex.*

Exercise 88. *At a chess tournament, there are 18 competitors. In each round, the contests are paired off, and play against each other once. Show that, after 8 rounds, there are three contestants, no pair of which has played against each other.*

Solution 88. *Draw a graph with two vertices (vertices corresponding to players) connected if they have not played each other, with all vertices having degree 9. We claim that this graph contains at least one 3-cycle. Assume the contrary; then let the vertices be $v_1, v_2, \ldots, v_9$ and $w_1, w_2, \ldots, w_9$ as their immediate neighbors. The neighbours of w_1 must be the v_i; this holds for $w_2, \ldots, w_9$ as well. Thus, each vertex v_i is adjacent to all vertices w_i, and each w_i is adjacent to each v_i, hence the graph is a complete bipartite graph. Thus, in each round, all players are paired, but this is a contradiction. Therefore, the graph must have a 3-cycle.*

Extension Exercise 37. *Let there be n points in the plane such that the distance between every pair of points is at least 1. Prove that there are at most 3n pairs of points with a distance of exactly 1.*

Extension Solution 37. *Interpret this in graph-theoretic terms: consider the points as vertices, with vertices adjoined by an edge iff their distance in the plane is 1. At any point, draw a circle of radius 1. Since the points are at least 1 apart, the circle has at most six points. Hence, an upper bound for the degree of any given vertex is 6, with no more than $\frac{6n}{2} = 3n$ edges, as desired.*

Extension Exercise 38. *Find the smallest positive integer n such that, in any set of n irrational numbers, there exist three numbers with all pairwise sums of those numbers irrational.*

Extension Solution 38. *Assume $n = 4$. Then among any three of $\pm\sqrt{2}$, $1 \pm \sqrt{2}$, there are two whose sum is rational. Thus, $n \geq 5$. Construct a graph G with vertices corresponding to the numbers. In G, two vertices are adjacent iff their corresponding numbers sum to a rational number. Then G does not contain C_3; if $x + y$, $y + z$, and $z + x$ are rational, then all of x, y, z will be rational as well. Similarly, C_5 is not a subgraph of G. Hence, G is bipartite. One of the partitions contains three vertices; the sum of any two of them is irrational.*

Extension Exercise 39. *In Podunk School District, there are 3 schools, each of which has n students. Every student knows exactly $n + 1$ students from the other two schools. Prove that we can find one student from each of the schools such that the three students all know each other (pairwise).*

Extension Solution 39. *Let the schools be A, B, C. Let each student be a vertex, and let an edge indicate whether the students know each other. For each vertex $v \in A$, let $NA(x)$ be the set of $a \in A$ adjacent to v. Let $m = \max\{NY(x) \mid x \in Y\} = |NA(b)|$, where $b \in B$. Then b is adjacent to m vertices in A, as well as with some $c \in C$. But as $|NB(c)| \leq m$, $|NA(c)| \geq n + 1 - m$. $|NA(b)| = m$, however, so $NA(c) \cap NA(b)$ contains some vertex (WLOG assume a). It follows that a, b, c all know each other.*

Extension Exercise 40. *Prove that any coloring of the edges of K_6 using only the colors red and blue will necessarily result in there being two monochromatic triangles (triangles with all three edges the same color).*

Extension Solution 40. *Call a pair of adjacent edges a monochromatic pair if they have the same color. Each vertex spawns at least four monochromatic pairs at each vertex, which yields at least 24 monochromatic pairs overall. Each monochromatic triangle has three such pairs, while a non-monochromatic triangle has only one pair.*

Chapter 11 Review Exercises

Exercise 89. *(2018 HMMT November Guts Round) Consider an unusual biased coin, with probability p of landing heads, probability $q \leq p$ of landing tails, and probability $\frac{1}{6}$ of landing on its side (i.e. on neither face). It is known that if this coin is flipped twice, the likelihood that both flips will have the same result is $\frac{1}{2}$. Find p.*

Solution 89. We have $p + q = \frac{5}{6}$, and $p^2 + q^2 + \left(\frac{1}{6}\right)^2 = \frac{1}{2}$. Solving the system yields $(p, q) = \boxed{\left(\frac{2}{3}, \frac{1}{6}\right)}$, as $p > \frac{5}{12}$.

Exercise 90.

(a) *(2014 HMMT November Guts Round) If you flip a fair coin 1000 times, what is the expected value of the product of the number of heads and the number of tails?*

(b) *If you flip a fair coin k times, where $k \in \mathbb{N}$, what is the expected value of the product of the number of heads and the number of tails?*

(c) *If you choose a positive integer n uniformly at random from $[1, 2019]$, and then flip a fair coin n times, what is the expected value of the expected value of the product of the number of heads and the number of tails?*

Solution 90.

(a) The expected value of the product is the product of the expected values, i.e. $500^2 = \boxed{250,000}$.

(b) By the same logic as above, our answer is $\boxed{\dfrac{k^2}{4}}$.

(c) This is just the average of the above for $k \in [1, 2019]$, which is $\frac{1}{8076} \sum_{i=1}^{2019} i^2 = \frac{1}{8076} \cdot \frac{2019 \cdot 2020 \cdot 4039}{6} = \boxed{\dfrac{505 \cdot 4039}{6}}$.

Exercise 91. *Call an ordered triple of positive integers (a, b, c) quadratically satisfactory if the quadratic equation $ax^2 + bx + c = 0$ has two real solutions. Let $a, b, c \in \{1, 2, 3, 4, 5\}$. Compute the probability that (a, b, c) is quadratically satisfactory.*

Solution 91. We require $b^2 - 4ac > 0 \implies b^2 > 4ac$; perform casework on b. The answer is $\boxed{\dfrac{4}{25}}$.

Exercise 92. *(1990 USAMO P1) A certain state issues license plates consisting of six digits (from 0 through 9). The state requires that any two plates differ in at least two places. (Thus the plates $\boxed{027592}$ and $\boxed{020592}$ cannot both be used.) Determine, with proof, the maximum number of distinct license plates that the state can use.*

Solution 92. Consider the set of n-digit long license plates. By Pigeonhole, we have at most 10^{n-1} distinct plates. Indeed, there are exactly 10^{n-1} distinct plates. For the first $n - 1$ digits, any digits are permissible; for the final digit, take the sum of the other digits mod 10. Then if two plates have the same first $n - 1$ digits, then they must be identical in the last digit as well, and if two plates differ in any of the first $n - 1$ digits, they will differ in the last digit as well. Thus, the answer is $\boxed{10^5}$.

Exercise 93.

(a) *How many cycles does the complete graph K_6 have?*

(b) In terms of n, compute the number of cycles in the complete graph K_n.

Solution 93. Let $p \leq n$ be a positive integer, with C_p a length-p cycle such as $2 \to 3 \to 4 \to \ldots \to (p-1) \to 2$. Fix the first/last vertex, and consider the remaining vertices. We have $\dfrac{(p-1)!}{2}$ cycles that may result from permuting those vertices. Among the n vertices in K_n, we can choose p of them in $\binom{n}{p}$ ways. Thus, the number of cycles in K_n is given by the sum

$$\boxed{\sum_{p=3}^{n} \binom{n}{p} \frac{(p-1)!}{2}}$$

which evaluates to $\boxed{197}$ at $n = 6$ in particular.

Exercise 94. *(2012 HMMT November Guts Round)* Alice generates an infinite decimal by rolling a fair 6-sided die with the numbers 1, 2, 3, 5, 7, and 9 infinitely many times, and appending the resulting numbers after a decimal point. What is the expected value of her number?

Solution 94. Let X_n be her n^{th} number. The resulting number is $\sum_{k=1}^{\infty} \dfrac{X_k}{10^k}$. By linearity of expectation, the answer is the sum of the expected value of the rolls divided by 10^k. This is $\dfrac{9}{2} \cdot \left(\dfrac{1}{10} + \dfrac{1}{100} + \ldots \right) = \boxed{\dfrac{1}{2}}$.

Exercise 95. A restaurant has a 4.2 star average rating, based on 15 reviews, on Yowl, the hot new online review service. Ratings are integers from 1 to 5, inclusive. If it received no 1- or 2-star ratings, how many distributions of star ratings are possible? (The order in which the reviews were left is irrelevant.)

Solution 95. The sum of star ratings is 63 stars, which means the restaurant "lost 12 stars among 15 reviews. In each review, it has lost no more than 2 stars. We can split this problem up into 2 cases:

Case 1: All reviews are 4- or 5- stars, meaning any star loss is at most 1. There are $\binom{15}{12} = \binom{15}{3} = 455$ ways for this to be the case.

Case 2: Some reviews are 3-stars, meaning star loss is 2. If there is a single 3-star review, there are $\binom{15!}{10!4!1!} = 15 \cdot 7 \cdot 13 \cdot 11$ ways for this to happen. With 2 3-star reviews, we have 8 4-star reviews, and so $\dfrac{15!}{8!5!2!}$ permutations. With 3 3-stars, we have $\dfrac{15!}{3!6!6!}$ ways, with 4 3-stars, we have $\dfrac{15!}{4!4!7!}$, with 5 3-stars, we have $\dfrac{15!}{5!2!8!}$, and with 6 3-stars, the rest are 5 stars and the number of permutations is $\binom{15}{6}$. Hence, we can straightforwardly compute the total from here (perhaps using Pascals Identity).

Exercise 96. *(2018 Fall OMO)* Patchouli is taking an exam with $k > 1$ parts, numbered Part $1, 2, \ldots, k$. It is known that for $i = 1, 2, \ldots, k$, Part i contains i multiple choice questions, each of which has $(i+1)$ answer choices. It is known that if she guesses randomly on every single question, the probability that she gets exactly one question correct is equal to 2018 times the probability that she gets no questions correct. Compute the number of questions that are on the exam.

Solution 96. Testing small cases, we find that the probability that Patchouli gets one question right is k times the probability that she gets none right. (We can prove this via induction; as a starting point, consider $k+1$.) Hence $k = 2018$, and the number of questions is $1 + 2 + 3 + \ldots + 2018 = 2019 \cdot 1009 = \boxed{2,037,171}$.

Exercise 97. A string of l letters has exactly 360 distinguishable permutations. Compute all possible values of l, and prove that no others exist.

Solution 97. *Clearly,* $\frac{6!}{2!} = 360$, *so* $l = 6$ *is a possibility. Also* $\frac{10!}{360} = 10,080 = 7! \cdot 2!$, *so* $l = 10$ *works as well. No other possibilities exist (hint: show any decomposition of* $\frac{l!}{360}$ *would have prime factors summing to more than* l*).*

Exercise 98. *(2018 Berkeley Math Tournament) Alice and Bob play a game where they start from a complete graph with n vertices and take turns removing a single edge from the graph, with Alice taking the first turn. The first player to disconnect the graph loses. Compute the sum of all n between 2 and 100 inclusive such that Alice has a winning strategy. (A complete graph is one where there is an edge between every pair of vertices.)*

Solution 98. *If the graph is a tree, then the next move will disconnect it. Else, there is a move that does not disconnect it (i.e. preserves it). Each complete graph on n vertices has* $\binom{n}{2}$ *edges. Furthermore, any tree on n vertices has* $n - 1$ *edges, so optimal play leads to the removal of* $\binom{n}{2} - (n-1) + 1$ *nodes, i.e. Alice wins if* $\binom{n}{2} - n = \frac{n(n-3)}{2}$ *is odd and Bob wins if it is even. With some basic modular arithmetic, we can derive the answer of* $\boxed{2575}$.

Exercise 99. *(1990 AHSME) A subset of the integers* $1, 2, \cdots, 100$ *has the property that none of its members is 3 times another. What is the largest number of members such a subset can have?*

Solution 99. *Note that we can include integers from 34 to 100, as long as no integer from 11 to 33 (inclusive) is also included. Hence, there are at least 67 members of the maximal set. From 1 to 10, we can include everything except 3 (as its inclusion would then require us to remove both 1 and 9, which is sub-optimal) to get 9 more solutions* $\implies$ $\boxed{76}$ *in total.*

Exercise 100. *(2017 ARML Local Individual Round #5) ABCDEFG is a pyramid with a hexagonal base. Compute the number of distinct ways all seven vertices of ABCDEFG can be colored one of either red, blue, or green such that no vertices that share an edge are identically colored.*

Solution 100. *Consider the top vertex. With its color fixed, there are only 2 possibilities for the configurations of the colors of the base vertices, since they must alternate and none can be the same color as the top vertex. There are 3 choices for the color of the top vertex, so there* $2 \cdot 3 = \boxed{6}$ *color configurations in total.*

Exercise 101. *(2014 AMC 12B) The number 2017 is prime. Let* $S = \sum_{k=0}^{62} \binom{2014}{k}$. *What is the remainder when S is divided by 2017?*

Solution 101. *Note* $2014 \equiv -3 \pmod{2017}$, *so* $\binom{2014}{k} \equiv (-1)^k \cdot \binom{k+2}{2} \pmod{2017}$. *Thus the sum collapses to an alternating series of triangular numbers, which in turn allows us to find a pattern based on whether n is even (this case) or odd. Here,*

$$\sum_{k=1}^{n} (-1)^k \cdot \binom{k+2}{2} = \left(\frac{n}{2} + 1\right)^2 \pmod{2017} \equiv \boxed{1024} \pmod{2017}$$

Exercise 102. *(2017-2018 USAMTS Round 1 P2) After each Goober ride, the driver rates the passenger as 1, 2, 3, 4, or 5 stars. The passenger's overall rating is determined as the average of all of the ratings given to him or her by drivers so far. Noah had been on several rides, and his rating was neither 1 nor 5. Then he got a 1 star on a ride because he barfed on the driver. Show that the number of 5 stars that Noah needs in order to climb back to at least his overall rating before barfing is independent of the number of rides that he had taken.*

Solution 102. *Let r be Noahs original Goober rating. Then if we let m be the minimum number of rides that Noah needs to take to restore his original rating, m satisfies $\frac{5m+1}{m+1} \geq r \implies m \geq \frac{r-1}{5-r}$. Hence $m = \lceil \frac{r-1}{5-r} \rceil$, and the desired result is apparent.*

Extension Exercise 41. *(1964 IMO P4) Seventeen people correspond by mail with one another - each one with all the rest. In their letters only three different topics are discussed. Each pair of correspondents deals with only one of these topics. Prove that there are at least three people who write to each other about the same topic.*

Extension Solution 41. *We invoke the following lemma:*

Lemma. *In K_6 colored with two colors, there is a monochromatic triangle.*

Proof. Consider an arbitrary vertex and its edges. By Pigeonhole, there are at least 3 of the 5 edges emanating from that vertex of the same color. If any of those 3 segments are that color, we have established our monochromatic triangle. If they are all the other color, then we also have a monochromatic triangle. $\square$

Henceforth, draw the graph corresponding to the problem and observe that there exists a monochromatic triangle on the connected graph with 17 vertices colored with three colors. Consider some vertex v, as well as the other vertices it is connected with. By Pigeonhole, at least 6 vertices are connected by v that are of a single color. If any of the edges between these vertices are in that same color, QED. Otherwise, there is still a monochromatic triangle in those 6 vertices, so we are still done.

Extension Exercise 42. *(1967 IMO P6) In a sports contest, there were m medals awarded on n successive days ($n > 1$). On the first day, one medal and $\frac{1}{7}$ of the remaining $m-1$ medals were awarded. On the second day, two medals and $\frac{1}{7}$ of the now remaining medals were awarded; and so on. On the nth and last day, the remaining n medals were awarded. How many days did the contest last, and how many medals were awarded altogether?*

Extension Solution 42. *At the beginning of day d, let the number of medals left be m_d. Then $m = m_1$ and $m_{k+1} = \frac{6}{7}(m_k - k)$ for $k < n$, $m_n = n$.*

We compute $m = 1 + 2 \cdot \frac{7}{6} + 3 \cdot \left(\frac{7}{6}\right)^2 + \ldots$ which is a traditional arithemtico-geometric series. We have $n = 6$ and $m = 36$.

Extension Exercise 43. *(2018 Stanford Math Tournament) Consider a game played on the integers in the closed interval $[1, n]$. The game begins with some tokens placed in $[1, n]$. At each turn, tokens are added or removed from $[1, n]$ using the following rule: For each integer $k \in [1, n]$, if exactly one of $k-1$ and $k+1$ has a token, place a token at k for the next turn, otherwise leave k blank for the next turn. We call a position static if no changes to the interval occur after one turn. For instance, the trivial position with no tokens is static because no tokens are added or removed after a turn (because there are no tokens). Find all non-trivial static positions.*

Extension Solution 43. *Consider a token at position k. For k to remain constant, it must have one neighbor. WLOG let this neighbor be at position $k+1$, so that $k-1$ is empty. Then $k-1 = 0 \implies k = 1$ or $k-2$ is non-empty. Then $k-3$ must also be occupied, which implies $k = 3i + 1 \equiv 1 \pmod 3$ for some $i \in \mathbb{N}$. Similarly, $n = k + 1 + 3j$ for some $j \in \mathbb{N}$. The desired interval $[1, n]$ is then of length $3(i+j) + 2$, implying that all static positions depend on whether $n \equiv 2 \pmod 3$.*

Extension Exercise 44. *(2017 ARML Local Individual Round #10) Let S be a set of 100 points inside a square of side length 1. An ordered pair of not necessarily distinct points (P, Q) is bad if $P \in S$, $Q \in S$ and $|PQ| < \frac{\sqrt{3}}{2}$. Compute the minimum possible number of bad ordered pairs in S.*

Extension Solution 44. *By Pigeonhole, there must exist a bad pair among any five points in the square. Henceforth, consider a graph with vertices at the 100 points in S, and an edge connecting any two good (non-bad) vertices. This graph has no complete subgraph consisting of five vertices, and so the graph has the vertices partitioned into 4 subgraphs of 25 vertices each, connecting each vertex to each vertex in a different subgraph (much like a bipartite graph, except with 4 partitions). Hence, there are at least $4 \cdot 25^2 =$ $\boxed{2500}$ bad pairs.*

Extension Exercise 45. *(2018 HMMT February Guts Round) Michael picks a random subset of the complex numbers $\{1, \omega, \omega^2, \ldots, \omega^{2017}\}$ where ω is a primitive 2018th root of unity and all subsets are equally likely to be chosen. If the sum of the elements in his subset is S, what is the expected value of $|S|^2$? (The sum of the elements of the empty set is 0.)*

Extension Solution 45. *Consider $a, -a$ in that set. Let s be the sum of the other complex numbers in the set of primitive roots of unity; then it follows that $\dfrac{(s+a)(\overline{s-a}) + 2s\overline{s} + (s-a)(\overline{s+a})}{4} = s\overline{s} + \dfrac{a\overline{a}}{2} = s\overline{s} + \dfrac{1}{2}$. Repeating this for all the other pairs yields $\boxed{\dfrac{1009}{2}}$.*

Extension Exercise 46. *(1975 USAMO P5) A deck of n playing cards, which contains three aces, is shuffled at random (it is assumed that all possible card distributions are equally likely). The cards are then turned up one by one from the top until the second ace appears. Prove that the expected (average) number of cards to be turned up is $\dfrac{n+1}{2}$.*

Extension Solution 46. *We can use induction on n, but we can also consider the second ace in position k. If we draw the deck backwards, then we draw the second ace in position $n - k + 1$. Thus, the expected position of the second ace is $\dfrac{k + n - 1 - k}{2} = \dfrac{n+1}{2}$.*

Extension Exercise 47. *(2014 IMO P2) Let $n \geq 2$ be an integer. Consider an $n \times n$ chessboard consisting of n^2 unit squares. A configuration of n rooks on this board is peaceful if every row and every column contains exactly one rook. Find the greatest positive integer k such that, for each peaceful configuration of n rooks, there is a $k \times k$ square which does not contain a rook on any of its k^2 squares.*

Extension Solution 47. *We claim the answer is $k = \lceil \sqrt{n} \rceil - 1$. We then show that each $n \times n$ chessboard with a peaceful rook configuration contains a valid $k \times k$ square. Since $k < n$, we have a set S of k consecutive columns, one of which has the first rook. If we also consider the $n - k + 1$ squares in S with dimensions $s \times s$. Of these squares, only 1 actually has the rook. Moreover, each of the $k - 1$ other rooks in S can only make k of the squares in S have a rook. Since $n - k + 1 > k(k-1) + 1$ by Pigeonhole; there is a $k \times k$ square in S not covered by a rook in S (valid choice).*

It suffices to show that there is a chessboard without a $(k+1) \times (k+1)$ square. First position a rook at the upper-left corner, and then another rook $k + 1$ down and 1 right. Repeat this process until we reach the bottom of the board. Subsequently place a rook in the unoccupied column and the first free row. This configuration is peaceful, since $(k+1)^2 > n$ we satisfy the property of every $(k+1) \times (k+1)$ square in the chessboard is occupied by a rook, QED.

Chapter 12: Advanced Discrete Math

Section 12.1: Combinatorics and Probability: Part 3

Exercise 103. *A cube is painted on all six faces, with the color of each face chosen from six coats of paint: red, blue, yellow, orange, teal, and violet. Two color arrangements are considered to be identical if the cube can be rotated or reflected to obtain the other. Find the number of distinct color arrangements.*

Solution 103. *Initially, we consider the possible configurations of the cube independently from the potential colorings of each face. There are 6^6 possible colorings of the cube; it can either remain as is, rotate all faces by $90°$ or $180°$, rotate all vertices by $120°$, or rotate all edges by $180°$. (Due to symmetry, rotations of $270°$ and $360°$ are equivalent to rotations of $90°$ and $0°$, respectively.) Henceforth, suppose that we have a set X such that $|X| = 6^6$, upon which a rotation group G acts. We observe that $|G| \, 2^3 \cdot 3 = 24$, where the number of distinct face and edge rotation types is 3, and the number of vertex rotation types is 1. Thus, we partition G into five subsets G_i ($1 \leq i \leq 5$).*

G_1 is the identity subset, which contains the cube without any modifications to its faces, edges, or vertices. There are 6 choices for each face, and 6 colors, so G_1 has $\dfrac{6^6}{24}$ elements.

In G_2, all six $90°$ face rotations leave 3 faces unchanged, or equivalently 6^3 elements of X.

In G_3, each $180°$ face rotation leaves 4 faces unchanged, or 6^4 elements of $X \to |G_3| = \dfrac{3 \cdot 6^4}{24}$.

In G_4, a $120°$ vertex rotation leaves 2 faces unchanged $\to |G_4| = \dfrac{8 \cdot 6^2}{24}$.

Finally, in G_5, six edge rotations leave 6^3 elements of X unchanged, so that $|G_2 + G_5| = \dfrac{12 \cdot 6^3}{24}$.

Therefore, the number of rotationally distinct color arrangements is $\dfrac{6^6 + 3 \cdot 6^4 + 12 \cdot 6^3 + 8 \cdot 6^2}{24} = \boxed{2226}$.

Exercise 104. *(2016 HMMT February Combinatorics Test) Let $V = \{1, \ldots, 8\}$. How many permutations $\sigma : V \mapsto V$ are automorphisms of some tree? (Authors note: An automorphism of a tree T is a permutation $\sigma : V \mapsto V$ such that vertices $i, j \in V$ are edge-connected iff $\sigma(i)$ and $\sigma(j)$ are.)*

Solution 104. *Decompose into cycle types of σ. Note that, in each cycle of σ, all vertices are of equal degree. The tree, as well, has a degree sum of 14 across all its vertices (as it has 7 edges).*

For any permutation with a fixed point, let $1 \leq a \leq 8$ be that fixed point. Consider the tree consisting of the edges from a to the other seven vertices; this permutation is an automorphism. For any permutation with a 2-cycle and a 6-cycle, let a, b be elements of the 2-cycle. Similarly as above, it is not too hard to verify that σ is an automorphism of the tree.

We also have 2-2-4, 2-3-3, 3-5, 4-4, and 8 cycle types, but only 2-2-3, 8, 4-4, 3-5 are not automorphisms (briefly, argue by contradiction that the resulting permutation violates a property of a tree (e.g. it has no cycles)). Subtracting these (namely 1120, 5040, 1260, 2688 from 8! yields the answer, $\boxed{30,212}$.

Exercise 105. *(2019 AIME I) A moving particle starts at the point $(4,4)$ and moves until it hits one of the coordinate axes for the first time. When the particle is at the point (a, b), it moves at random to one of the points $(a-1, b)$, $(a, b-1)$, or $(a-1, b-1)$, each with probability $\dfrac{1}{3}$, independently of its previous moves. Compute the probability that it will hit the coordinate axes at $(0,0)$ is $\dfrac{m}{3^n}$.*

Solution 105. *Note that this probability is $\dfrac{1}{3}$ that of the particle reaching $(1,1)$, which can be computed using either recursion or weighted probabilities. Assume the particle takes only 3 moves to reach $(1,1)$ from*

361

$(4, 4)$; then all 3 moves must be diagonal movements (probability $\frac{1}{3^3} = \frac{27}{3^6}$. If it takes 4 moves, 2 moves are diagonals and 1 is up, 1 is right. There are 12 ways to arrange these movements, out of 3^4 choices total, so the probability here is $\frac{108}{3^6}$. For 5 moves, 1 is diagonal, 2 are right, and 2 are up $\implies \frac{90}{3^6}$. Finally, for 6 moves, we just have $\frac{\binom{6}{3}}{3^6} = \frac{20}{3^6}$, so the sum is $\frac{245}{3^6}$. Recall, however, that this is the probability that we reach $(1, 1)$, so we must divide by 3 to get $\frac{245}{3^7} \implies \boxed{252}$ as our final answer.

Exercise 106. *Show that for any sequence of integers $a_1, a_2, \ldots, a_n$,*
$$\prod_{1 \leq i < j \leq n} \frac{a_i - a_j}{i - j} \in \mathbb{Z}$$

Solution 106. *Sketch: Think about how this is related to the Vandermonde determinant, by definition. As the Vandermonde determinant can be expressed in the very similar form $\prod_{1 \leq i < j \leq n} (\alpha_j - \alpha_i)$ it must be an integer.*

Exercise 107. *Suppose that we want to color the edges of a tetrahedron with n colors. There is no limit on the number of times we may use each color. Up to rotation of the tetrahedron, how many different colorings are possible?*

Solution 107. *For this problem, we compute the n-color cyclic index of the face permutation group G of the tetrahedron: $a_k = \sum_{i=1}^{n} X_i^k$ and extract the coefficient. Then Burnsides Lemma is a special case of this method, with $\frac{1}{12}\left(n^4 + 8n^2 + 3n^2\right) = \frac{n^4 + 11n^2}{12}$.*

Exercise 108. *(2018 Berkeley Math Tournament) Consider a $2 \times n$ grid where each cell is either black or white, which we attempt to tile with 2×1 black or white tiles such that tiles have to match the colors of the cells they cover. We first randomly select a random positive integer N where N takes the value n with probability 2^{-n}. We then take a $2 \times N$ grid and randomly color each cell black or white independently with equal probability. Compute the probability the resulting grid has a valid tiling.*

Solution 108. *Let this probability be p_n; consider the cell at the end of the first row. If the cell beneath it is the same color as it, and there exists a valid tiling, then there is a valid tiling from covering those two cells with one 2×1 "cell of the same color. Hence, we are able to tile the resulting $2 \times (n-1)$ grid similarly. Otherwise, if a valid tiling were to exist, then there must be a pair of horizontal tiles of different colors spanning the first two columns, with a valid tiling over the $2 \times (n-2)$ grid that results. Therefore, $p_n = \frac{1}{2}p_{n-1} + \frac{1}{8}p_{n-2}$ with $p_0 = 1, p_1 = \frac{1}{2}$.*

By using generating functions, we can see that $P(x) = 1 + \frac{x}{2} + \sum_{n=2}^{\infty} p_n x^n$ or, after some algebra, $P(x) = \frac{1}{1 - \frac{x}{2} - \frac{x^2}{8}}$. We seek $\sum_{n=1}^{\infty} \frac{P_n}{2^n}$, which is $P(x) - 1$ for $x = \frac{1}{2}$, i.e. $\boxed{\frac{9}{23}}$.

Extension Exercise 48. *(2000 USAMO P4) Find the smallest positive integer k such that, if k squares of an $m \times n$ chessboard are painted black, then there will be three black squares whose centers form a right triangle with sides parallel to the edges of the board.*

Extension Solution 48. *We claim that $n \geq \boxed{1999}$. (For $n \leq 1998$, color any of the 1998 squares in the top row/left column, excluding the top-left corner.)*

Say a row/column is occupied if all of its squares are colored in. Then any of the remaining squares must share a row/column with one of the colored squares in an occupied row/column. Thus, these squares form a right triangle - contradiction.

Let m count the columns with 1 colored square. Then there are $1999 - m$ colored squares in the remaining columns; in each column, there are at least two colored squares, which form a triangle. Hence, each of the $1999 - m$ colored squares must be in different rows, which implies $m \geq 999$. If $m = 999$, then all 1000 rows can have exactly 1 colored square, but this leaves no place for the other 999; contradiction. We also get a contradiction for $m \geq 1000$; hence $n \geq 1999$ as claimed.

Extension Exercise 49. *(2018 Spring OMO) Let m and n be positive integers. Fuming Zeng gives James a rectangle, such that $m - 1$ lines are drawn parallel to one pair of sides and $n - 1$ lines are drawn parallel to the other pair of sides (with each line distinct and intersecting the interior of the rectangle), thus dividing the rectangle into an $m \times n$ grid of smaller rectangles. Fuming Zeng chooses $m+n-1$ of the mn smaller rectangles and then tells James the area of each of the smaller rectangles. Of the $\binom{mn}{m+n-1}$ possible combinations of rectangles and their areas Fuming Zeng could have given, let $C_{m,n}$ be the number of combinations which would allow James to determine the area of the whole rectangle. Given that*

$$A = \sum_{m=1}^{\infty} \sum_{n=1}^{\infty} \frac{C_{m,n} \binom{m+n}{m}}{(m+n)^{m+n}}$$

then find the greatest integer less than $1000A$.

Extension Solution 49. *Consider the bipartite graph $K_{m,n}$ with each vertex corresponding to the m rows and n columns of the rectangle. We draw an edge between two vertices iff we know what the area of the corresponding rectangle is. If we know three rectangle areas, then we must also know the fourth, as we can connect the rectangle centers. (For any length 3 path, we can connect the end two vertices.)*

This process ensures that $C_{m,n}$ is also the number of trees in $K_{m,n}$, and so $\sum_{m+n=S} C_{m,n} \binom{m+n}{m}$ is twice the number of trees on the graph with S vertices.

For some subset of the vertices which has cardinality m, there will be $C_{m,n}$ graphs mapped to $K_{m,n}$, so we have $C_{m,n} \binom{m+n}{m} = 2(m+n)^{m+n-2}$. Thus, we compute $A = \sum_{s \geq 2} \frac{2}{s^2} = \frac{\pi^2 - 6}{3}$; 1000 times this, rounded down, is $\boxed{1289}$.

Extension Exercise 50. *(2014 IMO Shortlist C5) Consider $n \geq 3$ lines in the plane such that no two lines are parallel and no three have a common point. These lines divide the plane into polygonal regions; let S be the set of regions having finite area. Prove that it is possible to colour $\lceil \sqrt{n/2} \rceil$ of the lines blue in such a way that no region in S has a completely blue boundary.*

Extension Solution 50. *Let L be the set of lines. Let $B \subseteq L$ be maximal, and color the lines in B blue, such that no region in $\mathcal{F}$ has a boundary that is entirely blue. We claim $|B| \geq \lceil \sqrt{n/2} \rceil$. Color all lines in L, not in B, red. Then call a point blue if two blue lines intersect at that point; it directly follows that the number of blue points is $\binom{k}{2}$.*

Consider the red lines l now. Since B is maximal, there exists a region $R \in \mathcal{F}$ with its only red side lying on a line l_i. R has at least three sides, so it has at least one blue point as a vertex.

As each blue point belongs to four regions, it associates with no more than 4 red lines as well. Hence $|R| \leq 4\binom{k}{2}$. But we also have $|R| = n - k$, so $n - k \leq 2k(k-1) \implies n \leq 2k^2 - k \leq 2k^2 \implies n \leq 2k^2$, as desired.

Extension Exercise 51. *(2016 IMO Shortlist C3) Let n be a positive integer relatively prime to 6. We paint the vertices of a regular n-gon with three colours so that there is an odd number of vertices of each colour. Show that there exists an isosceles triangle whose three vertices are of different colours.*

Extension Solution 51. For $k \in [1,3]$, define t_k as the number of isosceles triangles with vertices of exactly k colors. Assume that $t_3 = 0$. Then let v_1, v_2, v_3 be the number of vertices of each color. We want to compute the number of ordered pairs (T, E), where T is an isosceles triangle and E is an edge of T with differently-colored endpoints.

Since $t_3 = 0$ by assumption, every triangle in a pair must contain two different colors, and so the number of pairs is $2t_2$. On the contrary, if we pick any two vertices of different colors, then three isosceles triangles exist with those vertices as their own vertices. Note that these three triangles are all distinct from each other, so the number of pairs is $3(v_1v_2 + v_2v_3 + v_3v_1)$. But this is odd, while $2t_2$ is even; contradiction. Hence, $t_3 > 0$.

Extension Exercise 52. *(2016 IMO Shortlist C1)* The leader of an IMO team chooses positive integers n and k $(n > k)$, and announces them to the deputy leader and a contestant. The leader then secretly tells the deputy leader an n-digit binary string, and the deputy leader writes down all n-digit binary strings which differ from the leaders in exactly k positions. (For example, if $n = 3$ and $k = 1$, and if the leader chooses 101, the deputy leader would write down 001, 111 and 100.) The contestant is allowed to look at the strings written by the deputy leader and guess the leaders string. In terms of n and k, what is the minimum number of guesses needed to guarantee the correct answer?

Extension Solution 52. We claim that, if n is odd, the minimum is 1; and if n is even, the minimum is 2.

Let S be the leaders binary string, $\overline{S}$ the binary string differing in every place from S. The leaders strings are the same as those where the leaders string is $\overline{S}$ and $k \mapsto n - k$. Hence, we can assume WLOG $k \geq \frac{n}{2}$. (If $k = \frac{n}{2}$ exactly, $\overline{S} = S$, so the minimum cannot be 1.)

Now consider any string T differing from S in d digits, $d \in [1, 2k - 1]$. WLOG assume that the first d digits of S, T are distinct. Let U be the binary string that results from changing the first k digits of S. Then U is a string written by the leader. Further observe that T and U differ in $|d - k|$ places. Hence, T cannot be the desired string. Finally, by our earlier assumption $k \geq \frac{n}{2}$, when $n < 2k$, every string $T \neq S$ is different from S in no more than $2k$ digits. If $n = 2k$ exactly, then every string except $S, \overline{S}$ differs from S in no more than $2k$ places, as desired.

Section 12.2: Graph Theory: Part 2

Exercise 109. *Show that a tree with n vertices has $n - 1$ edges.*

Solution 109. *We prove this by induction. If $n = 1$, the graph cannot have any edges or there would be a loop - what with the vertex connecting to itself.*

If a tree T has $k + 1$ vertices, then we claim that it contains a vertex with a single edge connected to it. For contradiction, say we could start at any vertex, and follow edges marking each vertex. If we ever come to a marked vertex, then we inevitably run into an edge loop - contradiction!

However, since we implicitly assume that each vertex has degree ≥ 2, there is no reason we would have to stop at one, so we eventually run into a marked vertex, which is a contradiction. For a satisfactory construction, take the vertex v with a single edge connecting to it, and delete it and its edge from the tree. The new graph $T - v$ will have k vertices. It must be connected, since v was disconnected from all but one vertex, and all other vertices must be connected. $T - v$ is still a tree, since we did not remove any loops. By the inductive hypothesis, the new tree $T - v$ has $k - 1$ edges. Finally, reverting the whole process, we can additionally transform $T - v$ into T by way of adding one edge and one vertex, so T also satisfies the condition - QED.

Exercise 110. *Prove that a graph is bipartite iff it contains no cycles of odd length.*

Solution 110. *To show that a graph is bipartite, we equivalently must show that we can divide its vertices into subsets A, B such that every edge in the graph connects a vertex in set A to a vertex in set B. Take any vertex from the graph and assign it to set A. Follow every edge from that vertex and put all vertices at the other end in set B, making sure to erase all the vertices traversed. For all vertices in B, follow all edges back to A, once again erasing all the vertices traversed. Repeat this process until it is no longer possible to proceed.*

Note that this process never runs into a vertex that is already in one set that needs to be moved to the other; otherwise, there would be a cycle of odd length. (If the graph is not connected, there may still be vertices unaccounted for.) Repeat the process in the previous paragraph until all vertices are assigned to A or to B.

Exercise 111. *(1997 South Africa TST P5) Six points are joined pairwise by red or blue segments. Must there exist a closed path consisting of four of the segments, all of the same color?*

Solution 111. *Yes; assume the contrary. Let the vertices be a, b, c, d, e, f. From our first monochromatic triangle with vertices a, b, c, WLOG assume it is red. d, then cannot have 2 red edges into a, b, c, or we end up with a red C_4. (The same holds for e, f.)*

Also, if a, b both have blue edges to two other vertices, then we wind up with C_4 as well. Thus, the only possible configuration is a red matching between $\{a, b, c\}$ and $\{d, e, f\}$, with all other edges between those sets red. WLOG the matching is ad, be, cf. But then it would follow that de, ef are both red. We have a red C_4, namely $bfed$ in this case, so the proof is complete.

Exercise 112. *(2010 IMO Shortlist C2) On some planet, there are $2N$ countries ($N \geq 4$). Each country has a flag N units wide and one unit high composed of N fields of size 1×1, each field being either yellow or blue. No two countries have the same flag. We say that a set of N flags is diverse if these flags can be arranged into an $N \times N$ square so that all N fields on its main diagonal will have the same color. Determine the smallest positive integer M such that among any M distinct flags, there exist N flags forming a diverse set.*

Solution 112. *We claim* $\boxed{M = 2^{N-2} + 1}$.

Let M_N be the smallest satisfactory positive integer. We show that $M_N > 2^{N-2}$. If we consider all 2^{N-2} flags with their first squares yellow and their second squares below, this suffices to prove the lower bound.

We then proceed to induct upon $N = 4$ as the base case, which holds since we can decompose each flag into 2×2 squares. For the inductive step, assume that $N > 4$ and consider any $2^{N-2}+1$ length-N flags. We can arrange all of these flags into one flag of dimensions $(2^{N-2}+1) \times N$, which has a column of multiple colors since the flags are distinct. By Pigeonhole, this column has at least $2^{N-3}+1$ squares of a single color.

For the last step, we remove the upper-left corner of each flag. We have at least $2^{N-3}+1 \geq M_{N-1}$ length $n-1$ flags; by the inductive hypothesis, we have a rectangle with size $(N-1) \times N$, and want to supplant it with one more row. Depending on the color of the diagonal, we can accomplish this either way. In both cases, we end up with an $N \times N$ square flag, as desired.

Exercise 113. *Consider six points in the plane, no three of which are collinear. Prove that there are two triangles with among the six points such that the longest side length of one of them is the shortest side length of the other.*

Solution 113. *Draw a graph with vertices a, b, c, d, e, f, and WLOG color the longest side of each triangle red. If there is a monochrome red triangle, then its shortest side is also the longest side of another triangle. Thus it suffices to show that there is a monochromatic, red triangle. If vertex a is an endpoint of three red edges (for instance, ab, ac, ad), then one of the edges, perhaps bc, in $\triangle bcd$ must be red, which produces red $\triangle abc$. Thus, it suffices to prove that there is a vertex which is an endpoint of three red edges.*

Henceforth, consider the shortest side, WLOG ab. The vertices (along with the other four) form four triangles. Thus, there are 4 red edges with endpoints at a or b. If at least three of these edges have endpoint at a, then we know that a monochromatic red triangle will be form. Suppose that ac, ad, be, bf are all red; consider $\triangle bcd$. If the longest side of this triangle is cd, then $\triangle acd$ is red as well. Otherwise, there will be three red edges with common endpoint at b. Again, we will end up with a red triangle, as in all possible cases. Indeed, this triangle satisfies the problem statement as well, hence completing the proof.

Extension Exercise 53. *(2014 USA TST) Let n be an even positive integer, and let G be an n-vertex graph with exactly $\frac{n^2}{4}$ edges, where there are no loops or multiple edges (each unordered pair of distinct vertices is joined by either 0 or 1 edge). An unordered pair of distinct vertices $\{x, y\}$ is said to be amicable if they have a common neighbor (there is a vertex z such that xz and yz are both edges). Prove that G has at least $2\binom{n/2}{2}$ pairs of vertices which are amicable.*

Extension Solution 53. *WLOG assume that G has $2n$ vertices and n^2 edges. We want to show that G has at least $n(n-1)$ amicable pairs of vertices. Note that we seek vertices that are neighbors of the neighbors of x. Thus, x is in at least $\max(\deg(y)) - 1 \geq $ average $\deg(y) - 1$ edges, with y a neighbor of x. Furthermore, the average degree of a neighbor of x is given by*

$$\frac{1}{\deg(x)} \cdot \sum_{y \in N(x)} \deg(y),$$

so the number of amicable pairs with x is bounded below by

$$\frac{1}{\deg(x)} \cdot \sum_{y \in N(x)} \deg(y) - 1.$$

The sums (which we omit here for reasons of concision and flow) eventually simplify to one cornerstone observation: if $(a, b) \in E(G)$, then we can express the sum as

$$\sum_{(x,y) \in E(G)} \frac{\deg(x)}{\deg(y)} + \frac{\deg(y)}{\deg(x)}$$

By AM-GM, we bound this below by $2n^2$, from which the conclusion follows.

Extension Exercise 54. *There are n people in a room. Any group of m people ($m \geq 3$) in the room have a common friend, and moreover, this friend is unique for all such groups. Compute n in terms of m.*

Extension Solution 54. *Construct a graph as usual. If this graph G contains a subgraph K_i ($i \leq m$), then every vertex in K_i is adjacent to a common vertex, and so G must also contain K_{i+1} as a subgraph. By induction, $K_2 \in G$, which is true. Hence $K_{m+1} \in G$ is our inductive hypothesis.*

Let $A = \{a_1, a_2, \ldots, a_{m+1}\}$ be the set of vertices set of K_{m+1}. Assume the existence of vetex $b \neq a_i, i \in [1, m+1]$, with b adjacent to $a_m, a_{m+1} \in A$. Then a_m, a_{m+1} are common neighbors of $b, a_1, a_2, \ldots, a_{m-1}$, but this is a contradiction. Therefore, there are m vertices (a_i, $i \in [1, m]$) in A that are not adjacent to b. All of the vertices $b, a_3, a_4, \ldots, a_{m+1}$ have a common neighbor c. Then $c \neq a_i$ a priori; nevertheless, c is adjacent to $m-1 \geq 2$ vertices in A, which is, yet again, a contradiction, Conclusively, G is just $K_{m+1} = K_n$ ($n = m+1$).

Chapter 12 Review Exercises

Exercise 114. *Show that any tree with at least two vertices is bipartite.*

Solution 114. *This is by induction. The only tree on 2 vertices is clearly bipartite, which proves the base case. Then suppose that every tree on n vertices is bipartite; we show this for $n+1$ as well. We know that the vertices of T on $n+1$ vertices contain a leaf v. Hence, $T = T - v$ is itself a tree. By the inductive hypothesis, the vertices of T can be decomposed into disjoint X, Y with $X \cup Y = T$.*

Finally, let v have neighbor w in T. Then $w \in T$ also, and assume WLOG that $w \in X \subset T$. Then $X = X, Y = Y \cup \{v\}$ is a partition of T, demonstrating that T is bipartite.

Exercise 115. *(2019 AMC 12B) How many sequences of 0s and 1s of length 19 are there that begin with a 0, end with a 0, contain no two consecutive 0s, and contain no three consecutive 1s?*

Solution 115. *Any valid sequence that is n binary digits long begins with a 0, and must be followed either by 10 or 110. Hence, we can set up the recurrence relation $f(n) = f(n-3) + f(n-2)$, which yields $\boxed{65}$ as the answer.*

Exercise 116. *There is a group of 9 people gathered at an airport waiting on the plane for an international business meeting. Each of them can speak at most three languages. Among any three of them, at least two can speak a common language. Prove that there are three people who speak a common language.*

Solution 116. *In graph G, represent each person as a vertex; join two people with an edge iff they have a language in common. If we have a vertex v with $\deg v = 4$, since the person associated to v speaks at most three languages, two of their neighbors speak a language in common with v, and QED.*

Otherwise, the degree of every vertex is at most 3. Consider vertex v again. There are at least 5 vertices not connected to v; let one of these be vertex w. Since w has no more than three neighbors, let vertex $x \notin N(a) \cup N(b)$. Thus v, x, x are vertices such that there is no edge joining any two of them, a contradiction.

Exercise 117. *A computer has an exponentially distributed lifespan. Explain why it would follow that its failure rate at any given time would remain constant.*

Solution 117. *Because of the memoryless property, the length of time a component has functioned in the past has no bearing on its future behavior, so the probability that the component fails in the near future is always the same and doesn't depend on its current age.*

Exercise 118. *(2008 AIME I) Consider sequences that consist entirely of A's and B's and that have the property that every run of consecutive A's has even length, and every run of consecutive B's has odd length. Examples of such sequences are AA, B, and $AABAA$, while $BBAB$ is not such a sequence. How many such sequences have length 14?*

Solution 118. *Let a_n, b_n respectively denote the number of length-n sequences ending in A and B. If a sequence ends in A, then it was formed by appending two As to a length $n-2$ string. Similarly, with a string ending in B, it was either formed by appending one B to a length $n-1$ string, or by appending two Bs to a string of length $n-2$ ending in B. All of this allows us to set up the combined recurrence $a_n = a_{n-2} + b_{n-2}$, $b_n = a_{n-1} + b_{n-2}$. Hence, $a_{14} + b_{14} = \boxed{172}$.*

Exercise 119. *(2001 AIME I) A mail carrier delivers mail to the nineteen houses on the east side of Elm Street. The carrier notices that no two adjacent houses ever get mail on the same day, but that there are never more than two houses in a row that get no mail on the same day. How many different patterns of mail delivery are possible?*

Solution 119. *Let 1 represent a house receiving mail, and let 0 represent a house that does not receive mail. Then we want the number of 19-digit binary strings with no two consecutive 1s and no three consecutive 0s. The last two digits cannot be 11, so they can be 00, 01, or 10. Then let a_n, b_n, c_n be the numbers of n-digit strings ending in 00, 01, 10 accordingly. We can then set up recursive formulae using the constructions of a_n, b_n, c_n, so $a_{14} + b_{14} + c_{14} = \boxed{351}$.*

Exercise 120. *(1990 APMO) A graph G on n vertices satisfies all three of the following conditions:*

(i) *No vertex has degree $n - 1$.*

(ii) *For any 2 points A, B which are not adjacent, there exists exactly one C such that AC, BC are both edges.*

(iii) *There are no triangles.*

Prove that all vertices have the same degree.

Solution 120. *(Indeed, we can verify that $n \geq 5$ alongside the requested statement.)*

Take a point A, and suppose $\deg A = m$. Then points $B_1, B_2, B_3, \ldots, B_m$ are connected to A. No two distinct B_i can be adjoined, and a point $C \neq A$ cannot be joined to $B_i, B_j, i \neq j$. Hence, we have $\deg B_i - 1$ points of the form $C_{i,j}$ joined to B_i, with all $C_{i,j}$ distinct.

Henceforth, the only points that can be connected via an edge to $C_{i,j}$ are other points of that form, but $C_{i,j}$ cannot be connected to $C_{i,k}$, since there are no triangles. In addition, $C_{i,j}$ is joined to at least one point $C_{k,h}, k \neq i$. In fact, it must be connected to this point. Thus, $\deg C_{i,j} = m$. Extending this to a starting point of B_i (rather than A), all vertices have the same degree, as desired.

(Note: Suppose this common degree is m. Then there are $n = m^2 + 1$ points in all, but $m = 1$ does not work. $m = 2$ does, however, so $n \geq 5$.)

Exercise 121. *The pattern of sunny and rainy weather in Inconsistentown can be modeled by a Markov chain. If it is sunny one day, the probability that the next day will also be sunny is $\frac{4}{5}$. If it is rainy one day, the probability that the next day will also be rainy is $\frac{3}{5}$. What is the probability that it rains in Inconsistentown on Tuesday, given that it is sunny on Sunday?*

Solution 121. This is, of course, just $\frac{4}{5} \cdot \frac{1}{5} + \frac{1}{5} \cdot \frac{3}{5} = \boxed{\frac{7}{25}}$, *but the idea is to use a Markov chain. This exercise primarily offers you an opportunity to use, and work with, a Markov steady state chain to derive the obvious answer in a novel, inventive way! Using the concept of matrices, we can multiply the matrix corresponding to the probabilities of each event occurring with a column vector to obtain probabilities into the future.*

Exercise 122. *(2018 AIME I) Let $SP_1P_2P_3EP_4P_5$ be a heptagon. A frog starts jumping at vertex S. From any vertex of the heptagon except E, the frog may jump to either of the two adjacent vertices. When it reaches vertex E, the frog stops and stays there. Find the number of distinct sequences of jumps of no more than 12 jumps that end at E.*

Solution 122. *Yet more dynamic programming (recursion); the specific trick here is to draw a table of values for the paths diagonally, as it is a great visual representation of the task at hand! Specifically, our constraints are $p(x, y) = p(x - 1, y) + p(x, y - 1)$ for the recursion and that we start at the origin, $(0, 0)$. It then follows from constructing the table that the desired answer is $\boxed{351}$.*

Exercise 123. *(2002 USAMO P1) Let S be a set with 2002 elements, and let N be an integer with $0 \leq N \leq 2^{2002}$. Prove that it is possible to color every subset of S either blue or red so that the following conditions hold:*

1. *the union of any two red subsets is red;*

2. *the union of any two blue subsets is blue;*

3. *there are exactly N red subsets.*

Solution 123. *We prove inductively that any set with n elements can be satisfactorily colored in this manner for $0 \leq N \leq 2^n$. The base case ($n = 0$) is trivial.*

Now let $s \in S$, $|S| = k+1$, $S = S - s$. Then if $N \leq 2^k$, we can establish a coloring with s blue and the rest colored according to the conditions. If $N > 2^k$, then color all subsets of which s is a member red, and the rest such that S is colored according to the conditions. Then S, too, is colored satisfactorily, completing the inductive step and the proof.

Extension Exercise 55. *(2000 Putnam B2) Prove that*

$$\frac{\gcd(m,n)}{n} \binom{n}{m}$$

is an integer for all $n \geq m \geq 1$, $n, m \in \mathbb{Z}$.

Extension Solution 55. *By Bezouts Identity and the Euclidean Algorithm, we have that $\gcd(m,n)$ is a linear combination of m and n. Therefore, $\dfrac{\gcd(m,n)}{n}\binom{n}{m}$ is also a linear combination of $\dfrac{m}{n}\binom{n}{m} = \binom{n-1}{m-1}$ and $\binom{n}{m}$ and so is an integer.*

Extension Exercise 56. *(1956 Putnam B5) Show that a graph with $2n$ nodes and $n^2 + 1$ edges necessarily contains a 3-cycle, but that we can find a graph with $2n$ points and n^2 edges without a 3-cycle.*

Extension Solution 56. *We prove this by induction. For the base case - $n = 2$ - the result is obvious. Suppose it holds for n, and we consider a graph G with $2n+2$ vertices and $n^2 + 2n + 2$ edges. Take points P, Q connected by an edge in G. Now consider two cases: one where there are fewer than $2n + 1$ edges to P, Q, then removing P, Q results in a graph with $2n$ vertices, and $n^2 + 1$ edges, which contains a triangle (and, therefore, G does as well).*

Let S, T be disjoint sets with n points. Then connect each point in G to each point in H. The resulting graph has n^2 edges, and it does not contain any triangles.

Extension Exercise 57. *(2018 HMIC P5) Let G be an undirected simple graph. Let $f(G)$ be the number of ways to orient all of the edges of G in one of the two possible directions so that the resulting directed graph has no directed cycles. Show that $f(G)$ is a multiple of 3 if and only if G has a cycle of odd length.*

Extension Solution 57. *Denote by $\phi_q(G)$ the number of ways to color G using q colors. (This is the chromatic polynomial of G.) Pick an edge $e \in E$, and let G/e be G with the vertices on either side of e converging together. Hence, $f_G(q) = f_{G\backslash E}(q) + f_{G/E}(q)$. In particular, for the empty graph E_n with n vertices, we have $f_{E_n}(q) = q^n$.*

We also show that the number of ways to arrange the vertices of G such that they form a directed path (without forming directed cycles) is equal to $c_G = (-1)^n \cdot \phi_G(-1)$. Then $c_G = c_{G\backslash e} + c_{G/e}$ with $c_{E_n} = 1$.

Part 5: Number Theory

Chapter 13: Beginning Number Theory

Section 13.1: Factors and Divisibility

Exercise 124.

(a) How many factors does 2019 have?

(b) How many factors does 876 have?

(c) How many factors does $1^1 \cdot 2^2 \cdot 3^3 \cdots 10^{10}$ have?

Solution 124.

(a) $2019 = 3 \cdot 673 \implies \tau(2019) = (1+1)(1+1) = \boxed{4}$.

(b) $876 = 2^2 \cdot 3 \cdot 73 \implies \tau(876) = (2+1)(1+1)(1+1) = \boxed{12}$.

(c) The product has distinct prime factors 2, 3, 5, 7, namely in the factorization $(5^{10} \cdot 2^{10}) \cdot 3^{18} \cdot 2^{24} \cdot 7^7 \cdot (3^6 \cdot 2^6) \cdot 5^5 \cdot 2^8 \cdot 3^3 \cdot 2^2 = 7^7 \cdot 5^{15} \cdot 3^{27} \cdot 2^{50} \implies \boxed{182,784}$ factors.

Exercise 125. *Show that the square of a prime has exactly 3 positive integral factors.*

Solution 125. Let p be a prime; then p^2 has $2 + 1 = 3$ factors, since p is the only prime factor of p^2.

Exercise 126. *Prime factorize each of the following:*

(a) 54

(b) 97

(c) 1600

(d) 676

Solution 126.

(a) $54 = 3^3 \cdot 2$

(b) $97 = 97$ (prime).

(c) $1600 = 2^6 \cdot 5^2$

(d) $676 = 24^2 = (2^3 \cdot 3)^2 = 2^6 \cdot 3^2$.

Exercise 127. *What is the largest prime factor of $5904 = 10^4 - 8^4$?*

Solution 127. $10^4 - 8^4 = (10^2)^2 - (8^2)^2 = (10^2 + 8^2)(10^2 - 8^2) = 164 \cdot 36 = (41 \cdot 2^2) \cdot (3^2 \cdot 2^2) \implies \boxed{41}$.

Exercise 128. *Compute the smallest positive integer with exactly 19 divisors.*

Solution 128. *19 is prime, so the only way to express it as a product of exponents plus one is as $18 + 1$. Hence, the desired integer is* $\boxed{2^{18}}$.

Exercise 129. *(2016 AMC 8) The number N is a two-digit number. When N is divided by 9, the remainder is 1. When N is divided by 10, the remainder is 3. What is the remainder when N is divided by 11?*

Solution 129. *N ends in the digit 3, so we can look through numbers ending in 3. We find that $N = 73$, so the requested remainder is* $\boxed{7}$.

Exercise 130. *Describe the set of all positive integers with exactly 2019 positive integral factors.*

Solution 130. *Since $2019 = 3 \cdot 673 = 1 \cdot 2019$, we can either have p^{2018} for p prime, or $p_1^2 \cdot p_2^{672}$.*

Exercise 131. *If the positive integer $6N$ has twice as many factors as does N, how many factors can N have?*

Solution 131. *Since $6 = 2 \cdot 3$, we are not adding prime factors of 2 and 3 into the prime factorization of N (otherwise, there would be four times as many factors). Instead, if we had a factor of 2^2 and 3^1 ($3 \cdot 2 = 6$ to begin with, then making it into 2^3 and 3^2 yields $4 \cdot 3 = 12$, which indeed doubles the number of factors. Hence, N has any number of factors that is a multiple of 6.*

Exercise 132. *For $p \geq 3$ a prime number, show that the only positive integers with exactly p divisors are the perfect $(p-1)^{th}$ powers of a prime.*

Solution 132. *Same logic as an earlier exercise: a number with p divisors has a power of a^{p-1} for a prime by definition.*

Exercise 133. *Find the largest integer n for which $(3!)^n$ divides $2019!$*

Solution 133. *Since $3! = 3 \cdot 2$, we seek the largest power of 3 that goes into $2019! = (3 \cdot 673)!$ By Legendre's formula, $n \leq 673 + \lfloor \frac{673}{3} \rfloor + \lfloor \frac{673}{9} \rfloor + \ldots = 673 + 224 + 74 + 24 + 8 + 2 = \boxed{1005}$.*

Exercise 134. *(2018 ARML Local Individual Round #5) Compute the number of triples of consecutive positive integers less than 50 whose product is both a multiple of 20 and 18.*

Solution 134. *We have $\text{lcm}(20, 18) = 180 = 5 \cdot 3^2 \cdot 2^2$. Thus, one of the multiples must be a multiple of 5. We also need a multiple of 9, and a multiple of 4. We search for clusters where a multiple of 9 and a multiple of 5 are close, and find $(8, 9, 10), (18, 19, 20), (34, 35, 36), (35, 36, 37), (43, 44, 45), (44, 45, 46)$. The answer is* $\boxed{6}$.

Extension Exercise 58. *(1986 AIME) Let S be the sum of the base 10 logarithms of all the proper divisors (all divisors of a number excluding itself) of 1000000. What is the integer nearest to S?*

Extension Solution 58. *We have $1,000,000 = 10^6 = 5^6 \cdot 2^6$, so there are 49 factors; by symmetry, the sum is 147, minus 10^6 itself (as a number is not a proper divisor of itself), so the answer is* $\boxed{141}$.

Extension Exercise 59. *(2018 ARML Individual Round #1) Compute the greatest prime factor of N, where*
$$N = 2018 \cdot 517 + 517 \cdot 2812 + 2812 \cdot 666 + 666 \cdot 2018$$

Extension Solution 59. *We have $N = 517 \cdot 4830 + 666 \cdot 4830 = 4830 \cdot 1183 = (23 \cdot 7 \cdot 5 \cdot 3 \cdot 2) \cdot (13^2 \cdot 7)$, so the largest prime factor is* $\boxed{23}$.

Extension Exercise 60. *Let N be a positive integer with exactly 20 divisors. Does there exist a positive integer a such that $a \cdot N$ has exactly 19 divisors? If so, give at least one satisfactory ordered pair (N, a). If not, prove that no such ordered pair exists.*

Extension Solution 60. *For $a \cdot N$ to have 19 divisors, we must have $a \cdot N = p^{18}$ for some prime p. But if p is prime, we cannot introduce another prime factor while keeping a a positive integer, so no such ordered pair exists.*

Extension Exercise 61. *(2018 ARML Local Individual Round #9) Compute the number of positive integers N between 1 and 100 inclusive that have the property that there exist distinct divisors a and b of N such that $a + b$ is also a divisor of N.*

Extension Solution 61. *We can perform a casework bash on $a = 2, 3, 4, \ldots$, WLOG assuming $b \geq a$. Note that, if $a + b \geq 10$, it cannot be prime. Using this method, we get $\boxed{22}$.*

Extension Exercise 62. *(2013 Caltech-Harvey Mudd Math Competition) Determine all positive integers n whose digits (in decimal representation) add up to $\dfrac{n}{57}$.*

Extension Solution 62. *Note that, if n has four digits, $n \leq 57 \cdot 36 = 2052$. Searching manually reveals that n has three or fewer digits, so $n = \boxed{513}$ is the only solution.*

Extension Exercise 63. *(2016 ARML Individual Round #10) Compute the largest of the three prime divisors of $13^4 + 16^5 - 172^2$.*

Extension Solution 63. *Note that $13^4 = 169^2$, and $16^5 = 2^{20}$, so we get $2^{20} - (172 + 169)(172 - 169) = 2^{20} - 1023 = 2^{20} - 2^{10} + 1$. We can multiply by $2^{10} + 1$ to get $2^{30} + 1$, then apply Sophie-Germain to factor it as $1321 \cdot 793$ after simplification. Hence, the answer is $\boxed{1321}$.*

Section 13.2: Bases Other than Base 10

Exercise 135. *Convert each of the following to base 10:*

(a) 99_{11}

(b) 10101101_2

(c) 2019_{16}

(d) 345_{-6}

(e) $20321_{\sqrt{10}}$

(f) 10010.01001_i

Solution 135.

(a) $99_{11} = 9 \cdot 11 + 9 = 108_{10}$

(b) $10101101_2 = 2^7 + 2^5 + 2^3 + 2^2 + 2^0 = 109_{10}$

(c) $2019_{16} = 2 \cdot 16^3 + 1 \cdot 16 + 9 = 8217_{10}$

(d) $345_{-6} = 3 \cdot (-6)^2 + 4 \cdot (-6) + 5 = 89_{10}$.

(e) $20321_{\sqrt{10}} = 2 \cdot (\sqrt{10})^4 + 3 \cdot (\sqrt{10})^2 + 2 \cdot \sqrt{10} + 1 = (231 + 2\sqrt{10})_{10}$

(f) $10010.01001_i = i^4 + i^1 + i^{-2} = 1 + i + (-1) = i_{10}$.

Exercise 136. *In base b, the positive integer N has d digits. Describe the interval of possible values for N in base 10.*

Solution 136. *We have $b^{d-1} \leq N \leq b^d - 1$ by definition of place values.*

Exercise 137. *(2003 AMC 10A) A base-10 three digit number n is selected at random. What is the probability that the base-9 representation and the base-11 representation of n are both three-digit numerals?*

Solution 137. *The base-9 representation has three digits when $81 \leq n \leq 9^3 - 1 = 728$, and the base-11 representation has three digits for $121 \leq n \leq 11^3 - 1 = 1330$ (upper bound here is irrelevant). Hence, $121 \leq n \leq 728 \implies 608$ values of n out of 900, for a probability of $\dfrac{608}{900} = \boxed{\dfrac{76}{125}}$.*

Exercise 138. *(2015 AMC 10A) Hexadecimal (base-16) numbers are written using numeric digits 0 through 9 as well as the letters A through F to represent 10 through 15. Among the first 1000 positive integers, there are n whose hexadecimal representation contains only numeric digits. What is the sum of the digits of n?*

Solution 138. *Observe that $1000_{10} = 3E8_{16}$, so we seek 3-, 2-, and 1-digit numbers without letters. There are $400 - 1 = 399$ combinations for 3-digit numbers, which has a digit sum of $\boxed{21}$.*

Exercise 139. *(2013 AMC 10A) In base 10, the number 2013 ends in the digit 3. In base 9, on the other hand, the same number is written as $(2676)_9$ and ends in the digit 6. For how many positive integers b does the base-b-representation of 2013 end in the digit 3?*

Solution 139. *We essentially require 2013 to leave a remainder of 3 when divided by 3; i.e. for 2010 to be an even multiple of $b \geq 4$. $2010 = 67 \cdot 5 \cdot 3 \cdot 2$ has 16 factors, 3 of which are less than 4 (namely 1, 2, 3), so $\boxed{13}$ values of b are valid.*

Exercise 140. *Explain how we can determine a unique base-b representation for any $0 < b < 1$, $b \in \mathbb{R}$.*

Solution 140. *We can express any real number as the sum of quotients of powers of real numbers (in particular, with rational numbers; this holds true especially if b itself is rational). Consider the infinite geometric series formula.*

Exercise 141. *In base $-2i$, what is the representation of 63_{10}? (Hint: First consider the representation of -63_{10}.)*

Solution 141. Note that $(-2i)^6 = (-2)^6 \cdot i^6 = -64$, so we add $1 = (-2i)^0$ to get $\boxed{1000001_{-2i}}$.

Exercise 142. *Compute the number of base-10 positive integers that have the same number of digits in base 3 and base 5.*

Solution 142. Let this number of digits be d. In base 3, a d-digit number is between 3^{d-1} and $3^d - 1$ in base 10, and in base 5, the range is from 5^{d-1} to $5^d - 1$. Hence, we construct a table of values, with $d = 1$ yielding $[1, 2]$ and $[1, 4]$; $d = 2$ yielding $[3, 8]$ and $[5, 24]$; $d = 3$ yielding $[9, 26]$ and $[25, 124]$; and finally, $d = 4$ yielding $[27, 80]$ and $[125, 624]$. At this point, the overlap stops, so all the integers are $1, 2, 5, 6, 7, 8, 25, 26 \implies \boxed{8}$ integers.

Exercise 143. *What is the smallest positive integer b such that 2019 has 2 or fewer digits in base b?*

Solution 143. We require that $2019 < b^2 \implies b \geq \boxed{45}$.

Exercise 144. *Describe a formula for the number of trailing zeros of a positive integer N in base b, where $b \in \mathbb{N}$.*

Solution 144. In base b, if N has a prime factor of b^k, then N will have k trailing zeros. (This is entirely equivalent to Legendre's formula.)

Exercise 145. *(2015 ARML Team Round #3) A positive integer has the Kelly Property if it contains a zero in its base-17 representation. Compute the number of positive integers less than 1000 (base 10) that have the Kelly Property.*

Solution 145. Clearly, a positive integer $N < 1000_{10}$ cannot have more than 3 digits in base 17. If N has 3 digits, perform casework based on whether the 0 lies in the middle or right position. If N has 2 digits, the zero must be in the right position rather than the left. The final total is $\boxed{106}$ integers.

Exercise 146. *Find the base-10 three-digit number whose base-4 representation is the reverse of its base-5 representation.*

Solution 146. Both representations must contain the same number of digits. Try $abc_4 = cba_5$, set up the corresponding equations, and test different possibilities to find $\boxed{152_{10}}$.

Extension Exercise 64. *Call a base-10 positive integer (m, n)-palindromic if it is a palindrome in both base 4 and base 5. What is the set of $(4, 5)$-palindromic base-10 positive integers?*

Extension Solution 64. Let $N = aba_4 = cdc_5$. Then $N = 17a + b = 26c + d \implies N \equiv 0, 1, 2, 3 \pmod{17}, 0, 1, 2, 3 \pmod{26}$. Use CRT to finish from there.

Extension Exercise 65. *(2018 AIME I) The number n can be written in base 14 as $\underline{a}\,\underline{b}\,\underline{c}$, can be written in base 15 as $\underline{a}\,\underline{c}\,\underline{b}$, and can be written in base 6 as $\underline{a}\,\underline{c}\,\underline{a}\,\underline{c}$, where $a > 0$. Find the base-10 representation of n.*

Extension Solution 65. We have $n = 196a + 14b + c = 225a + 15c + b = 216a + 36c + 6a + c = 222a + 37c = 37(6a + c)$. Hence, $3a + b = 22c$, and so $c = 1$ with $3a + b = 22$ due to restraints on the digits. Then, in conjunction with $29a - 13b = 14$ from the first pair of equations, we get $(a, b) = (4, 10)$, so $n = \boxed{925}$.

Extension Exercise 66. *(2017 AIME I) A rational number written in base eight is $\underline{ab}.\underline{cd}$, where all digits are nonzero. The same number in base twelve is $\underline{bb}.\underline{ba}$. Find the base-ten number $\underline{abc}$.*

Extension Solution 66. Observe that $8a + b = 12b + b \implies 2a = 3b$. In addition, $\dfrac{c}{8} + \dfrac{d}{64} = \dfrac{b}{12} + \dfrac{a}{144} = \dfrac{3b}{32} \implies 8c + d = 6b$. We can either have $(a, b) = (3, 2)$ or $(6, 4)$ (not $(9, 6)$, as 9 is not a valid digit in base 8). $(6, 4)$ also does not work, so $a = 3, b = 2, c = 1, d = 4 \implies \boxed{321}$.

Extension Exercise 67. *(2016 HMMT November Guts Round) On the blackboard, Amy writes 2017 in base a to get 133201_a. Betsy notices she can erase a digit from Amy's number and change the base to base b such that the value of the number remains the same. Catherine then notices she can erase a digit from Betsy's number and change the base to base c such that the value still remains the same. Compute, in decimal, $a + b + c$.*

Extension Solution 67. We have $a^5 + 3a^4 + 3a^3 + 2a^2 + 1 = 2017 \implies a = 4$ by approximation/number sense. Hence, if Betsy wants to keep 5 digits of Amy's original number, she should erase a 3 to make the number $13201_6 = 2017$, since $6^4 = 1296$. Finally, Catherine should erase the last 3 to make the number 1201_{12}, so that $a + b + c = \boxed{22}$.

Extension Exercise 68. *(2013 USAMO P5) Given postive integers m and n, prove that there is a positive integer c such that the numbers cm and cn have the same number of occurrences of each non-zero digit when written in base ten.*

Extension Solution 68. WLOG let $m \geq n \geq 1$. Using the prime factorization of n, we have $c_1 \in \mathbb{N}$ with $c_1 n = 10^p n_1$, where $\gcd(n_1, 10) = 1$. If $k > p$, then $10^k c_1 m - c_1 n = 10^p t$, with $t > 0$ relatively prime with 10. Choose k sufficiently large for $t > c_1 m$; then we can find $b \in \mathbb{N}$ with $t \mid (10^b - 1) > 10 c_1 m$.

Since $c_1 m, n_1 < t$, the decimal expansions of the fractions $\dfrac{c_1 m}{t}$ and $\dfrac{n_1}{t}$ repeat after every b digits. We can also show that they are in fact just cyclic shifts of each other. Hence, $cm, c_2 n_1$ have the same number of occurrences of non-zero digits in base ten, as desired.

Section 13.3: Modular Arithmetic: Part 1

Exercise 147. *(2016 Berkeley Math Tournament) What is the sum of all positive integers less than 30 divisible by 2, 3, or 5?*

Solution 147. *The sum of evens is $2+4+6+\ldots+30 = 2(1+2+3+\ldots+15) = 240$, the sum of multiples of 3 is $3(1+2+3+\ldots+10) = 165$, and the sum of multiples of 5 is $5(1+2+3+\ldots+6) = 105$. But using PIE, we must subtract the multiples of 6, which sum to $6(1+2+3+4+5) = 90$, the multiples of 10, which sum to $10(1+2+3) = 60$, and the multiples of 15, which sum to $15(1+2) = 45$. We do not add back 30, as we must exclude 30 from consideration. Hence our final tally is $240 + 165 + 105 - 90 - 60 - 45 = \boxed{315}$.*

Exercise 148. *For how many positive integers $n \in [1, 100]$ does 7 divide $11^n \pm 1$?*

Solution 148. *Note that $11 \equiv 4 \pmod{7}$, and we can list out the cycle of 4^n by hand: $4, 2, 1, 4, 2, 1, \ldots$. Since $4^n \not\equiv 6 \pmod 7$ for any n, $11^n - 1 \equiv 0 \pmod 7$ when $n \equiv 0 \pmod 3$, which is true for $\boxed{33}$ integers between 1 and 100, inclusive.*

Exercise 149. *Compute the remainder when*
$$\sum_{k=1}^{2019} k!$$
is divided by 7.

Solution 149. *Note that from $7!$ and beyond, all $k!$ will be multiples of 7 by definition of the factorial. Computing the sum of the remaining factorials is not difficult: $6! + 5! + 4! + 3! + 2! + 1! = 873 \equiv \boxed{5} \pmod{7}$*

Exercise 150. *(2017 Berkeley Mini Math Tournament) Determine the smallest integer x greater than 1 such that x^2 is one more than a multiple of 7.*

Solution 150. *This occurs when $x \equiv \pm 1 \pmod 7$, i.e. $x \geq \boxed{6}$.*

Exercise 151. *(1964 IMO P1)*

(a) *Find all positive integers n for which $2^n - 1$ is divisible by 7.*

(b) *Prove that there is no positive integer n for which $2^n + 1$ is divisible by 7.*

Solution 151.

(a) *We require $2^n \equiv 1 \pmod 7$, which happens for all $n \equiv 0 \pmod 3$.*

(b) *We require $2^n \equiv 6 \pmod 7$, but the cycle repeats as $2, 4, 1, 2, 4, 1, \ldots$, so this never occurs.*

Exercise 152.

(a) *What is the units digit of 9^{19}?*

(b) *What is the units digit of 7^{2019}?*

(c) *What is the **tens** digit of 7^{2019}?*

(d) *What is the units digit of $1^2 + 2^3 + 3^4 + \ldots + 9^{10}$?*

Solution 152.

(a) *Note that $9^1 = 9$, $9^2 = 81$, $9^3 = 729$, and so forth. The units digits oscillate between 9 and 1, so 9^{19} has a unit digit of $\boxed{9}$.*

(b) *The pattern for powers of 7 is $7, 9, 3, 1, \ldots$, and $2019 \equiv 3 \pmod 4$, so $7^{2019} \equiv \boxed{3} \pmod{10}$.*

(c) We now want the cycle wrt mod 100, i.e. $07, 49, 43, 01, 07, \ldots$ The period is still 4, so $7^{2019} \equiv 43$ (mod 100) with a tens digit of $\boxed{4}$. (As a sanity check: make sure that this does not contradict the result of the above part!)

(d) This is equal to $1 + 8 + 1 + 4 + 5 + 6 + 1 + 8 + 1 \equiv \boxed{5}$ (mod 10).

Exercise 153. *(2000 AMC 12)* Mrs. Walter gave an exam in a mathematics class of five students. She entered the scores in random order into a spreadsheet, which recalculated the class average after each score was entered. Mrs. Walter noticed that after each score was entered, the average was always an integer. The scores (listed in ascending order) were 71, 76, 80, 82, and 91. What was the last score Mrs. Walters entered?

Solution 153. The sum of the first three numbers is a multiple of 3, so write the numbers mod 3, which yields 2, 1, 2, 1, 1. Clearly, we must add the three 1s, so those go first. The next average is divisible by 4, forcing 71 to be next and $\boxed{80}$ to be last.

Exercise 154. *(2016 Berkeley Math Tournament)* Let $g_0 = 1$, $g_1 = 2$, $g_2 = 3$, and $g_n = g_{n-1} + 2g_{n-2} + 3g_{n-3}$. For how many $0 \leq i \leq 100$ is it that g_i is divisible by 5?

Solution 154. Consider the entire sequence mod 5. We have $1, 2, 3, 0, 2, 1, 0, 3, 1, 2, 3, 0, 2, 1, 0, 3, \ldots$ so the period of the cycle is 8, and $g_i \equiv 0 \pmod 5 \iff i \equiv 3, 6 \pmod 8$. As such, $\boxed{25}$ values of i satisfy $5 \mid g_i$.

Exercise 155. For a positive integer n, let s_n be the sum of the first n positive integers, and let S_n be the sum of the first n perfect squares. Find the remainder when

$$\sum_{k=1}^{2018} S_k - s_k$$

is divided by 1000.

Solution 155. We have $S_k = \dfrac{k(k+1)(2k+1)}{6}$ and $s_k = \dfrac{k(k+1)}{2} = \dfrac{k(k+1)(3)}{6}$, so $S_k - s_k = \dfrac{k(k+1)(2k-2)}{6} = \dfrac{k(k+1)(k-1)}{3} = \dfrac{k^3 - k}{3}$. Applying the sum formulas for powers up to the cube yields $\boxed{690}$.

Exercise 156. *(2011 AMC 10B)* What is the hundreds digit of 2011^{2011}?

Solution 156. It suffices to compute $11^{2011} \pmod{1000}$. Using the Binomial Theorem, $11^{2011} = (1 + 10)^{2011} = 1 + \binom{2011}{1} \cdot 10^1 + \binom{2011}{2} \cdot 10^2 + \ldots$ where all terms afterward reduce to zero since $10^3 = 1000$. It follows that $11^{2011} \equiv 1 + 2011 \cdot 10 + 2011 \cdot 1005 \cdot 100 \equiv 1 + 110 + 500 = 611$, which has a hundreds digit of $\boxed{6}$.

Exercise 157. *(2016 Berkeley Math Tournament)* What are the last two digits of $9^{8^{7^{\cdot^{\cdot^{\cdot^{2}}}}}}$?

Solution 157. Note that the powers of 9 cycle as follows: $01, 09, 81, 29, 61, 49, 41, 69, 21, 89, 01, \ldots$ starting from 9^0 (with period 10). Hence, it suffices to compute $8^{7^{\cdot^{\cdot^{2}}}} \pmod{10}$. The cycle of the powers of 8 with respect to modulo 10 is $8, 4, 2, 6, \ldots$ with period 4. Again it suffices to compute $7^{6^{\cdot^{\cdot^{2}}}}$ modulo 4. But since $7^n \equiv 3^n \equiv 3 \pmod 4$ for odd n and $1 \pmod 4$ for even n, and 6 raised to any power is clearly even, the power of 7 reduces to 1 mod 4, and so 8 to that power is just 8, with $9^8 \equiv \boxed{21} \pmod{100}$.

Exercise 158. *(2017 AIME I)* For a positive integer n, let d_n be the units digit of $1 + 2 + \cdots + n$. Find the remainder when

$$\sum_{n=1}^{2017} d_n$$

is divided by 1000.

Solution 158. *We list out the values of d_n until we reach a stopping point where the values begin to repeat: $1, 3, 6, 0, 5, 1, 8, 6, 5, 5, 6, 8, 1, 5, 0, 6, 3, 1, 0, 0, \ldots$ so the period is 20. In each "block" of 20, the sum of the d_n's is 70, so in the first 2020 terms, the sum will be 7070. Hence, we subtract off the last three terms (which sum to 1) to get 7069 $\implies$ $\boxed{69}$. (Note that it is not sufficient to shortcut this problem by simply applying the triangular sum formula!)*

Exercise 159. *(2011 AIME I) Let R be the set of all possible remainders when a number of the form 2^n, n a non-negative integer, is divided by 1000. Let S be the sum of the elements in R. Find the remainder when S is divided by 1000.*

Solution 159. *By symmetry, note that we need only compute the first few terms of R, since the terms will eventually cancel out to zero. Since we can apply CRT wrt mod 8 and mod 125, it suffices to consider 1, 2, 4 as the "first" terms, leading immediately to the answer of $\boxed{7}$ (mod 1000).*

Exercise 160. *For what positive integral values of n is $2^n + 3^n + 6^n$ a multiple of 7?*

Solution 160. *The cycle of powers of 2 repeats every 3 terms wrt mod 7, that of powers of 3 repeats every 6 terms, and that of powers of 6 repeats every 2 terms, so we consider the sequence's pattern every $lcm(2, 6, 4) = 12$ terms (when all the cycles will align). Some computation reveals that $n \equiv 2, 4, 8, 10 \pmod{12}$.*

Extension Exercise 69. *What is the rightmost nonzero digit of 20^{18} in base 18?*

Extension Solution 69. *This is just the remainder when 20^{18} is divided by 18 (or, if zero, the remainder when divided by 18^2, and so on, which it will not be). This is equivalent to $2^{18} \pmod{18}$, which can be found to be $4 \cdot (2^4)^4 \equiv 4 \cdot (-2)^4 \equiv 4 \cdot -2 \equiv -8 \equiv 10 \pmod{18}$, which corresponds to the "digit" $\boxed{A}$ in base 18.*

Extension Exercise 70. *(2018 AIME I) Find the least positive integer n such that when 3^n is written in base 143, its two right-most digits in base 143 are 01.*

Extension Solution 70. *We require $3^n \equiv 1 \pmod{143^2}$. By CRT, $143^2 = 11^2 \cdot 13^2$, so $3^n \equiv 1 \pmod{121}$ and $3^n \equiv 1 \pmod{169}$. The first condition is easy: $5 \mid n$ just by listing the pattern by hand. As for the second congruence, note that $169 = 13^2$, and $3^3 = 13a + 1 \equiv 1 \pmod{13}$. Hence, we raise this to the 13th power to get $3^{39} \equiv 1 \pmod{169}$, since we want to find the first prime p such that $(13a + 1)^p \equiv 1 \pmod{169}$. Our answer is $lcm(5, 39) = \boxed{195}$.*

Extension Exercise 71. *(2011 USAJMO P1) Find, with proof, all positive integers n for which $2^n + 12^n + 2011^n$ is a perfect square.*

Extension Solution 71. *Clearly, $n = 1$ works; now consider the expression $\pmod{3}$. The expression reduces to $2^n + 3^n$, and all perfect squares are $0, 1 \pmod{4}$. But since $2^n \equiv 0 \pmod{4}$, and $3^n \equiv 1, 3 \pmod{4}$, n must be even. If $n = 2k$ for some integer $k \geq 1$, however, then this leads to a contradiction through algebra. Hence $n = 1$ is the sole solution.*

Extension Exercise 72. *What is the smallest positive integer n such that $2 \cdot 5^n$ is 1 less than a multiple of 49?*

Extension Solution 72. *We require $2 \cdot 5^n + 1 \equiv 0 \pmod{49}$, which implies that $7 \mid (2 \cdot 5^n + 1)$. Hence, $2 \cdot 5^n \equiv 6 \pmod 7$ and $5^n \equiv 3 \pmod 7$. By testing small values of n, we find that $n = 5, 11, 17, \ldots$ all work, and in fact, any $n \equiv 5 \pmod 6$ works as well.*

Now consider $2 \cdot 5^n \pmod{49}$. Beginning with $n = 0$, we can multiply 2 by 5 repeatedly (and simplify $\pmod{49}$ as necessary) to obtain $2, 10, 1, 5, 25, 27, 37, 38, 43, \ldots$ (where $5^6 \equiv 2 \cdot 5^8 \equiv 43 \pmod{49}$, since $2 \cdot 5^2 = 50 \equiv 1 \pmod{49}$), such that $2 \cdot 5^5 \equiv 27, 2 \cdot 5^{11} \equiv (2 \cdot 5^5) \cdot 5^6 \equiv 27 \cdot 43 \pmod{49} \equiv 34 \pmod{49}$, $2 \cdot 5^{17} \equiv 34 \cdot 43 \pmod{49} \equiv 41 \pmod{49}$, and $2 \cdot 5^{23} \equiv 41 \cdot 43 \pmod{49} \equiv -1 \pmod{49}$ (more easily seen by observing $41 \cdot 43 = 42^2 - 1$ by the difference of squares identity, and then noting that $42^2 = (6 \cdot 7)^2 = 6^2 \cdot 7^2 \equiv 0 \pmod{49}$). Hence, $n = \boxed{23}$ is the smallest n such that $2 \cdot 5^n + 1$ is a multiple of 49.

Extension Exercise 73. *(1989 AIME) One of Euler's conjectures was disproved in the 1960s by three American mathematicians when they showed there was a positive integer n such that $133^5 + 110^5 + 84^5 + 27^5 = n^5$. Find the value of n.*

Extension Solution 73. *Note that n must be even by a parity argument, and furthermore, Fermats Little Theorem guarantees that $n^5 \equiv n \pmod 5$. Hence $n \equiv 4 \pmod 5$ as well as $n \equiv 0 \pmod 3 \implies n \equiv 9 \pmod{15}$ by CRT. Thus, $n = \boxed{144}$ (since even $n = 159$ is much too large).*

Section 13.4: Studying Prime Numbers

Exercise 161. *If the digits of a positive integer sum to 69, show that the number cannot be prime.*

Solution 161. *As the sum of digits is a multiple of 3, the integer itself must be a multiple of 3. Hence it cannot be prime.*

Exercise 162. *Show that 2 and 3 are the only primes that differ by 1.*

Solution 162. *Assume otherwise. Then one prime would be odd and the other even (and ≥ 4), but then the even prime would have a prime factor of 2 and more than 2 divisors - contradiction.*

Exercise 163. *If p and q are both prime numbers, and $p > q > 2$, show that $p^2 - q^2$ is a multiple of 8.*

Solution 163. *We have $p^2 - q^2 = (p+q)(p-q)$. Since p, q are both odd, $p+q$ and $p-q$ are both even, and furthermore, either $p + q \equiv 0 \pmod 4$ or $p - q \equiv 0 \pmod 4$. Hence the product is equivalent to $0 \pmod 8$ as desired.*

Exercise 164. *Let p be a prime number. When is $p^2 + p + 1$ equal to the square of an integer?*

Solution 164. *We have $k^2 = p^2 + p + 1 = (p^2 + 2p + 1) - p = (p+1)^2 - p$, so p must be the difference of squares $(p+1)^2 - k^2 = (p+k+1)(p-k+1)$. However, if $p = 2$, then 7 is not a square, so p must be odd. Then its only factors are 1 and itself, which implies either $p + k = 0$ or $p - k = 0$. The former is absurd, while the latter implies $p = k$, which is also absurd. Hence no such primes exist.*

Exercise 165. *(2000 AMC 12) Two different prime numbers between 4 and 18 are chosen. When their sum is subtracted from their product, which of the following numbers could be obtained?*

- *21*
- *60*
- *119*
- *180*
- *231*

Solution 165. *Note that both primes are odd, so their product is odd and their sum is even; hence the number obtained must be odd - so we can rule out 60 and 180 right away. In addition, $13 \cdot 17 = 221 < 231$, so 231 is out as well. Indeed, $5 \cdot 7 - 5 - 7 = 23 > 21$, so 21 is out, leaving $\boxed{119}$ as the only viable answer choice (achieved for $11, 13$).*

Exercise 166. *Can the number $2^{2017} - 1$ be prime? What about $2^{2019} - 1$?*

Solution 166. *Yes, $2^{2017} - 1$ can theoretically be prime (although it isn't) by the Mersenne test; however, $2^{2019} - 1$ cannot be, as 2019 is composite.*

Exercise 167. *(2015 AIME I) There is a prime number p such that $16p + 1$ is the cube of a positive integer. Find p.*

Solution 167. *We have $16p + 1 = k^3$ for $k \in \mathbb{Z}$, i.e. $k^3 \equiv 1 \pmod{16}$. It follows that $k^3 \equiv 1 \pmod 4$, so $k \equiv 1 \pmod 4$ itself. Then $k = 4a + 1$, and substitution yields $16p = 64a^3 + 48a^2 + 12a \implies 4p = 16a^3 + 12a^2 + 3a$. Since $a = 4b$ for some $b \in \mathbb{Z}$, some more algebra yields $b = 1$, $a = 4$, and $p = \boxed{307}$.*

Exercise 168. *If $\gcd(a, b) = p$, where p is a prime number, what are the possible values of $\gcd(a^3, b)$?*

Solution 168. *Clearly, p is a possible value (in fact, it is always a value). p^3 is also possible (take $a = 2, b = 8$ for instance, with $p = 2$), as is p^2 (take $a = 3, b = 9, p = 3$). These are the only such values, since cubing the factors of a leads to a factor of p^3 in the case that $a > b$ (most optimistic).*

Exercise 169. *(1999 AIME) Find the smallest prime that is the fifth term of an increasing arithmetic sequence, all four preceding terms also being prime.*

Solution 169. *Clearly all terms are odd. The common difference must be at least 6 as well, and this leads to the sequence 5, 11, 17, 23, $\boxed{29}$.*

Exercise 170. Show that $19 \nmid (4n^2 + 4)$ for all $n \in \mathbb{N}$.

Solution 170. *If such an n were to exist with $19 \mid (4n^2 + 4)$, it would have to satisfy $n^2 \equiv 18 \pmod{19}$. But 18 is not a quadratic residue modulo 19, so no such n exists.*

Exercise 171. *(2005 AMC 12A)* Call a number prime-looking if it is composite but not divisible by $2, 3,$ or 5. The three smallest prime-looking numbers are $49, 77,$ and 91. There are 168 prime numbers less than 1000. How many prime-looking numbers are there less than 1000?

Solution 171. *Apply complementary counting and PIE to split the integers from 1 to 1000 into 1, and sets of numbers divisible by 2, 3, and 5, primes not including 2, 3, or 5, and prime-looking numbers. We have that the cardinality of the prime-looking set is $1000 - 165 - 1 - |S_2 \cup S_3 \cup S_5|$ (where S_2, S_3, S_5 are the sets of multiples of 2, 3, 5 respectively). By PIE, $|S_2 \cup S_3 \cup S_5| = 500 + 333 + 200 - 166 - 66 - 100 + 33 = 734$, so the size of the prime-looking set is $\boxed{100}$.*

Exercise 172. Show that $5 \mid (7^{4k+2} + 11^{3k+1})$ for all $k \in \mathbb{N}$.

Solution 172. *Observe that, when 7 is raised to a power that is equivalent to 2 mod 4, it will always be congruent to 4 mod 5; similarly, 11 raised to 1 mod 3 will always be congruent to 1 mod 5. Their sum is thus a multiple of 5.*

Extension Exercise 74. Prove that there are infinitely many primes of the form $4k + 1$, $k \in \mathbb{N}$.

Extension Solution 74. *For all n even, all prime divisors of $n^2 + 1$ are equivalent to 1 (mod 4), since any $p \mid n^2 + 1$ must satisfy $n^2 \equiv -1 \pmod{p} \implies \left(\dfrac{-1}{p}\right) = 1$ (see §15), which is equivalent to $p \equiv 1 \pmod{4}$.*

Then assume there are k primes p_1 through p_k equivalent to 1 (mod 4). Then we can arrive at a contradiction (similarly to Euclids infinitude of all primes argument) by considering $(2p_1 p_2 p_3 \cdots p_k)^2 + 1$.

Extension Exercise 75. Show that, for all positive integers $n \geq 2$, $n^4 + 4$ is composite.

Extension Solution 75. *This is an immediate application of Sophie-Germain. Without knowing that identity, one may still derive it (at least partially) by completing the square; i.e. $n^4 + 4n^2 + 4 = 4n^2 \implies (n^2 + 2)^2 = (2n)^2 \implies (n^2 + 2n + 2)(n^2 - 2n + 2) = 0$.*

Extension Exercise 76. Show that if p is prime and is congruent to 3 (mod 4), then

$$\left(\frac{p-1}{2}\right)! \equiv \pm 1 \pmod{p}$$

(Hint: Use Wilson's Theorem.)

Extension Solution 76. *For some integer r, we have $p - r \equiv -r \pmod{p}$, which implies $r \equiv -(p-r) \pmod{p}$. Then for uniqueness, $r \leq p - r \implies 2r \leq p \implies r \leq \dfrac{p}{2}$. Hence, $1 \leq r \leq \dfrac{p-1}{2}$.*

Subsequently, setting $r = 1, 2, 3, \ldots$ yields $1 \equiv -(p-1), 2 \equiv -(p-2), \ldots, \dfrac{p-3}{2} \equiv \dfrac{p+3}{2}, \dfrac{p-1}{2} \equiv \dfrac{p+1}{2}$. There are $\dfrac{p+1}{2}$ pairs, so $(p-1)! = (-1)^{\frac{p-1}{2}} \cdot \left(\dfrac{p-1}{2}!\right)^2$. By Wilson's Theorem, the RHS is congruent to -1 (mod p). If $p \equiv 3 \pmod{4}$, set $p = 4t + 3$ for some $t \in \mathbb{Z}$. Then $\dfrac{p-1}{2} = 2t + 1$ odd, so $(-1)^{\frac{p-1}{2}} = -1$, which implies that $\dfrac{p-1}{2}! \equiv \pm 1 \pmod{p}$ upon taking the square root.

Extension Exercise 77. *(2002 AMC 12A)* Several sets of prime numbers, such as $\{7, 83, 421, 659\}$ use each of the nine nonzero digits exactly once. What is the smallest possible sum such a set of primes could have?

Extension Solution 77. *The even digits cannot be the units digit of a prime (with the exception of 2), so the sum is at least $40 + 60 + 80 + 1 + 2 + 3 + 5 + 7 + 9 = 207$. Indeed, the construction $\{2, 5, 7, 43, 61, 89\}$ works, as does $\{2, 3, 5, 41, 67, 89\}$.*

Extension Exercise 78. *Prove that there always exists at least one prime number between n and $1 + n!$, for $n \geq 2$ a positive integer. (You may use Bertrand's Postulate, which guarantees the existence of a prime number between k and $2k$ for all natural numbers $k \geq 2$.)*

Extension Solution 78. *Given $n > 2$, for every $x \in \mathbb{Z}$ such that $1 < x < n+1$, we have $x \mid n!$, and furthermore, $x \nmid (n! - 1)$. So either $n! - 1$ is prime or there exists a prime number $p \geq n + 1$ such that $p \mid (n! - 1)$. In either case, there exists prime p with $n + 1 \leq p \leq n! - 1$, and so the weaker statement in this exercise also holds.*

Section 13.5: Introduction to Diophantine Equations

Exercise 173. *Solve the Diophantine equation $20x - 19y = 45$.*

Solution 173. *We have $7(20) - 5(19) = 45$, and if we increase x by 19, we increase y by 20 as well to compensate. Hence the solution set is* $\boxed{\{(x,y) \mid x = 7 + 19t, y = 5 + 20t, x, y \in \mathbb{Z}, t \in \mathbb{R}\}}$

Exercise 174.

(a) *Find all ordered pairs of positive integers (x, y) such that $x^2 - y^2 = 20$.*

(b) *Find all ordered pairs of positive integers (x, y) such that $x^2 - y^2 = 2019$.*

Solution 174.

(a) *We have $(x+y)(x-y) = 20 \implies x - y = 2 \implies (x, y) = (6, 4)$ only, as we must consider the pairs of factors whose product is 20 (both factors must not only be integers, but also must be of the same parity).*

(b) *As above, $(x+y)(x-y) = 2019 = 3 \cdot 673 = 1 \cdot 2019$, so $x + y = 673, x - y = 3 \implies (x, y) = (338, 335)$ or $x + y = 2019, x - y = 1 \implies (x, y) = (1010, 1009)$.*

Exercise 175. *Find the sum of all solutions for x in the system of equations $xy = 2$, $x + y = 3$.*

Solution 175. *Clearly $(x, y) = (1, 2), (2, 1)$. We can verify that these are the only solutions, since $y = \frac{2}{x} \implies x + \frac{2}{x} = 3 \implies x^2 - 3x + 2 = 0 \implies x = 1, 2$. The sum is* $\boxed{3}$.

Exercise 176. *(2015 Berkeley Mini Math Tournament) Given integers a, b, c satisfying $abc + a + c = 12$, $bc + ac = 8$, $b - ac = -2$, what is the value of a?*

Solution 176.

Exercise 177. *(2016 AMC 10A) How many ways are there to write 2016 as the sum of twos and threes, ignoring order? (For example, $1008 \cdot 2 + 0 \cdot 3$ and $402 \cdot 2 + 404 \cdot 3$ are two such ways.)*

Solution 177. *We can write $2016 = 1008 \cdot 2 + 0 \cdot 3 = 1005 \cdot 2 + 2 \cdot 3 = 1002 \cdot 2 + 4 \cdot 3 = \ldots = 3 \cdot 2 + 670 \cdot 3 = 0 \cdot 2 + 672 \cdot 3$, for a total of $\boxed{337}$ ways (decreasing the coefficient of the 2 by three in order to increase the coefficient of the 3 by two).*

Exercise 178. *(2011 AMC 10B) In multiplying two positive integers a and b, Ron reversed the digits of the two-digit number a. His erroneous product was 161. What is the correct value of the product of a and b?*

Solution 178. *Note that $161 = 23 \cdot 7$, so the correct product should be $32 \cdot 7 = \boxed{224}$.*

Exercise 179. *Compute the prime number p such that $n^2 = 19p + 1$, where n is a positive integer.*

Solution 179. *We have $n^2 \equiv 1 \pmod{19} \implies n \equiv 1, 18 \pmod{19}$. $n = 1$ doesn't work, but $n = 18 \implies p = \boxed{17}$.*

Exercise 180. *Find all ordered pairs (a, b) of positive integers such that $a^2 - a + b - b^2 = 36$.*

Solution 180. *Rearranging, we have $(a^2 - b^2) + (b - a) = 36 \implies (a-b)(a+b) + (b-a) = 36 \implies (a-b)(a+b-1) = 36$. Then we filter through the factors of 36, obtaining $(a, b) = (7, 3), (8, 5), (19, 18)$.*

Exercise 181. *Find all ordered pairs of integers (x, y) for which $4^y - x^2 = 615$.*

Solution 181. *Clearly $y \geq 5$, since $x^2 \geq 0$ for all $x \in \mathbb{Z}$. Then x^2 is odd $\implies x$ is odd. It follows that $x^2 \equiv 1 \pmod 4$. But $615 \equiv 7 \pmod{16}$, so we require $x \equiv 3 \pmod 4$ for $615 + x^2$ to be divisible by 16, and hence by $4^2 \mid 4^y$ - contradiction. Hence there are no solutions.*

Exercise 182. *Find all rational ordered pairs (x, y) that satisfy the equation $5x^2 + 2y^2 = 1$.*

Solution 182. The number of solutions is equal to the number of minimal solutions to $5x^2 + 2y^2 = z^2$, $x, y, z \in \mathbb{Z}$. Taking this mod 5 yields $2y^2 = z^2$. But 2 is a quadratic non-residue modulo 5, so there are no solutions.

Exercise 183. Find all positive integers n such that $n \pm 76$ are both perfect cubes.

Solution 183. The difference of cubes between $n+76$ and $n-76$ is 152, i.e. $152 = a^3 - b^3 = (a-b)(a^2+ab+b^2)$ for some integers a, b with $a > b$. Since $152 = 8 \cdot 19 = 4 \cdot 38 = 2 \cdot 76$, we obtain $a = 6, b = 4$. Then $n + 76 = 6^3 = 216$ and $n - 76 = 4^3 = 64 \implies n = \boxed{140}$. This is indeed the only possibility, which is not too hard to verify using the factorization of 152.

Exercise 184. How many ordered pairs of positive integers (m, n) are there such that $mn + 2m + 2n = 176$?

Solution 184. Using Simon's Favorite Factoring Trick, we get $(m + 2)(n + 2) = 180 = 5 \cdot 3^2 \cdot 2^2 \implies 18$ factors, but $m+2, n+2 \geq 3$ by default. Hence, we exclude the factors 1, 2, 90, 180, for a total of $18 - 4 = \boxed{14}$ ordered pairs.

Exercise 185. Solve the equation $\dfrac{3}{x} - \dfrac{7}{y} = \dfrac{1}{5}$ over $\mathbb{Z}$.

Solution 185. We have $\dfrac{3y - 7x}{xy} = \dfrac{1}{5} \implies 15y - 35x = xy \implies xy + 35x - 15y = 0 \implies (x - 15)(y + 35) = -525$ by SFFT. Since $525 = 7 \cdot 5^2 \cdot 3$, we can use this to determine all solutions.

Exercise 186. *(2015 AMC 10A)* The zeroes of the function $f(x) = x^2 - ax + 2a$ are integers. What is the sum of the possible values of a?

Solution 186. The roots are of the form $\dfrac{a \pm \sqrt{a^2 - 8a}}{2}$, so $\sqrt{a^2 - 8a}$ is an integer $\implies a = -1, 0, 8, 9 \implies \boxed{16}$.

Exercise 187. *(2008 AMC 12B)* A rectangular floor measures a by b feet, where a and b are positive integers with $b > a$. An artist paints a rectangle on the floor with the sides of the rectangle parallel to the sides of the floor. The unpainted part of the floor forms a border of width 1 foot around the painted rectangle and occupies half of the area of the entire floor. How many possibilities are there for the ordered pair (a, b)?

Solution 187. The outer area is just ab, but the inner area is $(a-2)(b-2)$. In addition, $ab = 2(a-2)(b-2) \implies 2ab - 4a - 4b + 8 = ab \implies ab - 4a - 4b + 8 = 0$. By Simon's Favorite Factoring Trick, $(a-4)(b-4) = 8$, which yields the solutions $(a, b) = (5, 12), (6, 8)$, i.e. $\boxed{2}$ possibilities.

Extension Exercise 79. *(1996 AIME)* The harmonic mean of two positive integers is the reciprocal of the arithmetic mean of their reciprocals. For how many ordered pairs of positive integers (x, y) with $x < y$ is the harmonic mean of x and y equal to 6^{20}?

Extension Solution 79. The harmonic mean of x and y is given by $\dfrac{2xy}{x+y}$, so $xy = (x+y)(3^{20} \cdot 2^{19})$. By SFFT, we have $(x - 3^{20} \cdot 2^{19})(y - 3^{20} \cdot 2^{19}) = 3^{40} \cdot 2^{38}$, which has 1599 factors, one of which is $3^{20} 2^{19}$ (its square root). As $x < y$, the answer is $\dfrac{1599 - 1}{2} = \boxed{799}$.

Extension Exercise 80. Find all pairs of positive integers (a, b) with $a > b$ and $a^2 + b^2 = 40,501$.

Extension Solution 80. They are $(201, 10), (199, 30)$, not too hard to determine by hand. Indeed, notice that $40,501 = 200^2 + 501 = 200^2 + 200 + 200 + 101$, which lends itself to the difference of squares identity.

Extension Exercise 81. *(2008 AIME I)* There exist unique positive integers x and y that satisfy the equation $x^2 + 84x + 2008 = y^2$. Find $x + y$.

Extension Solution 81. Completing the square (in reverse?) yields $(x+42)^2 + 244 = y^2 \implies 244 = 122 \cdot 2$ is a difference of squares. Since 244 only has six factors, this is the only possibility, with $y = 62, x + 42 = 60 \implies x = 18$. Hence, $x + y = \boxed{80}$.

Extension Exercise 82. *(1984 AHSME) The number of distinct pairs of integers (x, y) such that $0 < x < y$ and $\sqrt{1984} = \sqrt{x} + \sqrt{y}$ is*

Extension Solution 82. $\sqrt{1984} = 8\sqrt{31}$, and so the solutions must be of the form $a\sqrt{31} + b\sqrt{31}$ for $a + b = 8$, $0 < a < b$. There are $\boxed{3}$ distinct pairs satisfying this condition.

Extension Exercise 83. *(2005 AMC 12A) How many ordered triples of integers (a, b, c), with $a \geq 2$, $b \geq 1$, and $c \geq 0$, satisfy both $\log_a b = c^{2005}$ and $a + b + c = 2005$?*

Extension Solution 83. We have $b = a^{c^{2005}}$. Performing casework on c, we have three cases:

Case 1: $c = 0$. We have $b = 1 \implies (2004, 1, 0)$.

Case 2: $c = 1$. Then $a = b$ and we get $(1002, 1002, 1)$.

Case 3: $c \geq 2$. There is no way to satisfy the condition here.

Hence, the answer is $\boxed{2}$.

Extension Exercise 84. *(2011 AIME I) For some integer m, the polynomial $x^3 - 2011x + m$ has the three integer roots a, b, and c. Find $|a| + |b| + |c|$.*

Extension Solution 84. By Vieta's, $a + b + c = 0$ and $ab + bc + ac = -2011$. Let $a = -(b + c)$; WLOG assume $|a| \geq |b| \geq |c|$. It follows that $a^2 = bc + 2011$. Thus, $|a| \geq 45$. If we plug in reasonable values for a, then we find that $a = 49, b = 39, c = 10$ work perfectly. The answer is $49 + 39 + 10 = \boxed{98}$.

Section 13.6: Special Functions in Number Theory

Exercise 188. What is the value of $\left\lceil \dfrac{2+2^3}{3+2^2} \right\rceil$?

Solution 188. $\left\lceil \dfrac{2+2^3}{3+2^2} \right\rceil = \lceil \dfrac{10}{7} \rceil = \boxed{2}$.

Exercise 189. Either prove or disprove that $\lfloor -x \rfloor = -\lfloor x \rfloor$ for all $x \in \mathbb{R}$.

Solution 189. Any non-integer value for x is a counterexample; e.g. $x = -1.5$; $\lfloor -(-1.5) \rfloor = 1$, $-\lfloor x \rfloor = -(-2) = 2$.

Exercise 190. Prove or disprove that $\lfloor nx \rfloor = n\lfloor x \rfloor$ for all positive integers $n > 2$.

Solution 190. Any x for which $x - \lfloor x \rfloor > \dfrac{1}{n}$ is a counterexample; for instance, $n = 3, x = 3.5$.

Exercise 191. Show that, for all $x \in \mathbb{R}$, $\lfloor 2x \rfloor = \lfloor x \rfloor + \lfloor x + \dfrac{1}{2} \rfloor$.

Solution 191. Case 1: $x - \lfloor x \rfloor < \dfrac{1}{2}$. In this case, let $x = \lfloor x \rfloor + a$. Then $\lfloor 2x \rfloor = \lfloor 2\lfloor x \rfloor + 2a \rfloor = 2\lfloor x \rfloor$ and $\lfloor x + \dfrac{1}{2} \rfloor = \lfloor x \rfloor$.

Case 2: $x - \lfloor x \rfloor \geq \dfrac{1}{2}$. This case is similar except the LHS is increased by 1 and $\lfloor x + \dfrac{1}{2} \rfloor$ is also increased by 1 to compensate.

Hence, proved.

Exercise 192. (2003 AMC 10B) Compute
$$\lfloor \sqrt{1} \rfloor + \lfloor \sqrt{2} \rfloor + \lfloor \sqrt{3} \rfloor + \cdots + \lfloor \sqrt{16} \rfloor.$$

Solution 192. Up to 3, the floors are 1. From 4 to $3^2 - 1 = 8$, they are 2. From 9 to 15, they are 3, and for 16 only, the floor is 4. Hence, the sum evaluates to $1 \cdot 3 + 2 \cdot 5 + 3 \cdot 7 + 4 = \boxed{38}$.

Exercise 193. Show that $\lfloor \sqrt{n^2 + n} \rfloor = n$ for all $n \in \mathbb{N}$.

Solution 193. We have $n^2 < n^2 + n < (n+1)^2 = n^2 + 2n + 1$, so $n < \sqrt{n^2 + n} < n + 1 \implies \lfloor \sqrt{n^2 + n} \rfloor = n$, as desired.

Exercise 194. Evaluate the sum
$$\sum_{k=1}^{\infty} \left\lfloor \dfrac{30}{k} \right\rfloor.$$

Solution 194. Listing the first few terms yields $30 + 15 + 10 + 7 + 6 + 5 + 4 + 3 \cdot 3 + 2 \cdot 5 + 1 \cdot 15 = \boxed{111}$.

Exercise 195. In how many trailing zeros does 2019!!! end? (n!!! denotes the triple factorial; i.e. $n \cdot (n-3) \cdot (n-6) \cdots$).

Solution 195. We have $2019!!! = 2019 \cdot 2016 \cdot 2013 \cdot 2010 \cdots 9 \cdot 6 \cdot 3$. We count the number of factors of 5, which are the multiples of 15 from 15 to $2010 = 15 \cdot 134$, inclusive. Hence, the answer is $134 + 26 + 5 + 1 = \boxed{166}$.

Exercise 196. (2014 Berkeley Math Tournament) For a positive integer n, let $\phi(n)$ denote the number of positive integers between 1 and n, inclusive, which are relatively prime to n. We say that a positive integer k is total if $k = \dfrac{n}{\phi(n)}$, for some positive integer n. Find all total numbers.

Solution 196. The total numbers are $\boxed{1, 2, 3}$. First prime factorize n, then use the formula for the totient function. (In this particular case, testing smaller cases suffices.)

Exercise 197. *For a real number x, let $\lfloor x \rfloor$ be the largest integer less than or equal to x. For example, $\lfloor 1 \rfloor = 1$ and $\lfloor 2.99 \rfloor = 2$. The area under the graph of $y = \lfloor x^2 - 2x \rfloor$ between $x = 0$ and $x = 5$ can be written in the form $N - R$, where N is a positive integer and R is a sum of radicals. Find N.*

Solution 197. The answer is $\boxed{49}$; the area consists of 14 unit intervals through graphing the function, namely $x \in [1 + \sqrt{a+1}, 1 + \sqrt{a+2})$ where $f(x) = a$. We have
$$\sum_{a=0}^{14} a(\sqrt{a+2} - \sqrt{a+1})$$
which telescopes and collapses into the integer part of 49.

Exercise 198. *(2018 AMC 10A) What is the greatest integer less than or equal to*
$$\frac{3^{100} + 2^{100}}{3^{96} + 2^{96}}?$$

Solution 198. Let this value be x. Then $3^{100} + 2^{100} = x(3^{96} + 2^{96})$. If we plug in $81 = 3^4$ for x, we obtain $3^{100} + \frac{81}{16} 2^{96} > x$, so $x < 81 \implies \boxed{80}$.

Exercise 199. *(1983 AIME) Let a_n equal $6^n + 8^n$. Determine the remainder upon dividing a_{83} by 49.*

Solution 199. Note that $a_{83} = (7-1)^{83} + (7+1)^{83}$; using the Binomial Theorem demonstrates that half the terms cancel, and that all terms except for the last one are divisible by 49. Hence, the remainder is $\boxed{35}$.

Exercise 200. *(2015 AMC 10B/12B) Cozy the Cat and Dash the Dog are going up a staircase with a certain number of steps. However, instead of walking up the steps one at a time, both Cozy and Dash jump. Cozy goes two steps up with each jump (though if necessary, he will just jump the last step). Dash goes five steps up with each jump (though if necessary, he will just jump the last steps if there are fewer than 5 steps left). Suppose the Dash takes 19 fewer jumps than Cozy to reach the top of the staircase. Let s denote the sum of all possible numbers of steps this staircase can have. What is the sum of the digits of s?*

Solution 200. Essentially, we just need to solve the equation $\lceil \frac{s}{2} \rceil - 19 = \lceil \frac{s}{5} \rceil$. By casework with $\lceil \frac{s}{2} \rceil = \frac{s}{2}$ or $\frac{s+1}{2}$, we can determine that $s = 64, 66$ in the first case, and $s = 63$ in the second case. Summing yields $193 \implies \boxed{13}$.

Exercise 201. *What fraction of the positive integers up to 20192019 are relatively prime with it? (Note: $10,001 = 137 \cdot 73$.)*

Solution 201. $\phi(20192019) = \phi(2019 \cdot 10001) = \phi(673 \cdot 137 \cdot 73 \cdot 3) = 20192019 \cdot \frac{672}{673} \cdot \frac{136}{137} \cdot \frac{72}{73} \cdot \frac{2}{3}$ (no, we won't bother to simplify this!)

Extension Exercise 85. *Let $f(x) = \left\lfloor \frac{x}{7} \cdot \left\lfloor \frac{37}{x} \right\rfloor \right\rfloor$, where $1 \leq x \leq 45$ is an integer. How many distinct values does $f(x)$ assume?*

Extension Solution 85. *Sketch:* Consider $\lfloor \frac{37}{x} \rfloor$ by itself. Clearly, for $x > 37$, the only possible value is 0. For $19 \leq x \leq 37$, $f(x) = \lfloor \frac{x}{7} \rfloor$, and so forth. What is the significance of 7 and 37 in particular in this problem?

Extension Exercise 86. *(2011 AMC 12B) For every m and k integers with k odd, denote by $\left[\frac{m}{k}\right]$ the integer closest to $\frac{m}{k}$. For every odd integer k, let $P(k)$ be the probability that*
$$\left[\frac{n}{k}\right] + \left[\frac{100-n}{k}\right] = \left[\frac{100}{k}\right]$$
for an integer n randomly chosen from the interval $1 \leq n \leq 99!$. What is the minimum possible value of $P(k)$ over the odd integers k in the interval $1 \leq k \leq 99$?

Extension Solution 86. *If* $\lfloor \frac{n}{k} \rfloor + \lfloor \frac{100-n}{k} \rfloor = \lfloor \frac{100}{k} \rfloor$ *as given, then it follows that* $\lfloor \frac{n-k}{k} \rfloor + \lfloor \frac{100-(n-k)}{k} \rfloor = \lfloor \frac{100}{k} \rfloor$. *We consider* $1 \leq n \leq k$ *as a repeating outcome. Hence, we consider* $P(k)$ *for* $1 \leq n \leq k$. *Split the cases into* $k > \frac{200}{3}, n < \frac{k}{2}, \frac{k}{2} < n < k$. *At* $k = 67$, *we attain the minimum of* $\boxed{\frac{34}{67}}$.

Extension Exercise 87. *(2010 AIME I) For each positive integer n, let $f(n) = \sum_{k=1}^{100} \lfloor \log_{10}(kn) \rfloor$. Find the largest value of n for which $f(n) \leq 300$.*

Extension Solution 87. *For 100 terms to add up to 300, we need each term's average value to be close to 3. Hence, $n \approx 100$. Testing values near 100 yields $n = \boxed{109}$, which works, but $f(110) > 300$. (Here we use the properties of the floor function to "split" into cases that we can work with more easily.)*

Extension Exercise 88. *(2015 AIME I) For each integer $n \geq 2$, let $A(n)$ be the area of the region in the coordinate plane defined by the inequalities $1 \leq x \leq n$ and $0 \leq y \leq x \lfloor \sqrt{x} \rfloor$, where $\lfloor \sqrt{x} \rfloor$ is the greatest integer not exceeding $\sqrt{x}$. Find the number of values of n with $2 \leq n \leq 1000$ for which $A(n)$ is an integer.*

Extension Solution 88. *By graphing the function, we can more clearly observe that it consists of trapezoids with x-values ranging from a^2 to $(a+1)^2$. The topmost part is the line $y = ax$, with each trapezoids width being $3, 5, 7, \ldots$ When a is odd, $A(n)$ increases by some integer value plus $\frac{1}{2}$. When it is even, $A(n)$ will increase by an integer value. n is an integer iff $a \equiv 0 \pmod{4}$, it will not be an integer if $a \equiv 2 \pmod{4}$. When a is odd, every other value is integral. Thus, we have reduced the problem to computing the number of areas where $a \equiv 0 \pmod{4}$, i.e. the sum $(5^2 - 4^2) + (9^2 - 8^2) + (13^2 - 12^2) + \ldots + (29^2 - 28^2) = 231$, and the number of areas where a is odd, i.e. the sum $\frac{(2^2 - 1^2) + (4^2 - 3^2) + \ldots + (30^2 - 29^2) + (1000 - 31^2)}{2} = 252$. Summing these yields the answer of $\boxed{483}$.*

Extension Exercise 89. *Prove Hermite's Identity:*

> **Theorem 17.5: Hermite's Identity**
>
> $$\lfloor na \rfloor = \lfloor a \rfloor + \lfloor a + \frac{1}{n} \rfloor + \ldots + \lfloor a + \frac{n-1}{n} \rfloor$$

Extension Solution 89. *A key observation should be made here as a preface: $\lfloor x \rfloor, \lfloor x + \frac{1}{n} \rfloor, \ldots, \lfloor x + \frac{n-1}{n} \rfloor$ can assume only two values, namely $\lfloor x \rfloor, \lfloor x + 1 \rfloor$. Then note that $\lfloor x \rfloor \leq \lfloor x + \frac{1}{n} \rfloor \leq \ldots \leq \lfloor x + \frac{n-1}{n} \rfloor$, the minimum of which is $\lfloor x \rfloor$, and the maximum of which is still less than or equal to $\lfloor x \rfloor + 1$.*

Proceeding with the proof itself: where $x \in \mathbb{Z}$, the result is clear, so assume otherwise. Then $\{x\} \in (0, 1)$ strictly, and by our prior observation, there exists i with $\lfloor x \rfloor = \lfloor x + \frac{1}{n} \rfloor = \ldots = \lfloor x + \frac{i-1}{n} \rfloor$ as well as $\lfloor x + \frac{i}{n} \rfloor = \lfloor x + \frac{i+1}{n} \rfloor = \ldots = \lfloor x + \frac{n-1}{n} \rfloor = \lfloor x \rfloor + 1$.

It then follows that $\lfloor x \rfloor + \lfloor x + \frac{1}{n} \rfloor + \lfloor x + \frac{2}{n} \rfloor + \ldots + \lfloor x + \frac{n-1}{n} \rfloor = n \lfloor x \rfloor + n - i$. i then satisfies $\{x\} + \frac{i-1}{n} < 1$ and $\{x\} + \frac{i}{n} \geq 1$, so $\frac{n-i}{n} \leq \{x\} < \frac{n-i+1}{n}$.

Hence, we arrive at $\lfloor nx \rfloor = n \lfloor x \rfloor + n - i$. Hence the sum reduces cleanly to $\lfloor nx \rfloor$, as desired.

Chapter 13 Review Exercises

Exercise 202. *What is the remainder when 2018^2 is divided by 2019?*

Solution 202. This is $(-1)^2 \equiv \boxed{1} \pmod{2019}$.

Exercise 203. *This book became available for pre-order on March 14, 2019 (π Day!), a Thursday. What day of the week is the day that falls exactly 314 days after π Day 2019?*

Solution 203. Since $314 \equiv 6 \pmod 7$, 314 days after Thursday, March 14, 2019 falls on a $\boxed{\text{Wednesday}}$.

Exercise 204. *Let m and n be two odd numbers with $m > n$. What is the largest integer that is a factor of all positive integers of the form $m^2 - n^2$?*

Solution 204. $m^2 - n^2 = (m+n)(m-n)$, so $m+n$ and $m-n$ will definitely collectively have a factor of 8.

Exercise 205. *Let a_n be the sum $1 + 2 + 4 + \ldots + 2^n$. Compute the remainder when $a_1 + a_2 + a_3 + \ldots + a_{2019}$ is divided by 100.*

Solution 205. We have $a_n = 2^{n+1} - 1$, so $\sum_{i=1}^{2019} a_i = (2^{2021} - 4) - 2019 \equiv \boxed{29} \pmod{100}$.

Exercise 206. *Compute the units digit of $\sum_{i=1}^{100} (i!)^2$.*

Solution 206. At 5!, the first point where a factorial ends in at least one trailing zero, the factorials become irrelevant. All we need to compute is $1!^2 + 2!^2 + 3!^2 + 4!^2 = 1 + 4 + 36 + 576 \equiv \boxed{7} \pmod{10}$.

Exercise 207. *(1997 AIME) Sarah intended to multiply a two-digit number and a three-digit number, but she left out the multiplication sign and simply placed the two-digit number to the left of the three-digit number, thereby forming a five-digit number. This number is exactly nine times the product Sarah should have obtained. What is the sum of the two-digit number and the three-digit number?*

Solution 207. Let x be the two-digit number and y the three-digit number. We have $1000x + y = 9xy \implies 9xy - 1000x - y = 0$. SFFT implies $(9x - 1)\left(y - \dfrac{1000}{9}\right) = \dfrac{1000}{9}$, and so $(9x - 1)(9y - 1000) = 1000$. $9x - 1$ is between 89 and 890, so we consider the factors of 1000. As it turns out, $x = 14, y = 112$ works $\implies \boxed{126}$.

Exercise 208. *(2009 AIME I) Call a 3-digit number geometric if it has 3 distinct digits which, when read from left to right, form a geometric sequence. Find the difference between the largest and smallest geometric numbers.*

Solution 208. The smallest geometric number is 124, and the largest is 964. Hence our answer is $\boxed{840}$.

Exercise 209. *(2016 ARML Local Individual Round #9) The integer $Z = 104,060,001$ is the product of three distinct prime numbers. Compute the largest prime factor of Z.*

Solution 209. We have $Z + 400 = 104,060,401 = 101^4 \implies Z = 101^4 - 20^2 = (101^2)^2 - 20^2 = (10,201 + 20)(10,201 - 20) = 10,221 \cdot 10,181 = 3 \cdot 3407 \cdot 10,181$. Clearly, $\boxed{10,181}$ is the largest prime factor.

Exercise 210. *Determine the number of distinct ordered pairs of positive integer solutions (m, n) to the equation $20m + 19n = 2019$.*

Solution 210. We have $(m, n) = (100, 1)$, as well as $m = 81, 62, 43, 24, 5 \implies \boxed{6}$ ordered pairs (use Euclidean Algorithm/Bezout's Identity to prove this rigorously).

Exercise 211. *(2014 AIME II) The repeating decimals $0.\overline{abab}ab$ and $0.\overline{abcabc}abc$ satisfy*

$$0.\overline{abab}ab + 0.\overline{abcabc}abc = \frac{33}{37},$$

where a, b, and c are (not necessarily distinct) digits. Find the three digit number abc.

Solution 211. We have $\dfrac{10a+b}{99} + \dfrac{100a+10b+c}{999} = \dfrac{33}{37} = \dfrac{891}{999}$; simplifying and equating coefficients yields $c = 7, a = b = 4 \implies \boxed{447}$.

Exercise 212. Compute the remainder when 7^{2018} is divided by 650.

Solution 212. By CRT, we can (and should) consider both mod 13 and mod 50. Mod 13, the pattern is $1, 7, 10, 5, 9, 11, -1, -7, -10, -5, -9, -11, 1, \ldots$ with period 12. Mod 50, the pattern is $1, 7, -1, -7, 1, \ldots$ with period 4. Hence, the patterns align/overlap every 12 terms. As such, $7^{2016} \equiv 1 \pmod{650}$, with $7^{2018} \equiv \boxed{49} \pmod{650}$.

Exercise 213. *(2000 AIME I)* Find the least positive integer n such that no matter how 10^n is expressed as the product of any two positive integers, at least one of these two integers contains the digit 0.

Solution 213. If a factor of 10^n has a 2 and 5 in its prime factorization, then that factor ends in 0. From $n = 1$, we continue onward until $n = \boxed{8}$, our desired result.

Exercise 214. *(1994 AIME)* The increasing sequence $3, 15, 24, 48, \ldots$ consists of those positive multiples of 3 that are one less than a perfect square. What is the remainder when the 1994th term of the sequence is divided by 1000?

Solution 214. Each term is of the form $k^2 - 1 = (k+1)(k-1)$, so either $3 \mid (k+1)$ or $3 \mid (k-1)$. This holds for $n \equiv -1, 1 \pmod 3$. As 1994 is even, $n \equiv 1 \pmod 3$ and will be the 997th term in the sequence. Hence $n = 2992 \implies n^2 - 1 \equiv \boxed{63} \pmod{100}$.

Exercise 215. *(2013 ARML Individual Round #9)* Compute the smallest positive integer base b for which 16_b is prime and 97_b is a perfect square.

Solution 215. We require $b + 6$ to be prime and $9b + 7 \equiv 7 \pmod 9$ to be a perfect square. Clearly b is odd. Also note that all perfect squares are congruent to either 0 or 1 mod 4. Since b is odd, $9b + 7$ is even, and therefore a multiple of 4. This means that not only is b odd, it is 1 (mod 4). (Subsequently, $b + 6 \equiv 3 \pmod 4$.) The answer is $b = \boxed{53}$.

Exercise 216. *(2016 AMC 12B)* The sequence (a_n) is defined recursively by $a_0 = 1$, $a_1 = \sqrt[19]{2}$, and $a_n = a_{n-1} a_{n-2}^2$ for $n \geq 2$. What is the smallest positive integer k such that the product $a_1 a_2 \cdots a_k$ is an integer?

Solution 216. It suffices to list everything through until we hit $\boxed{17}$, but perhaps a more efficient way would be to observe that the sum of logs must be an integer. Then multiply all logs by 19 to ensure that the sum is actually a multiple of 19. (What do you notice? Can you derive a recursive relationship?)

Exercise 217. Let a_n be the tens digit of 7^n. What is the remainder when $\displaystyle\sum_{k=1}^{2018} a_k$ is divided by 1000?

Solution 217. Notice the familiar cycling trait: $01, 07, 49, 43, 01, \ldots$. Hence, in each "block" of 4, $\{0, 4, 4, 0\}$, with $a_0 = 0$, the sum is 8, and so the sum up to a_{2016} is 4032. Two more terms yields a sum of $4036 \equiv \boxed{36} \pmod{1000}$.

Extension Exercise 90.

(a) *(2016 AMC 12A)* How many ordered triples (x, y, z) of positive integers satisfy $\text{lcm}(x,y) = 72$, $\text{lcm}(x,z) = 600$ and $\text{lcm}(y,z) = 900$?

(b) *(2016 AMC 12B)* There are exactly 77,000 ordered quadruplets (a, b, c, d) such that $\gcd(a,b,c,d) = 77$ and $\text{lcm}(a,b,c,d) = n$. What is the smallest possible value for n?

Extension Solution 90.

(a) *(2016 AMC 12A) Prime factorizing each of 72, 600, and 900, we get that $72 = 3^2 \cdot 2^3, 600 = 5^2 \cdot 3 \cdot 2^2$, and $900 = 5^2 \cdot 3^2 \cdot 2^2$. Designate by x, y, z the quantities $2^a 3^b 5^c, 2^d 3^e 5^f, 2^g 23^h 5^i$. After setting up maximum configurations, we deduce that $\max(d, g) = 2, a = 3$. Since $\max(b, h) = 1$, e must be 2. After a long road, we are left with $\max(b, h) = 1, \max(d, g) = 2$. Using these to our advantage, we have $3 \cdot 5 = \boxed{15}$ possible ordered pair configurations.*

(b) *(2016 AMC 12B) Let $a = 77A$, and define b, c, d analogously (so that $\gcd(a, b, c, d) = 1$ as a natural consequence). For each prime power in the prime factorization of $N = \dfrac{n}{77}$, one of the prime factorizations of a, b, c, d has that same prime power, at least one has $p^0 = 1$, and all have p^m for all $0 \le m \le k$.*

Define by $f(k)$ the number of ordered 4-tuples (m_1, m_2, m_3, m_4) such that $0 \le m_i \le k$, the largest is k, and the smallest is 0. Then $77,000 = f(k_2)f(k_3)f(k_5)\cdots$.

There are 14 4-tuples consisting only of 0 and k, $36k - 36$ of them that include 3 different values, and $12(k-1)(k-2)$ that include four different values. Thus, $f(k) = 14 + 12(k^2 - 1)$, and $f(1) = 14, f(2) = 50, f(3) = 110$. Immediately, we note that $14 \cdot 50 \cdot 110 = 77,000$. Hmm

Indeed, this is quite interesting, for the prime factorization of N actually uses the exponents of 1, 2, 3. Then assign the large exponents to smaller prime bases to get $N = 2^3 3^2 5^1 = 360, n = 360 \cdot 77 = \boxed{27,720}$.

Extension Exercise 91. *(2016 AMC 12B) For a certain positive integer n less than 1000, the decimal equivalent of $\dfrac{1}{n}$ is $0.\overline{abcdef}$, a repeating decimal of period of 6, and the decimal equivalent of $\dfrac{1}{n+6}$ is $0.\overline{wxyz}$, a repeating decimal of period 4. What is n?*

Extension Solution 91. *Let $\dfrac{1}{n} = 0.\overline{abcdef}$. Then $n \mid 10^6 - 1$, and so must be $n+6$ of $10^4 - 1$. We observe that $n = \boxed{297}$ is a solution, as it is a factor of both 9999 and 999,999.*

Extension Exercise 92. *(1968 IMO P2) Find all natural numbers x such that the product of their digits (in decimal notation) is equal to $x^2 - 10x - 22$.*

Extension Solution 92. *Let $x = \overline{d_1 d_2 d_3 \cdots d_n}$ in base 10. Then $x \ge d_1 \cdot 10^{n-1}$, and the product of the digits of x is less than or equal to $d_1 \cdot 10^{n-1}$, with equality iff x consists of a single digit. Hence the product of digits of x is no greater than x itself. Thus, $x^2 - 10x - 22 \le x \implies x^2 - 11x - 22 \le 0$. This holds for all $x \in [1, 12]$. However, for $x \in [1, 11]$, $x^2 - 10x - 22 < 0$. With $x = 12$, on the other hand, $12^2 - 12 \cdot 10 - 22 = 2$, which is indeed the product of digits of 12. Therefore, $\boxed{12}$ is the single solution.*

Extension Exercise 93. *(1975 IMO P4) When 4444^{4444} is written in decimal notation, the sum of its digits is A. Let B be the sum of the digits of A. Find the sum of the digits of B. (A and B are written in decimal notation.)*

Extension Solution 93. *Note that $4444^{4444} < 10000^{4444} = 10^{17,776}$, so 4444^{4444} has fewer than 17,776 digits. Hence, $A < 9 \cdot 17,775 = 159,975$. The digit sum of A is highest at $A = 99,999$, so $B \le 45$. (The maximal sum of digits of B is then 12, achieved at $B = 39$.)*

Using the notion of digital roots, we have $4444^{4444} \equiv A \equiv B \pmod 9$. Where $1 \le x \le 9$, we want to compute $4444^{4444} \equiv x \pmod 9$. By taking advantage of the cyclic pattern, we have $4444^{4444} \equiv \boxed{7} \pmod 9$.

Extension Exercise 94. *Show that $\lim\limits_{n \to \infty} \dfrac{\tau(n)}{\phi(n)} = 0$.*

Extension Solution 94. *Let $n = p_1^{k_1} p_2^{k_2} p_3^{k_3} \cdots p_n^{k_n}$. By definition, $\tau(n) = (k_1+1)(k_2+1)(k_3+1)\cdots(k_n+1)$, whereas $\phi(n) = n \cdot \left(1 - \dfrac{1}{p_1}\right) \cdot \left(1 - \dfrac{1}{p_2}\right) \cdots \left(1 - \dfrac{1}{p_n}\right)$. $\phi(n)$ dominates $\tau(n)$ with an extra n term, so the limit is zero.*

Chapter 14: Intermediate Number Theory

Section 14.1: Modular Arithmetic: Part 2

Exercise 218. *For all positive integers $1 \leq n \leq 2019$, what is the largest value of $2019 \pmod{n}$?*

Solution 218. Let this value be m. Then $n \mid (2019 - m)$, and $m < n$. $n = 1010$ works, giving $\boxed{1009}$ as the maximal value. (Anything larger is absurd.)

Exercise 219. *(2016 A-Star Math Tournament) Compute the remainder when $56666\underbrace{\ldots}_{2016 \text{ sixes}}6666$ is divided by 17.*

Solution 219. We can consider just the 5, then continually multiply by 10 and add 6. This yields $5, 5, 5, \ldots$ infinitely, since $5 \cdot 10 + 6 = 56 \equiv 5 \pmod{17}$. Hence, the answer is just $\boxed{5}$.

Exercise 220. *What is the remainder when 149162536496481100121 (the perfect squares from 1^2 to 11^2 concatenated together) is divided by 99?*

Solution 220. By Chinese Remainder Theorem, we want the remainders upon dividing the number by 9 and 11. By 9, the sum of digits is $74 \equiv 2 \pmod 9$. By 11, the difference in the alternating sums of digits is $43 - 31 = 12 \equiv 1 \pmod{11}$, so the number is congruent to $\boxed{56} \pmod{99}$.

Exercise 221.

(a) *(2017 AMC 10B/12B) Let $N = 123456789101112\ldots4344$ be the 79-digit number that is formed by writing the integers from 1 to 44 in order, one after the other. What is the remainder when N is divided by 45?*

(b) *If we write all of the positive integers from 1 to 19, inclusive, in consecutive order, and call the resulting number N, what is the remainder when N is divided by 330?*

(c) *(2017 AIME I) Let $a_{10} = 10$, and for each integer $n > 10$ let $a_n = 100a_{n-1} + n$. Find the least $n > 10$ such that a_n is a multiple of 99.*

Solution 221.

(a) By CRT, consider $N \pmod 5$ and $N \pmod 9$. $N \equiv 4 \pmod 5$ and the sum of the digits of N is 270 $\implies N \equiv 0 \pmod 9$. Hence $N \equiv \boxed{9} \pmod{45}$.

(b) $330 = 11 \cdot 5 \cdot 3 \cdot 2$. $N \equiv 7 \pmod{11}, 4 \pmod 5, 1 \pmod 3, 1 \pmod 2$, so $N \equiv \boxed{139} \pmod{330}$.

(c) We have $a_n + a_{n-1} + \ldots + a_{10} = 100(a_{n-1} + \ldots + a_{10}) + (n + \ldots + 10) \implies a_n = 99(a_{n-1} + \ldots + a_{10}) + \frac{(n+10)(n+9)}{2}$. Hence, $99 \mid a_n \implies 99 \mid \frac{(n+10)(n-9)}{2}$, at which point we can bust out our good friend, CRT! (Hi, CRT!) Thus, $n \geq \boxed{45}$ using casework on which of the two terms is a multiple of 9, and which is a multiple of 11.

Exercise 222. *(2018 Berkeley Math Tournament) Find*

$$\sum_{i=1}^{2016} i(i+1)(i+2) \bmod 2018$$

Solution 222. We can expand the summand out to get $i(i^2 + 3i + 2) = i^3 + 3i^2 + 2i$, and then use linearity of the sum to split it into three sums of each individual term. Use the sum formulae up to $\sum i^3$ to conclude that the answer is $\boxed{0}$, since the first term contains a factor of 2018, and the other two cancel wrt $\pmod{2018}$.

Exercise 223. *(2018 Berkeley Math Tournament) What is the remainder when $201820182018\ldots$ [2018 times] is divided by 15?*

Solution 223. Let N be the number in question. By CRT, we want to examine both N (mod 3) and N (mod 5). The sum of digits is $2018 \cdot 11 \equiv 2 \cdot 2 \equiv 1$ (mod 3), and clearly, $N \equiv 3$ (mod 5). Hence, $N \equiv \boxed{13}$ (mod 15).

Exercise 224. For how many positive integers $x \leq 100$ is $3^x - x^2$ divisible by 5?

Solution 224. Consider x (mod 5), and let $N = 3^x - x^2$. For $x \equiv 0$ (mod 5), $N \equiv 1$ (mod 5). For $x \equiv 1$ (mod 5), $N \equiv 2$ (mod 5). $x \equiv 2, 4$ (mod 5) $\implies N \equiv 0$ (mod 5), but $x \equiv 3$ (mod 5) $\implies N \equiv 3$ (mod 5). Hence the answer is $100 \cdot \dfrac{2}{5} = \boxed{40}$.

Exercise 225. Show that $111\ldots 111$ with 91 ones is composite.

Solution 225. Call this number N. Then note that
$$9N = 10^{91} - 9 \implies N = \frac{10^{91} - 9}{9} = \frac{(10^7)^{13} - 9}{9} = \frac{(10^7 - 1)((10^7)^{12} + (10^7)^{11} + \ldots + 1)}{9}$$
which is the product of $\dfrac{10^7 - 1}{9}$ and a number that is clearly an integer; hence, proved.

Exercise 226. Compute the remainder when $1 \cdot 1! + 2 \cdot 2! + 3 \cdot 3! + \ldots + 2015 \cdot 2015!$ is divided by 2017. (Recall that 2017 is prime.)

Solution 226. In general, we have $\sum_{i=1}^{k} i \cdot i! = (k+1)! - 1$ (prove this by induction), so $\sum_{i=1}^{2015} i \cdot i! = 2016! - 1$. By Wilson's Theorem, $2016! \equiv -1 \equiv 2016$ (mod 2017), and so our answer is $\boxed{2015}$.

Exercise 227. (2012 Stanford Math Tournament) Define a number to be boring if all the digits of the number are the same. How many positive integers less than $10,000$ are both prime and boring?

Solution 227. A prime number can only be boring if its single digit is 1 (this is called a repunit) or if it has one digit. They are 2, 3, 5, 7, and 11, so the answer is $\boxed{5}$.

Exercise 228. (2014 HMMT February Guts Round) Find the number of ordered quadruples of positive integers (a, b, c, d) such that a, b, c, and d are all (not necessarily distinct) factors of 30 and $abcd > 900$.

Solution 228. $abcd > 900 \implies \dfrac{30}{a} \cdot \dfrac{30}{b} \cdot \dfrac{30}{c} \cdot \dfrac{30}{d} < 900$. There are $\binom{4}{2}^3 = 216$ solutions to $abcd = 30^2 = 5^2 \cdot 3^2 \cdot 2^2$, so our answer is $\dfrac{8^4 - 216}{2} = \boxed{1940}$.

Exercise 229. (2014 HMMT February Team Round) Compute
$$\sum_{k=0}^{100} \left\lfloor \frac{2^{100}}{2^{50} + 2^k} \right\rfloor.$$

Solution 229. Denote $\dfrac{2^{100}}{2^k + 2^{50}}$ by a_k. Then for $k \in [0, 49]$, note that $a_k + a_{100-k} = 2^{50}$ by symmetry, but $a_k, a_{100-k} \notin \mathbb{Z}$ for $k \neq 50$, so $\lfloor a_k \rfloor + \lfloor a_{100-k} \rfloor = 2^{50} - 1$. Hence our answer is $50(2^{50} - 1) + 2^{49} = \boxed{101 \cdot 2^{49} - 50}$.

Exercise 230. (2004 AIME II) Let S be the set of integers between 1 and 2^{40} whose binary expansions have exactly two 1's. If a number is chosen at random from S, the probability that it is divisible by 9 is $\dfrac{p}{q}$, where p and q are relatively prime positive integers. Find $p + q$.

Solution 230. Note that $2^3 \equiv -1$ (mod 9) $\implies 2^6 \equiv 1$ (mod 9), i.e. $2^{6n+3} \equiv 1$ (mod 9). The solutions are of the form $2^a + 2^b$, i.e. $2^x(2^{6n+3}) + 1 \equiv 0$ (mod 9). Casework on x such that $0 \leq 3 + 6n + x \leq 39$ yields $\dfrac{37 + 31 + 25 + 19 + 13 + 7 + 1}{\binom{40}{2}} = \dfrac{133}{780} \implies \boxed{913}$.

394

Exercise 231. *Let $S(n)$ be the sum of the digits of n. Prove that $S(n) \equiv n \pmod 9$.*

Solution 231. *Write $n = 10^k \cdot d_k + 10^{k-1} \cdot d_{k-1} + \ldots + 10^2 d_2 + 10 d_1 + d_0$, then take each term mod 9 to get $n = d_k + d_{k-1} + \ldots + d_2 + d_1 + d_0$, as desired.*

Exercise 232. *Find all ordered pairs of integers (m,n) such that $m^4 + n! = 2016$.*

Solution 232. *Note that $n \leq 6$. We conclude that $(m,n) = (6,4)$ is the only solution, since $7^4 = 2401$ is the next cube above $6^4 = 1296$.*

Extension Exercise 95. *(2005 USAMO P2) Prove that the system*

$$x^6 + x^3 + x^3 y + y = 147^{157}$$
$$x^3 + x^3 y + y^2 + y + z^9 = 157^{147}$$

has no solutions in integers x, y, and z.

Extension Solution 95. *It suffices to prove that no solution exists $\pmod{19}$. Note that $157 \equiv 5 \pmod{19}$, so $\text{ord}_{19}(5) = 9$. It follows that $147^{157} \equiv -5^4 \equiv 2 \pmod{19}$, and $157^{147} \equiv -8 \pmod{19}$. Rewriting the system and using Fermats Little Theorem, $z^9 \equiv 0, \pm 1$, so $(x^3 + y + 1)^2 \equiv -4, -5, -6 \pmod{19}$. But these are all quadratic nonresidues mod 19, which is a contradiction.*

Extension Exercise 96. *Prove that if $n > 4$ is composite, then $(n-1)!$ is a multiple of n.*

Extension Solution 96. *Note that $(n-1)! = (n-1) \cdot (n-2) \cdot (n-3) \cdots 2 \cdot 1$, so it will contain all the prime factors of the integers from 1 to $n-1$, inclusive. If n is composite, it consists of at least two prime factors less than n, hence they must be contained in the expansion of $(n-1)!$*

Extension Exercise 97. *(2000 IMO Shortlist N4) Find all triplets of positive integers (a,m,n) such that $a^m + 1 \mid (a+1)^n$.*

Extension Solution 97. *The answer is whenever $\boxed{a=1}$ or $\boxed{m=1}$ or $\boxed{(2,3,k), k \geq 1}$.*

Clearly, the first two cases work. For the last case, we first show m must be odd.

Suppose we have $p \mid a^m + 1$ for $p > 2$ prime. Then $p \mid (a+1)^n \implies p \mid a+1$. But then $a \equiv -1 \pmod p$, so $a^m + 1 \equiv (-1)^m + 1 \pmod p \implies m$ odd. If no p exists, then set $a^m + 1 = 2^k$ for some $k \in \mathbb{Z}$. But $a, m \geq 2$ by assumption, so then $a^m + 1 \geq 5$ and $k \geq 3$. Hence, $a^m \equiv -1 \pmod 4$. But this is a contradiction, since m is even and perfect squares cannot be congruent to $3 \pmod 4$. Thus, m is odd.

Suppose $p \mid m$; then $a^p + 1 \mid (a^m + 1) \implies a^p + 1 \mid (a+1)^n$. But p is odd, so $a + 1 \mid (a^p + 1)$ and $\dfrac{a^p + 1}{a+1} = a^{p-1} - a^{p-2} + \ldots - p + 1 \mid (a+1)^{n-1}$. Let $f(x)$ correspond to this expansion (with a replaced by x). We obtain $f(-1) = p$, hence $f(x) - p = (x+1)g(x)$ for some polynomial $g(x)$. Put $x = a$ to obtain $f(a) = (a+1)g(a) + p$. If q prime divides $f(a)$, then $f(a) \mid (a+1)^{n-1}$ implies that $q \mid (a+1)$. Thence $q \mid p$, so $q = p$ by construction. Thus, $f(a) = p^k$ for $k \geq 1$. As $f(a) \mid (a+1)^{n-1}$, $p \mid (a+1)$.

We also show $k = 1$, so that $f(a) = p$. If $p^2 \mid (a+1)$, $f(a) = (a+1)g(a) + p \implies p^2 \nmid f(a)$. Otherwise, $a + 1 = pm$, $p \nmid m$. So $f(a) = \dfrac{a^p + 1}{a + 1} \equiv p \pmod{p^2}$. Then $f(a) = p$.

Since $a^2 - a \geq 2, a^4 - a^3 \geq 8$ and $a^{2r} - a^{2r-1} \geq 2$ for $r \geq 2$. Thus, $f(a) \geq 11$ for $p = 5$, which shows that $p = 3$ is our only possibility. Hence $a^2 - a + 1 = 3 \implies a = 2$.

Finally, m must be 3, as $2^9 + 1, 2^{27} + 1, 2^{81} + 1, \ldots$ are not powers of 3. Thus, $9 \mid (2+1)^n$, so $n \geq 2$. This completes the proof.

Extension Exercise 98. *Find all ordered pairs of integers (m,n) such that $m^4 - n! = 2016$.*

Extension Solution 98. *For all $n \geq 5$, $120 \mid n!$; note that $2016 \equiv 36 \pmod{120}$. Hence, for $n \geq 5$, $m^4 \equiv 36 \pmod{120}$. But this congruence has no integer solutions by CRT. Indeed, even among trivial possibilities, there are no solutions, as $6^4 = 1296$, $7^4 = 2401$.*

Extension Exercise 99. *(1976 USAMO P3) Determine all integral solutions of $a^2 + b^2 + c^2 = a^2 b^2$.*

Extension Solution 99. *If $a^2 = 0$, then $b^2 = c^2 = 0$ as well (and same for $b^2 = 0$). Otherwise, if $a^2 \neq 0$ and $b^2 \neq 0$, then necessarily $c^2 \neq 0$.*

Consider $a^2 \pmod 4$. If $a^2 \equiv 0 \pmod 4$, then $b^2, c^2 \equiv 0 \pmod 4$ as well (since all perfect squares are $0, 1 \pmod 4$). Let $a = 2a_1, b = 2b_1, c = 2c_1$; then $a_1^2 + b_1^2 + c_1^2 = 4a_1^2 b_1^2$. But by infinite descent, there are no solutions in this case. Similarly, if $a^2 \equiv 1 \pmod 4$, since $b^2 \neq 0 \pmod 4$, we must have $b^2 \equiv 1 \pmod 4$, and so $c^2 \equiv 3 \pmod 4$ - contradiction. Hence the only solution is $(a, b, c) = (0, 0, 0)$.

Section 14.2: Primality Theorems

Exercise 233. *(2016 ARML Local Individual Round #6) Compute the remainder when $100!$ is divided by 103.*

Solution 233. *By Wilson's Theorem, $102! \equiv -1 \equiv 102 \pmod{103} \implies 101! \equiv 1 \pmod{103}$. It remains to find the value of $x \in [0, 102]$ such that $101x \equiv 1 \pmod{103}$. As $101 \equiv -2 \pmod{103}$, $x = \boxed{51}$.*

Exercise 234. *What is the remainder when 3^{2019} is divided by 19?*

Solution 234. *By Fermat's Little Theorem, $3^{19} \equiv 3 \pmod{19}$, so $3^{18} \equiv 1 \pmod{19} \implies 3^{2016} \equiv 1 \pmod{19}$. In turn, $3^{2019} \equiv 3^3 = 27 \equiv \boxed{8} \pmod{19}$.*

Exercise 235. *Compute $2^{20} + 3^{30} + 4^{40} + 5^{50} + 6^{60} \pmod{7}$.*

Solution 235. *By symmetry, note that $k^{20} + (7-k)^{20} \equiv 0 \pmod 7$; adding 1 to our expression produces this, so the answer is $\boxed{6}$.*

Exercise 236. *(2008 AMC 12A) Let $k = 2008^2 + 2^{2008}$. What is the units digit of $k^2 + 2^k$?*

Solution 236. *We have $k \equiv 4 + 6 \equiv 0 \pmod{10}$. Hence, $k^2 \equiv 0 \pmod{10}$. Additionally, $k \equiv 0 \pmod 4$ as well, so $2^k \equiv 6 \pmod{10}$. This makes the answer $\boxed{6}$.*

Exercise 237. *(2002 AMC 12B) The positive integers $A, B, A-B$, and $A+B$ are all prime numbers. What is their sum?*

Solution 237. *Note that $A + B$ and $A - B$ are of the same parity, and so they must both be odd. If B were odd, then A would need to be even, but then $A + B = 4$ and $A - B = 0$; contradiction. Thus, B is even $\implies B = 2$, and A is odd. Then $A - 2, A, A + 2$ are consecutive odd primes, which implies $A = 5$ (as $3, 5, 7$ is the only triple of consecutive odd primes). Thus, the sum, $3A + B = 17$.*

Exercise 238. *Let p be a prime number, and let k be a positive integer. Prove that the decimal representation of $\frac{k}{p}$ consists of $d \mid (p-1)$ repeating digits.*

Solution 238. *Consider $\frac{1}{p}$ in base b, with $\gcd(p, b) = 1$. We want to multiply the denominator/numerator by k so that the denominator becomes $b^e - 1$ for an exponent e.*

Consider $1, b, b^2, \ldots \pmod{n}$. This sequence clearly repeats, so b does have multiplicative order e. Thus, $b^e - 1 = kp$ for some k.

We conclude that $\frac{1}{p} = \frac{k}{kp} = \frac{k}{b^e - 1} = p^{-e} \cdot \frac{k}{1 - b^{-e}} = kb^{-e} + kb^{-2e} + \ldots$. Since $k < b^e$, there are fewer than e digits in base b. Hence this proves the statement.

One may also observe that the multiplicative group of units modulo n has $\phi(n)$ elements, so the multiplicative order of $e \bmod n$ of b must divide $\phi(n)$. (In particular, for prime p, $\phi(p) = p - 1$.)

Exercise 239. *Show that the converse of Fermat's Little Theorem is false.*

Solution 239. *A counterexample is $2^{10} - 1 = 341 \cdot 3$, with $2^{10} \equiv 1 \pmod{341}$. But 341 is not prime, hence this contradicts FLT.*

Exercise 240. *(2017 HMMT February Guts Round) At a recent math contest, Evan was asked to find $2^{2016} \pmod{p}$ for a given prime number p with $100 < p < 500$. Evan has forgotten what the prime p was, but still remembers how he solved it:*

- *Evan first tried taking 2016 modulo $p - 1$, but got a value e larger than 100.*
- *However, Evan noted that $e - \frac{1}{2}(p - 1) = 21$, and then realized the answer was $-2^{21} \pmod{p}$.*

What was the prime p?

Solution 240. Let $p = 2k+1$, $50 < k < 250$. Then $2016 \equiv k + 21 \pmod{2k}$, so p is a factor of 1995. Then we obtain $p = 191, 211$ from the factorization of 1995, and of these $p = \boxed{211}$ is the correct solution (using quadratic reciprocity).

Extension Exercise 100. *(1989 IMO P5) Prove that for each positive integer n, we can find a set of n consecutive integers such that none of them is a power of a prime number.*

Extension Solution 100. Consider the set $\{N!^2 + 2, N!^2 + 3, \ldots N!^2 + N\}$. Note that $N!^2 + r$ is divisible by r (since $N!^2$ is), but $\dfrac{N!^2 + r}{r} = N! \cdot \dfrac{N!}{r} + 1 > 1$, but relatively prime with r. For each r, let prime $p_r \mid r$, so that $p_r \mid N!^2 + r$, but is not a power of p_r. Take $N = n + 1$ to obtain the desired result.

Extension Exercise 101. *(2003 IMO Shortlist N6) Let p be a prime number. Prove that there exists a prime number q such that for every integer n, the number $n^p - p$ is not divisible by q.*

Extension Solution 101. We have $\dfrac{p^p - 1}{p - 1} = p + p^2 + p^3 + \ldots + p^{p-1}$, which is congruent to $p + 1 \pmod{p^2}$. Thus we may obtain at least one prime divisor which is not congruent to $1 \pmod{p^2}$; let this divisor be q. Indeed, we show that this is the desired q.

Assume that there exists $n \in \mathbb{N}$ with $n^p \equiv p \pmod{q}$. Then $n^{p^2} \equiv p^p \equiv 1 \pmod{q}$. By FLT, $n^{q-1} \equiv 1 \pmod{q}$ since q is prime. Because $p^2 \nmid (q-1)$, we have $\gcd(p^2, q-1) \mid p$, so $n^p \equiv 1 \pmod{q} \implies p \equiv 1 \pmod{q}$. However, this leads to $1 + p + p^2 + \ldots + p^{n-1} \equiv p \pmod{q}$. This, in turn, implies that $p \equiv 0 \pmod{q}$, which is a contradiction.

Extension Exercise 102. *(2016 HMMT February Algebra Test) Define $\phi^!(n)$ as the product of all positive integers less than or equal to n and relatively prime to n. Compute the number of integers $2 \leq n \leq 50$ such that n divides $\phi^!(n) + 1$.*

Extension Solution 102. If $\gcd(k, n) = 1$, then there exists a unique value of k^{-1} between 1 and $n - 1$ such that $kk^{-1} \equiv 1 \pmod{n}$. Thus, if $k^2 \not\equiv 1 \pmod{n}$, then pair k with its inverse to get a product of 1.

On the other hand, if $k^2 \equiv 1 \pmod{n}$, then by symmetry, $(n - k)^2 \equiv 1 \pmod{n}$ as well; then $k(n - k) \equiv -k^2 \equiv -1 \pmod{n}$. Hence these k can also be paired up. When $n \neq 2$, there is no k with $k^2 \equiv 1 \pmod{n}$, $k \equiv 2 - k \pmod{2}$, so the product becomes $(-1)^{\frac{m}{2}}$, where m is the number of k with $k^2 \equiv 1 \pmod{n}$.

For p a prime, and i some positive integer, $k^2 \equiv 1 \pmod{p^i}$ has 2 solutions for p odd, 4 solutions in the case $p = 2$ and $i \geq 3$, and 2 solutions where $p = i = 2$. By CRT, we require $n = p^k, 2p^k, 4$. This yields $\boxed{30}$ as the answer.

Section 14.3: Symmetry and Quadratic Residues/Reciprocity

Exercise 241. Let $\left(\dfrac{a}{n}\right) = 1$ *(Jacobi symbol). Does it follow that a is a quadratic residue* (mod n)?

Solution 241. *No, this does not follow. To see why, split n up into at least two prime factors. Then the Legendre symbol may evaluate to -1 at both of these factors, but to $+1$ for the product of those factors.*

Exercise 242. *Calculate* $\left(\dfrac{1001}{2019}\right)$.

Solution 242. $\left(\dfrac{1001}{2019}\right) = \left(\dfrac{7}{2019}\right) \cdot \left(\dfrac{11}{2019}\right) \cdot \left(\dfrac{13}{2019}\right) = 1 \cdot 1 \cdot 1 = \boxed{1}$, *taking advantage of multiplicativity of the Legendre function (proven below).*

Exercise 243. *Show that* $\left(\dfrac{a}{p}\right)$ *(Legendre symbol) is a multiplicative function.*

Solution 243. *This follows from Euler's criterion; i.e. the identity* $(ab)^{\frac{p-1}{2}} = a^{\frac{p-1}{2}} b^{\frac{p-1}{2}}$.

Exercise 244. *Explain how we can test whether a number is prime using the Jacobi symbol.*

Solution 244. *Note that the Legendre symbol satisfies Euler's criterion; however, this does not necessarily hold for the Jacobi symbol. The relevance of this observation is that, if we test a number n for primality, if there is an integer a with $\gcd(a,n) = 1$ such that $\left(\dfrac{a}{n}\right) \neq a^{\frac{n-1}{2}}$ (mod n), then n cannot be prime. (This is known as the Solovay-Strassen primality test.)*

Exercise 245. *Find all solutions to the modular congruence $x^2 + 20x - 19 \equiv 0$ (mod 2017), or show that none exist.*

Solution 245. *Add 38 to both sides to get $(x+19)(x+1) \equiv 38$ (mod 2017). Note that this does not have any integer solutions, as 38 fails Euler's Criterion.*

Exercise 246. *Find all primitive roots* (mod 7).

Solution 246. *Recall that g is a primitive root* (mod n) *iff, for every integer a with $\gcd(a,n) = 1$, there is an integer k with $g^k \equiv a$ (mod n). Hence, we require $g^k \equiv a$ (mod 7). The values of g are* $\boxed{g = 2, 3, 4, 5}$, *i.e. the values that, when exponent-cycled, "cover" all values between 0 and $n-1$, inclusive.*

Extension Exercise 103. *Let g and h be primitive roots of p. Can gh be a primitive root of p?*

Extension Solution 103. *We claim this is not possible. If g, h are primitive roots, then $g^{p-1} \equiv h^{p-1} \equiv 1$, but then $g^{\frac{p-1}{2}} = h^{\frac{p-1}{2}} = -1$. Hence, $(gh)^{\frac{p-1}{2}} = g^{\frac{p-1}{2}} h^{\frac{p-1}{2}}$, so gh cannot be a primitive root.*

Extension Exercise 104. *(2018 Berkeley Math Tournament) Evaluate*

$$\prod_{j=1}^{50} \left(2\cos\left(\frac{4\pi j}{101}\right) + 1\right).$$

Extension Solution 104. *Define $n = \dfrac{2\pi}{101}$, $q = e^n$; then the term in the product is $q^{2j} + q^{-2j} + 1 = \dfrac{q^{3j} - q^{-3j}}{q^j - q^{-j}}$. 3 then "sends elements from 1 to 50 to a or $-a$, $a \in \{1, 2, 3, \ldots, 50\} = S$. If $3i \equiv 3j$ (mod 101) $\implies i \equiv j$ (mod 101), and similarly for their negatives, i, j cannot both be in S. Hence, $3S \equiv S$ (with respect to negatives). If $j \equiv -i$ (mod 101), $q^{3j} - q^{-3j} = -(q^{3i} - q^{-3i})$.*

Hence, the product is equal to the product of negative signs in $3S$. Note that the multiples of 3 are all positive, $3 \cdot 17, 3 \cdot 19, 3 \cdot 20, 3 \cdot 21, \ldots, 3 \cdot 33$ are all negative in S^c, and $3 \cdot 34, 3 \cdot 37, \ldots, 3 \cdot 50$ are all positive. Hence, there are $33 - 17 + 1 = 17$ negative signs and the product is $\boxed{-1}$.

Chapter 14 Review Exercises

Exercise 247. *For which prime numbers p is $\left(\dfrac{2^p}{p}\right) = 1$?*

Solution 247. *p must be odd; then we can take advantage of the multiplicative property to simply compute the symbol with an upper argument of 2. When this is equal to 1, p is a satisfactory prime; otherwise, it is not.*

Exercise 248. *(2002 AIME I) Harold, Tanya, and Ulysses paint a very long picket fence. Harold starts with the first picket and paints every h^{th} picket; Tanya starts with the second picket and paints every t^{th} picket; and Ulysses starts with the third picket and paints every u^{th} picket. Call the positive integer $100h + 10t + u$ paintable when the triple (h,t,u) of positive integers results in every picket being painted exactly once. Find the sum of all the paintable integers.*

Solution 248. *Note that $h, t, u \neq 1$. Furthermore, $h \neq 2$, as we will then have the third picket painted twice. If $h = 3$, $t, u \not\equiv 0 \pmod 3$; for fourth/fifth pickets to be painted, however, 333 does work. If $h = 4$, $t, u \equiv 0 \pmod 2$, but $u \not\equiv 2 \pmod 4$. Hence, $t \equiv 2 \pmod 4, u \equiv 0 \pmod 4$. Then $t = 2, u = 0$ for 5 and 6 to be paintable. Thus 424 is also paintable. The answer is $333 + 424 = \boxed{757}$.*

Exercise 249. *(2012 Stanford Math Tournament) Find the sum of all integers x, $x \geq 3$, such that 201020112012_x (that is, 201020112012 interpreted as a base x number) is divisible by $x - 1$.*

Solution 249. *Observe that $x \equiv -1 \pmod{x+1}$, and so $x^n \equiv 1 \pmod{x-1}$. Then the number in base x is congruent to its digit sum, i.e. 12 $\pmod{x-1}$. Then this reduces to $(x-1) \mid 12$, i.e. $x = 3, 4, 5, 7, 13 \implies \boxed{32}$.*

Exercise 250. *(2012 Winter OMO) Find the largest prime number p such that when $2012!$ is written in base p, it has at least p trailing zeroes.*

Solution 250. *We require $\lfloor \dfrac{2012}{p} \rfloor + \lfloor \dfrac{2012}{p^2} \rfloor + \ldots \geq p$ by Legendre's formula. Testing prime values near $44 < \sqrt{2012} < 45$ reveals that $p = \boxed{43}$ is indeed the value we seek.*

Exercise 251. *(2015 HMMT February Guts Round) Find the least positive integer $N > 1$ satisfying the following two properties:*

- *There exists a positive integer a such that $N = a(2a1)$.*
- *The sum $1 + 2 + \ldots + (N-1)$ is divisible by k for every integer $1 \leq k \leq 10$.*

Solution 251. *From the second condition, we write $16 \mid a(2a-1)(2a^2 - a - 1) \implies a \equiv 0, 1 \pmod{16}$. But $a = 1$ would result in $N = 1$, a contradiction to the problem statement. Hence, $a = 32$ is the minimum ($a = 16, 17$ also produce contradictions similarly), so $1 + 2 + \ldots + (N-1) = \boxed{2016}$.*

Exercise 252. *Show that an integer n has $\left(\dfrac{n}{p}\right) = 1$ for all prime numbers p iff n is a perfect square.*

Solution 252. *By definition of the quadratic residue/Legendre symbol, we require there to exist a solution to the congruence $k^2 \equiv n \pmod p$ for some $k \in \mathbb{Z}$. This is trivially true by construction, as we have assumed n is a perfect square. (In essence, this exercise boils down to knowing your definitions!)*

Exercise 253. *Show that there exists $n \in \mathbb{N}$ such that $n < \sqrt{p} + 1$ is a quadratic nonresidue modulo p.*

Solution 253. *Note that $n^2 \equiv (p-n)^2 \pmod p$, so all quadratic residues are determined by the values of the Legendre symbol up to $\sqrt{p} + 1$. Since every prime p has a quadratic non-residue, the desired result follows.*

Exercise 254. *(2014 Fall OMO) Find the sum of all positive integers n such that $\tau(n)^2 = 2n$, where $\tau(n)$ is the number of positive integers dividing n.*

Solution 254. We can write $n = 2k^2$, $k \in \mathbb{N}$. Then note that $\frac{\tau(n)^2}{n}$ is multiplicative (since $\tau(n)$ itself is multiplicative). Then the power of 2 in n is odd, so $g(2) = g(8) = 2$, but $g(2^i) < 2$ for $i \geq 5$. We can also verify that $g(p^{2j}) \leq 1$ (1 exactly when $j = 0, 1$ and $p = 3$). Hence, $g(n) = 2$ for $n = 2, 8, 18, 72 \implies \sum n = \boxed{100}$.

Exercise 255. *(2008 AIME II)* Find the largest integer n satisfying the following conditions:

(i) n^2 can be expressed as the difference of two consecutive cubes;

(ii) $2n + 79$ is a perfect square.

Solution 255. Let $m, m+1$ be the consecutive cubes in question. Completing the square, we get $(2m)^2 - 3(2m+1)^2 = 1 \equiv a^2 - 3b^2$, which is a Pell equation on the RHS. Then the solutions are of the form $(2 + \sqrt{3})^k = a_k + b_k\sqrt{3}$, with a even and b odd. By induction, a, b switch parities, so the solutions are $(0, 2), (1, 26), (2, 362), (3, 5042)$. Hence $2n = 362 \implies n = \boxed{181}$, and we are done.

Exercise 256. *(2015 Fall OMO)* Toner Drum and Celery Hilton are both running for president. A total of 2015 people cast their vote, giving 60% to Toner Drum. Let N be the number of "representative sets of the 2015 voters that could have been polled to correctly predict the winner of the election (i.e. more people in the set voted for Drum than Hilton). Compute the remainder when N is divided by 2017.

Solution 256. Let $m = 806$ people vote for Hilton, and let $n = 1209$ people vote for Drum. Then we can choose the voters with p more people voting for Drum than Hilton in $\sum_{i=0}^{m} \binom{m}{i}\binom{n}{i+p}$ ways, which by Vandermondes is $\binom{m+n}{m+p}$. Summing over all p yields $\sum_{m+1}^{m+n} \binom{m+n}{i}$. Since $m + n = 2015$,

$$\binom{m+n}{i} \equiv \binom{2015}{i} \equiv \frac{i+1}{2016}\binom{2016}{i+1} \equiv \frac{i+1}{2016} \cdot (-1)^{i+1} \equiv (i+1) \cdot (-1)^{i+1}$$

The sum finally simplifies to $-808 + 809 - 810 + \ldots - 2016 \equiv \boxed{605} \pmod{2017}$.

Extension Exercise 105. *(1998 IMO Shortlist N2)* Determine all pairs (a, b) of real numbers such that $a\lfloor bn \rfloor = b\lfloor an \rfloor$ for all positive integers n.

Extension Solution 105. The solutions are $\boxed{a = b}$, either $\boxed{a = 0}$ or $\boxed{b = 0}$, or $\boxed{a, b \in \mathbb{Z}}$. It is easy to check that these are, in fact, solutions.

Henceforth, show that no more exist. Assume $a, b \neq 0, a \neq b, a, b \notin \mathbb{Z}$. Then induction reveals that $\lfloor 2^n a \rfloor = 2^n \lfloor a \rfloor$ (the same holds for b). If we let $n = 1$, then we obtain $2\lfloor a \rfloor \leq 2a < 2\lfloor a \rfloor + 2$, so $\lfloor 2a \rfloor = 2\lfloor a \rfloor$ or $2\lfloor a \rfloor + 1$. In the latter case, $a\lfloor 2b \rfloor = b\lfloor 2a \rfloor = 2b\lfloor a \rfloor + b = 2a\lfloor b \rfloor + b$. Hence $\lfloor 2b \rfloor = 2\lfloor b \rfloor + \frac{b}{a} \implies a = b$, contradiction. So $\lfloor 2a \rfloor = 2\lfloor a \rfloor$ and $\lfloor 2b \rfloor = 2\lfloor b \rfloor$ (i.e. the result holds for $n = 1$).

Completing the inductive step, we have $a\lfloor 2^{n+1}b \rfloor = b\lfloor 2^{n+1}a \rfloor$. Repeat the same logic from before; we again obtain $a = b$, so contradiction. Hence, a, b must be integers, but this is also a contradiction, so there are no other solutions.

Extension Exercise 106. *(1998 IMO Shortlist N6)* For any positive integer n, let $\tau(n)$ denote the number of its positive divisors (including 1 and itself). Determine all positive integers m for which there exists a positive integer n such that $\tau(n^2)\tau(n) = m$.

Extension Solution 106. Let $n = p_1^{k_1} p_2^{k_2} \cdots p_r^{k_r}$. Then $\tau(n) = (k_1 + 1)(k_2 + 1)\cdots(k_r + 1)$. Let $m_i = k_i + 1$; then we must find all natural numbers N such that $N = \frac{2m_1 - 1}{m_1} \cdot \frac{2m_2 - 1}{m_2} \cdots \frac{2m_r - 1}{m_r}$. The numerator $\tau(n^2)$ is odd, so N must also be odd. We claim that all odd N have a representation in this form.

For $N = 1$, this is obvious. For every odd $N = 2k - 1$ with k odd, set $N = \dfrac{2k-1}{k} \cdot k$, with k having a representation of the desired form. Furthermore, if $N \equiv 3 \pmod{4}$, then this approach also works. However, it does not work for k even.

If this is the case, suppose $N = 2^t k - 1$, where k is odd. We then write

$$N = \frac{2^t(2^t-1)k - (2^t-1)}{2^{t-1}(2^t-1)k - (2^{t-1}-1)} \cdots \frac{2(2^t-1)k - 1}{(2^t-1)k}$$

completing the inductive step and the proof.

Extension Exercise 107. *(1994 IMO P4) Find all ordered pairs (m, n) where m and n are positive integers such that $\dfrac{n^3+1}{mn-1}$ is an integer.*

Extension Solution 107. Suppose $k = \dfrac{n^3+1}{mn-1} \in \mathbb{Z}$. Then $k(mn-1) = n^3+1$ and $k \equiv -1 \pmod{m}$. Let $k = jn - 1$ for $j \in \mathbb{N}$; then $n^3 + 1 = (mn-1)(jn-1) = mjn^2 - n(m+j) + 1$ which yields $n^2 = mjn - (n+j) \implies n^2 - mjn + m + j = 0$. This is a quadratic with a root of n; let p be the other root. Then $np = m + j$ by Vietas and $j, m, n, p = 2$ is a trivial solution.

Otherwise, we have $np > n + p, mj > m + j$, but this contradicts the equations $np = m + j, n + p = mj$. Thus, at least one of the variables equals 1.

If $m = 1$ or $j = 1$, WLOG assume $j = 1$. Then $np = m+1, n+p = m$; i.e. $(n-1)(p-1) = 2$ by SFFT, which leads to $(n,p) = (2,3), (3,2)$. Thus, $m = 5$ or $(m,j) = (1,5)$.

If $n = 1$ or $p = 1$, repeat the procedure above to obtain $n = 5$, $(n,p) = (1,5)$. This yields all solutions.

Extension Exercise 108. *(2004 IMO Shortlist N2) The function ψ from the set $\mathbf{N}$ of positive integers to itself is defined by the equality*

$$\psi(n) = \sum_{k=1}^{n}(k,n), n \in \mathbb{N}$$

where (k, n) denotes the greatest common divisor of k and n.

(a) Prove that $\psi(mn) = \psi(m)\psi(n)$ for every two relatively prime $m, n \in \mathbf{N}$.

(b) Prove that for each $a \in \mathbf{N}$ the equation $\psi(x) = ax$ has a solution.

(c) Find all $a \in \mathbf{N}$ such that the equation $\psi(x) = ax$ has a unique solution.

Extension Solution 108. Let $n \in \mathbb{N}$, and consider $k \in [1,n]$. Then $\gcd(k,n) \mid n$, and consider $d \mid n$ in general. If $\gcd(k,n) = \dfrac{n}{d}$, $k = \dfrac{nl}{d}$ for some $l \in \mathbb{N}$, and $\gcd(k,n) = \dfrac{n}{d} \cdot \gcd(l,d)$. Thus, $\gcd(l,n) = 1$ with $l \leq n$. Thus,

$$\psi(n) = \sum_{k=1}^{n} \gcd(k,n) = n \sum_{d \mid n} \frac{\phi(d)}{d}$$

For part (a), let $\gcd(m,n) = 1$. Then for each divisor $f \mid mn$, we can write $d = pq$, where $p \mid n, e \mid m$. Thus it follows that

$$\psi(mn) = mn \sum_{p \mid n, q \mid m} \frac{\phi(pq)}{pq}$$

$$= mn \sum_{p \mid n, q \mid m} \frac{\phi(p)}{p} \frac{\phi(q)}{q}$$

$$= \left(n \sum_{p|n} \frac{\psi(p)}{p}\right)\left(m \sum_{q|m} \frac{\phi(q)}{q}\right) = \psi(m)\psi(n)$$

as desired.

For part (b), let $n = p^k$ for p prime and $k \in \mathbb{N}$. From our preliminary work, we have $\frac{\psi(n)}{n} = \sum_{i=0}^{k} \frac{\phi(p^i)}{p^i} = 1 + \frac{k(p-1)}{p}$. $p = 2, k = 2a - 2$ yields $\psi(n) = an$ for $n = 2^{2a-2}$.

For part (c), observe that p prime $\implies \psi(p^p) = p^{p+1}$. Then $x = p^p \cdot 2^{2a_1 - 2}$ is a solution if a has an odd prime factor of p (where $a_1 p = a$). Assume henceforth that a is a power of 2, namely $a = 2^k$ for $k \in \mathbb{N}$. Then suppose $x = 2^r y \in \mathbb{N}$ such that $\psi(x) = 2^k \cdot x$. Then $2^{r+k} y = \psi(x) = (a+2)\psi(y)$. For y odd, $\psi(y)$ is also odd, with $\psi(y) \mid y$. We must have $y = 1$, so $r = 2a - 2$ and $x = 2^{2a-2}$ is unique. Hence, $\psi(x) = ax$ has a unique solution iff a is a power of 2.

Chapter 15: Advanced Number Theory

Section 15.1: Lifting the Exponent

Exercise 257. Show that $v_p(mn) = v_p(m) + v_p(n)$ for all primes p and positive integers m, n.

Solution 257. If $a = p^u \cdot m$ and $b = p^v \cdot n$ with $\gcd(m, p) = \gcd(n, p) = 1$, then $ab = p^{u+v} \cdot mn$ with $\gcd(mn, p) = 1$.

Exercise 258. Show that $v_p(m + n) = \min\{v_p(m), v_p(n)\}$ for all prime p, positive integers m, n, and whenever $v_p(m) \neq v_p(n)$.

Solution 258. *Hint:* If $u \leq v$ (using the same notation as above), then $a + b = p^u(m + p^{v-u}(n))$.

Exercise 259. Evaluate $v_2(2020^{2019} + 20^{2019})$.

Solution 259. $v_2(2020^{2019} + 20^{2019}) = v_2(20^{2019} \cdot (101^{2019} + 1)) = 4038 + 1 = \boxed{4039}$.

Exercise 260. Let n be a squarefree integer. Show that there does not exist a pair of positive integers (x, y) with $\gcd(x, y) = 1$ such that $(x + y)^3$ is a factor of $x^n + y^n$.

Solution 260. *For contradiction, assume that $(x+y)^3 \mid (x^n + y^n)$ with x, y relatively prime. Suppose that n is even. If there is an odd prime $p \mid (x+y)$, then $x^n + y^n \equiv 2x^n \pmod{p}$, so $p \mid x$ and $p \mid y$, which is a contradiction. As $x, y > 0$, $x + y$ must be a power of 2. In this case, x, y are odd (since $\gcd(x,y) = 1$), so $x^n \equiv y^n \equiv 1 \pmod 8$, and so $v_2((x+y)^3) \geq 3$, which is a contradiction since $v_2(x^n + y^n) = 1$.*

Henceforth, assume n is odd. Assuming the existence of an odd prime $p \mid (x + y)$, we have $3v_p(x + y) \leq v_p(x^n + y^n)$. By LTE, $3v_p(x+y) \leq v_p(x^n + y^n) = v_p(x+y) + v_p(n) \leq v_p(x+y) + 1$ as n does not have a perfect square in its prime factorization. Then we obtain $v_p(x+y) \leq \dfrac{1}{2}$, which is absurd.

Then it must be the case that $x + y$ is a power of 2, but this yields $v_2(x^n + y^n) = v_2(x+y)$ (hint: consider $\dfrac{x^n + y^n}{x + y}$). So, to summarize, we cannot have $(x+y)^3 \mid (x^n + y^n)$, since $v_2((x+y)^3) > v_2(x^n + y^n)$.

Exercise 261. Prove that 2 is a primitive root of 3^k for all $k \in \mathbb{N}$.

Solution 261. *We seek the smallest $n > 0$ with $2^n \equiv 1 \pmod{3^k}$. If $3^k \mid (2^n - 1)$, this implies $2^n \equiv 1 \pmod 3 \implies n$ is even. Then $n = 2m$ and $3^k \mid (4^m - 1)$. Then using LTE, $k \leq v_3(4^m - 1^m) = v_3(3) + v_3(m) = 1 + v_3(m)$. Hence, $v_3(m) \geq k - 1$ and $m \geq 3^{k-1}$. Thus, the smallest n is $2 \cdot 3^{k-1} = \phi(3^k)$, as desired.*

Exercise 262.

(a) Prove Kummer's Theorem:

> **Theorem 17.6: Kummer's Theorem**
>
> Given integers $n \geq m \geq 0$ and prime p,
> $$\nu_p\left(\binom{n}{m}\right)$$
> equals the number of carries when m is added to $n - m$ in base p.

(b) Generalize Kummer's Theorem to the multinomial coefficient. You do not need to prove this generalized form.

Solution 262.

(a) Write $\frac{n}{m} = \frac{n!}{m!(n-m)!}$, then apply Legendre's formula. Specifically, by Legendre's, we can write

$$v_p\left(\frac{n}{m}\right) = v_p((m+n)!) - v_p(m!) - v_p(n!)$$

$$= \frac{m+n-s_p(m+n)}{p-1} - \frac{m-s_p(m)}{p-1} - \frac{n-s_p(n)}{p-1}$$

$$= \frac{s_p(m) + s_p(n) - s_p(m+n)}{p-1}$$

which is the sum of coefficients in the below expansion of n.

(b) Write n in its base-p form, $n = n_0 + n_1 p + n_2 p^2 + \ldots + n_r p^r$; define $S_p(r) = n_0 + n_1 + n_2 + \ldots + n_r$. Then

$$v_p\left(\binom{n}{m_1, m_2, \ldots, m_k}\right) = \frac{1}{p-1}\left(\sum_{i=1}^{k} S_p(m_i) - S_p(n)\right)$$

Extension Exercise 109. *(2000 IMO P5)* Does there exist a positive integer n such that n has exactly 2000 prime divisors and n divides $2^n + 1$?

Extension Solution 109. We claim that such an n exists. For each $k \in \mathbb{N}$, define $n(k) \in \mathbb{N}$ with $n \mid 2^n + 1, 3 \nmid n$, and n has exactly k prime factors.

We have $n(1) = 3$; and we proceed by induction. n is odd, and since $2^{3n} + 1 = (2^n + 1)(2^{2n} - 2^n + 1)$ and $3 \nmid 2^{2n} - 2^n + 1$, so $3n \nmid 2^{3n} + 1$. For m odd, we have $2^{3n} + 1 \mid 2^{3mn} + 1$. Thus, for p prime with $p \nmid n, p \mid 2^{3n} + 1, 3np \mid (2^{3np} + 1, n(k+1) = 3np$. We claim that, for any $a > 2, a \equiv 2 \pmod{3}, \exists p$ such that $p \mid a^3 + 1, p \nmid a + 1$. Assume otherwise. Since $a^3 + 1 = (a+1)(a^2 - a + 1)$, each prime factor of the second factor is also a factor of $a + 1$. In turn, we have $a^2 - a + 1 = (a+1)(a-2) + 3$, we have $a^2 - a + 1$ a power of 3. Then $9 \nmid a^2 - a + 1$, but this is a contradiction.

Extension Exercise 110. If x and y are positive reals such that $x - y$, $x^2 - y^2$, $x^3 - y^3$, $\ldots$ are all positive integers, prove that x and y are themselves positive integers.

Extension Solution 110. As $x - y \in \mathbb{Z}$ and $x^2 - y^2 \in \mathbb{Z}$, $x + y \in \mathbb{Q}$. Then $x = \frac{(x-y)+(x+y)}{2}$ and $y = \frac{(x+y)-(x-y)}{2}$ are rational, which means that we may write $x = \frac{a}{z}, y = \frac{b}{z}$. Then minimize z such that $z^n \mid (a^n - b^n)$.

Let $p \mid z$, and note that $p \mid (a-b)$ as a result of $z \mid (a-b)$. Then $p \mid a \implies p \mid b$, but then z would not be minimal: contradiction. Hence $p \nmid a, b$, which primes the problem for an application of LTE. For $p > 2$ odd, write $n \leq v_p(z^n) = v_p(a^n - b^n) = v_p(a-b) + v_p(n)$. Then $p^n \leq n(a-b), \frac{p^n}{n} \leq a-b$. But since $\lim_{n \to \infty} \frac{p^n}{n} = \infty$, this is clearly impossible. If $p = 2$, on the other hand, $n \leq v_2(z^n) = v_2(a^n - b^n) = v_2(a-b) + v_2(a) + v_2(a+b) - 1$, so $\frac{2^{n+1}}{n} \leq (a-b)(a+b)$ which is also a contradiction. Hence $p \nmid z$, and $z = 1$ with $a, b \in \mathbb{N}$.

Extension Exercise 111. *(1990 IMO P3)* Determine all integers $n > 1$ such that $\frac{2^n + 1}{n^2}$ is an integer.

Extension Solution 111. We observe that $n = 3$ is a solution. To prove that it is the only solution, we apply LTE on $2^n + 1$ with $v_3(n) = k$ to get $v_3(2^n + 1) = 1 + k \geq v_3(n^2) = 2k$, which implies that $k = 1$. Hence n has a prime factor of 3, but not of any higher power of 3. Then let $n = 3t$, and assume $t \neq 1$ ($t = 1$ is trivial). Then we obtain a contradiction using CRT, namely $2^n + 1 \equiv 8^t + 1 \equiv 2 \pmod{7}$. (We are working with the 3-adic valuation because the prime p that serves as the base of the p-adic valuation $v_p(n)$ must the smallest prime divisor of n, and n cannot be even by a simple parity check.)

Extension Exercise 112. *(1991 IMO Shortlist P18) Find the highest degree k of 1991 for which 1991^k divides the number $1990^{1991^{1992}} + 1992^{1991^{1990}}$.*

Extension Solution 112. *The answer is* $\boxed{1991^{1991}}$.

For $h \geq 3$ odd and $k > 0$, we have $(1+h)^{h^k} + 1 + s_k h^{k+1}$, where $h \nmid s_k$. We induct upon k. For the base case, $k = 1$, the Binomial Theorem yields $(1+h)^h = 1 + h \cdot h + \dfrac{(h-1)h^3}{2} + O(h^3)$, so the result holds.

Letting $H = h^{k+1}$, we obtain $(1+h)^H = (1 + s_k h^{k+1})^h = 1 + h s_k h^{k+1} + \dfrac{(h-1)s_k^2 h^{2k+3}}{2} + O(h^{k+2})$, so the statement holds for $k+1$ as desired.

It follows that $v_{1991}(1992^n - 1) = 1991$ and $v_{1991}(1990^m + 1) = 1993$, so the answer is 1991^{1991}.

Section 15.2: Continued Fraction Representations and the Pell Equation

Exercise 263. *Find all integer solutions to the equation $x^2 - 7y^2 = 2$.*

Solution 263. *The smallest trivial solution is $(x,y) = (3,1)$, which generates a family of solutions (x_i, y_i) such that $x_i = x_{i-1}y_i + 7y_{i-1}y_1, y_i = x_{i-1}y_i + 7y_{i-1}x_1$.*

Exercise 264. *Let n be a nonzero integer. Show that if $x^2 - dy^2 = n$ has at least one solution (x,y), then it has infinitely many solutions.*

Solution 264. *By definition, the solution (x,y) generates an entire family of solutions $(x_1, y_1), (x_2, y_2), \ldots, (x_n, y_n)$. This essentially amounts to showing that the recurrence of a Pell equatio exists, which is just equivalent to showing that $(x_1 + y_1\sqrt{d})^2$ is an infinite sequence (this is trivial).*

Exercise 265. *What are the first three convergents of*

(a) e^2, *and*

(b) $\sqrt{5}$?

Solution 265. *Of e^2, they are given by $[7;2,1]$, i.e. $7, \dfrac{15}{2}, \dfrac{22}{3}$. Of $\sqrt{5}$, they are $2, \dfrac{9}{4}, 2 + \dfrac{4}{17} = \dfrac{38}{17}$.*

Exercise 266. *(2017 HMMT February Guts Round) Find the smallest possible value of $x+y$ where $x, y \geq 1$ and x and y are integers that satisfy $x^2 - 29y^2 = 1$.*

Solution 266. *The convergents to $\sqrt{29}$ are $5, \dfrac{11}{2}, \dfrac{16}{3}, \dfrac{27}{5}, \dfrac{70}{13}, \ldots$ Indeed, $70^2 - 29 \cdot 13^2 = 1$, so $(70 + 13\sqrt{29})^2 = 9801 + 1820\sqrt{19}$, and the answer is $9801 + 1820 = \boxed{11,621}$.*

Exercise 267. *Show that x is a Fibonacci number iff either $5x^2 + 4$ or $5x^2 - 4$ is a perfect square.*

Solution 267. *Lemma: If x, y are positive integers such that $y^2 - xy - x^2 = \pm 1$, then $(x,y) = (F_n, F_{n+1})$ for $n \geq 1$ (where F_n denotes the nth Fibonacci number with $F_1 = F_2 = 1$).*

Proof. Consider the cases $x > y, x = y, x < y$. The first case, $x > y$ is absurd, since $y^2 - xy - x^2 < -y^2$. If $x = y$, then $y^2 - xy - x^2 = \pm 1$, so $y = 1$ and $x = 1$ as a direct result. The lemma holds true here. Finally, if $x < y$, consider $(y - x, x)$, which also satisfies the conditions of the lemma. Since $(y-x) + x = y < x + y$, we conclude that $y - x$ and x must be consecutive Fibonacci numbers, which means that $(x, y) = (F_{n+1}, F_{n+2})$. □

From here, write the equation in y as $y^2 + (-x)y + (-x^2 \pm 1) = 0$. Then $y = \dfrac{x + \sqrt{5x^2 \pm 4}}{2}$. We claim that if $5x^2 \pm 4$ is a perfect square, y is an integer. We observe that y is integral iff $x + \sqrt{5x^2 \pm 4}$ is even. Note that $\sqrt{5x^2 \pm 4}$ has the same parity as x, so $x + \sqrt{5x^2 \pm 4}$ is always even as a result. Thus, x, y must be Fibonacci numbers. QED.

Exercise 268. *(1991 AIME) Find A^2, where A is the sum of the absolute values of all roots of the following equation:*

$$x = \sqrt{19} + \cfrac{91}{\sqrt{19} + \cfrac{91}{\sqrt{19} + \cfrac{91}{\sqrt{19} + \cfrac{91}{\sqrt{19} + \cfrac{91}{x}}}}}.$$

Solution 268. *We have $x = \sqrt{19} + \dfrac{91}{x}$, so $x^2 - x\sqrt{19} - 91 = 0 \implies x = \dfrac{\sqrt{19} \pm \sqrt{383}}{2} \implies A^2 = \boxed{383}$.*

Exercise 269. *Determine linear transformations onto the trivial solution $(3,2)$ of the Pell equation $x^2 - 2y^2 = 1$ that will produce more solutions for x and y.*

Solution 269. $(3,2)$ generates solutions according to the recurrence $x_i = x_{i-1}y_i + 7y_{i-1}y_1, y_i = x_{i-1}y_i + 7y_{i-1}x_1$, yielding $(17, 12), (99, 70), \ldots$. From here, we multiply x_i by 5 and add y_i to obtain x_{i+1}, and

Extension Exercise 113. *(1999 IMO Shortlist N3) Prove that there exist two strictly increasing sequences (a_n) and (b_n) such that $a_n(a_n + 1)$ divides $b_n^2 + 1$ for every natural n.*

Extension Solution 113. Let $d^2 \mid (c^2+1)$, and take $b = c(d^2+1) + d^3$; then $b^2 + 1 = c^2(d^2+1)^2 + 2cd^3(d^2+1) + d^6 + 1 = c^2(d^2+1)^2 + 2cd^3(d^2+1) + (d^2+1)(d^4 - d^2 + 1)$. Hence, $d^2 + 1 \mid (b^2+1)$. By assumption, it also follows that $d^2 \mid (b^2+1)$. Take $a = d^2$ as well. This allows us to derive the Pell equation $c^2 + 1 = 2d^2$, from which we have $(7, 5)$, which generates the solution family $(41, 29), (239, 169), \ldots$ which yields increasing sequences according the problem conditions.

Extension Exercise 114. *(2000 Putnam A2) Prove that there exist infinitely many integers n such that $n, n+1, n+2$ are each the sum of the squares of two integers.*

Extension Solution 114. Let $N = 2n^2(n+1)^2$, and $m = n^2 + n$; then $N = m^2 + n^2$ and $N + 2 = (m+1)^2 + (m-1)^2$. Also, $N + 1 = (n^2-1)^2 + (n^2+2n)^2$. (Hint: try $N-1$ instead of $N+1$.)

Extension Exercise 115. *Describe an algorithm to find the nth triangular number that is also a perfect square.*

Extension Solution 115. Recall that $T_n = \dfrac{n(n+1)}{2}$; then let $x^2 = \dfrac{n^2+n}{2} \implies 2x^2 = n^2 + n \implies (2n+1)^2 - 8x^2 = 1$. Let $u = 2n+1, v = 2x$ to obtain $u^2 - 2v^2 = 1$ which has trivial solution $(3, 2)$, hence generating $(17, 12), (99, 70), \ldots$ so $(n, x) = (1, 1), (8, 6), (49, 35), \ldots$

Extension Exercise 116. *Prove that the solutions to the Pell equation $x^2 - dy^2 = 1$ form a group under multiplication.*

Extension Solution 116. Recall that the elements of a ring are closed under multiplication, i.e. form a group, as they obey the other group axioms as well (proving this is left as an exercise to the reader). Hence, if (x_1, y_1) and (x_2, y_2) are solutions to the generalized Pell equation $x^2 - dy^2 = 1$, then we can undergo multiplication in $\mathbb{Z}[\sqrt{d}]$ such that $N((x_1 + y_1\sqrt{d})(x_2 + y_2\sqrt{d})) = N(x_1 + y_1\sqrt{d})(x_2 + y_2\sqrt{d}) = (x_1^2 - dy_1^2)(x_2^2 - dy_2^2) = 1$; this yields another solution, namely $(x_1x_2 + dy_1y_2, x_1y_2 + x_2y_1)$.

Extension Exercise 117. *(2004 IMO Shortlist A3) Does there exist a function $s: \mathbb{Q} \mapsto \{-1, 1\}$ such that if x and y are distinct rational numbers satisfying $xy = 1$ or $x + y \in \{0, 1\}$, then $s(x)s(y) = -1$? Justify your answer.*

Extension Solution 117. Yes; every rational number $r > 0$ can be written as a continued fraction

$$a_0 + \cfrac{1}{a_1 + \cfrac{1}{a_2 + \ldots}} = [a_0; a_1, a_2, \ldots]$$

As n depends solely on r, we can define a function $f(r) = (-1)^n$. (For $r < 0$, $f(r) = -f(-r)$, with $f(0) = 1$.)

Trivially, $f(x)f(y) = -1$ if $x + y = 0$, i.e. $x = -y$. Suppose that $xy = 1$; WLOG assume that $x > y > 0$. Then $x > 1$, so if $x = [a_0; a_1, a_2, \ldots]$, then $a_0 \geq 1, y = \dfrac{1}{x} = [0; a_0, a_1, a_2, \ldots, a_n]$. Hence, $f(r) = (-1)^n$, with $f(y) = (-1)^{n+1} \implies f(x)f(y) = -1$.

Henceforth, suppose that $x + y = 1$. Either $x, y > 0$ or $x > 0, y < 0$. In the first case, WLOG assume $x > \dfrac{1}{2}$; then $\exists a_2, a_3, \ldots, a_n \in \mathbb{N}$ such that $x = [0; 1, a_2, \ldots, a_n] = \dfrac{1}{1 + \frac{1}{t}}$, $t = [a_2, a_3, \ldots, a_n]$. $y = \dfrac{1}{1+t} = [0; 1+a_2, a_3, \ldots, a_n]$, so $f(x)f(y) = -1$ as desired. In the second case, $-y = [a_0; a_1, a_2, \ldots, a_n]$ and so $x = [1 + a_0; a_1, a_2, \ldots, a_n]$. Again $f(x)f(y) = -1$.

Section 15.3: Order and Cyclotomic Polynomials

Exercise 270. *Evaluate $\text{ord}_7 13$, $\text{ord}_5 2019$, and $\text{ord}_p 2^p$ for any prime p.*

Solution 270. $\text{ord}_7 13 = 2$, as $13^2 \equiv 1 \pmod{7}$. Similarly, $\text{ord}_5 2019 = 2$, since $2019^2 \equiv 1 \pmod{p}$. Finally, $\text{ord}_p 2^p = p - 1$ by experimentation.

Exercise 271. *Show that $\deg(\psi_n) = \phi(n)$.*

Solution 271. By definition, the cyclotomic polynomial ψ_n has the primitive n^{th} roots of unity as roots. Then by Fundamental Theorem of Algebra, it is the product of n $x - \xi$ terms, so its degree must be n.

Exercise 272. *Prove that n is a divisor of $\phi(a^n - 1)$ for all positive integers a, n.*

Solution 272. *Sketch:* First compute the order n of $a \pmod{a^n - 1}$. Then, apply Euler's Criterion, and show that Lagrange's Theorem applies. Hence, since the order of a subgroup divides the order of the group, the conclusion follows.

Exercise 273. *Prove that the nth roots of unity form a group under multiplication.*

Solution 273. Let $G = \{1, z, z^2, \ldots, z^{n-1}\}$ be the set of primitive roots of unity of z^n. G is indeed a group, as it has closure, associativity, the identity element 1, and the inverse (as well as commutativity). Specifically, $z^i z^j = z^{i+j} \in G$ wrt mod n for closure. The rest are left as an exercise for the readers own edification (as the proofs are markedly similar).

Exercise 274. *Let $n > 1$ be an odd integer. Show that $\psi_{2n}(x) = \psi_n(-x)$ for all x.*

Solution 274. Note that $\psi_{2n}(x)$ is just the cyclotomic polynomial with the roots of unity that are the negatives of those of $\psi_n(x)$. Hence, by symmetry, we have $\psi_{2n}(x) = (-1) \cdot (-1) \cdot \psi_n(-x) = \psi_n(-x)$ as desired.

Exercise 275. *Describe the relationship between the Euler totient function $\phi(n)$ and the primitive n^{th} roots of unity for any positive integer n.*

Solution 275. The totient function $\phi(n)$ gives the number of $1 \leq k \leq n$ such that $\gcd(k, n) = 1$, i.e. z^k is a primitive n^{th} root of unity.

Exercise 276. *Express $\psi_n(x)$ in terms of the Mobius function, $\mu(n)$.*

Solution 276. This is
$$\psi_n(x) = \prod_{d \mid n} \left(x^{\frac{n}{d}} - 1\right)^{\mu(d)}$$

(In short, show that each root of unity assumes the general form $\xi = (-1)^{\phi(p^k)}$ for a kth root of unity.)

Extension Exercise 118. *(2006 IMO Shortlist N5, modified) Find all integer solutions (x, y) to the equation $x + x^2 + x^3 + x^4 + x^5 + x^6 - y^5 = -2$.*

Extension Solution 118. We have $\dfrac{x^7 - 1}{x - 1} = y^5 - 1$. We show that this equation has no integer solutions.

If $x \in \mathbb{Z}$, $p \mid \dfrac{x^7 - 1}{x - 1}$ prime, then $p = 7, p \equiv 1 \pmod{7}$. To prove this, observe that $p \mid x^7 - 1, x^{p-1} - 1$, by FLT. Suppose that $7 \nmid (p-1)$; this leads to $\gcd(p-1, 7) = 1$, so that $\exists k, m \in \mathbb{Z}$ such that $7k + (p-1)m = 1$. We get a contradiction from expanding the fraction, hence $p \mid 7$ and $p = 7$ if $p \equiv 1 \pmod{7}$ does not.

As such, $d \equiv 0 \pmod{7}, d \equiv 1 \pmod{7}$. Now assume, for contradiction, that the original equation has some integer solution (x, y). Note that $y > 1$, and since $y - 1 \mid (y^5 - 1), y \equiv 1, 2 \pmod{7}$ (see above). If $y > 1$, $\dfrac{y^5 - 1}{y - 1} \equiv 5 \pmod{7}$, and in the latter case, $3 \pmod{7}$ - both contradictions.

Extension Exercise 119. *Prove that all cyclotomic polynomials have integer coefficients.*

Extension Solution 119. *Inducting upon n, $n = 1$ yields the trivial first cyclotomic polynomial $\psi_1(x) = x - 1$. For all $m < n$, we have*
$$\prod_{d|n} \psi_d(x) = x^n - 1$$
so by the induction hypothesis,
$$\prod_{d|n, d \neq n} \psi_d(x)$$
is monic with integer coefficients.

Extension Exercise 120. *For each $n \in \mathbb{N}$, prove that there are infinitely many primes $p \equiv 1 \pmod{n}$.*

Extension Solution 120. *Suppose otherwise. Then define c to be equal to $N!$ times the product of these finitely many primes, where N is a positive integer such that a prime divisor of $\psi_n(k)$ for some $k \in \mathbb{Z}$ is less than N and/or congruent to $1 \pmod{n}$. As the polynomial approaches infinity as $x \to \infty$, take M such that $\psi_n(c \cdot M) > 1$.*

For $n \geq 2, \psi_n(0) = 1$, so $\psi_n(c \cdot M)$ is relatively prime with c, but this is a contradiction.

Extension Exercise 121. *(2003 IMO P6) Let p be a prime number. Prove that there exists a prime number q such that for every integer n, the number $n^p - p$ is not divisible by q.*

Extension Solution 121. *If $p = 2$, then $q = 3$, so assume p is odd.*

Then consider $N = 1 + p + p^2 + \ldots + p^{p-1}$, which has p terms. p is odd $\implies$ N is odd, in addition to $N \equiv (p+1) \not\equiv 1 \pmod{p^2}$. Thus $q \mid N$, q prime.

As $N \equiv 1 \pmod{p}$, $p \nmid N \implies p \neq q$. If $p \equiv 1 \pmod{q}$, then $N \equiv p \pmod{q}$. This is a contradiction, however, as $N \equiv 0 \pmod{q} \implies q \mid p$. Hence $q \nmid (p-1)$.

Let $n^p \equiv p \pmod{q}$. $n \not\equiv 1 \pmod{q}$, so $q \nmid n$. Thus, $n \not\equiv 0, 1 \pmod{q}$. Raise both sides to the pth power, and obtain $p^p \equiv 1 \pmod{q}$. But by FLT, if we let $d = \gcd(q-1, p^2)$, then $n^d \equiv 1 \pmod{q}$ by order. Pick q so $q - 1 \nmid p^2$; but then $d = 1, p$ which implies $d \neq 1 \implies d = p$ which leads to a contradiction.

Extension Exercise 122. *Prove Wedderburn's Theorem:*

> **Theorem 17.7: Wedderburn's Theorem**
>
> *A finite division ring is a field.*

Extension Solution 122. *Let A be a finite domain. For $x \in A, x \neq 0$, we see that A is a group under multiplication.*

To prove it is a field, we induct upon the size of the group. We want to show that $q = 1$, where q^n is the order of $A \implies \text{ord} A = 1$. We can subsequently write $q^n - 1 = (q-1) + \sum \dfrac{q^n - 1}{q^d - 1}$ over x such that the order is $q^d - 1$ for any x. We can factor both the numerator and denominator in terms of the cyclotomic polynomials, namely $x^n - 1 = \prod_{m|n} \psi_m(x), x^d - 1 = \prod_{m|d} \psi_m(x)$.

We obtain $x = q$ and $\psi_n(q) \mid (q^n - 1), \dfrac{q^n - 1}{q^d - 1}$. Then $\psi_n(q) \mid (q - 1)$, and factorization over the complex numbers reveals that $|\psi_n(q)| = \prod |q - \xi|$ for roots of unity ξ.

Chapter 15 Review Exercises

Exercise 277. *Compute $v_3(2019^{4^5} - 1920^{5^4})$.*

Solution 277. *This is equal to $v_3(2019^{1024}) - v_3(1920^{625}) = \min(1024, 625) = \boxed{625}$.*

Exercise 278. *Prove that, for any $n \in \mathbb{N}$, we have*
$$\sum_{d|n} \phi(d) = n$$

Solution 278. *This is quite a classical proof, which proceeds at follows. Define $S_d = \{m \in \mathbb{Z} \mid 1 \le m \le n, \gcd(m,n) = d\}$. We have $\gcd(m,n) = d \iff \frac{m}{d}, \frac{n}{d} \in \mathbb{Z}$, so $|S_d| = \phi\left(\frac{n}{d}\right)$ and $\{1,2,3,\ldots,n\} = \bigcup_{d|n} S_d$. We then have $n = \sum_{d|n} |S_d| = \sum_{d|n} \phi\left(\frac{n}{d}\right) = \sum_{d|n} \phi(d)$ by the symmetry of ϕ, hence the result.*

Exercise 279. *(2005 Canadian MO P2) Let (a,b,c) be a Pythagorean triple, i.e., a triplet of positive integers with $a^2 + b^2 = c^2$.*

(a) *Prove that $\left(\frac{c}{a} + \frac{c}{b}\right)^2 > 8$.*

(b) *Prove that there does not exist any integer n for which we can find a Pythagorean triple (a,b,c) satisfying $\left(\frac{c}{a} + \frac{c}{b}\right)^2 = n$.*

Solution 279.

(a) *We have $\left(\frac{c}{a} + \frac{c}{b}\right)^2 = 2 + \left(\frac{a^2}{b^2} + \frac{b^2}{a^2}\right) + 2\left(\frac{a}{b} + \frac{b}{a}\right)$. By AM-GM, $\left(\frac{c}{a} + \frac{c}{b}\right)^2 > 8$.*

As $a,b,c \in \mathbb{N}$, $\frac{c}{a} + \frac{c}{b} = \frac{p}{q} \in \mathbb{Q}$. If $n = \frac{p^2}{q^2} \in \mathbb{Z}$, $\sqrt{n} \in \mathbb{Z}$.

(b) *Note that all Pythagorean triples can be written in the form $(2mn, m^2 - n^2, m^2 + n^2)$ for $m, n \in \mathbb{N}$. Thus one of a, b is even $\implies c$ is even as well. By modular arithmetic techniques, we can successfully argue that we run into an inevitable contradiction! Hence, there are no integers n for a Pythagorean triple (a,b,c) exists such that $\left(\frac{c}{a} + \frac{c}{b}\right)^2 = n$.*

Exercise 280. *Describe the positive integers n such that*
$$\sum_{i=1}^{n} \sum_{j=1}^{n} (i-j)^2$$
is a perfect square.

Solution 280. *Expand out to get $n^2(n^2 - 1) = 6x^2$, then $\frac{n^2 - 1}{6}$ will be a rational square, which implies that it is in fact a square integer. Solve the Pell equation $n^2 - 6m^2 = 1$ to determine the appropriate n.*

Exercise 281. *(1986 USAMO P3) Define the root mean square of positive integers $a_1, a_2, a_3, \ldots, a_n$ to be*
$$rms(a_1, a_2, a_3, \ldots, a_n) = \sqrt{\frac{a_1^2 + a_2^2 + a_3^2 + \ldots + a_n^2}{n}}$$
Compute the smallest positive integer n such that $rms(1, 2, 3, \ldots, n)$ is also an integer.

Solution 281. *Let $M(n)$ be the RMS of the first n positive integers. Using $(x+1)^3 = x^3 + 3x^2 + 3x + 1$ yields $M(n) = \sqrt{\frac{2n^2 + 3n + 1}{6}}$, and so $(2n+1)(n+1) = 6k^2$ for $k \in \mathbb{Z}$. By Euclidean Algorithm, observe that $\gcd(2n+1, n+1) = 1$, so $2n+1, n+1$ are each proportional to a perfect square. As $2n+1$ is odd, we can either have $2n+1 = 3a^2, n+1 = 2b^2$, or $2n+1 = a^2, n+1 = 6b^2$. In the first case, $2n+1 = 3a^2 = 4b^2 - 1$, so $b = 13 \implies n = 337$ is the smallest b value. In the second case, for $b = 1, 2, \ldots, 8$, $12b^2 - 1$ is not a perfect square, but $n = 383$ is the minimum, which is contradicted by $n = \boxed{337}$.*

Exercise 282. *(2001 IMO Shortlist N5) Let $a > b > c > d$ be positive integers and suppose that $ac + bd = (b + d + a - c)(b + d - a + c)$. Prove that $ab + cd$ is not prime.*

Solution 282. *First note that $(ab+cd) - (ac+bd) = (a-d)(b-c) > 0$ by assumption, so $ab+cd > ac+bd$. Similarly, $ac + bd - (ad + bc) > 0 \implies ac + bd > ad + bc$. Thence, $ab + cd > ac + bd > ad + bc$.*

It then follows that $(a + b + c + d)(-a + b + c + d) = ac + bd \implies b^2 + bd + d^2 = a^2 - ac + c^2$. So $(ac + bd)(b^2 + bd + d^2) = (ab + cd)(ad + bc)$. Hence, we can obtain $(ac + bd) \mid (ab + cd)(ad + bc)$. As $ab + cd > ac + bd$ with $ab + cd$ prime, they share no common factors other than 1. In order to satisfy the problem condition, we require $(ac + bd) \mid (ad + bc)$, but this is a contradiction, since $ac + bd > ad + bc$. Therefore, $ab + cd$ is composite.

Exercise 283. *(2017 USAMO P1) Prove that there are infinitely many distinct pairs (a, b) of relatively prime positive integers $a > 1$ and $b > 1$ such that $a^b + b^a$ is divisible by $a + b$.*

Solution 283. *Let $a - 2n = 1, b = 2n + 1$, with a, b coprime and $a, b > 1$. Note that $a^2 = 4n^2 - 4n + 1 \equiv 1 \pmod{n}$, and similarly $b^2 \equiv 4n^2 + 4n + 1 \equiv 1 \pmod{n}$. Hence $a^b + b^a = a \cdot (a^2)^n + b \cdot (b^2)^{n-1} \equiv a + b \equiv 0 \pmod{4n}$. As $a + b = 4n$, all such pairs work.*

Extension Exercise 123. *(2016 IMO Shortlist N3) Define $P(n) = n^2 + n + 1$. For any positive integers a and b, the set*
$$\{P(a), P(a + 1), P(a + 2), ..., P(a + b)\}$$
is said to be fragrant if none of its elements is relatively prime to the product of the other elements. Determine the smallest size of a fragrant set.

Extension Solution 123. *(Note: Here we use (x, y) as a shorthand for $\gcd(x, y)$.)*

The answer is $\boxed{6}$. Observe that $(P(n), P(n+1)) = 1$, $(P(n), P(n+2)) = 1$ for $n \not\equiv 2 \pmod 7$, $(P(n), P(n+2)) = 7$ for $n \equiv 2 \pmod 7$, and $(P(n), P(n+3)) = 1$ for $n \not\equiv 1 \pmod 7$, $(P(n), P(n+3))$ is a multiple of 3 for $n \equiv 1 \pmod 3$. (Checking these will be an exercise.)

For contradiction, assume that a fragrant set exists with ≤ 5 elements, namely $P(a), P(a + 1), P(a + 2), P(a + 3), P(a + 4)$. Consider $P(a + 2)$; it is coprime with $P(a + 1), P(a + 3)$. Hence we can assume WLOG that $(P(a + 1), P(a + 3)) = 1$. Then $a \equiv 2 \pmod 7$, and for the resulting set to be fragrant, we require $(P(a), P(a+3)), (P(a+1), P(a+4)) > 1$. Then by our third initial observation, $a, a+1 \equiv 1 \pmod 3$ - but this is absurd.

Finally, we can construct a size-6 fragrant set using CRT. This can be done by taking $a \equiv 7 \pmod{19}$ and checking all three conditions. Hence we are done.

Extension Exercise 124. *(2017 IMO Shortlist N5) Find all pairs (p, q) of prime numbers with $p > q$ for which the number*
$$\frac{(p+q)^{p+q}(p-q)^{p-q} - 1}{(p+q)^{p-q}(p-q)^{p+q} - 1}$$
is an integer.

Extension Solution 124. *Outline of proof/solution. We claim the only pair is $\boxed{(3, 2)}$. Let $N = (p + q)^{p-q}(p - q)^{p+q} - 1$, with $\gcd(N, p + q) = \gcd(N, p - q) = 1$. Eliminate the -1 in the numerator to get $((p + q) \cdot (p - q)^{-1})^{2q} \equiv 1 \pmod N$, and split the problem into three cases: $q \geq 5, q = 2, q = 3$. In case 1, $q \geq 5$, consider $r \mid N$ prime, $r \geq 3$. Then consider the multiplicative order modulo r, and by FLT, the order divides $r - 1$. We can observe that all prime divisors of N are either q or $1 \pmod q$ from another application of FLT, which shows that $p \nmid N$. Note that N is the product of two consecutive odd numbers, and so both should be equivalent to 0 or 1 modulo q - contradiction.*

In case 2, $q = 2$, we have $(p + 2)^{p-6}(p - 2)^{p+2} \leq 1$. So $p \leq 4$, and for $p = 3$, we indeed have the solution $(p, q) = (3, 2)$ by computation.

In case 3, we have $64(p+3)^{p-9}(p-3)^{p+3} \leq 1$, which invites the same approach as above (no solutions).

Extension Exercise 125. *Prove that all elements of the infinite sequence*

$$10001, 100010001, 1000100010001, \ldots$$

are composite.

Extension Solution 125. *Consider $S_k = 1 + 10^4 + 10^8 + \ldots + 10^{4k-4}$. Note that*

$$S_{km} = S_k \left(1 + 10^{4k} + 10^{8k} + \ldots + 10^{4k(m-1)}\right)$$

and furthermore, we can write $S_k = \dfrac{10^{4k} - 1}{10^4 - 1}$ with $r = 10^4$ (geometric series).

For k odd, we have $(10^{k-1} + 10^{k-2} + \ldots + 1)(10^{k-1} - 10^{k-2} + \ldots + 1)(10^{2k-2} - 10^{2k-4} + \ldots + 1)$ so S_k is composite for k odd. The composite-ness of S_k then depends on whether $10^4 + 1$ is prime. We find that it is not, since $10^4 + 1 = 10,001 = 137 \cdot 73$. Hence the sequence $(S_k)_{k \geq 1}$ contains no prime numbers.

Extension Exercise 126. *Using the properties of the Jacobi symbol, show that 561 is the smallest Carmichael number.*

Extension Solution 126. *We demonstrate that 561 is a pseudoprime wrt every base, with the Jacobi symbol evaluated with 561 in its bottom argument indicating that it can be prime (not necessarily that it is).*

Indeed, we have $561 = 51 \cdot 11$, verifying that it is not actually prime (as a sanity check). But FLT says that $2^{560} \equiv 1 \pmod{561}$ (and the same goes for a base of 5). Furthermore, all numbers smaller than 561 either are prime, or fail the pseudoprime test. This is because it fails the FLT test for at least one prime factor.

Extension Exercise 127. *If a and n are integers, then $\left(\dfrac{a}{n}\right) = 1$ does not necessarily guarantee that a is a quadratic residue modulo n. Approximately how often does the Jacobi symbol incorrectly identify quadratic residues?*

Extension Solution 127. *As a (hopefully) motivating start to thinking about this problem, but not as a rigorous proof, we will propose the following general approach. For n an odd integer, let k be the number of distinct prime factors it has, such that $n = \prod_{i=1}^{k} p_i^{a_i}$. If $\gcd(a, n) = 1$, a is a QR mod n iff a is a QR mod $p_i^{a_i}$ for all i by CRT.*

Pick $a \in (\mathbb{Z}/n\mathbb{Z})^{\times}$ uniformly at random, which yields a is a QR mod p_i with probability $\dfrac{1}{2}$. Thus by CRT, a is a square modulo all the p_i with probability $\dfrac{1}{2^k}$.

We also have that $\left(\dfrac{a}{n}\right) = \prod_{i=1}^{k} \left(\left(\dfrac{a}{p_i}\right)\right)^{a_i}$, so n is a square $\iff \left(\dfrac{a}{n}\right) = 1$ (probability $\dfrac{1}{2}$).

Bibliography

Aloupis, Greg, et al. "Classic Nintendo Games Are (Computationally) Hard." SpringerLink, Springer, Cham, 1 July 2014,

link.springer.com/chapter/10.1007/978-3-319-07890-8_4.

Barbosa, Andre Luiz. "A Human-Checkable Four-Color Theorem Proof." *Arxiv.org,* Arxiv.org, 2010,

arxiv.org/pdf/1708.07442.pdf.

Bing, R. H. "Necessary and Sufficient Conditions That a 3-Manifold Be S3." Annals of Mathematics, vol. 68, no. 1, pp. 1737.

Bix, Robert. *Conics and Cubics: A Concrete Introduction to Algebraic Curves,* Second Edition

Crandall, Richard, et al. A SEARCH FOR WIEFERICH AND WILSON PRIMES. Jan. 1997,

www.reed.edu/physics/faculty/crandall/papers/S0025-5718-97-00791-6.pdf.

Gasarch, William I. *Guest Column: The P=?NP Poll.* University of Maryland, 2002,

www.cs.umd.edu/~gasarch/papers/poll.pdf.

Glaisher, J.W.L. *Periods of Reciprocals of Integers Prime to 10.* 1878.

Hin, Lau Chi. Muirhead's Inequality. *Mathematical Excalibur,* Mathematical Excalibur, 2006,

www.math.ust.hk/excalibur/v11_n1.pdf.

Jiang, Albert Xin, and Kevin Leyton-Brown. A Tutorial on the Proof of the Existence of Nash Equilibria. Department of Computer Science, University of British Columbia, University of British Columbia,

www.cs.ubc.ca/ jiang/papers/NashReport.pdf.

Lantz, Frank, and Bennett Foddy. Universal Paperclips. 9 Oct. 2017.

Lehmer, Derrick Henry, et al. Guide to Tables in the Theory of Numbers. *The National Academies Press,* 1941,

www.nap.edu/catalog/18678/guide-to-tables-in-the-theory-of-numbers.

Martello, Silvano. *An Enumerative Algorithm for Finding Hamiltonian Circuits in a Directed Graph.* University of Bologna, Italy, Mar. 1983, www.dcs.gla.ac.uk/~pat/jchoco/lds/hcp/p131-martello.pdf.

Ohana, R. Andrew. *A Generalization of Wilsons Theorem*. 3 June 2009, University of Washington.

Razborov, Alexander A, and Steven Rudich. "Natural Proofs." Proceedings of the Twenty-Sixth Annual ACM Symposium on Theory of Computing, 1994, pp. 204213.

Stein, William A. Euler's Proposition. Euler's Proposition, 2007,

wstein.org/edu/2007/spring/ent/ent-html/node52.html.

"Sum of Interior Angles of an n-Sided Polygon. Sum of Interior Angles of an n-Sided Polygon,

www.qc.edu.hk/math/Junior%20Secondary/interior%20angle.htm.

Tao, Terence. Bounded Gaps between Primes (Polymath8) a Progress Report. *WordPress,* WordPress, 11 July 2013,

terrytao.wordpress.com/2013/06/30/bounded-gaps-between-primes-polymath8-a-progress-report.

Tao, Terence. Finite Time Blowup for an Averaged Three-Dimensional Navier-Stokes Equation. WordPress, WordPress, 6 Feb. 2014,

terrytao.wordpress.com/2014/02/04/finite-time-blowup-for-an-averaged-three-dimensional-navier-stokes-equation/.

Weintraub, Steven H. SEVERAL PROOFS OF THE IRREDUCIBILITY OF THE CYCLOTOMIC POLYNOMIALS. www.lehigh.edu/~shw2/c-poly/several_proofs.pdf.

Glossary

p-adic valuation: Of a nonzero integer n, the largest power of a prime p that divides n.

p-series: A type of series of the form $\sum \frac{a}{p^k}$, which converges for $k > 1$ and diverges otherwise.

z-score: The number of standard deviations that a data point is away from the mean.

a priori: Relating to deductive reasoning, as opposed to information that is derived empirically.

abelian: Describing a commutative group G in which $ab = ba$ for all $a, b \in G$.

absolute value: The positive difference between a real number and zero on the number line.

algebra: The field of mathematics in which variables are used to represent numerical quantities, taking its name from the Arabic *al-jabr,* meaning "restoration."

altitude: The "height" of a triangle, which is perpendicular to the opposing side of the triangle.

AND: A logical conjunction/gate that requires both parts of the statement to be true to hold true.

angle: The space (measured in either degrees or radians) between two intersecting lines, measured at the point of intersection.

arithmetic: The field of mathematics that studies numbers, their properties, and their manipulation.

axiom: A mathematical truth that we take as fact and which is self-evidently true.

barycentric coordinates: Ordered 3-tuples corresponding to points within the interior of a triangle. Each coordinate describes the relative mass of the point with respect to each vertex.

Bayes' Theorem: A theorem that describes the probability of an event in terms of conditional probability based on prior knowledge.

bijection: Also known as a *one-to-one correspondence,* a bijection is a function between two sets that is both one-to-one (injective) and onto (surjective).

binomial: An expression consisting of exactly two terms.

Boolean: A true/false binary value, often used in truth tables.

bound: An upper bound of a set is a number greater than or equal to all elements of the set; the lower bound is defined similarly.

box-and-whisker plot: A plot visually depicting the five-number summary of a data set, with the "box" representing the interquartile range (IQR) and the "whiskers" representing the minimum and maximum of the data.

Cartesian plane: The standard xy-plane in 2 dimensions, where each point is represented by an ordered pair (x, y). In dimensions greater than or equal to 3 (namely, in n dimensions), each point is represented by an ordered n-tuple.

casework: The procedure of considering disjoint cases to outcome separate outcomes that can be combined to obtain a final probability or number of ways.

centroid: The point in the interior of a polygon whose coordinates are the arithmetic means of the coordinates of the vertices.

Ceva's Theorem: In a triangle $\triangle ABC$, we have concurrence iff the product of the signed lengths formed by the cevians is 1.

circle: The set of points equidistant (at distance r, where r is the *radius* of the circle) from the center of the circle.

circumcircle: The unique circle of a polygon that passes through all vertices of the polygon.

combinatorics: The study of enumeration of objects in a finite set (i.e. counting).

complex number: A number of the form $a + bi$, where $a \in \mathbb{R}$ and $b \in \mathbb{R}$, with $i = \sqrt{-1}$.

conjecture: An inference formed from incomplete or insufficient evidence to form a theory.

constant: A quantity that does not change; i.e. a quantity not associated with a variable.

contradiction: An absurd conclusion that shows that the alternative to the desired statement must be false, and hence that the desired statement must be true.

convergence: A property of a series that eventually tends to some finite, real value.

converse: Of the statement $p \implies q$, the converse is $q \implies p$ (i.e. we switch the order of the parts of the statement).

coordinate: A value in a set of values that depicts an exact position of a point in the plane.

decimal: A numerical quantity with digits in base 10 (0-9).

degree: (1) The highest exponent in the terms of a polynomial. (2) The number of vertices adjacent to a vertex in a graph.

derivative: The rate of change of a function; i.e. the slope of the tangent line at a point.

dependent event: An event whose outcome depends on the outcome of another event.

diagonal: A segment entirely contained within the interior of a polygon that connects two vertices, but is not an edge.

differential equation: Any equation in terms of derivatives.

Diophantine equation: A multi-variable equation to which we seek integer solutions.

discrete mathematics: The analysis of mathematical structures that are not continuous, but rather are given by individual points (more formally, by countable sets).

disjoint: With regard to sets, having no elements in common.

divergent: A property of a series that either tends towards (positive or negative) infinity, or to no value at all (i.e. the limit of the partial sums of the series does not exist).

divide: To be a factor of another number.

domain: The set of input values for a function that produce a value in the range of the function.

dynamic programming: A computer programming and mathematical optimization method that proves very useful in recursion problems by breaking it down into branching sub-problems.

edge: A line segment joining two vertices (of a polygon, polyhedron, or graph in particular).

eigenvalue: A value associated with a matrix such that multiplying it with the matrix yields the same result as multiplying it with its associated eigenvector. The eigenvalue is a root of the matrix's characteristic polynomial.

eigenvector: A column vector associated with a matrix such that multiplying it with the matrix yields the same result as multiplying it with the corresponding eigenvalue (see above).

ellipse: An oval shape sharing similar properties with the circle. Ellipses are traced out by the points equidistant from two foci.

equation: A statement that two numerical quantities (either in terms of constants or variable) are equal to each other.

equivalence class: A subset such that all pairs of elements in the equivalence class have an equivalence relation (i.e. reflexivity, symmetry, and transitivity).

error: The difference between an approximation and the true value (also called the *tolerance, margin of error*).

Euclid's lemma: If a prime number p divides ab, then either $p \mid a$ or $p \mid b$ (or both).

Euclidean algorithm: A tool used to more effectively compute the greatest common divisor of two integers.

Euler's Criterion: A method for determining whether an integer is a quadratic residue mod a prime p.

Euler's formula: The formula $F + V - E = 2$ for 3-dimensional polyhedra, where F is the number of faces, V is its number of vertices, and E is its number of edges.

excircle: A circle lying outside a triangle that is internally tangent to the extensions of two of the sides, and tangent to the other side. The vertex to which the excircle is opposed is the vertex after which that excircle is named (so, for instance, the A-excircle is opposite vertex A, and hence tangent to the extensions of $\overline{AB}$ and $\overline{AC}$).

exists: Denoted by $\exists$, the exists quantifier says literally that "there exists a quantity ... such that."

expansion: The full form of a quantity written with all terms (and no ellipses or sum notation).

face: A flat surface comprising a geometric solid (polyhedron).

Fact 5: The fifth of the nine facts of Euclidean geometry (which we have listed in §9.5); also referred to as the *Incenter-Excenter Lemma*.

factor: A divisor of a number n; i.e. a number such that k times the number is equal to n for some integer k. (Under this definition k is usually positive, though it can be negative as well.)

Fermat's Last Theorem: The famous statement that there exist no solutions to the equation $a^n + b^n = c^n$ in integers (a, b, c) for $n \geq 3$. Proven by Andrew Wiles in 1995.

Fermat's Little Theorem: The statement that, for any prime p and integer a, $a^p \equiv a \pmod{p}$ (and, hence, $a^{p-1} \equiv 1 \pmod{p}$ in its alternate form). Unrelated to Fermat's Last Theorem.

Fibonacci sequence: The sequence $1, 1, 2, 3, 5, 8, \ldots$ where $F_1 = F_2 = 1$, and for all $n \geq 3$, $F_n = F_{n-1} + F_{n-2}$. A related sequence is the *Lucas sequence*, with $L_1 = 2$ and $L_2 = 1$, again with $L_n = L_{n-1} + L_{n-2}$ for $n \geq 3$.

field: A set on which addition, subtraction, multiplication, and division are well-defined operations, and behave as they do on the real numbers.

five-number summary: The description of a data set that includes its minimum, first quartile, median, third quartile, and maximum.

floor: The largest integer less than or equal to a real number. (A similar definition is of the *ceiling*, which is the smallest integer greater than or equal to a real number.)

for all: The quantifier $\forall$, which is interpreted as "for all quantities ..."

formula: A mathematical rule or relationship that is expressed symbolically using constants and/or variables.

fraction: A quotient of two numerical quantities, of the form $\dfrac{a}{b}$.

function: A mathematical rule or relationship that is written in terms of numerical quantities and symbols.

functional equation: An equation expressed in terms of functions, where the unknown represents a function.

game: An interaction involving *players* that is equipped with rules, strategy sets, and outcomes (*payoffs,* usually expressed in utils) for each player, depending on the actions of each player.

generating function: A function that represents a sequence as a power series.

geometry: The branch of math that deals with the measurement, relationships between, and properties of lines, points, angles, solids, and surfaces, along with several other constructs.

golden ratio: The quantity $\phi = \dfrac{1 + \sqrt{5}}{2} \approx 1.618\ldots$, which is crucial in the Fibonacci sequence, as well as the inherently beautiful symmetry of the *golden spiral* in nature (see the front cover of this book for a prime example!)

graph: An ordered pair $G = (V, E)$, where V is a set of vertices and E is a set of edges connecting the vertices in V together.

group: A set equipped with a binary operation $\cdot$ that satisfies the group axioms (closure, associativity, invertibility, and the existence of an identity).

Hamiltonian: Can refer to anything associated with Sir William Rowan Hamilton, but usually refers to a *Hamiltonian path* - a path which visits each vertex exactly once - in a graph-theoretic context.

harmonic series: The series $1 + \frac{1}{2} + \frac{1}{3} + \ldots$, which diverges (as a p-series with $p = 1$ exactly).

Heron's formula: A formula commonly used to find the area of a triangle in terms of its side lengths a, b, c, namely $\sqrt{s(s-a)(s-b)(s-c)}$, where $s = \frac{a+b+c}{2}$ (the *semi-perimeter*).

histogram: A display of data in a set with the data grouped in ranges (or *bins*) on the x-axis.

Hockey-Stick Identity: A special case of Vandermonde's Identity, which can be used to sum series of binomial coefficients and simplify computations involving binomials massively.

homogeneous: Describes an expression in which all terms have the same degree.

Horizontal Line Test: A test that can be applied to the function of a graph to determine whether it is one-to-one (injective).

hypercube: The analogue of a cube in dimensions 4 and higher.

hypothesis: A proposed explanation for a natural phenomenon that must be tested through further experimentation and/or observation to either be disproven or to be advanced to theory status following subsequent replication and verification.

identity: An element similar to $1 \in \mathbb{R}$ such that adding/multiplying the element with another element in a group yields the same element.

iff: Short for "if and only if," which is a logical conjunction that holds true when both parts of the statement have the same truth value.

image: The subset of the codomain of a function that represents the output from the function's domain (also called the *range*, which is a more familiar term.

implication: A logically valid conclusion that can be drawn from a series of statements.

incenter: The center of the inscribed circle (incircle) of a polygon.

independent event: An event that does *not* rely on other events; its probability remains constant (it is *memoryless*).

induction: A method of proof that first proves a base case, then assumes the statement holds true for some general n (the *inductive hypothesis*) and, subsequently, proves the statement for $n+1$.

inequality: A relation that holds between two unequal values (for a strict inequality; non-strict inequalities involve $\geq$ or $\leq$).

inferential statistics: The branch of statistics that deals with conclusions by inferring properties of a population.

infinite descent: A proof method - a sub-method of proof by contradiction, traditionally used to solve Vieta jumping problems - in which we show that the existence of an integer solution would imply the existence of a smaller solution, which would contradict the implied minimality.

injectivity: The one-to-one property of a function; more specifically, it never maps distinct elements in its domain to the same element in its range. (That is, the graph of the function passes the HLT.)

inscribe: To fit a shape perfectly inside of another shape, such that we have tangent side lengths.

integral: Loosely speaking, the area under the curve of a function (or the reverse operation of the derivative in the case of the indefinite integral). More rigorously, we can define the definite integral as the limit of the Riemann sum of a function, and the indefinite integral via the Fundamental Theorem of Calculus.

interval: A continuous set of values in the real numbers with the property that any number that lies between two numbers in the set is itself included in the set.

interval of convergence: the set of all values of x for which a power series converges. Can be determined using the Ratio Test.

invariant: An object which remains unchanged under some operation. We say that this object is invariant under that operation.

inverse: A function that "reverses" or "undoes" another.

involution: A function f such that, for all x in the range of f, $f(f(x)) = x$. Essentially, f is its own inverse.

isogonal conjugate: With respect to a point P in the interior of a triangle $\triangle ABC$, the point formed from reflecting the lines $\overline{PA}$, $\overline{PB}$, and $\overline{PC}$ about the angle bisectors of the triangle.

Jacobi symbol: A generalization of the Legendre symbol to *all* odd integers, not just primes.

Jensen's Inequality: An inequality on convex functions f that is highly useful in probability theory.

kernel: See *null space*.

Kronecker-Weber Theorem: The statement that some cyclotomic field contains every finite abelian extension of the rational numbers.

lattice point: A point with coordinates that are all integers.

Law of Cosines: The equation $a^2 + b^2 - 2ab\cos C = 0$, which generalizes the Pythagorean Theorem to all angles C (note that this reduces to the Pythagorean Theorem when $C = 90°$; i.e. when $\triangle ABC$ is right).

Law of Large Numbers: The statement that, as the number of trials of an experiment approaches infinity, the ratio of the number of successes to the number of trials will approach the probability of success.

Law of Quadratic Reciprocity: A law that allows for the simple calculation of the Legendre symbol, and hence whether or not a number is a QR with respect to some prime p.

Law of Sines: The statement that, in any triangle $\triangle ABC$ and side lengths a, b, c opposite vertices A, B, C

respectively, $\frac{\sin A}{a} = \frac{\sin B}{b} = \frac{\sin C}{c}$.

Legendre symbol: A symbol defined such that $\binom{q}{p} = 1$ if q is a QR modulo p, and -1 otherwise.

limit: A statement in the middle of a proof that is used to prove a key statement within the proof.

line: A geometric figure that is perfectly straight, has zero thickness, and extends infinitely in both directions; is uniquely determined by any two points in the plane.

linear algebra: The study of linear equations and their solutions, as well as of their relation with matrices.

linearly (in)dependent: Vectors in a set are linearly independent iff no vector can be expressed as a linear combination of the others; otherwise, the set is linearly dependent.

logic gate: An implementation of a Boolean structure, which takes the Boolean variable (true/false) as input and outputs another binary value.

Maclaurin series: A Taylor series (power series) centered about $a = 0$.

manifold: A topological space that resembles Euclidean space pointwise.

map: A map is often taken to mean a continuous function between a domain and a codomain, but it can also refer to a morphism (isomorphism or homeomorphism).

Markov chain: A stochastic model that depicts events whose probabilities depend only on the outcomes of the previous events.

Markov decision process: An ordered 4-tuple consisting of the finite set of states, the finite set of actions, the transition probability, and the reward.

Markov property: See *memorylessness*.

mass: A scalar value assigned to each point in a polygon that indicates its distance relative to each vertex of the polygon. Analogous to the center of mass/gravity in a physical system.

matching: In a graph, a matching is a set of edge with no common vertices.

matrix: A matrix is a rectangular array of entires with special properties.

mean: The mean (in this context, arithmetic mean) of a set of numbers is the sum of the numbers divided by the number of numbers in the set.

median: The median is the element at the "center" of the set, position-wise when the elements of the set are arranged in increasing order. That is, if the number of elements in the set is odd, the median is the element with the same number of elements greater than and less than it. If the number of elements in the set is even, the median is the arithmetic mean of the two middlemost elements.

memorylessness: The property of a probabiity distribution to "forget" its past behavior, in that the probability of a future outcome is not influenced by the past outcomes.

Menelaus' Theorem: Given a transversal line that intersects a triangle at three points (one of these intersection points being at an intersection of one of the sides), the signed product of distances between the triangle's vertices and the intersection points is -1.

metric: A distance function $d : X \mapsto X$ equipped with positive-definiteness ($\forall x, y, z \in X, d(x, y) \geq 0$), equality at 0 ($d(x, y) = 0 \iff x = y$), symmetry ($d(x, y) = d(y, x)$), and satisfaction of the Triangle Inequality (i.e. $d(x, z) \leq d(x, y) + d(y, z)$).

mode: The most frequent element(s) in a data set. May not exist, in which case the set necessarily has all elements distinct.

modular arithmetic: A congruence relation over the integers in which integers may be "congruent" to each other modulo some other integer m; i.e. the two integers leave the same remainder upon division by m.

modulus: The integer m in the above definition.

Modus Ponens: The rule of logic that asserts that, if $p \implies q$ holds true, and if we know that p is true, then q must also be true.

Modus Tollens: If $p \implies q$ holds, but q is false, then p must also be false.

monic: Describing a polynomial with leading coefficient 1.

monomial: An expression consisting of a single term.

monotonicity: The property of a function to increase or decrease over its interval, never switching between them.

Monte Carlo: A simulation method in which we seek the largest possible number of data points to attain the closest possible approximation.

Multinomial Theorem: A generalization of the Binomial Theorem to multiple arguments in the bottom.

multiple: A multiple of a number is the result of multiplication of that number by an integer.

multiplicity: The number of times a member of a set appears in that set.

Nash equilibrium: A strategy set between multiple players that maximizes each player's relative payoff, such that no player is motivated to unilaterally change his/her strategy.

natural proof: As in the P/NP section, a natural proof is one that shows that two complexity classes differ.

negation: The "opposite" of a mathematical statement A, in that it inverts the truth value (denoted $\neg A$).

nine point circle: In $\triangle ABC$, the unique circle passing through the midpoints of each side, the feet of each altitude, and the midpoints of the line segments from each vertex to the triangle's orthocenter.

node: See *vertex*.

norm: A function defined over a normed vector space that essentially takes the "length" of a vector. Id est, the norm is the magnitude of the vector.

normal: Perpendicular (especially with regards to a plane or curve).

NOT: The logical operator which serves as a negation.

NP: The class of problems in non-deterministic polynomial time, in constrast to problems solvable in polynomial time.

null hypothesis: The default hypothesis in a statistical experiment, that there exists no statistically significant difference between specified populations, and that any difference that does come about is purely due to experimental error or as the result of a fluke. Accepted iff the p-value (the likelihood that there is a statistically significant difference) is 0.05 or greater, else the alternative hypothesis H_a is accepted. (Note that 0.05 is not set in stone, but it is by far the most common threshold used in the scientific and statistical literature.)

null space: With respect to a matrix A, the set of column vectors v such that $Av = \mathbf{0}$.

number: An arithmetic value/quantity that represents the degree of presence of an object, and can be used in calculation and counting, as well as to depict cardinal order.

number theory: Fundamentally, the study of numbers and their interactions; in particular, the positive integers.

object: Anything with a formal definition, and which is conducive to logical reasoning and rigorous proof.

Occam's Razor: The principle that the simplest answer is the best, or most likely/plausible, answer.

open: In describing a set (or ball), *open* refers to the property of the set that is related to the openness of an interval of numbers in the real line.

operations research: A branch of discrete math that utilizes optimization methods to improve decision-making.

operator: A mapping that acts upon elements of a set (or space) to output elements in another set/space.

optimization: The act of determining the solution with the greatest payoff that also minimizes the required resources.

or: The logical gate that outputs a value of true if at least one of its arguments holds true. (Id est, it only outputs false if *all* of its arguments are false.)

oracle: Another name for a black box with unknown inner workings, or, more formally, a *queueing node* with unknown servicing procedures.

order: In particular, with regard to a prime number, the order of $a \bmod p$ (denoted $\text{ord}_p(a)$) is the smallest positive integer k such that $a^k \equiv 1 \pmod{p}$.

ordered pair: A set of two numbers (x, y) whose order is significant in some way (i.e. $(x, y) \neq (y, x)$ unless $x = y$).

orthocenter: The intersection of internal perpendicular bisectors of a triangle.

orthogonality: The property of vectors to have an inner product of zero; roughly relates to their perpendicularity from a geometric perspective.

P/NP problem: The unsolved problem of whether any problem that is easy to verify is also easy to solve forwards.

paradox: A seemingly self-contradictory or otherwise absurd statement that may nevertheless be true, or perhaps hard to see the fault in.

parallel: Of lines, planes, or hyperplanes, never intersecting, and having the same distance between them at all points.

parallelepiped: A generalization of the rectangular prism with each face a parallelogram.

Parallelogram Law: The statement that the sum of the squares of the side lengths of a parallelogram equals the sum of the squares of the diagonals of the parallelogram.

parity: The even-ness or odd-ness of an integer.

Pascal's Triangle: The famous triangular array of binomial coefficients, beginning with 1 at the top and along the sides, and having each element below two other elements equal to the sum of those two elements.

path: A route traversing through vertices in a graph.

perimeter: The length around the exterior (along the edges of) a shape, as measured in units.

period: The smallest length of an interval after which the values of a function begin to repeat indefinitely.

permutation: An arrangement of the elements of a set with respect to order.

perpendicular: At an angle of 90° with respect to another curve, line, plane, or hyperplane.

phi function: Euler's totient function, which yields the number of positive integers less than or equal to n that are relatively prime to n (i.e. its gcd with n is equal to 1).

Pick's Theorem: In any polygon in the plane, the sum of the number of interior lattice points, and half the number of boundary lattice points, minus 1, gives the area of the polygon.

PIE: The Principle of Inclusion-Exclusion (or Inclusion-Exclusion Principle), which requires that we add the sizes of all subsets of odd cardinality, and subtract the sizes of all subsets of even cardinality, to obtain the desired size of intersection.

Pigeonhole Principle: The statement that, if there are more pigeons than holes, then at least one hole must contain multiple pigeons.

point: A zero-dimensional location on a line or in a plane, hyperplane, or any higher dimension of geometry. It has no size, no width, no length, and no depth.

polar: In conjunction with the pole, the polar has a relationship with the power of a point with respect to a circle.

polar coordinates: The coordinates in a 2-dimensional coordinate system in which we express each Cartesian coordinate as the combination of a radius r and an angle measure θ.

pole: In conjunction with the polar, the pole yields several useful properties (known collectively as the properties of *poles and polars*).

polygon: A figure in the plane with at least three straight sides and angles.

polyhedron: A three-dimensional solid with multiple plane faces.

polynomial: An expression consisting of the sum of multiple terms consisting of variables.

postulate: See *axiom*.

power: Aside from the very familiar definition of *exponent, power* may also refer to the context of the *power of a point* theorem, wherein it is the value $s^2 - r^2$, for s the distance from the point to a circle, and r is the circle's radius.

power series: An expression of an infinite series in the form $\sum_{n=0}^{\infty} a_n x^n$, where the a_n are constants.

prime: A positive integer that is evenly divisible only by 1 and itself (i.e. has exactly 2 positive factors, and 1 *proper* positive factor).

Prisoner's Dilemma: In game theory, a prototypical example of a game where the pretense of rationality on the part of each player seems to contradict the final outcome.

probability: A real number between 0 and 1, inclusive, that represents the likelihood of an event occurring.

projective plane: The Euclidean plane additionally equipped with the notions of points and lines at infinity; typically denoted $\mathbb{RP}^2$ in 2 dimensions.

proof: An argument that *rigorously* and firmly establishes the absolute truth of a statement, without any room for uncertainty.

proof by contradiction: A proof that assumes the opposite of the desired statement, then goes on to show that this assumption inevitably leads to an absurdity or other falsehood.

proof by exhaustion: A proof method that exhausts all possible cases via numerical computation and casework.

proposition: A rigid statement - an expression or assertion - that is not in any way subjective.

Pythagorean Theorem: The classical statement that, for all right-angled triangles $\triangle ABC$ with side lengths a, b, c (with c the hypotenuse length and $C = 90°$), we have $a^2 + b^2 = c^2$.

quadrant: One of the four quarters of the Cartesian coordinate plane. Quadrant I is the upper-right corner, Quadrant II is the upper-left corner, Quadrant III is the lower-left corner, and Quadrant IV is the lower-right corner.

quadratic: A polynomial (or expression in general) with degree 2.

quadratic residue: Modulo p, an integer q that is congruent to a perfect square.

quadrilateral: A polygon with exactly four edges.

quartic: A polynomial of degree 4.

quartile: The data points at the 25th (first), 50th (second/median), and 75th percentiles (third) of the data set.

query: In queueing theory, a request for service or information.

queue: In informal terms, a "line" for service by a queueing node.

queueing node: A sort of "service station" where tasks/customers can be completed/serviced.

radian: A unit of angle measure defined such that π rad $= 180°$.

radical axis: With respect to two circles, their radical axis is the set of points at which tangents drawn to both circles have the same length.

radius: The distance of each point on a circle to the circle's center.

radius of convergence: half the length of the interval of convergence.

random variable: A variable in a probability distribution whose values are randomly distributed in some fashion.

random walk: A stochastic process in which each step is randomly determined.

range: The set of outputs realized by a function.

rank: Of a matrix, the dimension of the vector space spanned by the columns of the matrix; i.e. the number of nonzero rows in its (reduced) row echelon form.

ratio: The quotient of one value with another.

rational number: A number that can be written in the form $\frac{a}{b}$, where a and b are integers such that $\gcd(a, b) = 1$.

Rational Root Theorem: For an arbitrary polynomial $P(x)$, each rational root $\frac{p}{q}$ satisfies the constraints that $p \mid a_0$ (the constant term) and $q \mid a_n$ (the leading coefficient).

real number: A value representing a quantity on a continuous spectrum that can be depicted along the real number line.

recursion: The process of using previous outputs of a function or other sysstem to determine future outputs of the function/system.

reductio ad absurdum: A form of argument that disproves a conclusion by showing that it leads to an absurdity. See *proof by contradiction*.

relation: A relationship between sets that has the properties of functions. It can also be used in defining equivalence relations.

residue class: A complete set of integers that are congruent modulo n for some $n \in \mathbb{N}$.

Riemann Hypothesis: One of the most widely-known unsolved mathematical problems, the Riemann hypothesis posits that all non-trivial zeros of the Riemann zeta function lie at the negative even integers, and on the critical strip $\text{Re}(z) = \frac{1}{2}$ for $z \in \mathbb{C}$.

ring: A set equipped with two binary operators (additional and multiplication) that satisfy a set of ring axioms, similar to the group axioms.

root: Of some polynomial $P(x)$, a value x for which $P(x) = 0$. (Also known as a *zero* of $P(x)$.)

root of unity: A complex number z for which $z^n = 1$ for some positive integer n. (In this case z is known as an nth root of unity.)

sample: A collection of data points that is sufficiently large and meaningful to be conducive to statistical analysis.

scalar: A real, numerical quantity (in contrast to a *vector*).

scalene: Describes a triangle whose side lengths, and angles, all differ.

standard deviation: An approximate measure of the average "distance" of each element in a data set from the mean of the set.

self-adjoint: Describes an operator/linear map T on a complex vector space V, $T : V \mapsto V$, such that $\langle Tv, w \rangle = \langle v, Tw \rangle$ for all vectors $v, w \in V$.

sequence: An enumerated list of objects, in which repetition is permitted.

series: The sum of the elements of a sequence (either finite or infinite).

Shoelace Theorem: A method of area-finding for a polygon in the coordinate plane, given its coordinates.

similarity: Two shapes are similar if their side lengths are in direct proportion with each other and the corresponding angles congruent; then we can also say that there is a *homothety* between the shapes.

simplex: The higher-dimensional analogue of a triangle and tetrahedron from 2 and 3 dimensions, respectively.

singular: Describes a matrix with determinant zero (and which is hence non-invertible).

smoothness: The infinite-differentiability property of a function.

spiral similarity: A homothety equipped with a rotation about a given center.

statement: A declarative sentence with clear meaning that is definitively either true or false (and not both).

states: A recursive probability method that expresses the probability of a dependent event in terms of the probabilities of previous events in the sequence.

statistics: A branch of applied mathematics concerned with the collection, interpretation, and presentation of data.

stem-and-leaf plot: A data plot that writes each data point as the sum of some larger increment and some smaller increment, with a stem of larger increments and a leaf of smaller increments (e.g. $12 = 1 \mid 2$).

stochastic: Describes a probability/data model in which the outcomes are randomly determined and the variables are random.

strategy: An option that a player has in a game, associated with a payoff (potentially depending on other players' strategies).

strong induction: A variant of induction in which we induct upon *all* previous cases of 1 through n to prove the case $n+1$, not just n itself.

subgraph: A subgraph of $G = (V, E)$ is a graph H whose vertices are in V and whose edges are in E.

subtend: To form an angle at a particular point.

surjection: A function whose image is equal to its codomain; that is, a function where nothing in the codomain is "left out" of the range.

symmedian: The isogonal conjugate of the centroid.

symmetry: The inherent property of an object to have similar properties about some frame of reference that inverts those points with equal distance from the frame.

tangent: A straight line or plane that touches another curve at a single point.

Taylor series: A power series centered about $x = a$ for some real number a, namely an expansion of the form $\sum_{n=0}^{\infty} \frac{f^{(n)}(a)}{n!}(x-a)^n$, where $f^{(n)}(a)$ denotes the nth derivative of f evaluated at $x = a$.

theorem: A proposition that is not inherently self-evident, but is a truth that has been established by axioms.

theory: A well-tested, widely accepted explanation of a natural phenomenon that has advanced beyond hypothesis status through repeated experimentation and verified results.

topology: The branch of mathematics dealing with geometric properties of spatial relations under which continuity and shape/size (along with other properties) are invariant.

totient: See *phi function*.

transversal: A line cutting through two (usually parallel) other lines.

tree: An undirected graph in which any two vertices are connected by exactly one path.

triangle: A shape with three edges that has all angles summing to 180° and whose side lengths satisfy the Triangle Inequality (see below).

Triangle Inequality: In any triangle, the longest side length is strictly less than the sum of the other two side lengths.

trilinear coordinates: For a point in the interior of a triangle, the trilinear coordinates of that point reflect its relative distances from each edge of the triangle.

Trivial Inequality: The inequality $x^2 \geq 0$ for all real numbers x.

truth table: A table that lists the Boolean true/false outputs of a logic gate for all possible initial inputs specified.

Turing machine: An abstract machine that uses a strip of tape to position its head over a cell and writes a symbol accordingly.

twin prime: A prime such that two more or two less than the prime is also a prime number.

uniform continuity: We say a function f is uniformly continuous iff we can guarantee that $f(x)$ and $f(y)$ are sufficiently close for any choices of x and y, with this distance not depending on x or y.

union: The *union* of sets is the set consisting of all elements in at least one of the sets. (We can define the intersection, then, as the set consisting of all elemetns in *all* of the sets.)

universal set: The set which serves as the base on which to operate in terms of the Venn diagram (drawn as a rectangle).

util: The unit of measurement of perceived utility.

utility: A concept of perceived benefit to the end user, measured in arbitrary units of *utils*.

value: Either a numerical constant, or the result of a numerical calculation.

variable: An unknown (usually represented by a letter) that we aim to solve for.

variance: The square of the standard deviation of a data set, which can be used to determine how far the spread of the data is in general.

vector: A quantity in terms of both direction and magnitude (length), often used to determine relative positions of points in space.

vertex: The common endpoint of two or more rays or line segments.

vertical angles: Angles on opposite sides of two intersecting lines that sum to $180°$.

Vertical Line Test: The test that determines whether or not the graph of a relation also represents a function. If there exists any vertical line in the plane passing through multiple points on the curve of the relation, the relation fails the test and is therefore not a function. Otherwise, it is a function.

Vieta's formulas: Formulas used to determine the *symmetric sums* of polynomials; i.e. the sums of products of the roots of a degree-n polynomial, especially for $n \geq 3$ where we cannot apply a direct formula.

Vieta jumping: The method traditionally used to prove the existence of only a specific class of solutions to Diophantine equations, which utilizes infinite descent as the crux.

weight: A non-negative real quantity assigned to non-negative real numbers; the sum of the weights of the numbers in the weighted AM-GM inequality must be 1 for the inequality to work.

Wilson's Theorem: For any prime p, $(p-1)! \equiv -1 \pmod{p}$.

WLOG: *Without loss of generality;* placed in a proof to indicate that an assumption does not change the outcome of the proof or restrict it in any way.

Yang-Mills Theory: The basis for the Yang-Mills existence and mass gap Millennium problem, which seeks a "smallest" particle that obeys the Wightman axioms (see §16).

Zermelo-Frankel set theory: Equipped with the axiom of choice, Zermelo-Frankel (ZFC) set theory is an instrumental tool in solving Hilbert's first problem.

Zero: See *root*.

Zero-sum game: A game in which the sum of the payoffs for all players, regardless of their chosen strategies, is zero.

Index

$-\frac{1}{12}$, 236
G-congruent, 283
G-equidecomposable, 283
L-function, 237, 248
γ, *see also* Euler-Mascheroni constant
h-cobordism theorem, 227
k-algebra, 257
k-cycle, 263
n-ary, 256
n-manifold, 225
n-modal, 20
n-sphere, 225, 227
p-adic, 256
p-adic valuation, 194
p-value, 28
xy-plane, 21
z-score, 37
Arithmetica, 265
Donkey Kong Country, 231
Legend of Zelda, The, 231
Metroid, 231
Pokémon, 231
Super Mario Bros., 231
Tetris, 231
Universal Paperclips, 69
a priori, 98
3-manifold, 225
3-sphere, 225

a priori, 28
Aaronson, Scott, 232
Abel Prize, 266
abelian, 253
abelian group, 240, 250
absolute value, 194
abstract, 120
Achilles, 283
Additive Law, 32
Additive Rule, 36
aerodynamics, 245
AI, 97
air flow, 245
algebra
    polynomial, 257
algebraic curve, 253, 258
algebraic cycle, 238
algebraic geometry, 185, 238, 257
algebraic number field, 253
algebraic number theory, 252
algebraic variety, 238, 257
algebrization, 232
almost all, 241
analysis, 75
analytic, 253
analytic theory, 200
analytical continuation, 234
AND, 232
annulus, 227
anomalous cancellation, 278
anti-commute, 251
Apéry, Roger, 235
Aristotle, 283
arithmetic, 120
    axioms, 253
arrival trate, 115
Artin, Emil, 258
assignment, 236
associativity, 97
Atiyah, Michael, 228
attention filter, 19
auxiliary line, 268
auxiliary point, 268
axiom of choice, 253, 282

backward induction, 63
Bacon number, 84
Bacon, Kevin, 84
Baker, Theodore, 232
ball, 6, 12
balls-and-urns, *see also* stars-and-bars
Banach space, 12
Banach-Tarski paradox, 282
bar graph, 22
base, 130
base rate fallacy, 35
Basel problem, 235

basis, 240
basis point, 248
Battle of the Sexes, 66
Battleship, 231
Bayes' Theorem, 35, 115
Bayesian inference, 115
Bayesian search theory, 115
bell curve, 38
Bender, 238
benefit, 60, 61
Bernoulli, 96
Bernoulli number, 235
Bernstein, Sergei, 259
Bertrand's postulate, 262
BI, *see also* backward induction
Big O notation, 112
big-O notation, 229
bijection, 75, 98, 168
bijective, 4
bijectivity, 4
bimodal, 20
binary, 130
binary operation, 97
Bing, R. H., 228
binomial coefficient, 49, 75
binomial coefficients, 51
Binomial Theorem, 46, 140, 174, 194, 269
black box, *see also* queueing node, *see also* oracle
blowup time, 247
Bolibrukh, Andrey A., 259
Boolean, 231, 232
Borsuk, Karol, 228
boson, 250
Bostrom, Nick, 69
boundary condition, 253
boundary value problem, 259
bounded, 227
box plot, *see also* box-and-whisker plot
box-and-whisker plot, 26
bracket, 22
Brahmagupta, 205
branch, 61, 258
Breuil, 249
Brody, 238
Brun's constant, 261
Brun's theorem, 261
BSDC, *see also* Birch and Swinnerton-Dyer
buffer, 113
    size, 114
bundle, 250
Burnside's Lemma, 97
Burnside, William, 98
butterfly effect, 76

calculus of variations, 253, 254
Cambridge, MA, 243
CAP, *see also* computer-assisted proof
cardinality, 4, 8, 10, 44
Carmichael number, 175
Cartesian plane, 258
Cartesian product, 9
casework, 44
Cauchy, 98
    distribution, 38
Cauchy-Schwarz Inequality, 25
Cayley-Hamilton Theorem, 279
CCT, *see also* computational complexity theory
CDF, *see also* cumulative distribution function
ceiling, 156
Central Limit Theorem, 40
chaos theory, 76
characteristic, 278
characteristic equation, 95
characteristic polynomial, 77
charge, 252
Chinese postman problem, 107
Chinese Remainder Theorem, 123, 165
chromodynamics, 97
cipher, 233
circuit, 109
Clay Institute, 231, 250
closed, 226, 245
closed ball, 12
closed set, 12
closure, 227
co-NP, 233
coalition, 61
    empty, 61
    grand, 61
Coates, John, 248
codomain, 34
coefficient of determination, 25
cohomology
    ring, 240
    singular, 240
cohomology class, 239, 242
cohomology group, 240
Collatz, Lothar, 263
collection, 3, 19
coloring arguments, 88
combination, 48
combinatorics, 44, 75, 94
commutative ring, 278
commute, 251
compact, 227, 254
complement, 233
complementary counting, 53
complete metric space, 12

433

completeness, 260
complex analysis, 234
complex number, 133
complex plane, 260
complexity class, 229
computational complexity, 177
computational complexity theory, 229
computer-assisted proof, 259
conditional probability, 115
conjecture
- Berry, 238
- Bing-Borsuk, 228
- Birch and Swinnerton-Dyer, 248
- Collatz, 120, 263
- Cramér's, 237
- Elliott-Halberstam, 262
- epsilon, *see also* Ribet's Theorem
- even Goldbach, 238
- Goldbach, 120, 238
- Hilbert-Pólya, 237
- Hilbert-Smith, 254
- Hodge, 238
- integral Hodge, 244
- odd Goldbach, 238
- Poincaré, 225, 237
- Polignac's, 262
- prime gap, 237
- Shimura-Taniyama, 256
- smooth Poincaré, 228
- Taniyama-Shimura, 249
- twin prime, 120, 145, 261
- weak Poincare, 228

connected, 226
conservation
- of energy, 246
- of mass, 246
- of momentum, 246

constant map, 240
continued fraction representation, 200
continuity equation, 246
continuous, 32, 104, 115
continuum hypothesis, 253
contractible space, 240
control theory, 260
convergents, 203
convex set, 227
Cook-Levin theorem, 231
cooperate, 66
cooperative game, 61
coordination game, 68
correlation, 24
- correlation coefficient, 24

correspondence, 13
countable, 3

covariance, 25, 251
CPT theorem, 252
Cramér, 237
critical line, 237
critical point, 241
critical value, 241
Cruise, Tom, 84
cryptographic hashing, 233
cryptography, 183, 248
cryptosystem, 183
current, 245
curve
- component, 258
- modular, 249

cycle, 85, 263
- induced, 88
- simple, 87

cyclotomic polynomial, 180, 211, 213

Davis, Martin, 255
de Morgan's Law, 7, 45
de Rham, 243
- cohomology, 244

de Weger, 263
decimal, 122
decimal expansion, 170
decision, 60
decision problem, 233
decision-maker, 60
decision-making function, 112
defect, 66
degenerate, 200
degree, 84
Dehn invariant, 254
Dehn, Max, 254
density, 246
dependent, 32
derivative, 277
difference of squares, 273
differentiability, 254
differentiable, 239
differential equation, 246
- linear, 254
- partial, 246, 259

differential form, 242, 245
dihedral angle, 254
Dijkstra's Algorithm, 108
Diophantine, 255
Diophantine analysis, 151
Diophantine equation, 75, 197, 253
- linear, 151

Diophantine Equations, 151
Diophantus, 151

Dirichlet Box Principle, *see also* Pigeonhole Principle
Dirichlet series, 234, 237, 248
discount factor, 113
discrepancy, 232
discrete, 21
disjoint, 4, 14, 36
disk, 6, 260
distance function, 12, 259
distinct, 3
distinguishable, 99
distribution
    fair, 63
    unfair, 63
divisibility
    tests, 122
divisor, *see also* factor
divisor function, 159
dominant strategy, 66
double-tailed, 40
DP, *see also* dynamic programming, *see also* dynamic programming
dropout rate, 114
dynamic programming, 113, 260
dynamics, 258

edge, 84
edge-preserving, 104
Einstein problem, 259
Einstein, Albert, 244
    field equations, 250
    laws of general relativity, 250
    theory of relativity, 252
    theory of unification, 244
Eisenstein's Lemma, 188
electromagnetism, 250
ellipses, 4
elliptic curve, 248, 266
    modular, 249
Ennio de Georgi, 259
Ennio de Gorgi, 253
enumerative geometry, 257
environmental science, 97
equinumerous, 4
equipotent, 4
equivalence class, 14
equivalence relation, 283
Erdős number, 84
Erdős, Paul, 261, 263
Erdős-Bacon number, 84
error, 29
    type I error, 29
    type II error, 29
Euclid, 127, 267

Euclid's lemma, 127
Euclid-Euler Theorem, 147
Euclidean Algorithm, 141
Euclidean algorithm, 200
Euler, 105, 205, 235
Euler product, 248
Euler's 4-square theorem, 190
Euler's Criterion, 184, 189
Euler's formula, 90
Euler's totient function, 157
Euler's totient theorem, 175
Euler-Lagrange equation, 259
Euler-Mascheroni constant, 235, 255
event, 32, 44
exact form, 245
existence, 127, 166, 245, 250
expected value, 33
extensionality, 3
extensionality property, *see also* extensionality
extensive form game, 62

factor, 121
false, 231
false proof, 276
faster-than-light travel, 252
Fermat number, 145
Fermat prime, 145
Fermat pseudoprime, 175
Fermat psuedoprime, 175
Fermat's Last Theorem, 120, 174, 233, 248, 263, 265
Fermat's Little Theorem, 138, 176
Fibonacci sequence, 76
field, 189, 257
field dynamics, 250
Fields Medal, 228
finance, 97
financial engineering, 112
five-number summary, 20, 26
fix, 156
fixed point, 99
floor, 156
flow, 245
FLRW model, 227
FLT, *see also* Fermat's Little Theorem, 174
FLT Rearrangement Lemma, 174
Foddy, Bennett, 69
force
    electromagnetic, 250
    strong, 250
    weak, 250
form
    extensive form, 61
    normal form, 61

four-color problem, 88
Fourier transform, 251
fraction, 122
fractional part, 156
FreeCell, 231
Freedman, Michael, 228
frequency polygon, 21
freshman's dream, 194, 278
Frey curve, 266
Frey, Gerhard, 266
Friedmann, Alexander, 227
Frobenius, 98
FTL, *see also* faster-than-light travel
full-measure, 241
function
    absolute value, 235
    analytic, 234
    automorphic, 254, 260
    complex conjugate, 235
    elementary, 235
    exponential, 235
    gamma, 235
    logarithmic, 235
    signum, 275
    trigonometric, 235
functional analysis, 252
functional equation, 266
fundamental group, 228
fundamental theorem, 185
Fundamental Theorem of Arithmetic, 126
Fundamental Theorem of Orders, 211

Gödel's incompleteness theorem, 254
Galois, 244
    representations, 244
Galois group, 256
gambler's ruin, 76, 94
game
    asymmetric, 65
    normal form, 64
    sequential, 62
    simultaneous, 62
    symmetric, 65
    ultimatum, 63
game theory, 97
Gasarch, William, 232
gauge theory, 250
    gauge field, 250
    gauge group, 250
Gauss' Lemma, 184
Gauss, Carl, 120, 177, 259
Gaussian distribution, 38
generation, 253
    of algebras, 257

geodesic, 253, 254
geometric distribution, 96
geometric series, 147
George David Birkhoff, 89
Germain, Sophie, 265
Gill, John, 232
GIMPS, *see also* Great Prime Search, 148, *see also* Great Prime Search
Glaisher, 213
gluon, 250
golden ratio, 255
golden theorem, 185
Grandi's series, 236
graph, 84
    d-regular, 86
    acyclic, 87
    bipartite, 87
    chromatic number, 88
    chromatic polynomial, 89
    color class, 89
    complete, 86
    connected, 86
    directed, 84, 108
    forest, 87
    leaf, 87
    mixed, 108
    regular, 86
    tree, 87
    undirected, 84, 108
graph theory, 84, 104
Grassmannian, 257
Great Prime Search, 125, 262
Gross-Zagier Theorem, 249
Grothendieck, 244
group, 97, 169, 250
    continuous, 253
    differential group, 253
    finite, 248
    modular, 266
    monodromic, 254, 259
group action, 98
group theory, 97
growth rate, 230

Hölder continuous, 259
half-plane
    lower, 266
    upper, 266
Hamilton, Richard S., 228
Hamiltonian, 238
Hamiltonian cycle, 109
Hamiltonian path, 109
Hamiltonian path problem, 109
Hardy, G. H., 284

harmonic form, 242
harmonic series, 234
Harnack, Carl Gustav Axel, 258
heat transfer, 245
Heesch, Heinrich, 259
Hermitian operator, 238
Heron's formula, 274
hex, *see also* hexadecimal
hexadecimal, 131
Hilbert space, 251
Hilbert's 23 problems, 252
Hilbert, David, 237, 252
Hippasus, 267
histogram, 22
Hockey Stick Identity, 53
Hockey-Stick Identity, 75, 78
Hodge class, 243
Hodge cycle, 238
Hodge theory, 244
homeomorphism, 104, 225
homogeneous, 152, 256
homology, 227
homology class, 239
homotopy, 240
homotopy theory, 228
howler, 278
hypercube, 88
hypergeometric distribution, 80
hyperplane, 241
hypersphere, 65, 225
hypothesis, 28
    alternative hypothesis, 28
    hypothesis testing, 28, 29
    null hypothesis, 28
hypothesis testing, 28

identity element, 97
image, 240, 245
implied edge, 109
incircle, 273
Inclusion-Exclusion Principle, 45
independent, 32
independent events, 35
indistinguishable, 99
induced mapping, 241
induction, 45, 278
inertia, 246
inferential statistics, 27
infinite series, 236
information technology, 112
injective, 4
injectivity, 4, 240
integrable, 115
integral
    definite, 277
    indefinite, 277
integration
    constant of, 277
integration by parts, 277
interactive proof system, 232
International Congress of Mathematicians, 243
interquartile range, 20
intersection, 8
invariance, 250
invariant, 253, 257
    ring, 257
inverse, 104
inverse function, 226
IP, 232
IQR, *see also* interquartile range, *see also* interquartile range
irreducibility, 219
isomorphism, 104

Jacobi symbol, 185
Jakobshe, Włodzimierz, 228

Königsberg, bridges of, 105
Kakuro, 231
kernel, 240
Kolmogorov's axioms, 254
Kronecker-Weber Theorem, 256
Kronecker-Weber theorem, 253
Kummer, Ernst, 265
Kuratowksi subgraph, 104
Kuratowski's Theorem, 104
Kuratowski, Kazimierz, 86

Lagarias, Jeffrey, 263
Lagrange, 205
Lagrange Theorem, 179
Lagrange's Theorem, 205, 212
Lagrangian, 250
Lantz, Frank, 69
Law of best approximation, 208
law of conservation of energy, 282
Law of Large Numbers, 96
law of reciprocity, 255
Law of Sines, 274
least squares, 24
Lebesgue integral, 260
Lebesgue integration, 260
Lebesgue, Henri, 260
Lefschetz, 243
Lefschetz Theorem, 243
left-tailed, 40
Legendre, 196
Legendre symbol, 183
Legendre's formula, 196

Lehmer, 213
Lehmer's Theorem, 175
Lemaître, Georges, 227
Leonardo da Vinci, 271
Leray, Jean, 247
Lie group, 250
Lie, Sophus, 254
Lifting the Exponent, 176
lifting the exponent, 194
Light Up, 231
limit, 75, 96, 230
limit cycle, 253, 258
line graph, 21
linear combination, 227
linear regression, 24
linear subspace, 257
linearity, 230
linearity of expectation, 25
linguistics, 245
Little's Law, 115
logic gate, 232
logistic distribution, 38
longest path problem, 231
Lorentz group, 251
LSRL, *see also* least squares
LTE, *see also* Lifting the Exponent
Lucas-Lehmer, 148
ludology, 60

M-curve, 258
Möbius function, 214
Möbius inversion, 214
Machine learning, 234
machine translation, 234
Malaysia 370, 115
manifold, 226, 238
    compact, 239
    complex, 239
    Kähler, 242
    orientability, 239
    smoothness, 239
    submanifold, 242
manufacturing, 112
mapping, 98, 104
Markov, 112
Markov chain, 76, 94, 112
Markov decision process, 112
Markov property, 96
Martello's algorithm, 109
mass gap, 250, 252
matching, 86
    induced, 88
mathematical object, 3
mathematical physics, 252

Matiyasevich's theorem, 255
Matiyasevich, Yuri, 255
matrix, 61, 240, 279
    characteristic polynomial, 280
    determinant, 279
    invertible, 280
    nilpotent, 280
    singular, 280
MDP, 112, *see also* Markov decision process
mean, 20, 26
    arithmetic mean, 20
    geometric mean, 20
measure, 241
median, 20, 26
megaprime, 148
memorylessness, 95
Mersenne number, 145
Mersenne prime, 145
meteorology, 234
method, 97
metric, 32, 253, 259
    p-metric, 12
    2-metric, 12
    discrete metric, 12
    maximum metric, 12
    supremum metric, 12
    taxicab metric, 12
metric space, 11
Millennium Prize Problems, 225, 252, 265
Minesweeper, 231
Minkowski, 4
Minkowski sum, 4
mixed strategy, 64
mod, *see also* modulo
mode, 20, 26
modular arithmetic, 137, 165
modular congruence, 137
modular equivalence, 137, 185
modular form, 244, 266
modular inverse, 140
modularity theorem, *see also* Taniyama-Shimura, 266
modulo, 137
modulus, 137
monic, 256
Monte Carlo, 96, 116
Montgomery, Hugh, 237
Morse, Marston, 260
Morse, Samuel, 260
MRDP theorem, *see also* Matiyasevich's theorem
Muller, 238
multiple, 121
Multiplicative Law, 32
Multiplicative Rule, 36

multivariate polynomial, 238

Nagata, Masayoshi, 257
Nash Equilibrium, 67
Nash equilibrium, 66
Nash's existence theorem, 67
Nash, John, 67
Nash, John Forbes, 253, 259
natural proofs, 232
natural science, 245
Navier-Stokes, 245, 249, 252
NE, *see also* Nash equilibrium
    mixed strategy, 67
    pure strategy, 67
    strong, 67
    weak, 67
neighborhood, 225
Nerode, Anil, 233
Newston, Dr. Maby Winton, 258
Newton's second law, 245
node, 61, *see also* vertex
Noether, Emmy, 260
non-abelian, 250
non-cooperative game, 61
norm, 194, 251
    p-adic norm, 194
normal, 38
normal distribution, 38
normed vector space, 12, 237
NOT, 232
NP, 109
NP completeness, 231
NP-complete, 109
NP-hard, 109, 231
NPC class, 231
null hypothesis, 39
number
    algebraic, 120, 253, 255
    complex, 122
    irrational, 122
    rational, 255
    transcendental, 120, 253, 255
number theory, 120
numerical analysis, 248
Nurikabe, 231

object, 3
Occam's Razor, 234, 260
OCR, *see also* text recognition
OEIS, *see also* Online Encyclopedia of Integer Sequences
Online Encyclopedia of Integer Sequences, 284
open ball, 12
open set, 12

operations research, 109, 112
operator
    self-adjoint, 251
operator theory, *see also* spectral theory, *see also* spectral theory
optimization problem, 109
OR, *see also* operations research, 232
oracle, *see also* black box, 232
orbit, 97
Orbit-Stabilizer Theorem, 98
order, 84, 169, 211, 229, 248
ordered pair, 84
orthogonality, 241
Osterwalder, 251

P/NP problem, 112, 125, 229, 249
Pólya, George, 98, 237
paperclip maximizer, 69
paradox, 276
    antinomy, 278
    arrow paradox, 284
    dialetheia, 278
    dichotomy paradox, 283
    falsidical, 278
    fletcher's paradox, *see also* arrow paradox
    interesting number paradox, 284
    veridical, 278
Pardon, John, 254
parity, 121, 252
parity function, 264
particle
    spin, 251
partition, 14
Pascal's Identity, 52
Pascal's Triangle, 51
path, 85, 226
    induced, 88
    simple, 87
path-connected, 226
payoff, 61
    punishment, 66
    reward, 66
    sucker, 66
    temptation, 66
payoff matrix, 62, 64
Pell equation, 200, 205
Pell, John, 205
percentile, 20
Perelman, Grigori, 225
perfect information, 62
period, 165
periodic point, 170
periodic sequences, 168
periodicity, 211, 259, 263

asymptotic, 171
eventual, 170
permutation, 48, 104
PH, 233
phi function, *see also* Euler's totient function
photon, 250
physics, 234
PIE, *see also* Inclusion-Exclusion Principle
Pierre de Fermat, 265
Pigeonhole, 190
Pigeonhole Principle, 47, 211
planar graph, 104
PMF, *see also* probability mass function
Poincaré group, 251
Poincaré homology sphere, 227
Poincaré, Henry, 225, 258
pointwise, 247
policy, 113
    optimal policy, 113
policy modeling, 112
pollution, 245
polyhedron, 253, 259
polynomial, 235, 253, 279
population, 27
positive-definite, 258
Power of a Point, 275
Prandtl number, 246
pressure, 245, 246
primality test, 177, 185
prime
    irregular, 265
    regular, 265
prime factor, 124
prime factorization, 124
prime gap, 262
prime number, 120, 144
PrimeGrid, 262
primtive root, 211
Prisoner's Dilemma, 65
    generalized, 66
probability, 32, 44, 75, 94
    conditional probability, 35
    experiment, 32
    probability distribution, 37
    transitional, 251
probability density function, 97
probability distribution, 95
probability distribution function, 116
probability theory, 234
Product Rule, 274
projective geometry, 257
projective space, 240
PSD, *see also* population standard deviation
PSPACE, 232

psychology, 245
Ptolemy's Theorem, 272
punctured space, 245
pure mathematics, 234
Putnam, Hilary, 255
Pythagorean Theorem, 265

QCD, *see also* quantum chromodynamics
quadratic field, 248
quadratic form, 253, 256
quadratic reciprocity, 177
quadratic residue, 138, 183
quantization, 238
quantum chromodynamics, 250
quantum field theory, 250
    constructive, 250
quantum mechanics, 238
quark, 250
quartile, 20
    first quartile, 20
    second quartile, *see also* median
    third quartile, 20
query, 232
queue length, 114
queue theory, 109
queueing node, 113
queueing theory, 113

radiative forcing, 97
Ramanujan, 284
random process, *see also* stochastic process
random variable, 34
random walk, 76, 94
    one-dimensional, 77
rank, 248
rational decision, 60
rational number, 120
Razborov, Alexander, 232
real line, 12
rearrangement, 49
reciprocity theorem, 253
recursion, 75, 76, 94
recursive probability, 112
reflexivity, 14, 137
regular problem, 259
Reinhardt, Karl, 259
relation, 13, 254, 260
    binary relation, 13
    equivalence relation, 13, 14
    reflexive relation, 13
    symmetric relation, 13
    transitive relation, 13
relative position, 258
relatively prime, 235

relativization, 232
repeating, 200
residue class, 168
residue system, 168
resource, 230
resources, 177
revenue mangaement, 112
Reynolds number, 246
RH, *see also* Riemann Hypothesis
Ribet's Theorem, 266
Ribet, Ken, 266
Ricci flow, 228
Riemann, 260
Riemann Hypothesis, 228, 229, 234, 249
Riemann hypothesis, 253
    generalized, 249
Riemann sphere, 260
Riemann zeta function, 234, 248
Riemannian structure, 242
right-tailed, 40
ring, 194, 240, 253
    commutative ring, 194
Robertson, Howard, 227
Robinson, Julia, 255
roots of unity, 169
rotation, 99
round, 156
route inspection problem, 107
Rubik's Cube, 231
Rudich, Steven, 232

sample, 27
sample space, 32
sampling, 27
    convenience, 27
    non-probability, 27
    probability, 27
    quota, 27
    self-selection, 27
    simple random, 27
    stratified random, 27
    systematic random, 27
Sard's Theorem, 241
SAT problem, 231
satisfiability, 231
scalar, 279
scalar product, 251
scheduling policy, *see also* service discipline
Schonefeld, 237
Schrader, 251
Schubert calculus, *see also* Schubert's enumerative calculus
Schubert cell, 257
Schubert enumerative calculus, 253

Schubert's enumerative calculus, 257
scientific method, 28
SD, *see also* standard deviation
Selbert, 237
self-adjoint, 237
semiperimeter, 273
separable, 251
Serre, Jean-Pierre, 266
service discipline, 114
set, 3, 44
    computably enumerable, 255
set theory, 3, 9
SFFT, *see also* Simon's Favorite Factoring Trick
Shimura, Goro, 256, 266
shortcut map, 263
shortest path problem, 107
sieve, 151
Sieve of Eratosthenes, 144
significant, 28
    statistically significant, 28
Simon's Favorite Factoring Trick, 152
Simons, 263
simply connected, 227
simulation, 97
single-tailed, 40
singular chain complex, 240
singularity, 259
Smale, Stephen, 228
smooth, 240
smooth embedding, 240
smoothness, 245
snake-in-a-box, 88
Solovay, Robert, 232
sophomore's dream, 278
space, 232, 250
space-like separation, 251
spectral theory, 237, 252
sphere packing, 253, 259
spherical geometry, 254
split, 219
SSD, *see also* sample standard deviation
stabilizer, 98
standard deviation, 21
    population standard deviation, 21
    sample standard deviation, 21
standard normal distribution, 37, 38
stars-and-bars, 75
state space, 251
states, 94
statistics, 19, 234, 245
    descriptive statistics, 19, 26
Steiner, 263
stem-and-leaf plot, 26
stochastic model, 112

stochastic process, 76
strategy, 60
    A100, 69
    B100, 69
    BEAT LAST, 69
    GENEROUS, 69
    GREEDY, 69
    MINIMAX, 69
    mixed, 60
    pure, 60
    RANDOM, 69
    TIT FOR TAT, 69
    totally mixed, 60
strategy set, 60
    finite, 60
stratum, 28
Streater, 251
stress, 245
string theory, 244
strong induction, 127
Student's t, 38
subatomic particles, 250
subdivsion, 104
subgraph, 85, 88
    induced, 88
subsequence, 263
subset, 5, 8
subspace, 251
Sudoku, 231
sum, 4
superset, 12
support, 251
surjective, 98
surjectivity, 241
symmetry, 38, 97, 137, 183
    global, 250
    local, 250
symplectic structure, 242

Taniyama, Yutaka, 256, 266
Tao, Terence, 247
Tate-Shafarevich group, 249
taxicab number, 284
Taylor series, 248
Taylor, Richard, 266
temperature, 246
tensor product, 254
terminating, 200
tessellation, 259
tetrahedron, 86
text recognition, 234
theory, 28
Thue-Morse sequence, 260
tiling

tile-transitive, 259
    anisohedral, 253, 259
    isohedral, 259
time
    constant, 230
    factorial, 230
    linear, 230
    logarithmic, 230
    polynomial, 230
time complexity, 229
time reversal, 252
topological manifold, 254
topological space, 225, 254
topology, 226, 238, 260
    differential, 260
torsion cohomology class, 244
tortoise, 283
torus, 247
totient, 157
transivitity, 137
transportation, 234
transportation modeling, 112
traveling salesman problem, 109, 234
tree, 61
Triangle Inequality, 208
trimodal, 20
true, 231
Tunnell's Theorem, 249
turbulence, 245
Turing machine, 232
Turing, Alan, 148
twin prime, 145
Twin Prime Search, 262
twin primes, 261

unification, 250
uniform continuity, 260
uniform convergence, 260
uniformization, 260
unilateral, 67
union, 8
uniqueness, 127, 134, 166
universal set, 8
util, 61
utility, 34
    function, 34
    util, 34

vacuum, 251
vacuum energy, 227
value, 113
Vandermonde Identity
    Chu-Vandermonde Identity, 80
    Rothe-Hagen identity, 80

Vandermonde's Identity, 78
variable
    dependent variable, 22
    independent variable, 22
variance, 37
variational problem, 253, 259
vector
    velocity, 246
vector field, 253
vector space, 257
velocity, 246
Venn diagram, 8, 45, 89
vertex, 84, 104
Vertical Line Test, 22
viscosity, 245

Walker, Arhur, 227
weather, 245
weight, 94, 107
well-defined, 3
Whitehead manifold, 228
Whitehead, J. H. C., 228

Whitney embedding theorem, 240
Wigderson, Avi, 232
Wightman, 251
Wightman axioms, 251
Wigner's theory of symmetry, 251
Wiles, Andrew, 248, 266
Wilson's Theorem, 138, 176, 221

Yang-Mills Existence and Mass Gap, 249, 250
Yang-Mills theory, 250
YMEMG, *see also* Yang-Mills Existence and Mass Gap

Zeno of Elea, 283
Zermelo-Fraenkel set theory, 253
zero
    non-trivial, 237
    trivial, 237
zero-sum game, 61
ZFC set theory, *see also* Zermelo-Fraenkel set theory

About the Authors

Dennis Chen

As of the time of publication, Dennis Chen is an eighth grader involved in competition math. He was given a rough introduction to competitive mathematics during fourth grade, when the elementary school "math team" held its tryouts. During fifth grade, his interest in math mostly dwindled, only taking the occasional mathleague test. However, in sixth grade, the wonderful middle school math team at his school and the AMC 8 sparked his interest in competitive mathematics again.

During the February of sixth grade, he participated in the MathCounts Chapter competition as an individual. While his "practice" before consisted of doing Alcumus and MathCounts Trainer during English class, being chosen for the Chapter team motivated him to improve much more. Somewhere along the way, he signed up for the AMC 10A, and nearly made AIME with a score of 106.5.

The fall of seventh grade starts. Motivated to have a good competitive year, Dennis practices competitive math. In the free time during the fall/winter before competition season, he starts a couple of projects. One of them was a blog where he posted problems based off of ideas that he had learned (from which a couple problems have been pulled into this book), and another project during the winter was teaching a math class, which is ongoing today. In the spring, he qualified for California's State MathCounts through his team. Though his contest results were unremarkable that year, his mathematical ability and outreach were greatly improved. Near the end of the competition season, he began drafting the backbone his work-in-progress geometry book.

In the fall of eighth grade, Dennis begins working on Proofs in Competition Math. He also works on a couple of articles. The eighth grade was mostly uneventful until the competition season began. With the AMC 10A/AMC 10B looming and his lack of preparation, he put everything else to rest while working to ensure an AIME qualification. After a couple of other competitions, including the MathCounts Chapter round and MathCounts State, Dennis has continued learning new things.

His other interests include cross country and track, physics, playing the trumpet, and beginning to play the flute. He also plays Super Smash Ultimate, primarily playing Pichu. His favorite numbers are anything divisible by 3 (though he gets extra happy when a number is divisible by 9), his favorite fruits are strawberries, and he has a completely rational hatred of gum.

Freya Edholm

A Silicon Valley native, Freya Edholm is currently taking a gap year before heading off to MIT, where she plans to study mathematics and chemical engineering. She qualified for the USAJMO and, for the past two years, was the board master of the San Francisco Bay Area A1 ARML team, which won the national championship in 2017. That year, she also achieved a near-perfect AMC 12 score of 141, placing her in the top $\sim$0.1% of contestants worldwide. In addition, Freya was one of the top 50 participants in the US National Chemistry Olympiad (USNCO). A linguistics enthusiast, she placed 11th in the North American Computational Linguistics Olympiad (NACLO) and qualified for the invitational round twice. Most recently, in December 2019, she scored in the top 500 on the William Lowell Putnam Competition.

Even though she is no longer eligible to participate in high school math competitions, Freya remains extensively involved with them. She writes problems for and edits math competition exams for mathleague.org, Berkeley Math Tournament, and Alpha Math Contest. She is also a coach for the SFBA ARML teams and a teaching assistant for math and chemistry classes on Art of Problem Solving, including the highly prestigious WOOT (Worldwide Online Olympiad Training) and ChemWOOT classes. Besides STEM, Freya enjoys learning foreign languages, specifically Latin and Ancient Greek. In what's left of her free time, she can be found hanging out with her friends (many of whom, like Alex, she met at math competitions or camps) or social dancing.

Her favorite number is 17, and her favorite color is turquoise (although it becomes pink at ARML).

Alex Toller

Alex Toller has been a passionate mathlete ever since his middle school days. When he was introduced to competitive math by his middle school math club coach, Alex was instantly amazed, and realized that there was so much more to math beyond the classroom. He had been a member of the Northern Nevada Math Club for six years, during which he attended a wide variety of competitions held at universities across the nation. He owes much of his love and passion for math to his experiences working with teammates, as well as to his best friends, who were also competitive mathletes in middle and high school.

In the eighth grade, Alex was the highest scorer at the Nevada MathCounts state competition and participated in the MathCounts National Competition. In high school, he scored in the top 2 percent on the nationwide AMC 12 examination, and scored 9 out of 15 on the next contest in the AMC series, the AIME. In addition, he has written the problems for four original math competitions, and co-organized three of them with the Northern Nevada Math Club. He is currently a sophomore mathematics major at the University of California, Berkeley, and is a problem writer for mathleague.org, as well as for the Berkeley Math Tournament. In December 2019, he scored in the top 500 on the William Lowell Putnam Competition. This is his debut work, with which he hopes to reach out to aspiring mathletes everywhere and of all backgrounds.

His favorite number is ϕ (the golden ratio), and his favorite color is light blue (but it used to be orange).